THE DELTA RECEPTOR

The Delta Receptor

EDITED BY
Kwen-Jen Chang
Ardent Pharmaceuticals, Inc.
Durham, North Carolina, U.S.A.

Frank Porreca
University of Arizona
Tucson, Arizona, U.S.A.

James H. Woods
University of Michigan
Ann Arbor, Michigan, U.S.A.

Marcel Dekker, Inc.   New York • Basel

Library of Congress Cataloging-in-Publication Data
A catalog record for this book is available from the Library of Congress.

ISBN: 0-8247-4031-9

This book is printed on acid-free paper.

Headquarters
Marcel Dekker, Inc., 270 Madison Avenue, New York, NY 10016, U.S.A.
tel: 212-696-9000; fax: 212-685-4540

Distribution and Customer Service
Marcel Dekker, Inc., Cimarron Road, Monticello, New York 12701, U.S.A.
tel: 800-228-1160; fax: 845-796-1772

Eastern Hemisphere Distribution
Marcel Dekker AG, Hutgasse 4, Postfach 812, CH-4001 Basel, Switzerland
tel: 41-61-260-6300; fax: 41-61-260-6333

World Wide Web
http://www.dekker.com

The publisher offers discounts on this book when ordered in bulk quantities. For more information, write to Special Sales/Professional Marketing at the headquarters address above.

Current printing (last digit):

10 9 8 7 6 5 4 3 2 1

PRINTED IN THE UNITED STATES OF AMERICA

Preface

The discovery of a receptor selective for opioids in 1973 heralded a heightened interest in opioid research into mechanisms of endogenous control of pain and in efforts to develop new analgesics. The initial simultaneous discovery of the opiate receptor by Candace Pert, Solomon Snyder, Lars Terenius, and Eric Simon rapidly led to a search for the identification and characterization of endogenous ligands for the opiate receptor and revealed the existence of a long-suspected endogenous system of pain regulation. In rapid succession, Hans Kosterlitz, John Hughes, and their colleagues discovered the enkephalins, C. H. Li discovered the endorphins, and Avram Goldstein identified the dynorphins. Interest in opiate research and mechanisms had never been higher.

Strong pharmacological data had indicated that the properties of opiate agonists could not be satisfactorily described based on evidence of a single opioid receptor. W. R. William Martin described significantly differing behavioral properties of opiates in the chronic spinal dog and postulated the existence of three distinct opiate receptors, which he termed mu, kappa, and sigma. Between 1977 and 1979, J. A H. Lord, Hans Kosterlitz, and their colleagues demonstrated differential activity profiles of [Leu^5]enkephalin, [Met^5] enkephalin, and morphine in isolated tissue assays. Using the mouse isolated vas deferens and the guinea pig isolated ileum preparations they proposed the existence of a receptor preferentially expressed in the mouse vas deferens that they termed the "delta" (for vas deferens) opioid receptor. Many subsequent radioreceptor binding and autoradiographic localization studies in vitro confirmed the existence of this receptor that was not preferential for morphine, now believed to act at a receptor termed the mu receptor by Martin and colleagues. The identification of a receptor for the endogenous enkephalins led to investigations of the physiological and pharmacological properties of these endogenous ligands and of the delta receptor itself. Understanding these issues is a journey that has been in progress for some 30 years.

Understanding the physiology and pharmacology of the delta receptor was limited initially by a lack of ligands suitable for in vivo studies. Although [Leu^5]enkephalin or [Met^5]enkephalin acted preferentially at the delta receptor, the selectivity of these straight-chain pentapeptides for the delta receptor over other opioid receptors was quickly found to be quite low. Additionally, and perhaps more importantly, these peptides lacked sufficient stability to be useful as a tool for the in vivo characterization of the properties mediated by activation of the delta receptor. Attempts to overcome these issues began with the inclusion of the D-enantiomer of constituent amino acids of the pentapeptide which produced less labile enkephalin derivatives, such as [D-Ala^2, D-Leu^5]enkephalin (DADLE) or Tyr-D-Ser-Gly-Phe-Leu-Thr (DSLET), making it possible for certain behavioral studies to be performed. However, these substances were still considerably labile in vivo. In the early 1980s, Victor Hruby, Henry Mosberg, and their associates developed the novel concept of introducing conformational constraints and discovered a class of cyclic penicillamine derivatives of enkephalin, which include [D-Pen^2, D-Pen^5]enkephalin (DPDPE), and [D-Pen^2, L-Pen^5]enkephalin (DPLPE). These peptides were important in that they increased selectivity for the delta receptor significantly and additionally gained a great deal of stability in vivo, allowing their use for in vivo studies.

This development was soon followed by the discovery of the deltorphins, peptides derived from the skin of the frog *Phyllomedusa sauvagei* by Vittorio Erspamer, Lucia Negri, and colleagues. The deltorphins showed superb selectivity for the delta receptor and became an important tool for in vivo characterization. A potentially significant consequence of the availability of these stable, selective delta receptor agonists was the pharmacological identification of two subtypes of the delta receptor. Also critical in the investigation of the receptor and its physiology was identification of peptidic ligands which showed high selectivity for the receptor but acted as antagonists. Here, Peter Schiller and his colleagues developed TIPP and TIPPpsi as ligands, which proved enormously important in the characterization of the receptor.

An important limitation of peptidic ligands is that of systemic bioavailability. Most of the data collected with the peptidic ligands described above came from direct injections into the brain or spinal cerebrospinal fluid or through in vitro studies of receptor function. In order to circumvent the problems inherent with peptides, the development of nonpeptidic agonists had to be undertaken. Kwen-Jen Chang and Robert McNutt reported a breakthrough in identification of a nonpeptidic structure with significant selectivity for the delta agonist. This compound, BW373U86, was shown to be a systemically active, delta antinociceptive agent and led to further important studies, which led to the identification of even more selective molecules. Silvia mu opioid agonist.

Calderon and Kenner Rice developed a series of compounds based on the structure of BW373U86. The chiral methylether derivative SNC-80 showed greater selectivity for the delta opioid receptor, but was apparently associated with a brief single episode of convulsant activity, seeming to indicate potential limitations in the therapeutic value of delta receptor agonists. Structurally similar compounds that did not bind to the delta receptor were also shown to produce similar convulsant activity, confusing the issue of whether convulsant activity was an effect associated with the delta receptor itself, or with the specific structure.

Studies in animals and primates with these highly selective delta agonists begin to reveal that unlike mu opioid agonists such as morphine, oxycontin, fentanyl, etc., agents acting at the delta receptor are unlikely to produce addictive liability and respiratory depression. In fact, delta agonists may actually counteract those side effects induced by mu opioids.

As important as the development of highly selective agonists for the delta receptor was the identification of selective nonpeptidic antagonists for the receptor. Working together, Aki Takemori and Phil Portoghese produced a series of molecules that have been used to define the receptor. Naltrindole, a selective, nonpeptidic and systemically available delta antagonist, became widely used to characterize the function of the receptor in vivo, and its radiolabeling led to many important studies characterizing the distribution and role of the receptor.

Perhaps the most important breakthrough of delta receptor biology came with the first cloning of the opioid receptor. Chris Evans and Brigitte Kieffer simultaneously reported the cloning of the delta receptor, the first one to be cloned, and this led to the confirmation of the existence of receptor in mouse and rat tissues. Henry Yamamura and his colleagues ultimately reported the identification of the human delta receptor. These studies also led to the important identification of distribution of the receptor in the nervous system initially through autoradiography and later through the elegant development of antibodies for the receptor by Robert Elde and Tomas Hokfelt. These, and other, investigators have extensively characterized the receptor in primary afferent fibers, in the spinal dorsal horn, and in the brain. Others confirmed the existence of the delta receptor in the submucous plexus of the gastrointestinal tract.

The understanding of the molecular and cellular signal transduction mechanism is well advanced for the delta receptor. The delta receptor belongs to the superfamily of the G-protein-coupled receptors (GPCR). Through the coupling of various G-proteins, the activation of the delta receptor can lead to the modulation of phospholipase C (PLC), adenylyl cyclase, ion channels, and mitogen-activating protein kinases (MAP kinase), and eventually a variety of cellular and neuronal functions including neurotransmitter release. The fate of the delta receptor in the cell membrane is also well studied by Ping

Yi Law, Horace Low, and colleagues. Similar to other G-protein-coupled receptors, upon the activation of delta receptor by its agonists, the receptor molecule can be internalized and degraded through endosomes, lysosomes, and proteosomes; some receptors may recycle back to the cell surface. Molecular components responsible for receptor trafficking have also been fully studied and documented in the literature.

The physiology and function of the delta receptor has slowly begun to emerge. It is now clear that activation of the receptor produces analgesia and antihyperalgesia. The latter seems especially important given changes in its trafficking and distribution during pathological pain states. Agonists at the delta receptor have been shown to act synergistically with those acting at the mu opioid receptor to produce enhanced states of antinociception with reduced side effect profiles. The co-administration of delta opioid agonists with mu opioid agonists inhibits the development of tolerance to the antinociceptive effect of mu opioid agonists.

Interactions between and among different types of opioid receptors have been documented in many in vivo and in vitro studies. The pharmacological significance of these interactions has also slowly emerged. Therapeutic indications beyond analgesia have emerged too. Delta agonists were recently shown by James Woods and his colleagues to possess antidepressant activity in animal models. The discovery of the presence of the delta receptor in cardiac myocytes led to the exploration by Garrett Gross and his colleagues of a cardioprotective role of the delta agonist against ischemic heart insults such as heart attacks. Other potential therapeutic applications are also implied for gastrointestinal disorders, bladder function, and immunomodulation. Availability of the pharmacophore structure of nonpeptide delta agonists such as BW373U86 and SNC80, and delta antagonists such as naltrindole, have facilitated the synthesis of a large number of new nonpeptide ligands. Explorations of the uses of these newly synthesized nonpeptide ligands in the previously mentioned potential therapeutic applications are underway. We are anticipating multiple major advances in the therapeutic applications of delta compounds in the future beyond analgesia.

This book is thus relevant to all with an interest in the delta receptor and receptor-related ligands, pharmacology, and physiology. We hope that it stimulates a broad readership in both the academic world and the pharmaceutical industry.

It would not have been possible to publish this book without the contributions of the authors of all the chapters, and we would like to express our thanks and gratitude to them.

Kwen-Jen Chang
Frank Porreca
James H. Woods

Contents

Contributors

Josue Alfaro-Lopez University of Arizona, Tucson, Arizona, U.S.A.

Michael J. Bishop, Ph.D. GlaxoSmithKline Research and Development, Research Triangle Park, North Carolina, U.S.A.

Iwona Bonney Department of Anesthesia, Tufts–New England Medical Center, Boston, Massachusetts, U.S.A.

David R. Brown, Ph.D. Pharmacology Section, Department of Veterinary PathoBiology and Mucosal and Vaccine Research Center, University of Minnesota, St. Paul, Minnesota, U.S.A.

Thomas Burkey University of Arizona, Tucson, Arizona, U.S.A.

Daniel B. Carr, M.D. Department of Anesthesia, Tufts–New England Medical Center, Boston, Massachusetts, U.S.A.

Kwen-Jen Chang, Ph.D. Ardent Pharmaceuticals, Inc., Durham, North Carolina, U.S.A.

Mary J. Clark, B.S. M.S. Department of Pharmacology, University of Michigan Medical School, Ann Arbor, Michigan, U.S.A.

D. J. Daniels Department of Medicinal Chemistry, College of Pharmacy, University of Minnesota, Minneapolis, Minnesota, U.S.A.

F. M. Décaillot Institut de Génétique et de Biologie Moléculaire et Cellulaire, Illkirch, France

Christopher J. Evans, Ph.D. Department of Psychiatry and Behavioral Sciences, Neuropsychiatric Institute, University of California at Los Angeles, Los Angeles, California, U.S.A.

Howard L. Fields, M.D., Ph.D. Departments of Neurology and Physiology and the Wheeler Center for the Neurobiology of Addiction, University of California at San Francisco, San Francisco, California, U.S.A.

Ryan M. Fryer, Ph.D. Abbott Laboratories, Abbott Park, Illinois, U.S.A.

Peter J. Gengo, Ph.D. Ardent Pharmaceuticals, Inc., Durham, North Carolina, U.S.A.

Elisa Giannini, Ph.D. Department of Human Physiology and Pharmacology, University La Sapienza, Rome, Italy

Ricardo Gomez-Flores Department of Microbiology and Immunology, Universidad Autónoma de Nuevo León, San Nicolás de los Garza, NL, Mexico

Garrett J. Gross, Ph.D. Department of Pharmacology and Technology, Medical College of Wisconsin, Milwaukee, Wisconsin, U.S.A.

Mary M. Heinricher, Ph.D. Departments of Neurological Surgery and Physiology and Pharmacology, Oregon Health and Science University, Portland, Oregon, U.S.A.

Keiko Hosohata University of Arizona, Tucson, Arizona, U.S.A.

Yoshiaki Hosohata University of Arizona, Tucson, Arizona, U.S.A.

Victor J. Hruby, Ph.D. Department of Chemistry, University of Arizona, Tucson, Arizona, U.S.A.

Brigitte L. Kieffer, Ph.D. Institut de Génétique et de Biologie Moléculaire et Cellulaire, Illkirch, France

Emily M. Jutkiewicz Department of Pharmacology, University of Michigan, Ann Arbor, Michigan, U.S.A.

Richard J. Knapp, Ph.D. Aventis Pharmaceuticals, Bridgewater, New Jersey, U.S.A.

Arnaud Lacoste Cousins Center for Psychoneuroimmunology, Neuropsychiatric Institute, University of California at Los Angeles, Los Angeles, California, U.S.A.

Josephine Lai, Ph.D. Departments of Pharmacology and Anesthesiology, College of Medicine, University of Arizona, Tucson, Arizona, U.S.A.

P. Y. Law, Ph.D. Department of Pharmacology, University of Minnesota Medical School, Minneapolis, Minnesota, U.S.A.

Horace Loh, Ph.D. Department of Pharmacology, University of Minnesota Medical School, Minneapolis, Minnesota, U.S.A.

Andrzej W. Lipkowski Medical Research Centre, Polish Academy of Sciences, Warsaw, Poland

Ewa Malatynska Pharmaceutical Research and Development, Johnson & Johnson, Spring House, Pennsylvania, U.S.A.

Robert W. McNutt, Ph.D. Ardent Pharmaceuticals, Inc., Durham, North Carolina, U.S.A.

Aleksandra Misicka Department of Chemistry, Warsaw University, Warsaw, Poland

Henry I. Mosberg, Ph.D. Department of Medicinal Chemistry, College of Pharmacy, University of Michigan, Ann Arbor, Michigan, U.S.A.

Minoru Narita, Ph.D. Department of Toxicology, Hoshi University School of Pharmacy and Pharmaceutical Sciences, Tokyo, Japan

Lucia Negri Department of Human Physiology and Pharmacology, University La Sapienza, Rome, Italy

S. Stevens Negus, Ph.D. McLean Hospital, Harvard Medical School, Belmont, Massachusetts, U.S.A.

Florence Noble Department of Molecular and Structural Pharmacochemistry, UFR des Sciences Pharmaceutique et Biologiques, Paris, France

Michael H. Ossipov, Ph.D. Departments of Pharmacology and Anesthesiology, College of Medicine, University of Arizona, Tucson, Arizona, U.S.A.

Kemal Payza, Ph.D. Molecular Pharmacology Department, AstraZeneca R&D Montreal, Ville St.-Laurent, Quebec, Canada

Hemal H. Patel Department of Pharmacology and Technology, Medical College of Wisconsin, Milwaukee, Wisconsin, U.S.A.

Frank Porreca, Ph.D. Departments of Pharmacology and Anesthesiology, College of Medicine, University of Arizona, Tucson, Arizona, U.S.A.

P. S. Portoghese, Ph.D. Department of Medicinal Chemistry, College of Pharmacy, University of Minnesota, Minneapolis, Minnesota, U.S.A.

William R. Roeske, M.D. College of Medicine, University of Arizona, Tucson, Arizona, U.S.A.

Bernard P. Roques Department of Molecular and Structural Pharmacochemistry, UFR des Sciences Pharmaceutique et Biologiques, Paris, France

Richard B. Rothman, M.D., Ph.D. Clinical Pharmacology Section, Addiction Research Center, National Institute on Drug Abuse, National Institutes of Health, Baltimore, Maryland, U.S.A.

Peter W. Schiller, Ph.D. Laboratory of Chemical Biology and Peptide Research, Clinical Research Institute of Montreal, Montreal, Quebec, Canada

Jo El J. Schultz, Ph.D. Department of Pharmacology and Cell Biophysics, University of Cincinnati, Cincinnati, Ohio, U.S.A.

Ping Sun, Ph.D. Department of Pharmacology, University of Minnesota Medical School, Minneapolis, Minnesota, U.S.A.

Tsutomu Suzuki, Ph.D. Department of Toxicology, Hoshi University School of Pharmacy and Pharmaceutical Sciences, Tokyo, Japan

Xuejun Tang University of Arizona, Tucson, Arizona, U.S.A.

DeWayne Townsend IV, D.V.M., Ph.D. Pharmacology Section, Department of Veterinary PathoBiology, University of Minnesota, St. Paul, Minnesota, U.S.A.

John R. Traynor, Ph.D. Department of Pharmacology, University of Michigan Medical School, Ann Arbor, Michigan, U.S.A.

Jennifer Tsang University of Arizona, Tucson, Arizona, U.S.A.

Todd V. Vanderah, Ph.D. Departments of Pharmacology and Anesthesiology, College of Medicine, University of Arizona, Tucson, Arizona, U.S.A.

Eva V. Varga College of Medicine, University of Arizona, Tucson, Arizona, U.S.A.

Richard J. Weber, Ph.D. Department of Biomedical and Therapeutic Sciences, University of Illinois College of Medicine at Peoria, Peoria, Illinois, U.S.A.

James H. Woods, Ph.D. Department of Pharmacology, University of Michigan, Ann Arbor, Michigan, U.S.A.

Heng Xu, Ph.D. National Institute on Drug Abuse, National Institutes of Health, Baltimore, Maryland, U.S.A.

Henry I. Yamamura, Ph.D. College of Medicine, University of Arizona, Tucson, Arizona, U.S.A.

THE DELTA RECEPTOR

1

History of Delta Receptors

Kwen-Jen Chang
Ardent Pharmaceuticals, Inc., Durham, North Carolina, U.S.A.

1 INTRODUCTION

The discovery of two endogenous opiate- or morphinelike peptides, Leu-enkephalin and Met-enkephalin, by Hughes and colleagues in 1975 [1] led to the explosive advances in our understanding of the functions of opioid receptors and the actions of opiates and opioids in the last two decades. These two enkephalins are pentapeptides with identical first four amino acids of H-Tyr-Gly-Gly-Phe-OH and either leucine (Leu) or methionine (Met) at the fifth position carboxy-terminus. These two peptides were also later found to be derived from three separate precursor proteins produced from three genes known as preproenkephalin (or preproenkephalin A), preproopio-melanocortin (or pre-POMC), and preprodynorphin (or preproenkephalin B) genes [see reviews 2,3]. These endogenous morphinelike peptides were collectively classified as endorphins.

Today, we know that there are three types of opioid receptors—mu, delta, and kappa receptors [see review 3]. Proenkephalin contains six copies of Met-enkephalins and one copy of Leu-enkephalin. Enkephalins, especially Leu-enkephalin, are believed to be selective to delta receptors. Opiomelano-cortin contains β-endorphin that has the Met-enkephalin at its amino terminus. β-Endorphin is a nonselective ligand for mu and delta receptors.

Prodynorphin contains three copies of Leu-enkephalin with carboxy-terminus extended polypeptides of various lengths known as dynorphin A (or dynorphin 1-17), dynorphin B (dynorphin 1–13), or α- and β-neoendorphin. These peptides derived from prodynorphin are selective to kappa receptors and can also be further broken down to Leu-enkephalin. The identification of the delta receptor (or the enkephalin receptor) was a direct consequence of the discovery of enkephalins. This chapter will review the major events that are important for the identification of delta receptors and the subsequent cloning of delta receptor genes, and eventually all other opioid receptor genes.

2 ENKEPHALIN IS MORE POTENT THAN MORPHINE IN MOUSE VAS DEFERENS

Portoghese was the first one to discuss the possibility of multiplicity of opioid receptors in 1965 [4]. In 1976 Martin's group first proposed three distinctive classes of mu, kappa, and sigma opioid receptors based on the different central pharmacological actions of morphine, nalorphine and benzomorphan drugs such as ketocyclazocine, ethylketocyclazocine, and *N*-allylnormetazocine (SKF 10,047) [5,6]. In 1977 Lord et al. [7] first provided evidence of the existence of distinctive delta opioid receptors, 2 years after the discovery of enkephalins. Those investigators demonstrated that Leu-enkephalin was more potent than morphine in inhibiting the electrically stimulated muscle contraction of the isolated mouse vas deferens. In contrast, morphine was more potent than Leu-enkephalin in the guinea pig ileum by acting on mu opioid receptors. This enkephalin selective receptor in the mouse vas deferens was postulated to be putative delta receptors, distinctive from the mu, kappa, and sigma receptors proposed by Martin's group.

This hypothesis of delta receptors in the mouse vas deferens was later further substantiated by the observation of the lack of cross-tolerance between a stable enkephalin analog [D-Ala^2, D-leu^5]enkephalin (DADLE, see below) and sulphentanyl (a mu agonist) in this tissue after chronic administration in vivo [8].

We know today that the mouse vas deferens does contain all three types of opioid receptors, delta, mu, and kappa receptors although this tissue is dominated by delta receptors. Delta agonists are far more potent than mu and kappa agonists in this tissue. However, highly selective and potent mu agonist actually interact preferentially with mu receptors in the mouse vas deferens and can only be antagonized by a mu antagonist but not a delta antagonist. This can result in confusing data and misinterpretation. The best way to study delta receptor activity is in the presence of high concentrations of highly selective mu-antagonist (i.e., CTOP cyclic[D-Phe-Cys-Tyr-D-Trp-Orn-Thr-Pen-Thr-NH_2]) and kappa antagonist (i.e., nor-binaltorphimine [nor-BNI]) to block out mu and kappa receptor activities. The mu receptor activity can be investigated in this tissue in the presence of high concentrations of highly selective delta (i.e., TIPP, H-Tyr-Tic-Phe-Phe-OH) and kappa (i.e., nor-BNI) antagonists that are used to block out delta and kappa receptors. Likewise, the kappa receptor activity can be investigated in the presence of high concentrations of highly selective delta (i.e., TIPP) and mu (i.e., CTOP) antagonists in the mouse vas deferens [9]. Today, the mouse vas deferens is the most convenient and popular tissue for the study of opioid actions.

In contrast to the mouse vas deferens, the guinea pig ileum contains predominately mu receptors and less kappa receptors, but no delta receptors. The function of mu and kappa receptors can be examined independently in the presence of a selective kappa antagonist (i.e., nor-BNI) and a mu antagonist (i.e., CTOP), respectively, similar to that described above for the mouse vas deferens [9]. The potency of mu and kappa agonists often exhibits higher potency in the guinea pig ileum than that obtained from the mouse vas deferens.

3 STABLE ENKEPHALIN ANALOGUES: [D-Ala^2, D-Leu^5] ENKEPHALIN (DADLE) AND [D-Ala^2, N-Me-Phe^4, Met(O)^5-ol]ENKEPHALIN (FK 33-824) ARE SELECTIVE DELTA AND MU AGONISTS, RESPECTIVELY

Naturally occurring endorphins and enkephalins are rapidly metabolized in vivo. Metabolically stable peptides are important in studies of the relevant systems. Substitution of the second and fifth amino acid with D-amino acids resulted in stable enkephalin analogues. Two metabolically stable enkephalin analogues described by Chang and colleagues [10,11] and Roemer and colleagues [12] in 1977 were pivotal in the characterization of delta receptors. D-Ala^2, D-Leu^5 substituted Leu-enkephalin (designated as DADLE), and D-Ala^2, N-Me-Phe^4 and Met(O)^5-ol-substituted Met-enkephalin (referred to as FK 33-824) were the first two metabolically stable analogues that were later described to be selective to delta and mu receptors, respectively. ^{125}I- or ^{3}H-labeled DADLE and FK 33-824 were used by many laboratories to further characterize opioid receptors and compared with 3H-labeled opiates such as naloxone, morphine, etorphine, or their derivatives [7,13–22].

An examination of the structure-activity relationship (SAR) of a series of enkephalin analogues and morphine derivatives in competing with the binding of radioisotope labeled DADLE and FK 33-824 or opiates revealed that DADLE binds to an opioid receptor selective to Leu-enkephalin and its analogues. In contrast, FK 33-824, morphine, or naloxone binds to opioid receptors with selectivity in favor of morphine and derivatives similar to mu opioid receptors previously described by Snyder's group [17], Simon's group [18], and Terenius [19] in 1973. Based on the biochemical receptor binding affinity, these enkephalin-selective opioid receptors were initially referred to as enkephalin receptors and morphine selective opioid receptors were referred to as morphine receptors. Since these enkephalin and morphine selective opioid receptors were similar to delta and mu receptors characterized in the pharmacological studies in vitro and in vivo, respectively, delta receptors were finally accepted as the same class of enkephalin opioid receptors, and mu receptors were accepted as the morphine opioid receptors.

4 ANATOMICALLY DIFFERENTIAL DISTRIBUTION OF DELTA- AND MU-BINDING SITES IN THE BRAIN

Using ^{125}I-labeled DADLE to localize delta-receptors and ^{125}I-FK 33-824 or [^{3}H]naloxone to localize mu receptors, Chang and colleagues [16] first demonstrated the differential distribution of delta and mu receptors in the regions of rat brain. Subsequently, Snyder's laboratories [23] confirmed this differential distribution in the brain by the light microscopic autoradiography of delta and mu receptors localized with ^{125}I-DADLE and ^{125}I-FK 33824, respectively.

In the rat front cortex, the highest concentrations of mu receptors are in the layers I and IV, whereas delta receptors have their highest concentration in layers II, III, and V. In the corpus striatum, mu receptors occur in high concentration in clusters and as a subcallosal streak, while delta receptors occur diffusely distributed and often in low concentration in mu receptor cluster areas. The hippocampus has a high concentration of mu receptors in the pyramidal cell layer and a relatively low concentration of delta receptors. The olfactory tubercle, the nucleus accumbens, and the amygdala have a relatively high concentration of delta receptors and very low levels of mu receptors. The most striking contrast is in the thalamus and hypothalamus, which contain high density of mu receptors but very few delta receptors. In the midbrain, several structures are enriched in mu receptors, including the inferior colliculi, periaqueductal gray, the median raphe, and the interpeduncular nucleus. The periaqueductal gray and median raphe have very low levels of delta receptors, while moderate concentrations occur in the inferior colliculi and interpeduncular nucleus. In contrast, the pontine nuclei have a relatively high concentration of delta receptors and few mu receptors. Other caudal areas display relatively high concentrations of both delta and mu receptors, including the nucleus tractus solitarious, vagal fibers, the nucleus ambiguous, and the sustantia gelatinosa.

The localization described above suggests possible functional difference for these two types of opioid receptors. High concentrations of mu receptors are found in areas that are relevant to pain sensation or stimulation-analgesia production, such as substantia gelatinosa, the periaqueductal gray, the median raphe, the dorsomedial thalamus, and layer IV of the cortex. The localization of mu receptors is consistent with mu receptors mediating a major portion of opiate-induced analgesia. Recent observations of the lack of potent analgesic or antinociceptive activity of systemic nonpeptide delta agonists BW 373U86 and SNC 80 [9] are consistent with the low concentration of delta receptors found above areas that are rich with mu receptors.

Many of the areas found to be high in delta receptors are part of the limbic system associated with the control of emotion and reward behavior. Such areas include olfactory tubercle, the nucleus accumbens, and the amygdala. However, recent demonstrations have shown that delta agonists may have antidepression activity in the forced swimming test [24; Chap. 20], and the lack of abuse or self-administrative activity in monkeys [25; Chap. 23] may point to the unknown functions of delta receptors in the limbic system.

The recent observation that delta agonists could mitigate the mu analgesic-induced respiratory depression [26] is consistent with the rich distribution of both mu and delta receptors in nucleus tractus solitarious and vagal fibers. This negative regulation of delta agonist on mu opioid analgesic-induced side effects may suggest an opposite regulatory function of delta receptor in the mediation of the actions of mu receptors.

5 NG108-15 AND N4TG1 CELLS CONTAIN DELTA RECEPTORS ONLY

Once the heterogeneous nature of opioid receptors was recognized in the brain tissue, scientists began to search for neuronal cells that contain homogeneous single type of opioid receptor. Three cultured neuroblastoma cell lines were described in the early 1970s to contain opioid receptors [15,27,28] without knowing the true nature of opioid receptor subtypes. Chang and Cuatrecasas [14] demonstrated in 1979 that indeed neuroblastoma cells N4TG1 and neuroblastoma-glioma hybrid cells NG-108-15 did contain homogenous high-affinity opioid receptors with selectivity similar to delta receptor described in the brain tissue. This was convincing evidence supporting the concept of multiple opioid receptors, particularly the delta receptors. These cultured neuronal cells that contain a single type of opioid receptor were also vital to our understanding of the cellular actions of delta opioids when highly selective delta ligands were not available in the 1970s and 1980s.

Both N4TG1 and NG108-15 neuroblastoma-glioma hybrid cells were later extensively used as model systems for the investigation of the cellular activities of opioids, particularly for the studies of morphine tolerance and dependence, and delta receptor downregulation at the cellular and molecular levels. A cDNA library was later prepared from NG108-15 cells and successfully used for the gene cloning of delta receptors. Today, most investigators will use CHO, HEK 293 cells that permanently expressed cloned cDNAs of delta, mu, kappa, or mixed opioid receptors for the actions at molecular and cellular levels for opioids that are selective or nonselective to different types opioid receptors.

6 DELTA RECEPTOR DOWNREGULATION

Initially, many laboratories [29–32] were unable to demonstrate the downregulation of opioid receptors after chronic treatment with morphine despite the development of tolerance and dependence. After the recognition that there were subtypes of opioid receptors selective to morphine and enkephalins, we decided to reinvestigate this phenomenon of downregulation using delta receptor–containing N4TG1 cells after the confirmation that in vivo chronic treatment of morphine did not induce receptor downregulation as described in vivo by others [33].

To our surprise, delta receptors were rapidly downregulated after the treatment of neuroblastoma cells with either DADLE, natural enkephalins and their analogues, or endorphins. However, as expected morphine treatment did not induce delta-receptor downregulation. Morphine, in fact, behaved as an antagonist in blocking the downregulation induced by opioid peptides, such as DADLE. This model of neuroblastoma cells was later used by many investigators to study the mechanisms of internalization, downregulation, degradation, recycling, and trafficking of delta receptors, and this area of progress will be discussed by Law in Chapter 5.

7 DISCOVERY OF DELTA AND MU RECEPTOR HIGHLY SELECTIVE PEPTIDE ANALOGS

Although above mentioned delta and mu receptor selective peptides and opiates were observed and useful for the identification of distinctive delta receptor from other opioid receptors, they suffered from the low selectivity and cross-actions at high doses or concentrations of these ligands, especially in in vivo pharmacological studies. The needs of having highly selective ligands for all types of opioid receptors were clear. In addition to the above mentioned supports for the existence of delta receptors, the subsequent discovery of many mu and delta receptor-selective enkephalin analogues added more additional evidence and facilitate studies of delta receptors and other opioid receptors.

In 1984, Cotton and colleagues described ICI 174, 864 (N,N-diallyl-Tyr-Aib-Aib-Phe-Leu-OH) as a highly selective antagonist for the delta receptors [34]. This peptide was later discovered by Hertz and colleagues [35,36] to be the first inverse delta agonist.

In 1983, Mosberg and colleagues [37] synthesized delta receptor highly selective cyclic peptide analogues of enkephalin such as bis-penicillamine (Pen) enkephalins, cyclic[D-Pen^2, D-Pen^5] and cyclic[D-Pen^2, L-Pen^5] enkephalin (referred to as DPDPE and DPLPE, respectively). The significance and the contribution of these analogues to our understanding of delta receptors in vivo and in vitro are discussed by Hruby and Mosberg in Chapter 10 and Ossipov et al. in Chapter 18 in this volume. Using DPDPE and deltorphins, Porreca and colleagues demonstrated the potential existence of delta receptor subtypes, delta 1 and 2 receptors.

In 1981, Chang and colleagues [38] discovered a mu receptor highly selective peptide, morphiceptin (H-Tyr-Pro-Phe-Pro-NH_2) and subsequently synthesized a potent and stable [D-Pro^4]morphiceptin (or PL 017) [39]. These highly selectively mu receptor agonists, morphiceptin and analogues, provided excellent tools to determine the portion of the binding to mu versus delta receptors for a given labeled ligand [40]. Almost at the same time, Kosterlitz and Paterson [41] described an enkephalin analogue, [D-Ala^2, N-Me-Phe^4, Gly-ol^5]enkephalin (DAMGO), that was also a highly selective agonist for mu receptors. Today, the [^{3}H]DAMGO is the most popular labeled ligand for studies of mu receptors.

The isolation of delta receptor-selective peptides from frog skin provided further understanding of delta receptors. Two peptides, deltorphin I and II were isolated by Erspamer and colleagues in 1989 from frog skin [42]. The most striking observations in deltorphins isolated from frog skin was that the D amino acid of D-Ala was found in the second position of the peptides. Negri and Giannini discuss the details on deltorphins in Chapter 11 (this volume). Today, [^{3}H]deltorphin II is the most commonly used delta agonist for delta receptor binding studies.

The latest advance in delta receptor-selective peptides was the synthesis of TIPP and TIPP(φ) (H-Tyr-Ticφ[CH_2NH]Phe-Phe-OH] and analogues by Schiller and colleagues in 1992 [43,44]. In contrast to DPDPE and deltorphins, TIPP and TIPP(φ) behave as antagonists in most pharmacological tests except in the modulation of mu agonist–induced respiratory depression. The availability of a highly selective delta antagonist provided a useful tool for further understanding the physiolocal and pharmacological actions of delta receptors in vivo. Schiller discusses this area of advance in Chapter 12 (this volume).

8 DISCOVERY OF NONPEPTIDE DELTA RECEPTOR–SELECTIVE LIGANDS: NALTRINDOLE, BW373U86, SNC 80, AND TAN 67

Peptides are known to have limited permeability in blood brain barrier and poor oral bioavailability. To study central effects of peptides, they need to be injected centrally either by intracerebralventriculer (ICV) or by intrathecal (IT) spinal injection. For a long period of time, there were only peptides available for delta receptors. It is also unclear whether or not effects produced by central injected peptides differ from systemically applied compounds. This has hampered progress of the understanding of the functions of delta receptors. This difficulty was finally overcome by the discoveries of delta receptor-selective nonpeptide antagonists such as naltrindole (NTI) in 1988 [45,46], agonists BW373U86, (±)-4-((α-R*)-α-(2S*,5R*)-4-allyl-2,5-dimethyl-1-piperazinyl)-3-hydroxybenzyl)-N,N-diethylbenzamide, in 1992 [9, 47], and TAN 67, (-)-2-methyl-4aα-(3-hydroxyphenyl)-1,2,3,4,4a,5,12,12aα-octahydroquinolino[2,3,3g]isoquiniline dihydrobromide, in 1994 [48]. A methyl ether derivative of (+)BW 373U86 (which was known as SNC 80) was later obtained by Rice and colleagues, and shown to have further improved selectivity for delta receptors [49]. The pharmacophore structure of BW373U86 is the basis of many new series of delta receptor–selective compounds described in patents recently filed by many pharmaceutical companies and institutes that are reviewed in Chapter 8.

The use of BW 373U86 by systemic administration generated many interesting surprises in the delta receptor pharmacology. BW 373U86 did not show antinociceptive efficacy in the traditional analgesic models such as rat hot plate and tail flick assays, rat tail pinch assay, mouse tail pinch assay [9], and in the warm-water tail withdrawal procedure in rhesus monkeys [25]. The compound did show activity in mouse hot plate and acetic acid–induced writhing tests [40,51]. This inconsistency of antinociceptive activity of BW 373U86 in various pain models suggests that delta receptors may only play a minor role in the regulation of pain perception. Delta agonists may not be a potent analgesic and cannot replace morphine in the management of acute moderate to severe pain. However, delta agonists were shown to be active in suppressing nociception in inflammatory and neuropathic chronic pain [this volume, Chapter 18 by Ossipov et al].

Furthermore, it was found that BW 373U86 produced a brief (lasting ~ 5–10 sec) and nonlethal, single-episode convulsion in mice [52]. Naltrexone at high doses (>10 mg/kg) and naltrindole produced dose-dependent shift in the potency of BW 373U86 to induce a convulsion. Midazolam completely eliminated convulsion induced by BW 373U86. Pretreatment with a single injection of BW 373U86 produced dose-related reduction in the capacity of BW 373U86 to induce a second convulsion. Recovery of sensitivity to BW 373U86 to induce a second convulsion did not return to control levels for up to 2 weeks after pretreatment with a single injection of 32 mg/kg of BW 373U86. It is interesting to note that this BW373U86-induced convulsion could also be attenuated by a mu agonist, fentanyl [50]. Despite the brief and nonlethal nature of convulsion, future clinical developments for systemically active delta agonists will have to overcome this convulsive effect.

The third interesting observation was that BW 373U86 did not produce physical dependence after chronic infusion of the compound, but attenuated the development and expression of morphine abstinence precipitated by naloxone in rats [53]. Continuous infusion of BW 373U86 by a subcutaneously implanted minipump did not induce any abnormal behavior. After 6 days of BW 373U86 infusion, naloxone at a high dose or naltrindole did not precipitate morphinelike abstinence syndrome. BW 373U86 did not induce abstinence syndromes or modulate morphine abstinence precipitated by naloxone in chronic morphine-treated rats. However, naloxone-precipitated abstinence syndromes in morphine-dependent rats were partially suppressed by BW 373U86 when the compound was infused simultaneously with chronic morphine treatment. In rhesus monkeys, BW 373U86 did not produce reinforcing effects in a self-administration procedure [25]. These data suggest that delta receptors do not appear to mediate reinforcing effects and physical dependence, and it is likely that delta agonists will not produce abuse liability. Negus will discuss this area of research on substance abuse [this volume, Chap. 23].

Finally, unlike mu agonists, which cause respiratory depression, BW 373U86 did not produce respiratory depression in rats [26] and monkeys [25]. In contrast, BW 373U86 and other delta agonists were demonstrated to reverse the respiratory depression induced by a mu agonist, alfentanil, in rats [26]. Furthermore, this discovery eventually led to the synthesis and development of a mixed delta/mu agonist such as DPI-3290 and DPI-125 as a safer analgesic [54,55]. Gengo and Chang will discuss this area of research [this volume, Chap. 14].

The availability of nonpeptide delta agonists and antagonists has recently advanced our understanding of delta receptors and their potential therapeutic indications. New potential therapeutic applications were uncovered recently by using these nonpeptide compounds. For instance, there are implications of delta agonists in the treatment of depression [23; this volume, Chap. 20] with (+)BW 373U86 and SNC 80, the cardioprotection effects of TAN 67 and BW 373U86 [56,57; this volume, Chap. 25, by Gross and colleagues], peripheral analgesia [this volume, Chap. 19, by Narita and Suzuki], the regulation of sexual ejaculation of BW 373U86 and SNC 80 [Chang and colleagues, unpublished works], the modulation of micturition response (unpublished observations), the regulation of motility and water mobilization in the gastrointestinal tracts [this volume, Chap. 24, by Townsend and Brown], and immunomodulation [this volume, Chap. 22, by Weber and Gomez-Flores].

9 CLONING OF DELTA RECEPTORS

In late 1992, the mouse delta receptor was the first opioid receptor gene to be cloned by two independent groups—Evans et al. [58], and Kieffer et al. [59]. Both groups used the expression cloning technique with a cDNA library prepared from the NG 108-15 mouse neuroblastoma-glioma hybrid cells. The cDNA library was transfected into COS cells for receptor expression. These cells that expressed delta receptors were subsequently identified and isolated by the receptor autoradiography with ^{125}I-labeled DADLE, a radioisotope label of a high radio specific activity and a high binding affinity, by Evans et al. [58], and [^{3}H]diprenorphine, a labeled opiate with a very high affinity for delta receptor and a very low nonspecific binding background, by Kieffer et al. [59; this volume, Chap. 4, by Décaillot and Kieffer]. The human delta receptor was subsequently cloned by Yamamura's group in 1994 [60; this volume, Chap. 3 by Knapp et al.]. Genomic structure of mouse delta receptor gene was identified [61]. Sun and Loh will discuss the transcriptional regulation of delta receptor expression [this volume, Chap. 7]. Following the cloning of the delta receptor cDNA, mu and kappa receptors were cloned by several laboratories using homology screening technique with a cloned delta receptor cDNA probe [62–64]. As expected, all opioid receptors including delta receptors belong to the members of the superfamily of G protein–coupled seven-transmembrane receptors (GPCR). Opioid receptors mediate opioid actions through interactions with a variety of GTP-binding proteins (G proteins) to modulate intracellular messenger systems or downstream effectors such as ion channels, adenylyl cyclase, phospholipase C, and/or MAP kinase. Clark and Traynor will review the complex interactions between opioid receptors and G-proteins [this volume, Chap. 6]. Finally, delta receptor knockout transgenic mice were also generated [65] for physiological and pharmacological studies.

10 CONCLUSION

The major events important for the advances of delta receptor biology are chronologically listed in Table 1 for reference.

TABLE 1 Major Events That Were Important for the Advances of Delta Receptor Biology

Event	Year	Authors
Discovery NG108-15 neuroblastoma-glioma hybrid	1974	Klee and Nirenberg [27]
Discovery of enkephalins	1975	Hughes et al. [1]
Proposal of putative delta receptors in mouse vas deferens	1976	Lord et al. [7]
Discovery of DADLE	1977	Miller et al. [10,11]
Discovery of FK 33824	1977	Roemer et al. [12]
Available of ^{125}I-DADLE	1978	Chang et al. [13,15]
Discovery of enkephalin high-affinity delta-binding sites	1979	Chang et al. [14]
Discovery of differential distribution of delta and mu receptors in brain	1979	Chang et al. [16]
Characterization delta receptors in N4TG1 and NG108-15	1979	Chang et al. [14,15]
Discovery of highly selective mu agonists, morphiceptin, and DAMGO	1981	Chang et al. [38,39], Kosterlitz [41]
Delta receptor downregulation in neuroblastoma cells	1982	Chang et al. [33]
Discovery of DPDPE	1983	Mosberg et al. [37]
Discovery naltrindole (NTI)	1988	Portoghese et al. [45,46]
Discovery of deltorphins	1989	Erspamer et al. [42]
Cloning cDNA of mouse delta receptors	1992	Evan et al. [58], Kieffer et al. [59]
Discovery of BW373U86	1992	Chang et al. [9]
Discovery TIPP and TIPP (φ)	1992	Schiller et al. [43,44]
Discovery of SNC 80, a methyl ether BW 373U86	1994	Calderon et al. [49]
Cloning of the gene of human delta receptors	1994	Knapp et al. [60]
Delta receptor knockout transgenic mice	2000	Filliol et al. [65]

The discovery of enkephalins has led to the development of the hypothesis of delta receptors. The availability of neuronal cells that contain single type of delta receptor has provided additional evidence to the concept of multiple opioid receptors and a simple system for the studies of opioids at cellular and molecular levels. The synthesis of many delta receptor-selective peptides strengthens the concept and provides tools for the identification and cloning of delta receptors, and in vitro delta receptor pharmacology. The systemically active nonpeptide delta agonists and antagonists have facilitated advances of the delta receptor pharmacology in vivo. The final cloning of all types of opioid receptors including delta receptors and the production of transgenic mice deleted with various types of opioid receptors should further enhance our understanding of the physiology and biology of opioid receptors, including delta receptors, and their endogenous ligands, enkephalins. It is anticipated that these advances will eventually lead to the development of new therapeutic applications for delta receptor compounds, agonists or antagonists, in the coming years.

REFERENCES

1. Hughes J, Smith TW, Kosterlitz HW, Fothergill LA, Morgan BA, Morris HR. Nature 1975; 258:577–579.
2. Hollt V. Annu Rev Pharmacol Toxicol 1986; 26:59–77.
3. Gutstein HB, Akil H. Opioid analgesics. In: Hardman JG, Limbird LE, Gilman AG, eds. Goodman & Gilman's: The Pharmacological Basis of Therapeutics. 10th ed. New York: McGraw-Hill, 2001:569–619.
4. Portoghese PS. J Med Chem 1965; 8:609–616.
5. Gilbert PE, Martin WR. J Pharmacol Exp Ther 1976; 198:66–82.
6. Martin WR, Eades CG, Thompson JA, Huppler RE, Gilbert PE. J Pharmacol Exp Ther 1976; 197:517–532.
7. Lord JA, Waterfield AA, Hughes J, Kosterlitz HW. Nature 1977; 267:495–499.
8. Schulz R, Wuster M, Kreness H, Herz A. Nature (London) 1980; 285:242–243.
9. Chang K-J, Rigdon GC, Howard JL, McNutt RW. J Pharmacol Exp Ther 1993; 267:852–857.
10. Miller RJ, Chang K-J, Cuatrecasas P, Wilkinson S. Biochem Biophys Res Commun 1977; 74:1311–1317.
11. Beddell CR, Clark RB, Hardy GW, Lowe LA, Ubatuba FB, Vane JR, Wilkinson S, Chang KJ, Cuatrecasas P, Miller RJ. Proc R Soc Lond B Biol Sci 1977; 198:249–265.
12. Roemer D, Buescher HH, Hill RC, Pless J, Bauer W, Cardinaux F, Closse A, Hauser D, Huguenin R. Science 1977; 268:547–549.
13. Miller RJ, Chang K-J, Leighton J, Cuatrecasas P. Life Sci 1978; 22:379–387.
14. Chang K-J, Cuatrecasas P. J Biol Chem 1979; 254:2610–2618.
15. Chang K-J, Miller RJ, Cuatrecasas P. Mol Pharmacol 1978; 14:961–970.

16. Chang K-J, Cooper BR, Hazun E, Cuatrecasas P. Mol Pharmacol 1979; 16: 91–104.
17. Pert CB, Snyder SH. Science 1973; 179:1011–1014.
18. Simon EJ, Hiller JM, Edelmand I. Proc Natl Acad Sci USA 1973; 70:1947–1949.
19. Terenius L. Acta Pharmacol Toxicol 1973; 33:377–384.
20. Pert CB, Snyder SH. Mol Pharmacol 1974; 10:868–879.
21. Pasternak GW, Snowman AM, Snyder SH. Mol Pharmacol 1975; 11:735–744.
22. Simon EJ, Groth J. Proc Natl Acad Sci USA 1975; 72:2404–2407.
23. Goodman RR, Snyder SH, Kuhar MJ, Young WS. Proc Natl Acad Sci USA 1980; 77:6239–6243.
24. Broom DC, Jutkiewicz EM, Folk JE, Traynor JR, Rice KC, Woods JH. Neuropsychopharmacology 2002; 26:744–755.
25. Negus SS, Butelman ER, Chang K-J, DeCosta B, Winger G, Woods JH. J Pharmacol Exp Ther 270:1025–1034
26. Su YF, McNutt RW, Chang K-J. J Pharmacol Exp Ther 1998; 287:815–823.
27. Klee WA, Nirenberg M. Proc Natl Acad Sci USA 1974; 71:3474–3477.
28. Law PY, Herz A, Loh HH. J Neurochem 1979; 33:1177–1187.
29. Pert CB, Pasternak GW, Snyder SH. Science 1973; 193:1359–1361.
30. Klee WA, Streaty RA. Nature 1974; 248:61–63.
31. Holt V, Dum J, Blasrg J, Schubert P, Herz A. Life Sci 1975; 16:4141–4145.
32. Simon EJ, Hiller JM. Fed Proc 1978; 37:141–146.
33. Chang K-J, Eckel RW, Blanchard SG. Nature 1982; 296:446–448.
34. Cotton R, Shaw JS, Miller L, Giles MG, Timms D. Eur J Pharmacol 1984; 97: 331–339.
35. Costa T, Hertz A. Proc Natl Acad Sci USA 1989; 86:7321–7325.
36. Costa T, Lang J, Gless C, Hertz A. Mol Pharmacol 1990; 37:383–394.
37. Mosberg HI, Hurst R, Hruby VJ, Gee K, Yamamura HI, Gallian JJ, Burks TF. Proc Natl Acad Sci USA 1983; 80:5871–5874.
38. Chang K-J, Killian A, Hazum E, Cuatrecasa P, Chang J-K. Science 1981; 212: 75–77.
39. Chang K-J, Wei ET, Killian T, Chang J-K. J Pharmacol Exp Ther 1983; 227:403–408.
40. Chang K-J. Psychopharmacol Bull 1981; 17:108–111.
41. Kosterlitz HW, Paterson SJ. Br J Pharmacol 1981; 73:299.
42. Erspamer V, Melchiorri P, Falconieri-Erspamer G, Negri L, Corsi R, Severini C, Barra D, Simmaco M, Kreil G. Proc Natl Acad Sci USA 1989; 86:5188–5192.
43. Schiller PW, Nguyen TM-D, Weltrowska G, Wilkes BC, Marsden BJ, Lemieux C, Chung NN. Proc Natl Acad Sci USA 89:11871–11875.
44. Schiller PW, Weltrowska G, Nguyen TM-D, Wilkes BC, Chung NN, Lemieux C. J Med Chem 1993; 36:3182–3187.
45. Portoghese PS, Sultana M, Nagase H, Takemori AE. J Med Chem 1988; 31: 281–282.
46. Portoghese PS, Sultana M, Takemori AE. Eur J Pharmacol 1988; 146:185–186.

47. Lee PHK, Mc Nutt RW, Chang K-J. A non-peptide delta-opioid receptor agonist BW 373U86 suppresses naloxone-precipitated morphine abstinence. In: Meeting of the International Narcotics Research Conference, Keystone, Colorado, USA, 1992, Abstract 34.
48. Nagase H, Wakita H, Kawai K, Endoh T, Matsuura H, Tanaka C, Takezawa Y. Jpn J Pharmacol 1994; 64(suppl 1):35.
49. Calderon SN, Rothman RB, Porreca F, Flippen-Anderson JL, McNutt RW, Xu H, Smith LE, Bilsky EJ, Davis P, Rice KC. J Med Chem 1994; 37:2125–2128.
50. O'Neill SJ, Collins MA, Pettit HO, McNutt RW, Chang K-J. J Pharmacol Exp Ther 1997; 282:271–277.
51. Wild KD, McCormick J, Bilsky EJ, Vanderah TW, McNutt RW, Chang K-J, Porreca F. J Pharmacol Exp Ther 1993; 267:858–865.
52. Comer SD, McNutt RW, Chang K-J, DeCosta BR, Mosberg HI, Woods JH. J Pharmacol Exp Ther 1993; 267:866–874.
53. Lee PHK, McNutt RW, Chang K-J. J Pharmacol Exp Ther 1993; 267:883–887.
54. Gengo PJ, Pettit HO, O'Neill SJ, Wei K, McNutt RW, Chang K-J, J Pharmacol Exp Ther. In press.
55. PJ Gengo, HO Pettit, O'Neill SJ, Su YF, McNutt RW, Chang K-J. J Pharmacol Exp Ther. In press.
56. Schultz JEJ, Hsu AK, Nagase H, Gross GJ. Am J Physiol 1998; 274:H909–H914.
57. Patel HH, Hsu A, Jeannine Moore, Gross GJ. J Mol Cell Cardiol 2001; 33:1455–1465.
58. Evans CJ, Keith DE, Morrison H, Magendzo K, Edwards RH. Science 1992; 258:1952–1955.
59. Kieffer BL, Befort K, Gaveriaux-Ruff C, Hirth CG. Proc Natl Acad Sci USA 1992; 89:12048–12052.
60. Knapp RJ, Malatynska E, Fang L, Li X, Babin E, Nguyen M, Santoro G, Varga E, Hruby VJ, Roeske WR, Yamamura HI. Life Sci 1994; 54:PL463–PL469.
61. Gylys KH, Tran N, Magendzo K, Zaki P, Evans CJ. Neuroreport 1997; 8:2369–2372.
62. Chen Y, Mestek A, Liu J, Hurley JA, Yu L. Mol Pharmacol 1993; 44:8–12.
63. Li S, Zhu J, Chen C, Chen Y-W, Deriel JK, Ashby B, Liu-Chen L-Y. Biochem J 1993; 295:629–633.
64. Meng F, Xie G-X, Thompson RC, Mansour A, Goldstein A, Watson SJ, Akil H. Proc Natl Acad Sci USA 1993; 90:9954–9958.
65. Filliol D, Gholand S, Chluba J, Martin M, Matthes HWD, Simonin F, Befort K, Gavereriaus C, Dierich A, LeMeur M, Valverde O, Maldonado R, Kieffer BL. Nat Genet 2000; 25:195–200.

2

Cloning of Delta Opioid Receptors

Arnaud Lacoste and Christopher J. Evans
University of California at Los Angeles, Los Angeles, California, U.S.A.

1 INTRODUCTION

Cloning of the opioid receptors did not come easy, and there were many false claims along the way. Efforts began in the mid-1980s in the wake of cloning of the opioid peptide precursors. However, it was not till 1992 that the delta receptor was first cloned and provided the critical probes leading to the characterization of the entire family of opioid receptors.

The history leading up to the discovery of opioid receptors, and the delta receptor in particular, is detailed elsewhere in this volume. Briefly, opioid receptors as specific sites of interaction with opioid ligands were first described in the early 1970s using crude binding of radiolabeled alkaloid opiates to various tissue homogenates [1]. The stereospecificity and saturation of opiate binding sites provided evidence of a limited number of interactive sites and made much less credible the view that opiates may be interacting with membrane lipids or in another nonspecific manner. At about the same time, behavioral experiments with multiple opioid drugs revealed heterogeneity in opioid-induced behaviors, suggesting the presence of multiple opioid receptors [2]. Following elegant experiments by John Hughes and Hans Kosterlitz [3] resulting in the discovery of the endogenous opioid peptides methionine enkephalin and leucine enkephalin, a search for a selective receptor for the enkephalins ensued. Binding assays and bioassays ranking opioid ligand activities revealed a high-affinity receptor for the enkephalins [4]. This receptor was christened the delta opioid receptor.

2 CHARACTERISTICS OF OPIOID RECEPTORS

Developing a strategy for the characterization of opioid receptors in the 1970s and early 1980s was not straightforward, given the lack of understanding of most other receptor systems. The fact that opioid receptors were membrane-associated was realized early since receptor binding was enriched in membrane preparations and the hydrophilic opioid peptides, which are membrane impermeable, bound to, and activated receptors in intact systems. However, whether the receptor consisted of a multisubunit complex similar to acetylcholine channels, a single membrane protein, an enzyme or a lipid derivative such as cerebroside sulphate (see arguments by Cho et al. [5]), was unknown and mattered greatly for the approaches pursued.

During the 1980s, the characterization of the nicotinic receptor channel, the insulin receptor kinase and G protein–coupled receptors such as rhodopsin and adrenergic receptors, greatly improved our understanding of receptor-mediated cellular signaling [6]. With regard to the characteristics of opioid receptors and indeed the final cloning of opioid receptors, the field owes much to the NG108-15 cell line developed in 1974 by Klee and Nirenberg [7]. The NG108-15 cell line was obtained by fusion of a rat glioma cell line (C6BU1) with a mouse neuroblastoma cell line (N-18TG2). This hybrid cell line has played a major role in elucidating the signaling and binding characteristics of opioid receptors and their likelihood of being G protein–coupled receptors.

Following experiments by Collier and Roy in 1973 [8] showing that opiates could inhibit prostaglandin E_1 or E_2 stimulation of adenylate cyclase in brain homogenates, inhibition of adenylate cyclase by opiate receptors in the NG108-15 cells was also demonstrated [9]. The dependence of GTP and sodium for the opiate inhibition of adenylate cyclase in NG108-15 cells showed an intimate relationship between GTP and signaling [10], and a role for GTP hydrolysis in cyclase signaling was demonstrated soon after [11]. Binding assays also provided evidence of opiate receptor association with guanyl nucleotides and several groups demonstrated sensitivity of opiate agonist binding to GTP and GTP analogues (for review see Snyder et al. [12]). Perhaps, the most powerful evidence that opioid receptors were likely G protein–coupled receptors was that pertussis toxin, which ADP-ribosylates and inactivates the alpha subunits of inhibitory G proteins, abolished opioid agonist actions in most systems including those coupled to adenylate cyclase, as well as calcium and potassium channels (reviewed by Law et al. [13] and Childers [14]).

By the late 1980s, evidence became heavily stacked in favor of the opioid receptors being members of the family of G protein–coupled receptors. This family of receptors was typified by rhodopsin and the adrenergic receptor, which were cloned in the early 1980s. Through the 1980s it became increasingly frustrating for opioid researchers to observe many other neurotransmitter and neuromodulator receptors cloned, but no opioid receptors.

3 APPROACHES TO CLONE OPIOID RECEPTORS

One of the main approaches to clone opioid receptors copied techniques used to identify beta-adrenergic receptor [15] and opioid peptide precursor transcripts [16]. This well-established strategy involved isolation and purification of the protein and determination of partial amino acid sequence by Edman degradation. Protein sequence information could then be used to design DNA probes, screen cDNA or genomic libraries, and identify nucleic acid stretches encoding the protein. The major issue was that the opioid receptor is membrane bound and required solubilization prior to isolation. Unfortunately, removing the receptor from its membrane environment either compromised or completely eliminated binding, an important characteristic required for purification. A review by Simon and Hiller in 1988 [17] revealed many issues which thwarted first attempts to purify opioid receptors, including the lack of a rich source of opioid receptors, the sensitivity of opioid binding to detergents, and the lack of tightly/irreversibly binding probes. This classical cloning approach did result in the identification of OBCAM, named because it bound to a morphine affinity column (Opiate Binding), and it belonged to a subfamily of proteins which includes the neural cell adhesion molecule (NCAM). Partial protein sequencing of purified OBCAM was used to generate DNA probes and isolate a cDNA clone encoding the entire protein [18]. Although OBCAM has been implicated in opioid actions (reviewed by Loh and Smith [19]), its function and relationship to opioid receptors are unclear. A clever twist to the classical cloning approach used a biotinylated derivative of the opioid ligand beta-endorphin [20], which was bound to the receptor prior to solubilization to provide a biotin tag facilitating purification. This strategy enabled partial Edman sequencing of the mu opioid receptor and provided a hook to clone the entire mu receptor in 1993.

A second strategy used cloning by homology. Although this approach undoubtedly identified opioid receptor clones, it did not reveal the identity of opioid receptors and claims were retrospective. The approach was based on the assumption that the opioid receptors would be G protein–coupled receptors, and as indicated above, earlier pharmacological evidence strongly supported this assumption. Using PCR with primers targeting conserved regions of G protein–coupled receptors or low-stringency hybridization with DNA probes encoding other G protein–coupled receptors, many orphan receptors have and continue to be identified. Besides technical concerns, including choice and preparation of appropriate probes, cDNA or genomic libraries, and starting tissue, the major hurdle of the approach was selection of an appropriate assay to determine if orphan clones were indeed opioid receptors. Three main tools provided effective screens for identifying orphan receptor clones: 1) expression in cell lines devoid of opioid receptors followed by pharmacological characterization; 2) correlation of novel transcripts with pharmacologically determined receptor distribution in various tissues and cell lines; and 3) coregulation with receptor binding. An excellent example of this approach discovered a novel somatostatin receptor (SSTR2B) from NG108-15 cells [21]. The strategy used degenerate primers from conserved transmembrane domains (TMDs) of other G protein–coupled receptors to amplify cDNA from NG108 cells. Amplified cDNA was then cloned and individual amplicons were sequenced. Novel G protein–coupled receptor-like sequences were identified and used to screen a NG108-15 cDNA library for full-length transcripts. Clones were then inserted into a mammalian expression vector, and expressed in COS cells where binding identified the cloned receptor as a somatostatin receptor. This very sound approach could also have resulted in cloning of the delta opioid receptor.

A third approach used subtractive hybridization combined with pharmacological regulation of opioid binding in attempts to enrich clones corresponding to the opioid receptor. This approach used cell culture models such as NGF-induced upregulation of delta opioid receptors in PC-12 cells [22] and delta receptor downregulation in NG108-15 cells. Unfortunately, some of these strategies were based on incorrect assumptions. For instance, we now know that downregulation of opioid binding does not require changes of opioid receptor mRNA levels (for review see Law et al., 2000 [13]). Though unsuccessful in the case of cloning opioid receptors, subtractive hybridization approaches and differential display have been exceptionally powerful for the identification of other proteins, such as those involved in development [23]. As with the homology approach mentioned previously, this strategy required authentication of any clones isolated. Clones were identified using this approach but DNA rearrangements and issues with marginal binding experiments compromised these studies and led to other premature claims of cloning of opioid receptors.

The fourth approach, which eventually proved successful, was expression cloning. In this strategy cDNA from a tissue or cell line expressing the receptor was cloned into a mammalian expression vector and transfected in to a cell line devoid of opioid receptors. Transfected cells could then be screened for opioid receptor binding. This approach was used by Xie et al. [24] to identify a G protein–coupled receptor with opioid-binding properties. The study used an oligo(dT)-primed human placental cDNA library cloned in the mammalian expression vector pME18S to express proteins in COS-7 cells. Intact transfected COS cells were screened for opioid receptors by binding dynorphin-32 and panning for cells with bound ligand using plates coated with a monoclonal antibody recognizing the C-terminus of dynorphin-32. After multiple cycles of panning, isolation of plasmids from enriched transfected COS cells by Hirt lysis, regrowing and retransfecting plasmids from enriched COS cells, a clone was identified and named hK1R. When hK1R was transfected into COS cells it was selected by the Dyn-32 panning assay, and binding of Dyn-32 could be disrupted by the alkaloid kappa agonist U-50488. Although hK1R was clearly a G protein–coupled receptor, it had close homology to neurokinin receptors, which couple to Gs and not to pertussis toxin–sensitive G proteins. Transfected hK1R cells bound many alkaloid and peptide opioid ligands, but binding selectivity and affinity were far below those expected from a cell with bona fied opioid receptors. There was some legitimate concern that the clone may have been truncated or unable to retain the original pharmacological properties in the transfected cell line. However, subsequent research has made these claims mute, and hK1R is now classified as an authentic neurokinin receptor.

One critical aspect of the expression cloning strategy is to identify a rich source of receptor and create cDNA that has reasonable representation of the coding region of the receptor. The source chosen for the initial cloning of the delta opioid receptor was the NG108-15 cell line. Creating cDNA with opioid receptor representation had an unpredicted pitfall, in which the Xie et al. study [24] unknowingly fell into. There are two common ways to generate cDNA: 1) using random oligomers, which prime from many different positions in the transcript, or 2) using oligo dT which generally prime only at the polyA tail. The problem with oligo-dT primed cDNA is that coding regions of mRNAs with extended 3′ untranslated regions may not be represented since reverse transcription rarely generates cDNA of >5 Kb in size. It was commonly considered that G protein–coupled receptors were encoded by simple mRNAs, and this was indeed the case of many G protein–coupled receptors. However, the opioid receptor mRNAs have extended 3′ untranslated regions, which can be >10 Kb and require random priming to be represented in the resulting cDNA.

4 CLONING OF THE DELTA OPIOID RECEPTOR

Over the same time frame, one group in France [25] and another in the United States [26], independently cloned the delta opioid receptor using remarkably similar strategies. Both approaches used the NG108-15 cell line to create a randomly primed cDNA library. Both groups also used expression cloning in COS cells and binding of a radiolabeled peptide ligand to identify the receptor protein. The major differences between the two approaches were in the screening: the U.S. group used iodinated D-Ala-D-Leu, enkephalin (DADLE) [27] binding to intact cells followed by autoradiography, while the French group screened pools of clones in a conventional binding assay using ^{3}H-labeled Tyr-D-Thr-Gly-Phe-Leu-Thr (DTLET) [28]. The clones identified by the two groups were identical. Comparison of binding affinities of a series of opioid and non-opioid ligands confirmed that the clone was a delta opioid receptor. Functional coupling to adenylate cyclase inhibition was also demonstrated [26]. Many subsequent experiments confirmed that the receptor cloned was indeed the delta opioid receptor.

5 CHARACTERISTICS OF THE CLONED DELTA RECEPTOR

The predicted amino acid sequence of the clone obtained from NG108-15 cells (Fig. 1) identified the delta opioid receptor as a member of the seven-transmembrane family of G protein–coupled receptors. The receptor had seven hydrophobic, predicted transmembrane domains (TMDs) and close homology to many other G protein–coupled receptors, including receptors for somatostatin, interleukin-8 and angiotensin. This structural homology confirmed the widely held notion that opioid receptors would indeed be members of this enormous family of receptors (the human genome has recently been calculated to contain approximately 950 G protein–coupled receptors [29]).

Since the clones were obtained from a mouse/rat hybrid cell line, an important question was whether the cloned delta receptor was from rat or mouse. Southern blot analysis using DNA from rat, NG108, and mouse showed that the clone was of murine origin. The Southern analysis also revealed that the coding sequence of the delta receptor from NG108-15 cells was likely to have arisen from multiple exons. Indeed, analysis of the delta gene (located on the proximal end of chromosome 1 in the human and chromosome 4 in mouse) reveals the presence of three protein coding exons with boundaries just following the first and fourth TMDs of the receptor (see Fig. 2 and Massotte and Kieffer [30] for review of gene structure). The mouse delta receptor has two predicted N-linked glycosylation sites at the N-terminus and several kinase sites in the second intracellular loop and C-terminal domain, which have been proposed to be involved in regulation of the receptor activity.

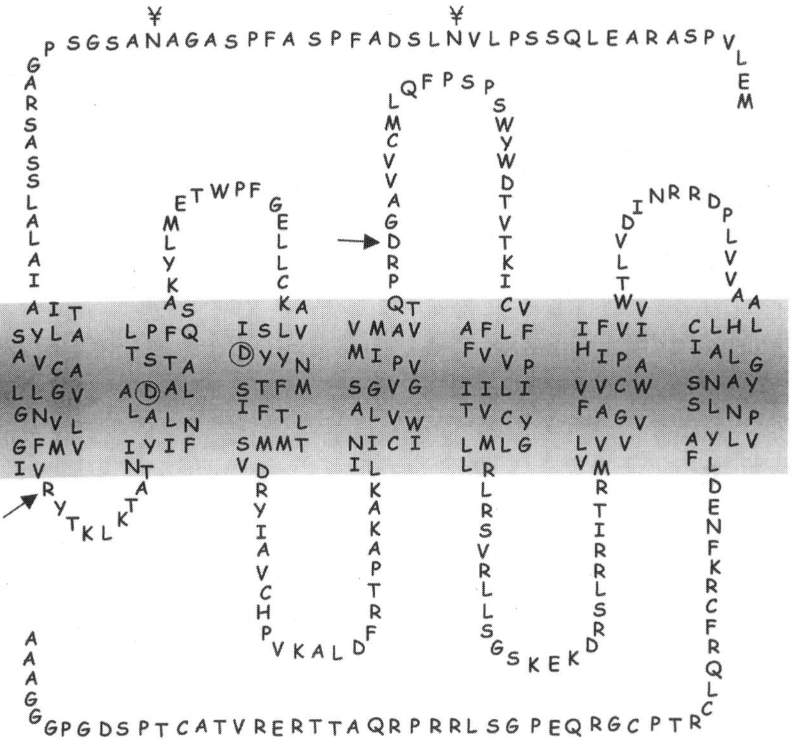

FIGURE 1 Predicted sequence and membrane topography of the murine delta opioid receptor. The mouse sequence was obtained from Genbank (accession No. L07271). Negatively charged amino acids within TMD 2 and 3 are circled and extracellular N-linked glycosylation sites are indicated (¥). Arrows indicate intron/exon boundaries in the mouse genome. (From Ref. 40.)

6 CLONING OF THE OPIOID RECEPTOR FAMILY

The cloning of the delta receptor set in motion a competitive race to identify other members of the opioid receptor family. Homologous orphan clones were quickly assessed for opioid receptor binding properties, which resulted in the identification of the kappa receptor and reassignment of an orphan clone as the delta opioid receptor [31]. PCR, genomic, and cDNA screens revealed the mu opioid receptor and an extremely abundant orphan member, named opioid receptor-like (ORL-1) receptor (reviewed by Massotte and Kieffer [30]).

```
                                                      ¥          ^   ¥
DELTA   ---------------------MEPAPS-AGAELQ-PPLFANASDAY-----PSAFPSAGANASGPPGARSASS
MU      MDSSAAPTNASNCTDALAYSSCSPAPS-PGSWVNLSHLDGNLSDPCG----PNRTDLGGRDSLCPP--TGSPS
KAPPA   --------MESPIQIFRG----EPGPTCAPSACL-PPNSSAWFPGWAEP--DSNGSAGSEDAQLEP---AHIS
ORL-1   -----------------------MEPL-FPAPFW-EVIYGSHLQGNLSLLSPNHSLLPPHLLLNAS-HGAFLP
                               *    :             .         .         .  .  .

                     1               ↓                   2
DELTA   LALAIAITALYSAVCAVGLLGNVLVMFGIVRYTKMKTATNIYIFNLALADALATSTLPFQSAKYLMETWPFGE
MU      MITAITIMALYSIVCVVGLFGNFLVMYVIVRYTKMKTATNIYIFNLALADALATSTLPFQSVNYLMGTWPFGT
KAPPA   PAIPVIITAVYSVVFVVGLVGNSLVMFVIIRYTKMKTATNIYIFNLALADALVTTTMPFQSTVYLMNSWPFGD
ORL-1   LGLKVTIVGLYLAVCVGGLLGNCLVMYVILRHTKMKTATNIYIFNLALADTLVLLTLPFQGTDILLGFWPFGN
            : * .:*  * . **.** ***: *:*:*******************:*.  *:***..  *:  ****

                  3                                                4
DELTA   LLCKAVLSIDYYNMFTSIFTLTMMSVDRYIAVCHPVKALDFRTPAKAKLINICIWVLASGVGVPIMVMAVTRP
MU      ILCKIVISIDYYNMFTSIFTLCTMSVDRYIAVCHPVKALDFRTPRNAKIINVCNWILSSAIGLPVMFMATTKY
KAPPA   VLCKIVISIDYYNMFTSIFTLTMMSVDRYIAVCHPVKALDFRTPLKAKIINICIWLLSSSVGISAIVLGGTKV
ORL-1   ALCKTVIAIDYYNMFTSTFTLTAMSVDRYVAICHPIRALDVRTSSKAQAVNVAIWALASVVGVPVAIMGSAQV
         *** *::********* ***   ******:*:***::***.**. :*: :*:. * *:* :*:.   .:. ::

        ↓                            5
DELTA   RDGA--VVCMLQFPSPSW-YWDTVTKICVFLFAFVVPILIITVCYGLMLLRLRSVRLLSGSKEKDRSLRRITR
MU      RQGS--IDCTLTFSHPTW-YWENLLKICVFIFAFIMPVLIITVCYGLMILRLKSVRMLSGSKEKDRNLRRITR
KAPPA   REDVDVIECSLQFPDDDYSWWDLFMKICVFIFAFVIPVLIIIVCYTLMILRLKSVRLLSGSREKDRNLRRITR
ORL-1   EDEE--IECLVEIPTPQD-YWGPVFAICIFLFSFIVPVLVISVCYSLMIRRLRGVRLLSGSREKDRNLRRITR
        .:     : * : :.     :*   .  **:*:*:*::*:*:* *** **: **:.**:****:****.******

                  6                              7
DELTA    MVLVVVGAFVVCWAPIHIFVIVWTLVDIDRRDPLVVAALHLCIALGYANSSLNPVLYAFLDENFKRCFRQLC
MU       MVLVVVAVFIVCWTPIHIYVIIKALVTIP-ETTFQTVSWHFCIALGYTNSCLNPVLYAFLDENFKRCFREFC
KAPPA    LVLVVVAVFVVCWTPIHIFILVEALGSTS-HSTAALSSYYFCIALGYTNSSLNPILYAFLDENFKRCFRDFC
ORL-1    LVLVVVAVFVGCWTPVQVFVLAQGLGVQP-SSETAVAILRFCTALGYVNSCLNPILYAFLDENFKACFRKFC
         :*****.. *: **:*::::::    *                :* ****.**.***:********* ***.:*

DELTA    RKPCGRPDPSSFSRAREATARERVTA-CTPSDGPGGGAAA-----------
MU       IPTSSNIEQQNSTRIRQNTRDHPSTA-NTVDRTNHQLENLEAETAPLP
KAPPA    FPLKMRMERQSTSRVRNTVQD-PAYL-RDIDGMNKPV-----------
ORL-1    CASALRRDVQVSDRVRSIAKD-VALACKTSETVPRPA-----------
              . : .   * *. .          .
```

FIGURE 2 Alignment and sequence comparison of the human opioid receptor family. Alignment was performed by T-COFFEE *(http://www.ch.embnet.org/software/TCoffee.html)* and the clustal format is presented. Sequences were obtained from Genbank (accession Nos.: human delta, NM_000911; mu, NM_ 000914; kappa, NM_000912; ORL-1, NM_000913). Identical amino acids (*), highly conserved (:), less conserved (.), or no conservation ( ) are indicated under the alignments. Consensus extracellular N-linked glycosylation sites (¥) are shown for the human sequence. Predicted TMDs regions are shaded and numbered 1–7. The ^ above the phenylalanine in the first extracellular domain of the delta is a polymorphism [41] and in some populations this is a cysteine residue (compare accession numbers U10504 and NM-000911). Arrows designate conserved intron/exon splice junctions among all four receptors.

Sequence homology among members of the opioid receptor family is depicted in Fig. 2. The highest conserved regions are the two intracellular loops, the C-terminal region adjacent to TMD 7, and TMDs 1, 2, and 4. The receptors are most highly divergent at the N- and C-terminal extremities as well as the third extracellular loop (between TMD 6 and 7). Interestingly, all members of the opioid receptor family identified in human and mouse exhibit conserved intron/exon boundaries just C-terminal to TMD 1 and TMD 4. Other features conserved among the opioid receptor family are the pattern of cysteine residues in extracellular loops, multiple N-linked glycosylation sites at the N-terminus, acidic residues in TMDs 2 and 3, and conserved kinase sites in the second intracellular loop (between TMD 5 and 6).

7 CLONING DELTA OPIOID RECEPTORS IN THE POSTGENOMIC ERA: COMPARATIVE APPROACHES AND EVOLUTIONARY PERSPECTIVES

Classical cloning approaches and comparative genomic analyses reveal that the existence of delta opioid receptors can be traced back at least 430 million years. Indeed, following cloning of the murine delta receptor from the NG108-15 cell line, complete delta receptor transcripts have been sequenced in distantly related vertebrates including mammals, such as human and rat as well as zebrafish (*Danio rerio*). Delta opioid receptors, together with other opioid receptors, can also be found in the recently sequenced genome of the pufferfish, *Takifugu rubripes* (see Fugu clone JGI-9982, JGI-18324, JGI-3781, and JGI-343). Partial delta opioid receptor cDNA sequences have been obtained from monkey (*Lagothrix lagotricha*) and pig (*Sus scrofa*), and, using PCR, genomic sequences have been identified in cow (*Bos taurus*), chicken (*Gallus domesticus*), bullfrog (*Rana catesbeiana*), stripped bass (*Morone saxatilis*), thresher shark (*Alopias vulpinus*), and Pacific hagfish (*Eptatretus stoutii*), demonstrating that this receptor is present in all vertebrate phyla [32].

High interspecies variation of delta receptor sequences are observed at the N-terminus and C-terminus as well as in the extracellular loops between TMD 4 and 5 and TMD 6 and 7 (Fig. 3). This conservation pattern is similar to that observed between members of the opioid receptor family. Analyses of gene structures also reveal identical intron/exon boundaries between evolutionary distant species, suggesting that the family of opioid receptors has evolved by gene duplication of a common ancestor that appeared before or during the emergence of early agnathans. Interestingly, the genome of the prochordate *Ciona intestinalis,* a representative of the closest invertebrate ancestor of all vertebrates, contains at least one opioid receptor-like sequence, suggesting that opioid receptors may have appeared during or before the Cambrian explosion, approximately 550 million years ago. Functional studies

```
                                         ¥        ^      ¥                             1
Human     --------------MEPAPSAGAELQPPLFANASDAYPSAFPSAGANASGPPGARSASSLALAIAITALYSAVCAVGLLGNVLVMFGIV
Rat       --------------MEPVPSARAELQFSLLANVSDTFPSAFPSASANASGSPGARSASSLALAIAITALYSAVCAVGLLGNVLVMFGIV
Mouse     --------------MELVPSARAELQSSPLVNLSDAFPSAFPSAGANASGSPGARSASSLALAIAITALYSAVCAVGLLGNVLVMFGIV
Zfish     MEPSVIPGA-----------DIPDLYSINPFNVTFPDDVLSFVPDGRNYTEPNPVKSRGIIIAISITALYSVICVVGLLGNILVMYGVV
Fugu      MEPVTVSGASGASGVSGAQLSLSDLFSVLPLNVSLPDEAAAERLQNSSRQEVPGRSAAGLIIAVSITALYSLICVVGLLGNILVMYGVV
                                .:*      * : .                   .: .: :*::****** :*.******:***:*:*

          ↓                 2                                      3
Human     RYTKMKTATNIYIFNLALADALATSTLPFQSAKYLMETWPFGELLCKAVLSIDYYNMFTSIFTLTMMSVDRYIAVCHPVKALDFRTPAK
Monkey                                                                                >PVKALDCRTPAK
Pig           >KTATNIYIFNLALADALATSTLPFQSAKYLMETWPFGELLCKAVLSIDYYNMFTSIFTLTMMSVDRYIAVCHPVKALDFRTPAK
Cow           >KTATNIYIFNLALADALATSTLPFQSAKYLMETWPFGELLCKVVLSIDYYNMFT<
Rat       RYTKLKTATNIYIFNLALADALATSTLPFQSAKYLMETWPFGELLCKAVLSIDYYNMFTSIFTLTMMSVDRYIAVCHPVKALDFRTPAK
Mouse     RYTKLKTATNIYIFNLALADALATSTLPFQSAKYLMETWPFGELLCKAVLSIDYYNMFTSIFTLTMMSVDRYIAVCHPVKALDFRTPAK
Chick         >KTATNIYIFNLALADALATSTLPFQSAKYLMETWPFGELLCKLVLSIDYYNMFT<
Frog          >KTATNIYIFNLALADALATSTLPFQSAKYLMETWPFGEVLCKVVLSIDYYNMFT<
Bass          >KTATNIYIFNLAFADALATSTLPFQSAKYLMSTWPFGEPLCKVVIAIDYYNMFT<
Zfish     RYTKLKTATNIYIFNLALADALATSTLPFQSTKYLMNTWPFGELLCKVVIAIDYYNMFTSIFTLTMMSVDRYIAVCHPVRALEFRTPIK
Fugu      RYTKMKTATNIYIFNLALADALATSTLPFQSAKYLMSTWPFGEVLCKLIIAIDYYNMFTSIFTLTMMSVDRYIAVCHPVRALGFRTPVK
Shark         >KTATNIYIFNLALADALATSTLPFQSAKYLMDTWPFGELLCKTVLSIDYYNMFT<
          ****:************:*************:****.****** *** :::****************************:** **** *

                      4                ↓                                   5
Human     AKLINICIWVLASGVGVPIMVMAVTR-PRDGAVVCMLQFPSPSWYWDTVTKICVFLFAFVVPILIITVCYGLMLLRLRSVRLLSGSKEK
Monkey    AKLINICIWVLASGIGVPIMVMAVTR-PRDGAVVCMLQFPSPSWYWDTVTKICVFLFAFVVPILIITVCYGLMLLRLRSVRLLSG<
Pig       AKLINICIWVLASGVGVPIMVMAVTR-PRDGAVVCMLQFPSPSWYWDTVTKICVFLFAFVVPILVITVCYGLMLLRLRSVRLLSGSKEK
Rat       AKLINICIWVLASGVGVPIMVMAVTQ-PRDGAVVCTLQFPSPSWYWDTVTKICVFLFAFVVPILIITVCYGLMLLRLRSVRLLSGSKEK
Mouse     AKLINICIWVLASGVGVPIMVMAVTQ-PRDGAVVCMLQFPSPSWYWDTVTKICVFLFAFVVPILIITVCYGLMLLRLRSVRLLSGSKEK
Zfish     AKIINVCIWILSSAVGVPIMIMAVTRVTNQNTTVCMLKFPDPDWYWDTVTKICVFIFAFVVPVLVITICYGLMILRLKSVRLLSGSKEK
Fugu      AKLINVLIWVVSSAIGVPIIVMAVTKVGNNGKIMCMLKFPDPDWYWDTVTKICVFIFAFVVPVLVITICYGLMILRLKSVRLLSGSKEK
          **:**: **:::*.:****::****:  .:. .:* *:**.*.************:*****:*:**:*****:***:************

                       6                                       7
Human     DRSLRRITRMVLVVVGAFVVCWAPIHIFVIVWTLVDIDRRDPLVVAALHLCIALGYANSSLNPVLYAFLDENFKRCFRQLCRKPCGRPD
Pig       DRSLRRITRMVLVVVGAFVVCWAPIHIFVIVWTLVDIDRRDPLVVAAL<
Rat       DRSLRRITRMVLVVVGAFVVCWAPIHIFVIVWTLVDINRRDPLVVAALHLCIALGYANSSLNPVLYAFLDENFKRCFRQLCRAPCGGQE
Mouse     DRSLRRITRMVLVVVGAFVVCWAPIHIFVIVWTLVDINRRDPLVVAALHLCIALGYANSSLNPVLYAFLDENFKRCFRQLCRTPCGRQE
Zfish     DRNMRRITRMVLVVVAAFIICWTPIHIFIIEKTLVDINQKNPFVIASWHLH-RTGYTNSSLNPVLYAFLDENFKRCFRDFCLPFRTRAD
Fugu      DRNMRRITRMVLVVVAAFIVCWTPIHIFIIVKTMVEIDTRNPFVIASWHLCIALGYTNSSLNPVLYAFLDENFKRCFRDFCLPCRTRME
          **.:***********.**::**:*****:*  *:*:*: ::*:*:*: **    **:*********************::*       :

Human     PSSFSRAREATARERVTACTPSDGPGGGAAA
Rat       PGSLRRPRQATARERVTACTPSDGPGGGAAA
Mouse     PGSLRRPRQATTRERVTACTPSDGPGGGAAA
Zfish     QSNLNRARNA-TREPVSVCALRIQERSRYD
Fugu      QNHLARGRNA-AREPVSVCSPTETGRKPA
           . : * *:* :** *:.*:
```

FIGURE 3 Alignment and amino acid sequence homology of delta opioid receptor sequences from different vertebrates. Alignment was performed and presented as described in Figure 2. Sequences were obtained from Genbank. Accession Nos.: human, NM-000911; rat, U00475; mouse, L11064; and zebrafish (Z.fish), NM-131258 and partial sequences (> <) for monkey, PC2218; and pig, U71149. The sequences for cow, bass, shark, and frog were from reference [32]. For explanation of symbols, see Figure 2.

are necessary to determine if this gene functions as an opioid receptor. To date, analyses of the genome of the prochordate *Ciona intestinalis*, the nematode *C. elegans* and the arthropods *Drosophila melanogaster* and *Anopheles gambiae* failed to reveal homologues of common vertebrate opioid receptor ligands.

Studies have demonstrated the presence of immunoreactive opioids in protostomes such as mollusks and a group has reported amino acid sequences for putative mollusc (mussel, *Mytilus edulis*) and annelid (leech, *Theromyzon tessulatum*) POMC, enkephalins, and dynorphins [33]. Unfortunately, corresponding DNA sequences have not been found. Interestingly, general gene database searches using these mollusk and annelid amino acid sequences reveal matching genes in vertebrates only, no significant invertebrate matches are found. Nucleic acid sequences would help confirm the existence of opioid ligands in mollusks and annelids, and they would clarify evolutionary processes that led to the presence of these ligands solely in certain protostomes. These mussel and leech POMC-like compounds were purified from the hemolymph, where they are thought to be present at high concentration [33], and mollusk immune cells respond in vitro to the presence of mammalian ACTH or beta endorphin [34]. Surprisingly, recent completion of EST projects on activated mollusk immunocytes did not reveal opioid receptor genes in these cells [35], which could support the hypothesis that in mussel and leech, opioid precursors are not processed to form opioid receptor ligands but instead form antibacterial peptides used for defense or immune signaling [36]. Opioid receptors are not found in other protostomes either. Automatic or semi-automatic annotations of the *C. elegans* and *D. melanogaster* genomes have labeled predicted gene products as putative opioid receptors but careful analysis of these sequences strongly suggests that they are related to other GPCR such as adrenergic or serotonin receptors. Taken together, these data show that *Ciona intestinalis* is the only invertebrate containing vertebrate opioid receptor-like sequences.

Ongoing genome projects in other deuterostomes such as the echinoderm *Strongylocentrotus purpuratus* will indicate if the presence of similar sequences can be traced further back in evolution (analysis of 76,000 bacterial artificial chromosome ends currently available did not reveal opioid receptor sequences in *S. purpuratus*). However, current information (see Fig. 4) on the evolution of opioid receptors is strikingly reminiscent of other vertebrate gene families with a common prototypical prochordate ancestor which diversified in vertebrates as a result of the two gene duplication events that occurred successively before and after the agnathan-gnathostome split [37]. Interestingly, cell signaling factors such as beta-arrestin, which acts downstream of opioid receptors, diversified during the same period from a prochordate ancestor [38], suggesting that main actors of the opioid receptor signaling pathway may have emerged early in vertebrate history. Surely future com-

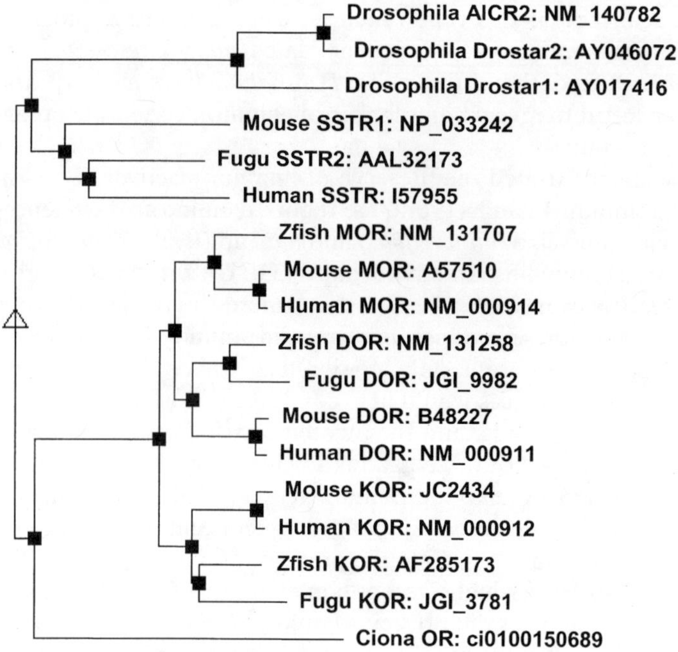

FIGURE 4 Phylogenetic analysis of opioid receptors. Phylogenetic analysis reveals that the opioid receptor identified in the genome of *Ciona intestinalis* is an orthologue of vertebrate opioid receptors. *C. intestinalis* OR is positioned in a clade containing vertebrate opioid receptors but outside of the DOR-MOR-KOR clade of the tree, indicating that the common ancestor of *C. intestinalis* OR and vertebrate opioid receptors predates a gene duplication that gave rise to DOR, MOR, and KOR receptors. BLAST searches revealed that somatostatin receptors were the closest relatives of *C. intestinalis* OR after vertebrate opioid receptors; however, phylogenetic analysis shows that these receptors cluster in an outgroup together with arthropod allatostatin receptors, the closest known invertebrate homologs of mammalian opioid/somatostatin receptors [42]. Abbreviations: OR, opioid receptor; DOR, delta opioid receptor; MOR, mu opioid receptor; KOR, kappa opioid receptor; SSTR, somatostatin receptor; ALCR, allatostatin C receptor; PAM, point accepted mutations. (△), indicates the root of the tree. The phylogenetic tree was constructed using MultiAlin [43].

parative genomic studies will enable more precise analyses of diversification mechanisms and evolutionary processes that gave rise to opioid signaling.

8 QUESTIONS REMAINING

Pharmacological subtypes of delta opioid receptors have been described. However, primary sequence analysis of transcripts has failed to provide an explanation for this pharmacological heterogeneity. This issue has been addressed in detail in other reviews such as Zaki et al. [39]. Although differential posttranslational modifications have not been ruled out, another possibility is that different protein complexes are formed either as a result of heterodimerization with other receptors, or via associated proteins such a G-proteins, arrestins or kinases. It is important to realize that the receptor is not an isolated protein floating alone in a membrane sea. The next challenge will be to decipher the actual complexes that regulate the delta opioid receptor activity, selectivity, trafficking, and signaling. Although the cloning was heralded as a breakthrough, understanding the structure and functioning of the delta opioid receptor in its cellular environment is just beginning.

Undoubtedly, further knowledge on opioid signaling processes will owe a lot to emerging technologies such as mouse genetics and RNA interference. Thus far, mutagenesis screens used so successfully in *Drosophila* and *C. elegans* to elucidate complex cell signaling pathways, could not be applied to opioid receptor biology because arthropod or nematode counterparts of mammalian opioid receptors have not been identified. The fact that mu, delta, and kappa opioid receptors are present in zebrafish provides hope that powerful forward genetics and functional genomic tools will be used in the near future to identify key regulatory mechanisms of opioid signaling pathways.

REFERENCES

1. Simon EJ, Hiller JM. The opiate receptors. Annu Rev Pharmacol Toxicol 1978; 18:371–394.
2. Gilbert PE, Martin WR. The effects of morphine and nalorphine-like drugs in the nondependent, morphine-dependent and cyclazocine-dependent chronic spinal dog. J Pharmacol Exp Ther 1976; 198:66–82.
3. Hughes J, Smith TW, Kosterlitz HW, Fothergill LA, Morgan BA, Morris HR. Identification of two related pentapeptides from the brain with potent opiate agonist activity. Nature 1975; 258:577–580.
4. Robson LE, Kosterlitz HW. Specific protection of the binding sites of D-Ala2-D-Leu5-enkephalin (delta-receptors) and dihydromorphine (mu-receptors). Proc R Soc Lond B Biol Sci 1979; 205:425–432.
5. Cho TM, Law PY, Loh HH. A proposed mode of action for narcotic agonists and antagonists. In: Loh HH, Ross DH, eds. Neurochemical Mechanisms of Opiates and Endorphins. New York: Raven Press, 1979:69–102.
6. Hollenberg MD. Mechanisms of receptor-mediated transmembrane signaling. Experientia 1986; 42:718–727.
7. Klee WA, Nirenberg M. A neuroblastoma times glioma hybrid cell line with morphine receptors. Proc Natl Acad Sci USA 1974; 71:3474–3477.
8. Collier HO, Roy AC. Morphine-like drugs inhibit the stimulation of E prostaglandins of cyclic AMP formation by rat brain homogenate. Nature 1974; 248:24–27.

9. Sharma SK, Nirenberg M, Klee WA. Morphine receptors as regulators of adenylate cyclase activity. Proc Natl Acad Sci USA 1975; 72:590–594.
10. Blume AJ, Lichtshtein D, Boone G. Coupling of opiate receptors to adenylate cyclase: requirement for Na^+ and GTP. Proc Natl Acad Sci USA 1979; 76: 5626–5630.
11. Koski G, Klee WA. Opiates inhibit adenylate cyclase by stimulating GTP hydrolysis. Proc Natl Acad Sci USA 1981; 78:4185–4189.
12. Snyder SH, Childers SR, Creese I. Molecular actions of opiates: historical overview and new findings on opiate receptor interactions with enkephalins and guanyl nucleotides. Adv Biochem Psychopharmacol 1979; 20:543–552.
13. Law PY, Wong YH, Loh HH. Molecular mechanisms and regulation of opioid receptor signaling. Annu Rev Pharmacol Toxicol 2000; 40:389–430.
14. Childers SR. Opioid receptor-coupled second messenger systems. Life Sci 1991; 48:1991–2003.
15. Dixon RA, Kobilka BK, Strader DJ, et al. Cloning of the gene and cDNA for mammalian beta-adrenergic receptor and homology with rhodopsin. Nature 1986; 321:75–79.
16. Dores RM, Akil H, Watson SJ. Strategies for studying opioid peptide regulation at the gene, message and protein levels. Peptides 1984; 5(suppl 1):9–17.
17. Simon EJ, Hiller JM. Solubilization and Purification of opioid binding sites. In: Pasternak GW, ed. The Opiate Receptors. Clifton, NJ: Humana Press, 1988: 165–194.
18. Schofield PR, McFarland KC, Hayflick JS, et al. Molecular characterization of a new immunoglobulin superfamily protein with potential roles in opioid binding and cell contact. EMBO J 1989; 8:489–495.
19. Loh HH, Smith AP. Molecular characterization of opioid receptors. Annu Rev Pharmacol Toxicol 1990; 30:123–147.
20. Eppler CM, Hulmes JD, Wang JB, et al. Purification and partial amino acid sequence of a mu opioid receptor from rat brain. J Biol Chem 1993; 268:26447–26451.
21. Vanetti M, Kouba M, Wang X, Vogt G, Hollt V. Cloning and expression of a novel mouse somatostatin receptor (SSTR2B). FEBS Lett 1992; 311:290–294.
22. Inoue N, Hatanaka H. Nerve growth factor induces specific enkephalin binding sites in a nerve cell line. J Biol Chem 1982; 257:9238–9241.
23. Dougherty JD, Geschwind DH. Subtraction-coupled custom microarray analysis for gene discovery and gene expression studies in the CNS. Chem Senses 2002; 27:293–298.
24. Xie GX, Miyajima A, Goldstein A. Expression cloning of cDNA encoding a seven-helix receptor from human placenta with affinity for opioid ligands. Proc Natl Acad Sci USA 1992; 89:4124–4128.
25. Kieffer BL, Befort K, Gaveriaux-Ruff C, Hirth CG. The delta-opioid receptor: isolation of a cDNA by expression cloning and pharmacological characterization. Proc Natl Acad Sci USA 1992; 89:12048–12052.
26. Evans CJ, Keith DE Jr, Morrison H, Magendzo K, Edwards RH. Cloning of a delta opioid receptor by functional expression. Science 1992; 258:1952–1955.
27. Miller RJ, Chang KJ, Leighton J, Cuatrecasas P. Interaction of iodinated enkephalin analogues with opiate receptors. Life Sci 1978; 22:379–388.

28. Zajac JM, Gacel G, Petit F, Dodey P, Rossignol P, Roques BP. Deltakephalin, Tyr-D-Thr-Gly-Phe-Leu-Thr: a new highly potent and fully specific agonist for opiate delta-receptors. Biochem Biophys Res Commun 1983; 111:390–397.
29. Takeda S, Kadowaki S, Haga T, Takaesu H, Mitaku S. Identification of G protein–coupled receptor genes from the human genome sequence. FEBS Lett 2002; 520:97–101.
30. Massotte D, Kieffer BL. A molecular basis for opiate action. Essays Biochem 1998; 33:65–77.
31. Yasuda K, Raynor K, Kong H, et al. Cloning and functional comparison of kappa and delta opioid receptors from mouse brain. Proc Natl Acad Sci USA 1993; 90:6736–6740.
32. Li X, Keith DE Jr, Evans CJ. Multiple opioid receptor-like genes are identified in diverse vertebrate phyla. FEBS Lett 1996; 397:25–29.
33. Stefano GB, Salzet M. Invertebrate opioid precursors: evolutionary conservation and the significance of enzymatic processing. Int Rev Cytol 1999; 187:261–286.
34. Duvaux-Miret O, Stefano GB, Smith EM, Dissous C, Capron A. Immunosuppression in the definitive and intermediate hosts of the human parasite Schistosoma mansoni by release of immunoactive neuropeptides. Proc Natl Acad Sci USA 1992; 89:778–781.
35. Gueguen Y, Cadoret JP, Flament D, et al. Immune gene discovery by expressed sequence tags generated from hemocytes of the bacteria-challenged oyster, *Crassostrea gigas*. Gene 2003; 303:139–145.
36. Salzet M. Invertebrate molecular neuroimmune processes. Brain Res Rev 2000; 34:69–79.
37. Escriva H, Manzon L, Youson J, Laudet V. Analysis of lamprey and hagfish genes reveals a complex history of gene duplications during early vertebrate evolution. Mol Biol Evol 2002; 19:1440–1450.
38. Nakagawa M, Orii H, Yoshida N, et al. Ascidian arrestin (Ci-arr), the origin of the visual and nonvisual arrestins of vertebrate. Eur J Biochem 2002; 269: 5112–5118.
39. Zaki PA, Bilsky EJ, Vanderah TW, Lai J, Evans CJ, Porreca F. Opioid receptor types and subtypes: the delta receptor as a model. Annu Rev Pharmacol Toxicol 1996; 36:379–401.
40. Augustin LB, Felsheim RF, Min BH, Fuchs SM, Fuchs JA, Loh HH. Genomic structure of the mouse delta opioid receptor gene. Biochem Biophys Res Commun 1995; 207:111–119.
41. Gelernter J, Kranzler HR. Variant detection at the delta opioid receptor (OPRD1) locus and population genetics of a novel variant affecting protein sequence. Hum Genet 2000; 107:86–88.
42. Kreienkamp HJ, Larusson HJ, Witte I, et al. Functional annotation of two orphan G-protein-coupled receptors, Drostar1 and -2, from *Drosophila melanogaster* and their ligands by reverse pharmacology. J Biol Chem 2002; 277: 39937–39943.
43. Corpet F. Multiple sequence alignment with hierarchical clustering. Nucleic Acids Res 1988; 16:10881–10890.

3

Cloning and Expression of the Human Delta Opioid Receptor

Richard J. Knapp
Aventis Pharmaceuticals, Bridgewater, New Jersey, U.S.A.

Ewa Malatynska
Johnson & Johnson, Spring House, Pennsylvania, U.S.A.

Eva V. Varga, William R. Roeske, and Henry I. Yamamura
University of Arizona, Tucson, Arizona, U.S.A.

Considerable evidence had been accumulated for the existence of multiple opioid receptor subtypes prior to the cloning of the cDNAs encoding the three major opioid receptor types—the mu, delta, and kappa opioid receptors [1,2]. Since our laboratory was working on the development of selective delta receptor agonists, including [D-Pen^2, D-Pen^4] enkephalin (DPDPE) and its halogenated analogues, we were primarily interested in the characterization of the delta opioid receptor and the identification of putative delta opioid receptor subtypes.

At least three hypotheses were proposed to explain pharmacological data suggesting delta receptor heterogeneity. The more conservative hypothesis was the existence of at least two distinct delta receptor proteins, encoded by two distinct mRNA species. A more imaginative hypothesis, proposed by

Rothman and Westfall [3,4], divided the delta receptor population into two components: one coupled to mu opioid receptors, and the other acting alone. The third hypothesis combined these ideas, by suggesting that the delta receptors that interact with the mu opioid receptors are different molecular entities from those that act alone [5,6]. A definitive test of the above hypotheses would be to identify two or more delta receptor proteins by molecular cloning. The cloning of the human delta opioid receptor [7] was the fortunate outcome of a larger project in our laboratory to clone of the cDNAs encoding putative delta opioid receptor subtypes.

Although the opioid receptors were perhaps the first neurotransmitter receptors to be characterized as molecular entities, the molecular cloning of the first opioid receptor cDNAs was elusive. The idea of a "receptive substance" was developed in the late 19th century by J.N. Langley, to explain the selectivity of drugs in certain tissues. While the concept of drug receptors was supported by pharmacological analyses in isolated tissue preparations, the molecular identity of the receptors was not known for a long time. The existence of opioid receptor molecules was confirmed only in the 1970s, by the work of Snyder [8], Simon [9], and Terenius [10], who independently used the technique of receptor radioligand binding, first suggested by Goldstein [11], to demonstrate the existence of a distinct, stereoselective opiate binding molecule in brain. These studies provided the first direct evidence that drugs could recognize a specific cellular component, with high affinity and a distinct tissue distribution. Further studies using this technique showed that receptor recognition was a property of many, though not all, drugs, and provided an essential tool to characterize their biochemical properties. Opioid receptors proved unsuitable for receptor isolation and biochemical characterization, largely because of their low density in neural tissue.

Our initial approach for the molecular cloning of delta receptor subtypes was to use the expression cloning technique, where cDNAs isolated from a suitable tissue are expressed in mammalian cells, and the presence of the desired receptor is detected by a selective radioligand. By that time, we had developed and characterized a novel radioiodinated ligand, [^{125}I-Phe^{3}]-DPDPE [12]. The high selectivity and specific activity, and low nonspecific binding of this radioligand made it ideal for expression cloning. However, before we could complete our work, two independent groups (Evans et al. [13] and Kieffer et al. [14]) applied this technique using other radioligands, to clone the mouse delta opioid receptor.

Both groups used expression cloning to screen cDNA libraries obtained from NG 108-15 mouse neuroblastoma-rat glioma hybridoma cells. These cells were a logical choice for delta opioid receptor cloning efforts, since they express high density of delta receptors and can be produced in great quantities in cell culture.

The delta opioid receptor protein described by Evans et al. [13] and Kieffer et al. [14] consists of 372 amino acids. Hydropathy analysis indicates seven putative α-helical transmembrane, domains in the cloned protein. The cloning of the mouse delta opioid receptor opened a new path to the molecular cloning of other opioid receptors by hybridization screening. Accordingly, the mouse mu [15] and kappa [16] opioid receptor types were cloned in rapid succession. Interestingly, one of the first reported kappa opioid receptor clones did not originate from homology screening using a cloned delta receptor probe, but instead resulted from efforts to clone somatostatin receptor types [17]. One of the cDNA clones identified in this work had low homology to other somatostatin receptors, but exhibited a relatively high homology to the just described mouse delta opioid receptor. Subsequently, this clone was tested for opioid ligand binding and was identified as a kappa opioid receptor.

The cloned mouse delta receptor sequence also facilitated the cloning of opioid receptor cDNAs from other animal species. Our cloning efforts were initially directed at finding a delta opioid receptor subtype, different from the cloned mouse delta opioid receptor type. We began by generating a cDNA probe to use for homology sceening. We used mRNA from NG 108-15 cells as template, and the primer pair: 5′-GGGTCTTGGCTTCAGGTGTCG-3′ (sense) and 5′-GCAGCGCTTGAAGTTCTCGTC-3′ (antisense) in a reverse transcriptase polymerase chain reaction (RT-PCR) to amplify a 467-bp fragment of the mouse delta opioid receptor cDNA. The PCR product was ligated into a pCR II cloning vector, amplified in *E. coli* cells, and labeled with [α-^{32}P]dCTP and random primers, to produce the desired radioactive hybridization probe. Since much of the work on delta receptor heterogeneity was done in mouse tissues, we subsequently used this probe to screen mouse brain cDNA libraries, to hopefully identify mouse delta opioid receptor subtypes. However, while we isolated several clones, none of the sequences were different from the previously published mouse delta opioid receptor. During this time we also started screening a human striatal cDNA library, and identified a single cDNA clone, designated 44-11. Since the human receptors are the ultimate pharmacological targets for drug development, and no human delta opioid receptor sequences had been described at that time, we decided to shift our attention to the cloning of the human delta opioid receptor.

A 0.7-kb human delta opioid receptor cDNA probe, corresponding to the 3′ end of the expected delta receptor sequence, was produced from the 1.6-kb cDNA 44-11 clone by digestion with EcoRI and Not I restriction enzymes. This probe was used to screen a human temporal cortex cDNA library. Three additional cDNA clones were isolated from this library. Each of these clones was ~1.0-kb long, and contained only partial fragments of the predicted receptor sequence. The breakthrough was the finding that one of the cloned fragments (designated 78-4) contained the elusive 5′ initiation codon, and that most of its 3′ sequence overlapped with that of the 44-11 clone (Fig. 1). However, an 84-bp stretch at the 3′ end of the 78-4 clone did not overlap with the sequence of the 44-11 cDNA clone. The 84-bp sequence was further analyzed, and was identified as a receptor fragment that was complementary to nucleotide positions 382–465 in the open reading frame of the cloned delta receptor, in the inverted orientation. This indicated that the 78-bp fragment in the 78-4 clone was an artifact of the library construction process. With this information, we turned to the problem of reassembling the two fragments to restore the full open reading frame.

We were able to reconstitute the full open reading frame by the identification of a unique, HincII restriction site, within the overlapping region of the two clones (Fig. 2). Furthermore, the presence of an ApaI restriction site in the untranslated 3′ sequence of the 44-11 clone allowed ligation of the HincII-ApaI double digestion product of the 44-11 clone into the corresponding sites of the pBluescript 78-4 clone, to assemble the complete open reading frame in the correct orientation. Restriction analysis with HincII and ApaI demonstrated that both sites were preserved in the ligation product.

The reconstituted 78x44 clone was then transferred into pREP10 and pcDNA3 expression vectors for expression in mammalian (COS-7, CHO, and others) cells. For initial characterization, the recombinant hDOR/pcDNA vector was transiently transfected into COS-7 cells. The presence of the human delta opioid receptor in the transfected cells was confirmed by radioligand binding studies, using [^{3}H] naltrindole, [^{3}H] [4′-Cl-Phe^4] DPDPE and [^{3}H][D-Ala^2, Glu^4]deltorphin as radioligands. Each of these delta-selective radioligands exhibited high affinity specific binding to transfected COS-7 cell membrane preparations. In contrast, very low specific binding was detected with either of the radiololigands in sham-transfected (empty pcDNA3 vector)

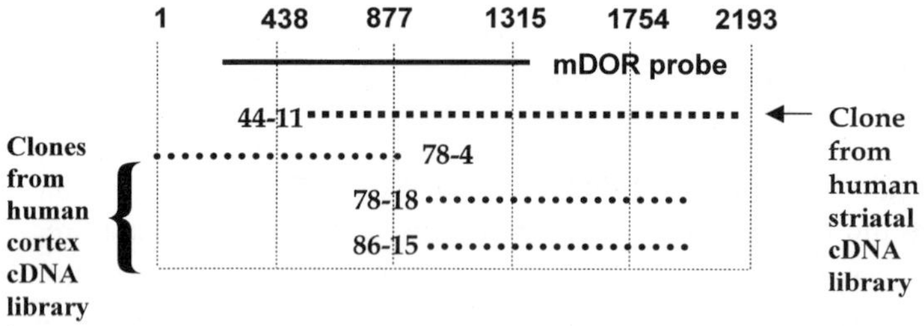

FIGURE 1 Positions of cloned cDNA sequences relative to the mouse delta opioid receptor open reading frame. The figure shows an alignment of the insert sequences for the four cDNA clones obtained from human brain cDNA libraries. Together the 44-11 and 78-4 cDNA inserts cover the entire open reading frame and include 3′ and 5′ flanking sequences.

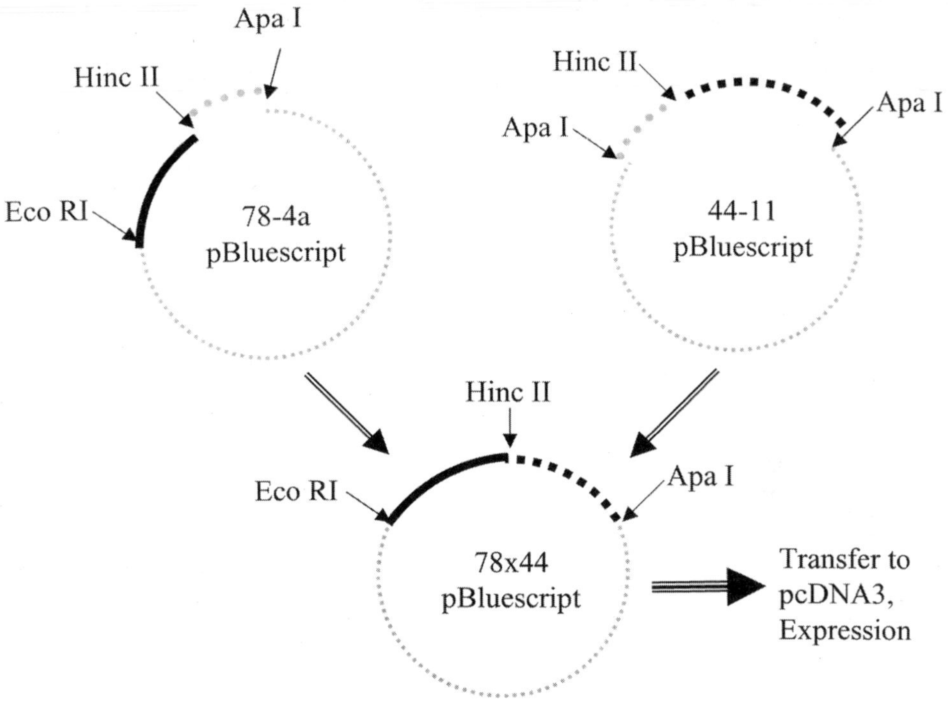

FIGURE 2 Assembly of human delta opioid receptor open reading frame. The figure illustrates how the human delta opioid receptor open reading frame was assembled from the 44-11 and 78-4 cDNA fragments using the HincII and ApaI restriction sites.

cell membranes. In addition, competition binding studies were also used to demonstrate that the expressed receptor exhibited the expected selectivity profile, with high affinity for delta-, but significantly lower affinities for mu- and kappa-selective opioid ligands (Table 1).

An initial controversy was encountered at this step, since the K_d and K_i values of the ligands tended to be higher in cells expressing the cloned opioid receptors than the previously determined values in brain membrane preparations, leading to an ambiguity regarding the identity of the cloned opioid receptors. Thus, the K_d values we obtained for [^{3}H]naltrindole binding to membranes from COS-7 cells transiently transfected with the human delta receptor clone ranged from 55 to 560 pM. Importantly, the obtained K_d values showed a strong correlation with the actual receptor concentrations in the particular batch of transfected cells used for the experiment (18–430 pM, depending on the transfection [31]). Ligand depletion, resulting in reduced

TABLE 1 Binding Affinities for Selective Opioid Receptor Ligands to Membrane Preparations from COS-7 Cells, Transiently Transfected with the Cloned Human Delta Opioid Receptor

Ligand[a]	Preferred receptor	Ki[b] (nM ± SEM)
NTB	δ	0.28 ± 0.11
BNTX	δ	2.30 ± 0.58
[D-Ala^2, Glu^4]Deltorphin	δ	8.70 ± 6.60
[4′-Cl-Phe]DPDPE	δ	2.80 ± 1.00
CTAP	μ	4100 ± 2100
U69593	κ	92000 ± 19000

[a] NTB: naltrenoxone-benzofuran, BNTX: benzylidene-naltrexone, DPDPE: [D-Pen^2-D-Pen^5] enkephalin, CTAP: D-Phe-Cys-Tyr-D-Trp-Arg-Thr-Pen-Thr-NH_2, U69,593: 5α,7α,8 β-(-)-N-methyl-N-[7-(-pyrrolidinyl)-l-oxaspiro (4,5)dec-8-y]-phenyl-benzeneacetamide.
[b] Ki: inhibition constant, calculated according to the Cheng-Prusoff equation [35].

apparent binding affinity, can be important when the free radioligand concentration is substantially less than the total radioligand concentration. Such radioligand depletion frequently becomes a problem when transfected cells are used, where the receptor density is usually much greater than that found in native tissues. Indeed, after extrapolation of the K_d versus B_{max} data to zero receptor concentration we obtained a corrected [^{3}H]naltrindole K_d value of 28 pM, consistent with the K_d value measured in rat brain homogenates [18].

Since at the time of the described studies it was thought that the two delta receptor antagonists naltriben (NTB) and 7-benzylidenenaltrexone (BNTX) were selective for the proposed delta-2 and delta-1 receptor subtypes, respectively [19], the relatively high affinity of NTB for the expressed receptors (as compared to BNTX, see Table 1) was thought to be consistent with the cloned human delta receptor being of the delta-2 opioid receptor subtype.

In summary, transient expression of the reconstituted 78-44 cDNA clone in COS-7 cells produced a pharmacological delta opioid receptor binding site in these cells. In order to confirm that the binding site indeed corresponds to a naturally occuring human delta opioid receptor sequence, PCR amplifications were performed, using human brain libraries from different sources as templates. The PCR primer pair contained the start and the stop codons of the 78x44 construct (sense PCR primer: 5′GGCCCCCTCCGCC-GGCGCC-3′, and antisense PCR primer: 5′TGAGGCGGCACGGCCAC-CGCCGGGACC-3′). Therefore, if a library contains a full-length cDNA corresponding to our reassembled clone, a 1.1-kb amplification product should be expected. Indeed, PCR amplification of a human occipital cortex cDNA library led to isolation of a 1.1-kb PCR product, with the nucleotide sequence identical to that of the 78x44 clone (unpublished results). This result indicates that a contiguous cDNA sequence, identical with the sequence the 78X44 ligation product is present in the human mRNA pool. Furthermore, after the publication of our report a similar cDNA sequence, encoding the human delta opioid receptor, was independently obtained by Simonin et al. [20].

The cloned human delta receptor open reading frame consists of 1116 nucleotides, encoding for a 372 amino acid protein (Fig. 3). The same number of amino acids were found in the rat and mouse delta opioid receptors. The cloned human delta receptor exhibits high sequence homology (93%) to both rat, and mouse delta opioid receptors, with most of the differences confined to the N- and C-termini. There are no amino acid differences in the seven putative transmembrane domain regions. There are substantially greater differences between the human delta receptor and the human mu and kappa opioid receptors. The primary amino acid sequence of the human delta opioid receptor displays the characteristics of a Group 1a G protein–coupled receptor, with seven α-helical transmembrane domains, extracellular N- and intracellular C-termini, several consensus extracellular glycosylation and intracellular phosphorylation sites, and a palmitoylation site in the C-terminal tail. In addition, most of the landmark amino acid residues, which are highly conserved in the biogenic amine and neuropeptide receptors, are also present in the human delta opioid receptor (Fig. 3).

In order to functionally characterize the human delta opioid receptor, the cDNA was also stably transfected into Chinese hamster ovary (CHO) cells, using a pREP10/hDOR recombinant expression vector. As expected, the K_d (139 ± 36 pM, range 71–196 pM), and B_{max} (968 ± 170 fmol/mg protein, range 790–1310 pM) values for [^{3}H]NTI showed significantly lower variability than in the transiently transfected COS-7 cells (K_d range 50–560; B_{max} range 900–3100 fmol/mg membrane protein) [21,22]. The stably transfected human delta opioid receptors were subsequently shown to mediate the inhibition of forskolin stimulated cAMP formation [22], stimulation of inositol lipid hydrolysis [23], and the phosphorylation of p42/44 MAP kinases [24]. Chronic treatment of the hDOR/CHO cells with delta opioid agonists on the other hand, was shown to lead to signal desensitization [22,23], receptor down-regulation [25] and adenylyl cyclase superactivation [22,23,26,27].

Distinct genes for delta opioid receptor subtypes however, have not been identified, reducing the probability that delta opioid receptor subtypes exist as distinct molecular entities. On the other hand, this does not eliminate the possibility that pharmacological delta opioid receptor subtypes can be produced by alternative splicing from the cloned delta receptor genomic sequence, as was found for the other opioid receptors [28]. However, there is no evidence for alternative splicing for the delta opioid receptor of this at

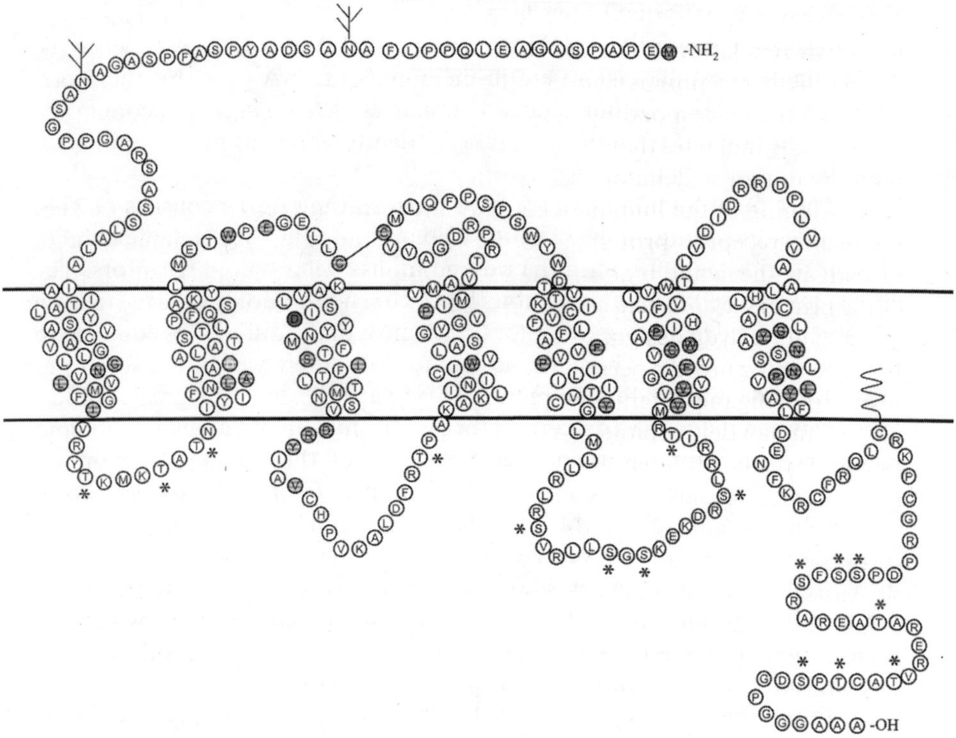

FIGURE 3 Putative amino acid sequence of the human delta opioid receptor. Dark gray circles indicate amino acid residues conserved in the rhodopsin family of G protein–coupled receptors; asterisks indicate putative intracellular phosphorylation sites. Putative N-terminal glycosylation (Asn), and C-terminal palmytoylation (Cys) sites are also marked.

present. Alternatively, as was previously suggested by Rothman and Westfall [3,4], the pharmacological delta receptor subtypes may originate from oligomerization of the delta opioid receptor with other (opioid or nonopioid) receptor proteins [29].

In summary, the cloning of the human delta opioid receptor has been a great advance in the molecular pharmacology of the opioid receptors that began with the cloning of the mouse delta receptor. This work provided recombinant cell lines to study the molecular pharmacology of ultimate target of therapeutic drug design—the human receptor. To illustrate the importance of studying the human receptors, species-specific differences have been observed earlier in the regulation of the opioid receptors [30]. Similarly, it was shown that Thr353 is crucially inportant in the molecular mechanism of the down-regulation of the mouse delta opioid receptor. However, the human delta opioid receptor is also downregulated by chronic agonist treatment, although the corresponding residue is an alanine in the human receptor [25]. This finding indicates that the molecular mechanism of the downregulation of the human mouse delta opioid receptor may be subtly different.

Cloning of the human delta opioid receptor also enabled us to characterize the receptor protein, and to perform site-directed mutagenesis to investigate the ligand binding [31] and functional [32] domains. Identification of the primary amino acid sequence of the human delta opioid receptor should also facilitate drug design, since molecular modeling of the tertiary structure of the receptor protein permits the identification and subsequent optimization of receptor/ligand interactions [33]. Recombinant cell lines expressing the cloned human delta opioid receptor are also sensitive model systems to study agonist specific differences in receptor signaling. Thus, we have recently demonstrated agonist-specific differences in the downregulation of the human delta opioid receptor [34]. Finally, the availability of recombinant model cell lines aids the characterization of the interactions between the receptors and a multitude of cytoplasmic proteins that regulate receptor activity and transmit signals into the interior of the cell [29,33] in a homogeneous cellular background. Better understanding of opioid receptor mediated cellular signaling mechanism will aid the future development of novel, more effective analgesic compounds with fewer side effects.

REFERENCES

1. Knapp RJ, Hunt M, Wamsley JK, Yamamura HI. In: London ED, ed. Imaging Drug Action in the Brain. Boca Raton: CRC Press, 1993:119–176.
2. Quock RM, Burkey TH, Varga E, Hosohata Y, Hosohata K, Cowell SM, Slate CA, Ehlert FJ, Roeske WR, Yamamura HI. Pharmacol Rev 1999; 51:503–532.
3. Rothman RB, Westfall TC. Mol Pharmacol 1982; 21:538–547.
4. Rothman RB, Westfall TC. Mol Pharmacol 1982; 21:548–557.
5. Jiang Q, Mosberg HI, Bowen WD, Porreca F. J Pharmacol Exp Ther 1990; 255:636–641.
6. Mattia A, Vanderah T, Mosberg HI, Porreca F. J Pharmacol Exp Ther 1991; 258:583–587.
7. Knapp RJ, Malatynska E, Fang L, Li X, Babin E, Nguyen M, Santoro G, Varga E, Hruby VJ, Roeske WR, Yamamura HI. Life Sci 1994; 54:PL463–PL469.
8. Pert CB, Snyder SH. Science 1973; 179:1011–1017.
9. Simon EJ, Hiller JM, Edelman I. Proc Natl Acad Sci USA 1973; 70:1947 1956.
10. Terenius L. Acta Pharmacol Toxicol (Copenh) 1973; 33:377–382.
11. Goldstein A, Lowney LI, Pal BK. Proc Natl Acad Sci USA 1971; 68:1742–1747.

12. Knapp RJ, Sharma SD, Toth G, Duong MT, Fang L, Bogert CL, Weber SJ, Hunt M, Davis TP, Wamsley JK, Hruby VJ, Yamamura HI. J Pharmacol Exp Ther 1991; 258:1077–1083.
13. Evans CJ, Keith DE, Morrison H, Magendzo K, Edwards RH. Science 1992; 258:1952–1955.
14. Kieffer BL, Befort K, Gaveriaux-Ruff C, Hirth CG. Proc Natl Acad Sci USA 1992; 89:12048–12052.
15. Chen Y, Mestek A, Liu J, Hurley JA, Yu L. Mol Pharmacol 1993; 44:8–12.
16. Minami M, Toya T, Katao Y, Maekawa K, Nakamura S, Onogi T, Kaneko S, Satoh M. FEBS Lett 1993; 329:291–295.
17. Yasuda K, Raynor K, Kong H, Breder CD, Takeda J, Reisine T, Bell GI. Proc Natl Acad Sci USA 1993; 90:6736–6740.
18. Fang L, Knapp RJ, Horvath R, Matsunaga TO, Haaseth RC, Hruby VJ, Porreca F, Yamamura HI. J Pharmacol Exp Ther 1994; 268:836–846.
19. Sofuoglu M, Portoghese PS, Takemori AE. Life Sci 1993; 52:769–775.
20. Simonin F, Befort K, Gaveriaux-Ruff C, Matthes H, Nappey V, Lannes B, Micheletti G, Kieffer B. Mol Pharmacol 1994; 46:1015–1021.
21. Malatynska E, Wang Y, Knapp RJ, Santoro G, Li X, Waite S, Roeske WR, Yamamura HI. NeuroReport 1995; 6:613–616.
22. Malatynska E, Wang Y, Knapp RJ, Waite S, Calderon S, Rice K, Hruby VJ, Yamamura HI, Roeske WR. J Pharmacol Exp Ther 1996; 278:1083–1094.
23. Rubenzik M, Varga E, Stropova D, Roeske WR, Yamamura HI. Mol Pharmacol 2001; 60:1076–1082.
24. Varga EV, Rubenzik MK, Sugiyama M, Stropova D, Roeske WR, Yamamura HI. Proceedings of the International Narcotic Research Conference, Pacific Grove, CA, 2002:41.
25. Okura T, Cowell SM, Varga EV, Burkey TH, Roeske WR, Hruby VJ, Yamamura HI. Eur J Pharmacol 2000; 387:R11–R13.
26. Varga EV, Stropova D, Rubenzik M, Waite S, Roeske WR, Yamamura HI. Eur J Pharmacol 1999; 364:R1–R2.
27. Varga EV, Rubenzik M, Grife V, Sugiyama M, Stropova D, Roeske WR, Yamamura HI. Eur J Pharmacol 2002; 451:101–102.
28. Zimprich A, Simon T, Höllt V. FEBS Lett 1995; 359:142–146.
29. Jordan BA, Devi LA. Nature 1999; 399:697–700.
30. Zhang F, Li J, Li JG, Liu-Chen LY. J Pharmacol Exp Ther 2002; 302:1184–1192.
31. Varga EV, Li X, Stropova D, Zalewska T, Landsman RS, Knapp RJ, Malatynska E, Kawai K, Mizusura A, Nagase H, Calderon SN, Rice K, Hruby VJ, Roeske WR, Yamamura HI. Mol Pharmacol 1996; 50:1619–1624.
32. Hosohata Y, Varga EV, Stropova D, Li X, Knapp RJ, Hruby VJ, Rice KC, Nagase H, Roeske WR, Yamamura HI. Life Sci 2001; 68:2233–2242.
33. Knapp RJ, Malatynska E, Collins N, Fang L, Wang JY, Hruby VJ, Roeske WR, Yamamura HI. FASEB J 1995; 9:516–525.
34. Okura T, Varga EV, Hosohata Y, Navratilova E, Cowell SM, Rice KC, Nagase H, Hruby VJ, Roeske WR, Yamamura HI. Eur J Pharmacol 2003; 459:9–16.
35. Cheng YC, Prusoff WH. Biochem Pharmacol 1973; 22:3099–3108.

4

In Vitro and In Vivo Mutagenesis: Insights into Delta Receptor Structure and Function

F.M. Décaillot and Brigitte L. Kieffer
Institut de Génétique et de Biologie Moléculaire et Cellulaire,
Illkirch, France

1 INTRODUCTION

The delta receptor was the first opioid receptor to be cloned in 1992, almost 20 years after opioid binding sites were discovered in the brain [1,2]. Cloning was achieved simultaneously by two independent teams, using an expression strategy [3,4]. Homology cloning techniques then delivered the whole opioid

---

Abbreviations: BW373U86, (±)-4-[(a-R*)-a-[(2S*, 5R*)-4-allyl-2,5-di-methyl-1-piperazinyl]-3-hydroxybenzyl]-N,N-diethylbenzamide; CFA, complete Freud's adjuvant; DAMGO, [D-Ala^2, MePhe^4,Gly-ol^5]; Delt I or II, Deltorphin I or II; DPDPE, cyclic[D-penicillamine^2, D-penicillamine^5] enkephalin; DSLET, [D-Ser^2, D-Leu^5]enkephalin-Thr; DTLET, Tyr-D-Thr-Gly-Phe-Leu-Thr; EL, extracellular loop; IL, intracellular loop; jom13, Tyr-c[D-Cys-Phe-D-Pen]OH; Leu-Enk, Leu-enkephalin; Met-Enk, Met-enkephalin; NTB, naltriben methanesulfonate; NTI, naltrindole; THC: Δ9-tetrahydrocannabinol; ICV, intracerebroventricular; ITH, intrathecal; SC, subcutaneous. Behavioral assays: HP, hot plate; TF, tail flick; TI, tail immersion; TIPP, Tyr-Tic-Phe-Phe-OH; TM, transmembrane domain; 3D, tridimensional; WT, wild-type.

receptor gene family. Gene encoding the previously proposed mu, delta and kappa receptors were isolated in several species, as well as the nonopioid homologous ORL-1 receptor [reviewed in 5]. The next experiments used receptor domain swapping, truncation, and site-directed mutagenesis to elucidate structure-activity relationships of opioid receptors at the cellular level. To date, receptor mutagenesis has provided a fairly clear view of the opioid binding site, as well as some clues on possible signaling and regulatory determinants, although the general picture of delta receptor activation is still missing. Genetic studies were initiated and mice lacking functional opioid receptor genes were created and studied behaviorally. Gene knockout in mice has brought unexpected and novel findings on delta receptor function in vivo, that should promote the search for novel delta ligands and therapeutic areas. Together, molecular approaches have uniquely addressed receptor function both at the membrane and in the complex environment of neural circuits. In this chapter we will briefly summarize some important findings that have stemmed from receptor in vitro and in vivo mutagenesis and told us how delta receptors operate at the protein level as well as within the brain.

Analyzing the consequences of receptor mutagenesis in cell culture systems or in a whole animal has strongly influenced our view of delta receptor signaling and physiology. Consequently, many of the data presented in this chapter will also be cited in other chapters across all sections covering receptor characterization, activation, and regulation at the cellular level (Part 1), receptor/ligand interactions (Part 2), and receptor function in vivo (Part 3). Our purpose here is to provide an overall picture on the molecular basis of delta receptor function.

2 IN VITRO MUTAGENESIS

cDNA sequence analysis confirmed that opioid receptors belong to the G protein–coupled receptor (GPCR) superfamily. Sequence alignment of the three subtypes highlighted conserved receptor domains. Among these, intracellular loops (IL) and transmembrane domains (TM) 2 and 3 are most conserved (86–100%), while extracellular loops (EL) 2 and 3, N- and C-terminal tails are most divergent (52–21%) [6]. Homology analysis, together with our knowledge from other GPCRs, suggested that (1) extracellular loops are involved in mu, delta, and kappa selectivity; (2) the transmembrane core contains a common opioid receptor binding pocket, as well key switches for receptor activation; (3) intracellular loops are responsible for G protein coupling that is similar across receptor subtypes; and (4) regulation of receptor activity by C-terminal determinants strongly differs among receptors. This was confirmed and refined by mutagenesis experiments. Tables 1, 2, and 3 summarize structural determinants for delta receptor function that have been identified by chimeric studies and site-directed mutagenesis in various host

TABLE 1 Study Delta Selectivity by Chimeric Constructs[a]

Chimera	Effect on binding
k(Nter-TM1)/d	No effect on DPDPE, DSLET, Delt II, NTI [18]
m(Nter-TM1)/d	No effect [19] or decreased on DPDPE (10X) and Delt II (10X) [7]
d(Nter-TM1)/m	Increased for DPDPE, Delt II (1000X) [7]
k(Nter-EL1)/d	Decreased for Leu-enk, DSLET (25X) [20]
m(Nter-EL2)/d	Decreased for DPDPE (6X) [19]
k(Nter-EL2)/d	Decreased for Met-Enk (15–20X), Delt II (> 100X), DPDPE (20X), jom 13 (10X) [20,21]
m(Nter-TM5)/d	No effect on DPDPE, Delt I [7] and DSLET [22]
m(Nter-TM7)/d	Abolished for DPDPE [40]
m(IL1-TM3)/d	Increased for DAMGO (30X) [19]
m(TM2-EL1)/d	Increased for DAMGO (affinity comparable to WT mu) [23]
m(TM5-Cter)/d	Abolished for DSLET (22); decreased for: DPDPE, Delt II (100-50X) [7]
k(TM5-Cter)/d	Abolished for Enks, DSLET, Delt II, jom 13 [20]
m(TM6-Cter)/d	Close to WT values for DPDPE, Delt II [7]
d(TM5)/m	Delt II, TIPP, NTB close to wt values except for DPDPE (10X) [7]
k(TM6-EL3)/d	Abolished for DPDPE, Enks and Delt II [21]
m(TM6-EL3)/d	Abolished for same ligands except for Enks (100X) [21]
d(TM6-EL3)/k	Increased for same ligands (to 100-10nM range) [21]
m(EL3)/d	Decreased for DPDPE, Delt II, SNC121, TAN67 [9]
m(EL3)/d	Decreased for NTI (300X) [11]
m(EL3-288-300)/d	Decreased for DPDPE, Delt II, SNC80 (400–1000X) [10]
m(EL3-291-300)/d	Decreased for DPDPE, SNC80 (>200X) [12]

[a] *Top part of table*: Progressive insertion of either mu (m) or kappa (k) sequences (in parentheses) into the N-terminal part of the delta receptor (/d) gradually disturbs delta ligand binding. Affinity decreases dramatically when EL3 is reached. Middle part: Chimera containing EL1 show molecular determinants responsible for the low DAMGO binding on delta receptor. *Bottom part:* Efforts concentrated on C-terminal half of the delta receptor confirm that EL3 is an important determinant for delta selectivity.

cell lines. Key amino acid residues for binding and activation are presented on a serpentine representation of the delta receptor as well as on a rhodopsin-based 3D model (Fig. 1). Many of these data have been reviewed recently [6,7], and below we discuss important features of delta receptor structure.

2.1 The Extended EL3: Delta Selectivity

In a first attempt to identify residues involved in subtype-specific binding, several chimeric opioid receptors were generated (reviewed in [8] and Table 1).

TABLE 2 Site-Directed Mutagenesis of the Delta Opioid Receptor: EL and TM Domains[a]

Location	Mutation	Effect on binding	Effect on activation
TM1	D95N	Decreased for DSLET, Delt II, Met-Enk (11–30X) and BW373U86 (300X) [18]	No change in cyclase response for all ligands [18]
EL1	K108N	Increased for DAMGO (50X) [16,17]	
EL3	W284A	Decreased for SNC80, DPDPE, Delt II (10–30X) [10]	
	W284K	Decreased for DADLE and DTLET (10–15X) [25]	MAPK activation by DADLE unaffected [25]
	W284L	Decreased for SCN80 (15X), no change or DPDPE [26]	Decreased EC50 of both ligands (10X) [26]
	W284E	Decreased for NTI (10X), Increase for nBNI, GNTI (8–42X) [15]	
	V296A	Decreased for SNC80, DPDPE, Delt II (4–5X) [10]	
	V297A	Decreased for same ligands (3–10X) [10]	
	m(291–300)d + W300L	Restored binding for DPDPE, SNC80, on chimera [12]	
Helical bundle	D128N	Decreased for all peptides (>200X) and alkaloids (20–100X) [32]	High basal activity [34]
	D128A	No change for peptides or alkaloids [32,33]	High basal activity [32,33], Nlx inverse agonist [33]
	D128K	Abolished for same ligands [33]	High basal activity, Nlx agonist [33]
	D128H	Not determined	High basal activity, Nlx antagonist [33]
	Y129A	Decreased for peptides (15–400X) [37]	High basal activity [34]
	Y129F	Decreased for DADLE, Delt II (10–90X) [37]	
	W173A	Decreased for DPDPE (14X) and Enks (20–43X) [37]	
	F222A	Decreased for DADLE and DPDPE (14X) [37]	
	W274A	Decreased for DPDPE and BW373U86 (10–15X); Nlx and bremazocine (50X) [37]	
	Y308F	Decreased for NTI (10X) [37], No change DSLET [39]	Low basal activity [34], unchanged IC_{50}; increased EC_{50} (10X), and decreased MAPK activation by DTLET [39]
	P276C	Abolished for diprenorphine [38]	
	I277C	Accessible to MTSEA alkylation [38]	
	H278C	Abolished for diprenorphine [38]	
	V280C	Accessible to MTSEA alkylation [38]	

[a] Location of the mutations is indicated in the left column. Nature of the mutation is provided using the single letter code. Effect of the mutation on binding affinity is indicated in fold changes. EC_{50} and IC_{50} reflect changes in agonist potency in [^{35}S]GTPγS and cyclase assays respectively. Overall, the selectivity of delta ligands is mainly dependent on EL3, and the helical bundle is involved in binding as well as activation of the receptor.

TABLE 3 Site-Directed Mutagenesis of the Delta Opioid Receptor: C-Terminal Tail[a]

C. mutation	Effect on signal and receptor regulation
ΔC-ter (S334)	Decreased internalization [48] and downregulation [49]
ΔC-ter (S344)	Decreased phosphorylation by Etorphine. β-Arrestin recruitment, internalization, and downregulation as WT [51]
S344A	Abolished phosphorylation and internalization by PMA, phosphorylation, and internalization by DPDPE as WT [47]
Any S/T after T352	Decreased internalization by DADLE [48]
T353A	Decreased downregulation by DADLE [49]
T358A/D	Decreased in phosphorylation [44,45], β-arrestin1 and 2 binding [46], and internalization [45] by DPDPE
T361A	Decreased in phosphorylation [44,45] and β-arrestin1 and 2 binding [46]
T361D	Phosphorylation as the WT [44]
S363G/A/D	Decreased in phosphorylation [44,45], β-arrestin1 and 2 binding [46], internalization [50], and desensitization [45,50] by DPDPE or Delt II
5 last S/T in A	Decreased in phosphorylation, β-arrestin recruitment, internalization, and downregulation by etorphine [51]

[a] C-terminal truncations (ΔC-ter, last residue indicated) and mutations of potential phosphorylation sites have been performed. Several aspects in the regulation of signaling are mediated by determinants of the C-terminal tail.

Most studies showed that EL3, together with the extracellular part of TM6 and 7 are important for delta ligand selectivity. Three groups replaced EL3 of the delta receptor by the equivalent portion of the mu receptor and found a dramatic decrease in affinities of DPDPE, Delt II, and NTI, as well as SNC 80, SNC 121, and TAN-67 [9–12]. Site-directed mutagenesis identified a stretch of hydrophobic residues in EL3 (Trp284, Val296, Val297; see Fig. 1) involved in DPDPE, Delt II, and SNC 80 recognition [10]. Binding rescue experiments also highlighted several determinants distributed in upper parts of TM5, TM6, and TM7, and flanking EL3 (Fig. 1). The reintroduction of Leu300 into the otherwise inactive m(EL3)/d chimera restored DPDPE and SNC80 binding, showing contribution of this particular residue [12]. An opioid binding site was created by inserting five residues from the delta receptor (Lys214, Ile277-His278-Ile279, and Ile304) into the nonopioid ORL receptor [13].

Together, the studies suggest that EL3 contributes to delta selectivity by enhancing the affinity of the receptor for delta ligands. Other mechanisms

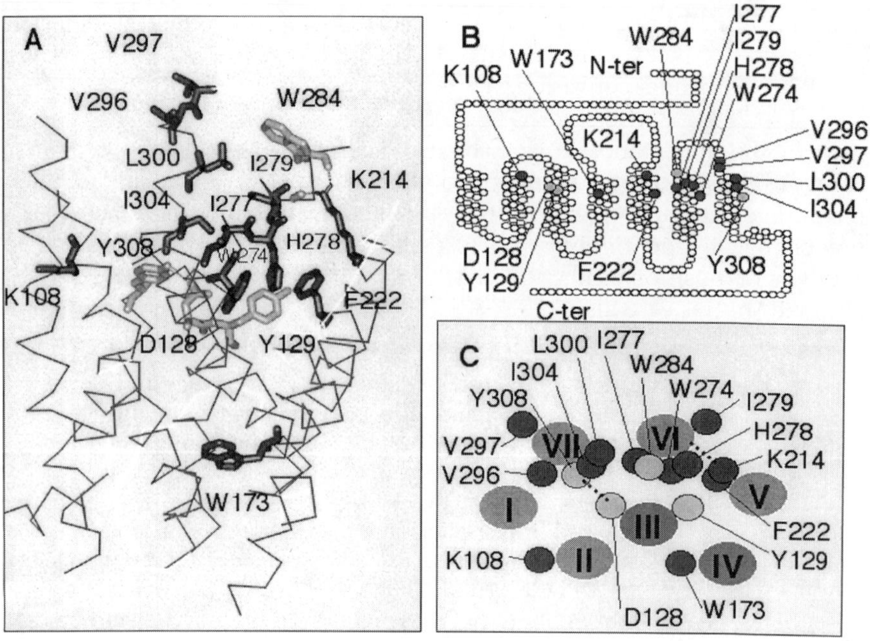

FIGURE 1 Important residues for delta receptor structure-activity. (A) Lateral view of a 3D model of the human delta opioid receptor. This model is based on x-ray crystallographic data from rhodopsin [91]. Helices are indicated as ribbons, side chains of amino acids implicated in binding (dark gray) or both binding and activation (light gray) are shown as sticks. Hydrophilic bonds are shown as dotted lines. (B) Position of important residues along the human delta opioid receptor sequence [34] using the same color code. (C) Scheme representing the receptor viewed from the extracellular face using the same color code.

can contribute to delta selectivity. In particular, some residues of the delta receptor could prevent binding of mu or kappa compounds [14,15]. As an example the insertion of mu EL1 into delta receptor (see Table 1), then the replacement of Lys 108 (delta) by Asn (mu) in EL1 [16,17] strongly increased the affinity of DAMGO (mu agonist) for the delta receptor.

2.2 Trp 284: A GPCR Key Residue for Binding and Receptor Activation

Much attention focused on Trp284 joining EL3 and TM6 (Table 2). In the delta receptor, mutation of Trp284 into Lys affects peptide binding [25], mutation into Leu modifies affinity for the small alkaloid SNC80 [26], and mutation into Glu decreases naltrindole binding [15]. Interestingly, a marked modification of agonist potency and efficacy was reported in these mutants that could not be explained by affinity changes, indicating that Trp284 also regulates functional activation of the receptor [26].

Trp 284 was also studied in other receptors, providing a "transversal" view of the importance of this particular amino acid across GPCRs. The equivalent residue in the kappa receptor (Glu297) is implicated in binding of the antagonist norBNI [27]. In the mu receptor, the equivalent Lys303 does not change ligand binding but strongly affects receptor activation [28]. Equivalent position (Gln286) in the nonopioid ORL receptor has been mutated. In this case also, the residue is not involved in agonist binding but activation by any nociceptin/orfaninFQ agonist is abolished [29]. Finally, this amino acid is also implicated in agonist potency of bombesin receptor [30]. In conclusion, Trp284 is important for delta ligand recognition and appears also as a key molecular determinant for agonist-induced signaling in delta and other GPCRs.

2.3 The Transmembrane Core: Bottom of the Opioid Binding Site and Center for Signaling

A panel of residues buried inside the helical bundle, implicated in the binding of many different opioid ligands, has been identified in early experiments [see 8]. Since then, the role of these residues has been investigated further in both binding and functional studies, and several novel molecular determinants have been reported (Table 2; Fig. 1).

The highly conserved residue Asp128 (TM3) was postulated to serve as a counterion for protonated ligands in biogenic amine GPCRs [31] and was therefore expected to anchor the charged nitrogen atom present in all opioid alkaloids and peptides. Results from mutagenesis experiments were complex. Whereas the Asn substitution strongly decreased the binding of many opioids, the Ala mutation was almost ineffective [32]. Another study showed that mutation of the same Asp into Lys decreased affinities for DPDPE and naloxone, and confirmed that the Ala mutation did not modify opioid binding [33]. A reasonable explanation was that Asp128 indeed interacts with the protonated amine of opioid ligands in the wild-type receptor, but that other determinants for opioid binding are prevailing, in particular a hydrophobic/aromatic pocket (see below). The mutated side chain of Asp128Ala could strengthen hydrophobicity of the binding site in a way that would compensate for the absence of ionic interaction. In contrast, Asn or Lys mutations would neither participate in the hydrophobic environment nor fulfill the counterion role, and sterically hinder the binding pocket. Additional data came from a two-dimensional mutagenesis study in the mu opioid receptor using mutations at the equivalent position (Asp147) and several morphine derivatives in which protonated nitrogen has been removed. The data supported the notion that Asp indeed represents a counterion for opioid ligands in the wild-type receptor [24].

The mutation Asp128 into Asn, Ala, Lys, and His produced constitutive activation of the delta receptor [33,34], suggesting a role for this residue in receptor signaling. Interestingly, naloxone had effects ranging from inverse agonist, agonist, and antagonist on the Ala, Lys, and His mutants, respectively, further confirming that these mutations influence the functional status of the receptor [33]. Asp128 is part of TM3 known to be central in the binding crevice of GPCRs [35]. In the 3D model, Tyr308 is forming an hydrogen bond with the Asp128, linking TM3 and TM7 (see Fig. 1C). Mutations of Asp128 most likely weaken this helix-helix interaction, inducing a more relaxed receptor conformation that more productively interacts with G proteins [34]. A movement of TM3 and TM7 apart from each other has been suggested to contribute to receptor activation in other GPCRs [36].

Other residues were tested by site-directed mutagenesis within the helical bundle [37]. Three-dimensional modeling identified an aromatic pocket encompassing TM3 to TM7 and located at mid-distance from extra- and intracellular faces of the receptor. The analysis of mutants suggested a limited contribution of the phenyl ring of F218 and Phe222 (TMV), a significant participation of indole rings from Trp173 (TM4) and Trp274 (TM6), and a prominent role of both hydroxyl and aromatic moieties of Tyr129 (TM3) and Tyr308 (TM7). This pocket was proposed to represent a general opioid-binding site [37] with a hydrophilic (Tyr, TM3/TM7) and a hydrophobic (Phe, Trp, TM4/TM5/TM6) part, possibly complementing amphiphilic opioid molecules. As for Asp128 mutants, the Tyr129 and Tyr308 mutants were later found to be constitutively active [34]. Therefore, in addition to forming the bottom of a general opioid-binding pocket, these residues represent a site where agonist binding translates into activation.

The binding site crevice was probed by cysteine accessibility mapping, where each amino acid from TM6 was replaced by a Cys residue and tested for reactivity to methanethiosulfonate ethylammonium [38]. The comparative study accross mu, delta, and kappa receptors showed a water-accessible surface on the extracellular face of the helix for all opioid receptors and located above the Pro kink (Ile277 and Phe280 to Leu286 in the delta receptor). The data were consistent with the notion of an opioid binding pocket penetrating the upper half of the helical bundle (see Fig. 1A).

2.4 The Intracellular Face

Mutagenesis of intracellular faces of transmembrane domains of mu receptors revealed molecular determinants for receptor activation within the helical bundle and located underneath the ligand binding pocket. Several mutants of Asp164 located in the Asp-Arg-Tyr motif at the TM3/IL2 interface showed constitutive activity [40], as well as the Thr279Lys mutant in TM6 [41]. It was suggested that an intrahelical salt bridge in TM3 between Asp164 and Arg165 of the conserved motif, stabilized by Thr279, would participate in the stabilization of the inactive state. More generally, mutations disrupting this "ionic lock" enhance spontaneous basal activity in other GPCRs [42,43]. Similar ionic interactions likely are important in delta receptor activation, but this has not been investigated yet.

Other mutants were created to study the regulation of delta receptor signaling, mostly within the C-terminal tail (summarized in Table 3). Among six Ser and Thr residues present in this region, Thr358 and Ser363 seem to play a major role in receptor phosphorylation, internalization, and/or desensitization [44–46]. Thr361 could also contribute, although in an indirect manner since the Ala but not the Asp mutation disrupts receptor phosphorylation [44].

3 GENE TARGETING IN MICE

Homologous recombination makes it possible to specifically mutate a gene of interest in a whole animal. Classical in vivo gene targeting strategies consist in deleting an essential exon, which leads to a nonfunctional gene. Using this approach, mice with a deficient delta opioid receptor gene have been generated by two groups. Zhu and colleagues [52] deleted the second coding exon, while Filliol et al. [53] deleted the first coding exon, including the ATG initiation codon.

In both cases animals with a deletion of two alleles (homozygous) showed no opioid binding, whether a delta 1 (DPDPE), a delta 2 (Delt I), or a general (NTI) delta radiolabeled ligand was used in homogenate [52,53] or autoradiographic [52,54] studies. Binding of the general opioid ligand bremazocine was also reduced [55]. In contrast, the binding of specific mu and kappa agonists was modified only subtly [54]. Similarly, expression patterns of preproenkephalin, preprodynorphin, and proopiomelanocortin genes did not seem modified [52,53], suggesting that the absence of delta receptors did not markedly alter the expression of remaining components of the opioid system during development. Both mouse lines were fertile and showed no obvious anatomical or growth deficit. The characterization of these mice and their comparison with mice lacking the mu or the kappa opioid receptor gene have been discussed in great detail recently [56].

3.1 Delta Compounds: A Complex Pharmacology

The pharmacology of delta ligands was investigated in opioid receptor knockout mice, and data from these compounds have been more difficult to understand than data obtained from mu or kappa agonists [56]. The analgesic activity of DPDPE and Delt, the prototypal delta agonists, was examined in delta receptor–deficient mice by Zhu et al. [52]. Surprisingly, the compounds remained active in the mutant mice when injected by ICV route (Table 4, left column). In the meantime, several groups had tested the delta agonists in mice lacking mu receptors and found decreased biological effects in many distinct experimental conditions (Table 4, right column). Obviously, and in view of previously postulated functional interactions between mu and delta receptors [57–59], data from both mu receptor– and delta receptor–deficient mice should be considered concurrently (pooled in Table 4).

A first comment is that experimental protocols and genetic backgrounds often vary among laboratories. Obvious discrepancies (as for example different results in mu receptor knockout mice following DPDPE ICV in the tail flick and hot plate tests across laboratories, see Table 4, right column upper part) occur for technical reasons and complicate the overall picture. Despite this methodological aspect, however, several conclusions can be drawn from Table 4:

1. The selectivity of currently used delta agonists may not be sufficient to avoid mu receptor activation in vivo. As an example, one study showed that DPDPE (selectivity delta/mu 100-fold) injected either ICV or ITH was less active in the mu receptor mutant than deltorphin (selectivity mu/delta 10,000-fold) [60]. This suggests that, in WT mice the less delta selective compound recruits mu receptors to produce analgesia in the tail flick and hot plate tests under their experimental conditions.

2. Independently from the ligand, some analgesic responses in vivo could be mediated by mu receptors, either because of strategic localization of this receptor in the nociceptive circuitry or because of receptor availability in the tissue. To illustrate this point, several experiments showed that both deltorphin and DPDPE analgesia was maintained in the delta receptor mutant [52] and decreased in the mu receptor mutant [60–62] when ICV route of administration was used, while ITH administration led to different results [63]. This strongly suggests that central mu receptors, located in the proximity of ventricles, represent targets for those opioid compounds, at least in response to thermal pain. These mu receptors also seem responsible for the nonanalgesic effects of deltorphin, since addictive [64] and respiratory [61] properties were abolished when the compound was injected ICV in MOR-deficient mice.

3. Delta receptors nevertheless mediate some delta agonist–induced analgesia, as suggested by reduced DPDPE and deltorphin activity in the DOR mutant after ITH applications [52], or enhanced antinociceptive activity in MOR mutants subjected to CFA inflammation [65]. Delta receptors also depress respiratory neurons in slice preparations [66] and mediate SNC80-evoked convulsions [67].

TABLE 4 Effects of Delta Compounds in Mice Lacking Delta and Mu Opioid Receptors[a]

Biological activity	Compound	Delta receptor KO (52)	Mu receptor KO
Analgesia to thermal pain	DPDPE ICV	Maintained	Decreased in TI [61] Decreased in TF [60,62] Maintained in TF [81] Decreased in HP [62] Maintained in HP [61,81]
	DPDPE ITH	Decreased	Decreased in TF [60] Maintained in TF [63]
	Deltorphin ICV	Maintained	Decreased in TI [61] Maintained in TF [60] Maintained in HP [61]
	Delt ITH	Decreased	Maintained in TF [60]
	BW373U86 ICV	Enhanced	
	BW373U86 ITH	No effect, as in WT	
	BW373U86 SC	Enhanced	
Analgesia to formalin	DPDPE ICV	Maintained	
Antihyperalgesia to CFA	DPDPE ITH		Enhanced [65]
	Delt ITH		Enhanced [65]
Analgesia to mechanical pain	DPDPE ICV		Decreased [82]
Analgesia to visceral pain	SNC80 SC		Decreased [83]
Convulsions	SNC80 SC	Abolished [67]	
Reward and physical dependence	Delt ICV		Decreased [64]
Respiratory depression	Delt ICV		Decreased [61]
Brain stem slices: firing of respiratory neurons	Delt		Maintained [66]
	DPDPE		Enhanced [66]
Vas deferens: inhibition of contractions	Delt I and II, DPDPE, BUBU, [D-Met]-Delt		Reduced [84]
Immunosuppression	NTI	Maintained, as well as in triple knockout [68]	

[a] Biological activities of prototypic delta agonists have been tested in delta receptor knockout mice essentially by Zhu et al. (left column) and in mu receptor knockout mice by several groups (right column).
Modes of injection: ICV, intracerebroventricular; ITH, intrathecal; SC, subcutaneous.
Behavioral assays: HP, hot plate; TF, tail flick; TI, tail immersion.
CFA, Complete Freud's adjuvant.

4. Beyond delta agonist selectivity and mu or delta receptor availability in vivo, the observation of decreased delta agonist efficacy in mice lacking mu receptors could also be explained by functional cooperation of mu and delta receptors. It is likely that some delta receptor–mediated effects require mu receptors for full activity (see Rothman and Xu, Chap. 21). Where in the brain, and whether this occurs between receptors located in the same neurons or on different neurons within neural circuits, remains to be determined.

5. The activity of a prototypical delta antagonist was also tested in vivo. The immunosuppressive effect of NTI was maintained in delta receptor knockout mice, as well as in mice lacking all three opioid receptors [68], opening the search for a novel molecular target for this widely used delta opioid compound.

In conclusion, the in vivo activity of available delta opioids is complex. DPDPE, or even Delt, administered ICV seems to recruit mu receptors and, from all the data, it appears that delta agonists often have mixed mu/delta activities. More selective delta agonists need to be produced to explore delta receptor pharmacology. The examination of nonanalgesic activities of delta ligands in opioid receptor knockout mice has been very informative: while the convulsive effect of SNC 80 seems indeed delta receptor mediated, the addictive activity of Delt most probably results from mu receptor activation and the immunosuppressive action of NTI is mediated by a nonopioid mechanism.

3.2 Delta Knockout Mice Behavior: Nociceptive and Emotional Responses

Behavioral alterations of delta receptor–deficient mice are shown in Table 5. A first observation is that delta receptor–deficient mice show significant hyperlocomotor activity [53]. This behavioral modification, however, did not seem to influence responses in other tests, which generally showed hyporeactivity of the animals.

Nociceptive thresholds have been examined in response to acute painful stimuli. First studies showed no change in responses to thermal, mechanical, visceral, and inflammatory pain [52,53]. A more extensive study with separate gender analysis later indicated modest but significant increased nociception in the tail pressure test, as well as in the late phase of the formalin test, but in females only [69]. These data, together with results using mice lacking mu and kappa receptors [see 56,69], suggest a modest opioid tone in modulating acute noxious information, with delta receptors contributing the least. In the future, models of chronic inflammation or neuropathic pain applied to the knockout animals may reveal a significant contribution of delta receptors in adaptive

TABLE 5 Behavioral Modifications in Mice Lacking Delta Receptors[a]

Behavior	Test	Phenotype
Locomotion	Actimetry	Increased [53]
Thermal pain	Tail immersion	Unchanged [53,69]
	Tail flick	Unchanged [52]
	Hot plate	Unchanged [53,69]
Mechanical pain	Tail pressure	Increased in females [69]
Visceral pain	Acetic acid/writhing	Unchanged [53]
Chemical and inflammatory pain	Formalin	Unchanged [52,53]
		Increased in females, late phase [69]
Anxiety	Elevated plus maze	Increased [53]
	Dark-light box	Increased [53]
		Reversed to WT levels after EtOH [73]
Depression	Forced swim	Increased [53]
Morphine analgesia	Tail flick	Unchanged [52]
Tolerance to morphine analgesia	Tail flick	Reduced [52,79]
Morphine dependence	Signs of withdrawal	Maintained [79]
Alcohol intake	Operant self-administration	Increased [73]
	Two-botte choice	Increased after EtOH training [73]
Cannabinoid hypolocomotion, hypothermia, and antinociception	Actimetry Rectal temperature Tail immersion	Acute and chronic effects unchanged [77]
Cannabinoid dependence	Signs of withdrawal	Paw tremor reduced; global score unchanged [77]
Cannabinoid reward	Conditioned place preference and aversion	Unchanged [77]

[a] These include changes in spontaneous behaviors, as well as responses to drugs of abuse.

responses to persistent pain. Good indications are the antihyperalgesic activity of delta compounds correlated to recruitment of delta receptors on the cell surface [70], or the enhanced antihyperalgesic efficacy of delta agonists in MOR-deficient mice [65] using the CFA model of inflammation.

Delta receptor knockout animals showed a strong and unexpected phenotype in assays for anxiety and depressive-like responses [53]. Mutant animals explored open arms of the plus maze and visited the aversive compartment of the light-dark box significantly less than their wild-type counterparts. These indications of increased levels of anxiety, paralleled with a similar phenotype in mice lacking preproenkephalin [71], suggest that delta receptors endogenously activated by enkephalin peptides have anxiolytic activity. Further, delta receptor knockout mice showed a depressivelike behavior, as indicated by increased immobility in the forced swim test. There is therefore a potential for delta receptors as a therapeutic target to treat affective disorders, a possibility that has been little investigated earlier [72]. This aspect of delta receptor function could also have implications in the treatment of drug abuse, particularly alcoholism, since mice lacking delta receptors consume more alcohol than wild-type [73] (see details below).

Finally, an important observation is that, although delta and mu receptors seem to cooperate in their antinociceptive activities (Sec. 3.1), this is generally not the case for other behaviors. In fact, mice lacking mu receptors showed responses opposing those of mice lacking delta receptors in most behavioral tests performed so far [see 74]. The mu receptor mutant exhibited hypolocomotion, decreased anxiety, and depressive-like behavior in a comparative study [53], and also did not self-administer alcohol, even after forced drinking [75]. Together, these studies confirm that delta receptors should definitely be considered very distinct from mu receptors functionally. This has been hampered in the past by the probable lack of delta agonist selectivity in vivo under some circumstances, and by the primary focus of pharmacological studies on pain physiology where indeed delta and mu receptors seem to ally their antinociceptive roles.

3.3 Delta Receptors and Chronic Drugs of Abuse

Mu receptor knockout mice showed no reponse to morphine [76], confirming definitely that mu receptors are the primary target for abused opiates and are critical in mediating opioid addiction. Interestingly also the MOR mutants showed no alcohol [75], THC [77], and nicotine [78] reward, suggesting that mu receptors are more generally key players in responses to drugs of abuse, presumably because of their prominent role in modulating neural circuits of reward and addiction. The implication of delta receptors in adaptations to chronic drug use is more complex, and responses of delta receptor knockout mice to drugs of abuse have been variable.

Morphine analgesia was unchanged in the delta receptor–deficient after an acute injection, demonstrating that the delta receptor is not required for morphine analgesia in vivo [52]. Interestingly, morphine analgesia was maintained after repeated injections, indicating that delta receptors contribute to the development of morphine tolerance as previously suggested by the pharmacology [52]. Also, tolerance did not develop in mice lacking the preproenkephalin gene, suggesting an enkephalin/delta receptor tone in adaptations to chronic morphine along nociceptive pathways [79]. Importantly also, the latter study showed that morphine withdrawal could be induced in both mutant mice that did not become tolerant to chronic morphine. This finding demonstrated that analgesic tolerance and physical dependence can be genetically dissociated, and that the enkephalin/delta receptor pair contributes to some, but not all, adaptations to chronic opiates within the endogenous opioid circuitry [79].

Responses of delta receptor–deficient mice to THC were thoroughly examined and found mainly unchanged [77]. Briefly, acute effects, including hypolocomotion, hypothermia, and analgesia, were unchanged and tolerance to these effects developed normally after repeated administration. Conditioned place preference and aversion to THC, which can be produced in mice following different administration protocols, were unchanged in mice lacking delta receptors while dramatic—and opposing—modifications were found for mice lacking mu and kappa receptors in those tests. One sign of THC withdrawal only was modified (Table 5), with a global withdrawal score unchanged. Together the data indicate that, in contrast to mu and kappa receptors, delta receptors barely interact with the cannabinoid system.

Delta receptor knockout mice were exposed to alcohol, and drinking was examined in the two-bottle choice and operant self-administration paradigms [73,80]. First, naive animals consumed alcohol as wild-type animals in the two-bottle choice. After training to self-administer alcohol, the mutant mice developed a stronger preference for alcohol and consumed more than wild-type mice in both tests. After alcohol experience also, their anxiety decreased and reversed to wild-type levels. This suggests that, in the case of alcohol, delta receptor activity may partly influence addictive behavior indirectly, by modulating anxiety-like responses. Whether this applies to self-consumption of other drugs of abuse remains to be determined.

4 PERSPECTIVES

In vitro site-directed mutagenesis has identified EL3 as a critical determinant of delta selectivity, as well as an amphiphilic binding pocket located on the extracellular face of the helical bundle encompassing TM3 to TM7. A few determinants for receptor activation have been found fortuitously, or from studies in other GPCRs, but no clear image has emerged on how the delta agonist triggers receptor signaling. In the future, understanding the entire sequence of events leading to receptor activation should benefit from more global approaches such as random mutagenesis strategies [85] (our unpublished results). Other key structural features of delta receptor function are domains for the postulated receptor dimerization [59,86,87] or intracellular signals for receptor trafficking and downregulation [88], and it is expected that these will be extensively examined in the future.

In vivo deletion of the delta opioid receptor gene in mice has brought to light the complex pharmacology of prototypic delta agonists, which recruit both mu and delta receptors in vivo. Mouse phenotyping has also revealed novel aspects of delta receptor function. The possibility of an anxiolytic and antidepressant delta tone deserves further investigations that may have implications in the field of emotional disorders and drug abuse. Also, the future examination of these mice in models of chronic pain should help in evaluating the therapeutic potential of novel more selective delta compounds in pain control. Finally, more sophisticated gene targeting experiments in mice will allow to identify the exact localization of functionally important delta receptors along neural circuits and help exploring the intriguing observation of delta receptor exocytosis under chronic morphine treatment [89] or the potential role of delta receptor spontaneous activity in vivo [90].

ACKNOWLEDGMENTS

We wish to thank ShiYi Yue and Philippe Walker for fruitful collaboration. The authors also wish to acknowledge support of AstraZeneca, the Human Frontier Science Program, the National Institute of Drug Abuse (2 P50 DA05010), the Centre National de la Recherche Scientifique, the INSERM, the Université Louis Pasteur, the Association de la Recherche pour le Cancer, the Institut UPSA de la Douleur, and the Mission Interministérielle de Lutte Contre la Drogue et la Toxicomanie.

REFERENCES

1. Barnard EA. Curr Biol 1993; 3:211–214.
2. Brownstein MJ. Proc Natl Acad Sci USA 1993; 90:5391–5393.
3. Evans CJ, Keith DE, Morrison H, Magendzo K, Edwards RH. Science 1992; 258:1952–1955.
4. Kieffer BL, Befort K, Gaveriaux-Ruff C, Hirth CG. Proc Natl Acad Sci USA 1992; 89:12048–12052.
5. Kieffer BI. Cell Mol Neurobiol 1995; 15:615–635.
6. Chaturvedi K, Christoffers KH, Singh K, Howells RD. Biopolymers 2000; 55:334–346.
7. Law PY, Wong YH, Loh HH. Biopolymers 1999; 51:440–455.
8. Befort K, Kieffer BL. In: Budd K, Hamann W, eds. Pain Reviews. London: Arnold, 1997:100–121.
9. Varga EV, Li X, Stropova D, Zalewska T, Landsmann RS, Knapp RJ, Malatynska E, Kawai K, Mizusura A, Nagase H, Calderon SN, Rice K, Hruby VJ, Roeske WR, Yamamura HI. Mol Pharmacol 1996; 50:1619–1624.
10. Valiquette M, Vu HK, Yue SY, Wahlestedt C, Walker P. J Biol Chem 1996; 271:18789–18796.

11. Li X, Varga EV, Stropova D, Zalewska T, Malatynska E, Knapp RJ, Roeske WR, Yamamura HI. Eur J Pharmacol 1996; 300:R1–R2.
12. Pepin M-C, Yue SY, Roberts E, Wahlestedt C, Walker P. J Biol Chem 1997; 272:9260–9267.
13. Meng F, Ueda Y, Hoversten MT, Taylor LP, Reinscheid RK, Monsma FJ, Watson SJ, Civelli O, Akil H. Mol Pharmacol 1998; 53:772–777.
14. Metzger TG, Ferguson DM. FEBS Lett 1995; 375:1–4.
15. Metzger TG, Paterlini MG, Ferguson DM, Portoghese PS. J Med Chem 2001; 44:857–862.
16. Fukuda K, Terasako K, Kato S, Mori K. FEBS Lett 1995; 373:177–181.
17. Minami M, Nakagawa T, Seki T, Onogi T, Aoki Y, Katao Y, Katsumata S, Satoh M. Mol Pharmacol 1996; 50:1413–1422.
18. Kong HY, Raynor K, Yano H, Takeda J, Bell GI, Reisine T. Proc Natl Acad Sci USA 1994; 91:8042–8046.
19. Fukuda K, Kato S, Mori K. J Biol Chem 1995; 270:6702–6709.
20. Meng F, Hoversten MT, Thompson RC, Taylor L, Watson SJ, Akil H. J Biol Chem 1995; 270:12730–12736.
21. Meng F, Ueda Y, Hoversten MT, Thompson RC, Taylor L, Watson SJ, Akil H. Eur J Pharmacol 1996; 311:285–292.
22. Wang WW, Shahrestanifar M, Jin J, Howells RD. Proc Natl Acad Sci USA 1995; 92:12436–12440.
23. Onogi T, Minami M, Katao Y, Nakagawa T, Aoki Y, Toya T, Katsumata S, Satoh M. FEBS Lett 1995; 357:93–97.
24. Li JG, Chen C, Yin J, Rice K, Zhang Y, Matecka D, de Riel JK, DesJarlais RL, Liu-Chen LY. Life Sci 1999; 65:175–185.
25. Chaturvedi K, Jiang X, Christoffers KH, Chinen N, Bandari P, Raveglia LF, Ronzoni S, Dondio G, Howells RD. Brain Res Mol Brain Res 2000; 80:166–176.
26. Hosohata Y, Varga EV, Stropova D, Li X, Knapp RJ, Hruby VJ, Rice KC, Nagase H, Roeske WR, Yamamura HI. Life Sci 2001; 68:2233–2242.
27. Hjorth SA, Thirstrup K, Grandy DK, Schwartz TW. Mol Pharmacol 1995; 47:1089–1094.
28. Bonner G, Meng F, Akil H. Eur J Pharmacol 2000; 403:37–44.
29. Mouledous L, Topham CM, Moisand C, Mollereau C, Meunier JC. Mol Pharmacol 2000; 57:495–502.
30. Sainz E, Akeson M, Mantey SA, Jensen RT, Battey JF. J Biol Chem 1998; 273:15927–15932.
31. Strader CD, Fong TM, Graziano MP, Tota MR. FASEB J 1995; 9:745–754.
32. Befort K, Tabbara L, Bausch S, Chavkin C, Evans C, Kieffer BL. Mol Pharmacol 1996; 49:216–223.
33. Cavalli A, Babey AM, Loh HH. Neuroscience 1999; 93:1025–1031.
34. Befort K, Zilliox C, Filliol D, Kieffer BL. J Biol Chem 1999; 274:18574–18581.
35. Lu ZL, Saldanha JW, Hulme EC. Trends Pharmacol Sci 2002; 23:140–146.
36. Meng EC, Bourne HR. Trends Pharmacol Sci 2001; 22:587–593.

37. Befort K, Tabbara L, Kling D, Maigret B, Kieffer BL. J Biol Chem 1996; 271: 10161–10168.
38. Xu W, Li J, Chen C, Huang P, Weinstein H, Javitch JA, Shi L, Kim de Riel J, Liu-Chen LY. Biochemistry 2001; 40:8018–8029.
39. Kramer HK, Andria ML, Kushner SA, Esposito DH, Hiller JM, Simon EJ. Brain Res Mol Brain Res 2000; 79:55–66.
40. Li J, Huang P, Chen C, de Riel JK, Weinstein H, Liu-Chen LY. Biochemistry 2001; 40:12039–12050.
41. Huang P, Li J, Chen C, Visiers I, Weinstein H, Liu-Chen LY. Biochemistry 2001; 40:13501–13509.
42. Ballesteros JA, Jensen AD, Liapakis G, Rasmussen SG, Shi L, Gether U, Javitch JA. J Biol Chem 2001; 276:29171–29177.
43. Scheer A, Costa T, Fanelli F, De Benedetti PG, Mhaouty-Kodja S, Abuin L, Nenniger-Tosato M, Cotecchia S. Mol Pharmacol 2000; 57:219–231.
44. Guo J, Wu Y, Zhang W, Zhao J, Devi LA, Pei G, Ma L. Mol Pharmacol 2000; 58:1050–1056.
45. Kouhen OM, Wang G, Solberg J, Erickson LJ, Law PY, Loh HH. J Biol Chem 2000; 275:36659–36664.
46. Cen B, Xiong Y, Ma L, Pei G. Mol Pharmacol 2001; 59:758–764.
47. Xiang B, Yu GH, Guo J, Chen L, Hu W, Pei G, Ma L. J Biol Chem 2001; 276:4709–4716.
48. Trapaidze N, Keith DE, Cvejic S, Evans CJ, Devi LA. J Biol Chem 1996; 271: 29279–29285.
49. Cvejic S, Trapaidze N, Cyr C, Devi LA. J Biol Chem 1996; 271:4073–4076.
50. Law PY, Kouhen OM, Solberg J, Wang W, Erickson LJ, Loh HH. J Biol Chem 2000; 275:32057–32065.
51. Whistler JL, Tsao P, von Zastrow M. J Biol Chem 2001; 276:34331–34338.
52. Zhu Y, King MA, Schuller AGP, Nitsche JF, Riedl M, Elde RP, Unterwald E, Pasternak GW, Pintar JE. Neuron 1999; 24:243–252.
53. Filliol D, Ghozland S, Chluba J, Martin M, Matthes H, Simonin F, Befort K, Gavériaux-Ruff C, Dierich A, LeMeur M, Valverde O, Maldonado R, Kieffer BL. Nat Genet 2000; 25:195–200.
54. Goody RJ, Oakley SM, Filliol D, Kieffer BL, Kitchen I. Brain Res 2002; 945: 9–19.
55. Simonin F, Slowe S, Becker JAJ, Matthes HWD, Filliol D, Chluba J, Kitchen I, Kieffer BL. Eur J Pharmacol 2001; 414:189–195.
56. Kieffer BL, Gavériaux-Ruff C. Prog Neurobiol 2002; 66:285–306.
57. Traynor JR, Elliot J. Trends Pharmacol Sci 1993; 14:84–85.
58. Rothman RB, Holaday JW, Porreca F. In: Herz A, ed. Handbook of Experimental Pharmacology, Opioids I. Berlin: Springer-Verlag, 1993:217–237.
59. Devi LA. Trends Pharmacol Sci 2001; 22:532–537.
60. Hosohata Y, Vanderah TW, Burkey TH, Ossipov MH, Kovelowski CJ, Sora I, Uhl GR, Zhang X, Rice KC, Roeske WR, Hruby VJ, Yamamura HI, Lai J, Porreca F. Eur J Pharmacol 2000; 388:241–248.
61. Matthes HWD, Smadja C, Valverde O, Vonesch J-L, Foutz AS, Boudinot E,

Denavit-Saubier M, Severini C, Negri L, Roques BP, Maldonado R, Kieffer BL. J Neurosci 1998; 18:7285–7295.
62. Sora I, Funada M, Uhl GR. Eur J Pharmacol 1997; 324:R1–R2.
63. Schuller AGP, King M, Zhang J, Bolan E, Pan Y-X, Morgan DJ, Chang A, Czick ME, Unterwald EM, Pasternak GW, Pintar JE. Nat Neurosci 1999; 2:151–156.
64. Hutcheson DM, Matthes HW, Valjent E, Sanchez-Blazquez P, Rodriguez-Diaz M, Garzon L, Kieffer BL, Maldonado R. Eur J Neurosci 2001; 13:153–161.
65. Qiu C, Sora I, Ren K, Uhl G, Dubner R. Eur J Pharmacol 2000; 387:163–169.
66. Morin-Surun MP, Boudinot E, Dubois C, Matthes HW, Kieffer BL, Denavit-Saubie M, Champagnat J, Foutz AS. Eur J Neurosci 2001; 13:1703–1710.
67. Broom DC, Nitsche J, Pintar JE, Woods JH, Traynor JR. J Pharmacol Exp Ther 2002; 303:723–729.
68. Gaveriaux-Ruff C, Filliol D, Simonin F, Matthes HWD, Kieffer BL. J Pharmacol Exp Ther 2001; 298:1–6.
69. Martin M, Matifas A, Maldonado R, Kieffer BL. Eur J Neurosci 2003; 17:1–8.
70. Cahill CM, Morinville A, O'Donnell D, Beauder A. Pain 2003; 101:199–208.
71. Konig M, Zimmer AM, Steiner H, Holmes PV, Crawley JN, Brownstein MJ, Zimmer A. Nature 1996; 383:535–538.
72. Tejedor-Real P, Mico JA, Smadja C, Maldonado R, Roques BP, Gibert-Rahola J. Eur J Pharmacol 1998; 354:1–7.
73. Roberts AJ, Gold LH, Polis I, McDonald JS, Filliol D, Kieffer BL, Koob GF. Alcohol Clin Exp Res 2001; 25:1249–1256.
74. Gavériaux-Ruff C, Kieffer B. Neuropeptides 2002; 36:62.
75. Roberts A, Mcdonald JS, Heyser CJ, Kieffer BL, Matthes HWD, Koob GF, Gold LH. J Pharmacol Exp Ther 2000; 293:1002–1008.
76. Kieffer BL. Trends Pharmacol Sci 1999; 20:537–544.
77. Ghozland S, Matthes HW, Simonin F, Filliol D, Kieffer BL, Maldonado R. J Neurosci 2002; 22:1146–1154.
78. Berrendero F, Kieffer BL, Maldonado R. J Neurosci 2002; 22:10935–10940.
79. Nitsche J, Schuller AGP, King MA, Zengh M, Pasternak GW, Pintar JE. J Neurosci 2002; 22:10906–10913.
80. Koob GK, Roberts AJ, Kieffer BL, Heyser CJ, Katner SN, Ciccocioppo R, Weiss F. In: Galante M, ed. Recent Developments in Alcoholism. New York: Plenum Press, 2002:263–281.
81. Loh HH, Liu H-C, Cavalli A, Yang W, Chen H-F, Wei L-N. Mol Brain Res 1998; 54:321–326.
82. Fuchs PN, Roza C, Sora I, Uhl G, Raja SN. Brain Res 1999; 821:480–486.
83. Sora I, Li XF, Funada M, Kinsey S, Uhl GR. Eur J Pharmacol 1999; 366:R3–R5.
84. Maldonado R, Severini C, Matthes HW, Kieffer BL, Melchiorri P, Negri L. Br J Pharmacol 2001; 132:1485–1492.
85. Parnot C, Bardin S, Miserey-Lenkei S, Guedin D, Corvol P, Clauser E. Proc Natl Acad Sci USA 2000; 97:7615–7620.
86. Portoghese PS. J Med Chem 2001; 44:2259–2269.

87. Filizola M, Weinstein H. Biopolymers 2002; 66:317–325.
88. Tsao PI, Von Zastrow M. J Biol Chem 2000; 275:11130–11140.
89. Cahill CM, Morinville A, Lee MC, Vincent JP, Collier B, Beaudet A. J Neurosci 2001; 21:7598–7607.
90. Costa T, Herz A. Proc Natl Acad Sci 1989; 86:7321–7325.
91. Palczewski K, Kumasaka T, Thori, Behnke CA, Motoshima H, Fox BA, Le Trong I, Teller DC, Okada T, Stenkamp RE, Yamamoto M, Miyano M. Science 2000; 289:739–745.

5

Delta Opioid Receptor Signaling and Trafficking

P. Y. Law
University of Minnesota Medical School, Minneapolis, Minnesota, U.S.A.

1 INTRODUCTION

From the initial hypothesis of Martin et al. [1] on the existence of multiple opioid receptors and the discovery of the endogenous opioid peptides [2–4], numerous studies have reported on the signaling and cellular control of these opioid receptors. In most instances, the molecular mechanism studies have been hampered by the presence of heterogeneous receptor or cell populations. The delta opioid receptor, with its expression in homogeneous clonal cell lines such as neuroblastoma x glioma NG108-15 hybrid cells [5] or in neuroblastoma N4TG2 [6], has been the best studied among the opioid receptors. With the reported cloning of the delta opioid receptor by Evans and Kieffer and their coworkers [7,8], followed by the cloning of mu and kappa opioid receptors [9–14], further understanding of the molecular mechanism of opioid receptor function was obtained by the heterologous expression of these cloned receptors in various systems and the subsequent receptor mutagenesis studies. The importance of the delta opioid receptor activity was demonstrated by absence of morphine tolerance in the delta opioid receptor null mice [15]. Thus, it is critical that the basis of the delta opioid receptor signaling and the cellular control of such signaling are fully understood. There are numerous review articles and also several chapters within this book that discuss the cloning and the structure–activity studies of the delta opioid receptor. In this chapter, we will review briefly the studies on the delta opioid receptor signal transduction processes, and the subsequent cellular regulation of the receptor activities.

2 DELTA OPIOID RECEPTOR SIGNALING

From the sequence analysis of the cloned delta opioid receptor, it is apparent that this receptor belongs to the rhodopsin subfamily of the super family of G protein–coupled receptors (GPCRs). One of the common features of GPCRs is that the receptor signal transduction is mediated via the heterotrimeric G proteins. The delta opioid receptor is prototypical "G_i/G_o-coupled" receptor because its signals are efficiently blocked by pertussis toxin (PTX), a bacterial toxin produced by *Bordetella pertussis* which ADP-ribosylates and inactivates the α subunits of G_i/G_o proteins ($G\alpha_{i/o}$ subunits). Like other GPCRs that utilize G_i/G_o subfamily members for signal transduction, the delta opioid receptors have long been known to inhibit adenylyl cyclases [5] and Ca^{2+} channels [16,17], as well as to stimulate K^+ channels [18] and to increase intracellular Ca^{2+} level [19]. More recently, the delta opioid receptors have been shown to regulate the mitogen-activated protein (MAP) kinase cascade [20,21]. The activation of multiple effectors presents myriad possibilities of signals cross talks and potentiation of the delta opioid receptor signals.

The promiscuity of the delta opioid receptor to regulate multiple effectors also is reflected in the ability of the receptor to induce GTP binding to all the G_i/G_o α subunits. By using either ^{32}P-azidoanilido GTP to photoaffinity label the Gα subunits or cholera toxin to ADP-ribosylate the G_i/Go α subunits after their dissociation from the βγ subunits, several reports have indicated that the delta opioid receptors could activate the G_i/G_o proteins with equal potency [22–24]. The ability of delta opioid receptor to activate G protein other than G_i/G_o was demonstrated by the coimmunoprecipitation of receptor with the recombinant G_z, which has 66% sequence homology with that of the G_i/G_o and which cannot be inactivated by PTX [25]. The ability of delta opioid receptor to interact with G_z presents a probable explanation for those observations that opioid receptor activities could not be blocked by PTX. We will review the delta opioid receptor activation of these G proteins and the mechanism for the subsequent effectors regulation separately.

2.1 Adenylyl Cyclase

One of the most studied effectors regulated by the delta opioid receptor is the adenylyl cyclase. Opioid agonist inhibition of the icAMP production has been demonstrated in both brain membranes [26] and in clonal cell lines such as the NG108-15 hybrid cells [5], and attributed mainly to the attenuation of the adenylyl cyclase activity. However, delta opioid receptor also could regulate the intracellular cAMP level by stimulating the phosphodiesterase activities [27]. Studies with Gα-specific antibodies suggested that G_{i2} mediates the PTX-sensitive delta opioid receptor inhibition of adenylyl cyclase activity in NG108-15 cells [28] while G_o mediates the mu opioid receptor inhibition of the adenylyl cyclase activity in SHSY5Y cells and brain membrane [29]. However, the overexpression of the G_z α-subunit with the delta opioid receptor resulted in PTX-insensitive agonist inhibition of the adenylyl cyclase activity [30]. Hence, the inhibition of the adenylyl cyclase by the delta opioid receptor is mediated by the G protein α subunits, either the $G_{i/o}$ or the G_z.

In addition to inhibiting the adenylyl cyclase activity, opioid receptor stimulated the adenylyl cyclase activity in brain membranes [31], F-11 neuroblastoma-sensory neuron hybrid cells [32], olfactory bulb [33], and spinal cord–ganglion explants [34]. Chronic agonist treatment also resulted in the superactivation of the adenylyl cyclase activity [5]. With the multiple subtypes of adenylyl cyclases, the opioid agonist stimulation of the enzymatic activity could be attributed to the $\beta\gamma$ subunits (G$\beta\gamma$) generated from the heterotrimeric G protein activation. Of the nine isoforms of mammalian adenylyl cyclases that have been cloned, G$\beta\gamma$ activates types 2, 4, and 7 adenylyl cyclases in the presence of GTP-bound $G\alpha_s$ [35–37]. Many classical inhibitory receptors (e.g.,α_2-adrenergic, dopamine-D_2, adenosine-A_1, and chemoattractant receptors) stimulate the type 2 adenylyl cyclase through the G$\beta\gamma$ released from activated PTX-sensitive G_i proteins [35,38]. Such coincidence signaling by the G$\beta\gamma$ and $G\alpha_s$ could be the basis for the delta opioid agonist–induced potentiation of the behavioral responses elicited by dopamine D_1 receptor agonists in mice [39]. The potentiation of the $G\alpha_s$ activity by G$\beta\gamma$ could be the reason for the subtype-specific superactivation of the adenylyl cyclase after chronic agonist treatment [40].

Others have suggested that the stimulation of the adenylyl cyclase is due to direct activation of $G\alpha_s$ by the delta opioid receptor. In dorsal root ganglion (DRG) neurons, the action potential duration is modulated by morphine in a bimodal fashion [41] where the cAMP-dependent excitatory effects are mediated by G_s-coupled opioid receptors. Treatment with GM1 ganglioside, but not with other gangliosides, rapidly converts the opioid receptors from an inhibitory to excitatory mode in the DRG neurons. Similar treatments with GM1 ganglioside allow the delta opioid receptor to stimulate cAMP formation in NG108-15 [42] and CHO [43] cells. Though it is possible for the delta opioid receptor to stimulate the adenylyl cyclase activity via a gangloside-dependent conversion mechanism, the direct interaction of the receptor with G's has not been demonstrated.

The complexity and versatility of the mammalian adenylyl cyclase system also allows for opioids to stimulate cAMP production. For instance, type 1 and 8 adenylyl cyclases are activated by Ca^{2+}/calmodulin while the basal activities of types 2, 4, and 7 adenylyl cyclases are elevated upon phosphorylation of the enzyme by PKC [36,37]. Since opioid receptors are capable of stimulating phospholipase C (PLC) and mobilizing intracellular Ca^{2+} (as discussed below), and that the third intracellular loop contains calmodulin binding motif with agonist regulating the calmodulin binding [44,45], the agonist activation of the delta opioid receptor could affect the Ca^{2+}/calmodulin and subsequently the PKC activation. Hence, it is not surprising to note that the opioid-induced elevation of basal cAMP level in SK-N-SH cells involves Ca^{2+} entry and calmodulin activation [46]. Whether the opioid receptor could stimulate or simply inhibit the intracellular cAMP production will depend greatly on the components present in the cellular environment.

2.2 Ion Channels

The ability of the opioid receptors to inhibit Ca^{2+} channels and hence affecting the neurotransmitters release has been established for more than a decade [16,17]. Delta opioid receptors have been shown to inhibit the L, N, P, and Q high voltage–activated Ca^{2+} channels in the postganglionic [47], neostriatal [48], and sensory [49] neurons, as well as clonal cell line models such as NG108-15 cells [16,50,51] and the neuronal F11 [52] cells. The cloned delta opioid receptor when expressed in the pituitary GH3 cells could inhibit the L-type Ca^{2+} channels [53]. The coupling of the delta opioid receptor to the L-type channels is less efficient than its regulation of the adenylyl cyclase activity [54]. Nonetheless, the delta opioid receptor regulation of the voltage-dependent Ca^{2+} channel is as complicated if not more than the receptor inhibition of the adenylyl cyclase activity. This is due to the heterogeneity of the Ca^{2+} channels involved. There are genes that encode for 10 α_1 (α_{1A}–α_{1I}, and α_{1S}) subunits, four β subunits, and three α_2-δ dimers [55]. The structural heterogeneity is complicated further by the presence of spliced variants of the α_1 subunits [56–60]. Since the α_1 subunit, with its 24 transmembrane domains, forms the voltage sensor and the ion pore of the Ca^{2+} channels, the composition of the channels with different combination of α_1, β, and the α_2-δ will determine the voltage dependence, current amplitude, and the kinetics of activation and inactivation.

In the past, G_α subunits were considered to mediate the opioid receptor inhibition of the Ca^{2+} activities. The involvement of $G_{\alpha o}$ was suggested initially by the injection of the α subunit into NG108-15 cells pretreated with PTX, thus restoring the delta opioid receptor activity [16]. This was later substantiated by the use of $G_{\alpha o}$-selective antisera [61]. Using antisense oligodeoxynucleotides to lower the Gα subunit content, Tang et al. [62] suggested that the opioid receptor regulation of the Ca^{2+} channels in ND8-47 neuroblastoma x dorsal root ganglion hybrid cells was mediated by the $Gi_{\alpha 2}$. However, it is apparent that G$\beta\gamma$ rather than the Gα subunit mediated the Ca^{2+} channels inhibition. GPCR mediated inhibition of the Ca^{2+} currents in rat sympathetic neurons [63] or in a heterologous expression system [64] can be mimicked by the G$\beta\gamma$ expression. The Q-X-X-E-R motif within the intracellular loop connecting domains I and II of the $\alpha 1$ subunit has been identified to be the G$\beta\gamma$ binding domain [63,65–68]. However, this QXXER motif located at the C-terminus of the α_{1D} subunit is not sufficient to confer the sensitivity to inhibitory G proteins [69]. The exact composition of the $\beta\gamma$ subunit appears to be critical in Ca^{2+} channel regulation. Intranuclear injections of rat superior cervical ganglion neurons with DNA encoding different Gβ subunits revealed that $G\beta_1$ and/or $G\beta_2$ subunits accounted for most of the voltage-dependent inhibition of N-type Ca^{2+} channels, while $G\beta_5$ produced weak inhibition and both $G\beta_3$ and $G\beta_4$ were ineffective [70]. Although the G$\beta\gamma$ subunits are responsible for mediating the inhibition of Ca^{2+} channels, the $G\alpha_o$ subunit remains important in the opioid receptor regulation of the channels. This is supported by the studies in which opioid inhibition of the Ca^{2+} channels was significant impaired in the DRG neurons obtained from $G\alpha_o$ knockout mice [71].

Opioid receptors, similar to many GPCRs, produce hyperpolarization at the postsynaptic membrane by activating K^+ channels, thereby preventing excitation or propagation of the action potentials. Electrophysiological studies in the rat locus coeruleus and substantia gelatinosa neurons have shown that delta opioid receptors can activate the G protein–gated inward rectifying potassium channel (K_G/GIRK/Kir3 channels) via PTX-sensitive G proteins [18]. K_G channels are tetramers formed by members of the Kir3/GIRK subfamily of inward rectifying K^+ channel subunits [72]. There are four Kir3 subunits, and three of these—the Kir3.1/GIRK1, Kir3.2/GIRK2, and Kir3.3/GIRK3—are expressed throughout the CNS [73,74]. The ability of delta opioid receptor to activate the K_G channels formed from these Kir3 subunits could be demonstrated by the heterologous expression of the subunits and receptor in *Xenopus* oocytes [75]. As a matter of fact, the K_G channels expressed in oocytes have been used to examine the mechanism of opioid receptor desensitization [76]. Using Kir3 knockout mice, it was demonstrated that the opioid inhibition of the firing rate and hyperpolarization of the locus ceruleus neurons was due to the activation of the K_G channels containing the Kir3.2 and Kir3.3 subunits [77]. Since active K_G could be formed by the various combinations of these Kir3 subunits, the delta opioid receptor could activate distinct populations of K_G channels with distinct characteristics.

Most likely, the delta opioid receptor activation of the K_G channels is mediated via the Gβγ subunits. Using glutathione S-transferase and different N- and C-terminal deletion mutants of Kir3.1 fusion proteins, two Gβγ-binding sites on Kir3 have been identified [78,79]. At the C-terminus of Kir3, the Gβγ-binding domain is composed of two segments [79], one of which contains the N-X-X-E-R motif that has been implicated to participate in Gβγ interaction [80]. The N-terminus interacts with the C-terminal domains to bind Gβγ in a synergistic fashion. There is evidence to support differential activation of Kir3.1 by different types of Gβ [81]. Such differences could be the basis for the observed differences in the three opioid receptors to activate K_G channels. The various opioid receptors may associate with distinct G proteins containing different Gβ subunits. However, the specificity of interaction with K_G is lost when Gβ is bound to Gγ, because K_G channel currents in *Xenopus* oocytes expressing Kir3.1 can be activated by different combinations of Gβγ [82]. Probably, opioid receptor could regulate the activities of the K_G channels via the PI3-kinase since PIP_2 has been reported to involve in Gβγ-induced activation of K_G channels [83].

2.3 Phospholipase C

Activation of phospholipase C (PLC) has been considered to be the property of the Gq coupled receptors [84]. Hence, the ability of the Gi/Go-coupled receptors, such as the delta opioid receptor, to activate PLC, has not been seriously considered. However, the ability of delta opioid agonist to stimulate the formation of IP3 and subsequent intracellular Ca^{2+} mobilization in NG108-15 were demonstrated [19,85]. The delta opioid receptor mediated activation of the PLCβ was reported also with spinal cord preparation [86]. Similar observations were noted with the cloned delta opioid receptor transfected in the Ltk^- cells [30]. Further, the delta opioid receptor activation of PLC appears to require the carboxyl tail of the receptor [87]. The significance of the opioid receptor stimulation of PLC was implicated in the attenuation of the antinociceptive response by antisense studies [88] or in the PLCβ3 knockout mice [89]. All this delta opioid receptor–mediated stimulation of PLC was blocked by PTX pretreatment. Since none of the PTX-sensitive Gα subunits can activate PLCβ by themselves [84], the opioid-induced stimulation of PLCβ appears to be mediated via the Gβγ subunits. This was demonstrated by blockade of the opioid response after the injection of Gβγ-binding peptide (QEHA) but not the Gq-binding peptide (QLKK) to the NG108-15 cells [90] or by the release of Gβγ from Gi2 or Go in the intestinal smooth muscle [91]. The activation of PLCβ3 by the Gβγ released from the activation of Gi/Go could account for the PTX sensitivity of this response. However, Gβγ subunits only potentiate the activities of PLCβ1 to β3 [84]. Only with the preactivation of the Gq-coupled receptor pathway could the Gi/Go-coupled receptor stimulation of PLC activity and subsequent Ca^{2+} mobilization be observed [92]. In neuroblastoma SHSY5Y cells, there is coincident signaling between the delta opioid receptor and the Gq-coupled m3-muscarinic receptor [93]. Delta opioid agonist–mediated increase in intracellular Ca^{2+} was observed only in the presence of muscarinic agonist. Whether the binding of the Gβγ subunits at the N terminal PH domain affecting the interaction of Gq α-subunit with the C2 domain of the PLCβ3 is the cause for such coincident signaling remains to be examined.

The delta opioid receptor also could stimulate the PLC activity and increase intracellular Ca^{2+} level in mechanism other than the activation of Gi/Go proteins. In a human neuroblastoma cell line, SK-N-BE, delta opioid receptors mobilize Ca^{2+} from intracellular ryanodine-sensitive stores which is independent of the PTX-sensitive G_i/G_o proteins [94]. Coexpressing the delta opioid receptor with G_{16}, a promiscuous G protein, allows for a PTX-insensitive stimulation of the PLC activity by the opioid agonist [95]. Thus, the ability of the opioid receptors to stimulate PLCβ is determined in part by the availability of complementary G proteins in any particular cell type.

2.4 MAPkinases Cascade

Opioid receptor has been reported to have modulating effects on the proliferation of cells range from neuroblastoma to endothelium [96,97]. Thus, similar to other GPCRs that regulate the cell growth and differentiation, opioid receptor could stimulate the mitogen-activated protein (MAP) kinase cascades. There are at least three sets of mammalian MAP kinases modules: the extracellular signal-regulated kinases, ERKs; the Jun N-terminal kinases, JNKs; and the p38 kinases. The ability of opioid receptor to stimulate the ERKs was demonstrated to be PTX sensitive in heterologous expression systems such as the CHO or Rat-1 fibroblast [20,21,98]. However, the direct demonstration of opioid receptor, specifically delta opioid receptor, regulating the Erk1/2 cascades in CNS has been difficult. At best, Erk1/2 activities in cortical neurons (layers II/III), median eminence, and amygdaloid and hypothalamic nuclei are diminished in rats with chronic morphine treatment [99], while acute morphine treatment has no effect on the Erk1/2 activity in these brain regions. On the other hand, morphine withdrawal produces a dramatic increase in Erk1/2 phosphorylation in somata and fibers of locus coeruleus, solitary tract, and hypothalamic neurons [99]. Since the in vivo effects of morphine have been shown to be mediated via the mu opioid receptor from the knockout mice studies [100], it is unlikely that delta opioid receptor has significant in vivo effects on the Erk1/2 activities.

The activation of Erk1/2 by opioid receptor has been shown to occur through the Gβγ subunits in a Ras-dependent manner [101]. However, whether the delta opioid receptor activation of the Erk1/2 requires the receptor being internalized is controversial. Coscia and coworkers suggested that delta opioid receptor internalization is a prerequisite for the agonist activation of the Erk1/2 [102]. However, several laboratories have since reported that opioid receptor activation of Erk1/2 did not depend on the receptor internalization [103,104]. Dominant negative dynamin mutant that would block the agonist-induced receptor internalization would not attenuate the Erk1/2 stimulation. Such data support the observation that morphine could stimulate the Erk1/2 activity but could not induce receptor internalization [105]. Thus, the delta opioid receptor stimulates the Erk1/2 activities prior to the receptor internalization processes.

In addition to the Gβγ subunit mechanism, the opioid receptor could increase the Erk1/2 activity via a transactivation process. Using receptor mutants that have impeded calmodulin binding activity, Belvecha et al. [106] demonstrated the activation of Erk1/2 by the mu opioid receptor involved the calmodulin-dependent transactivation of the EGF receptor. Such mechanism represents the participation of metalloproteases in producing the ligand for EGF receptor as in the case of β_2-adrenergic receptor [107]. Whether this is occurring with the neuronal delta opioid receptor remains to be demonstrated.

The stimulation of Erk1/2 by the Gβγ subunits could be modulated by the activation of disparate pathways by the subunits. Gβγ subunits can stimulate the γ-isoform of phosphoinositide 3-kinase (PI3K) [108]. The ability of opioid receptor to activate the PI3K signaling pathways was demonstrated clearly with the mu opioid receptors expressed in CHO cells. DAMGO activation of these receptors resulted in the activation of Akt or protein kinase B, the p70 S6 kinase, the 4E-BP1 and 4E-BP2, all downstream substrates of the PI3K [109]. Ability of delta opioid receptor to stimulate the Akt activity via a PTX-sensitive mechanism also can be observed in heterologous expression systems (unpublished observations). In addition, activation of the delta opioid receptor in Rat-1 fibroblasts resulted in the tyrosine phosphorylation of the p52 Shc adaptor protein [110] and also the activation of the p70 and p85 S6 kinases [111]. The stimulation of these effectors within the MAP kinase cascades by the delta opioid receptor activation and via the Gβγ subunits could provide a strong mitogenic signal for the opioids to regulate cell growth. This could be the basis for the reported delta opioid receptor potentiation of the T cells or neuroblastoma proliferation [96,112].

3 DELTA OPIOID RECEPTOR TRAFFICKING

With the multiple effectors that are activated by the delta opioid receptor, the cellular control of the signals generated represents a complicated picture needs to be resolved. Using β_2-adrenergic receptor as the model, Lefkowitz and coworkers have championed a mechanism for the cellular control of GPCR activities [113]. In the proposed mechanism, agonist binding to the receptor results in the rapid phosphorylation of the receptor by protein kinases including the G protein–coupled receptor kinases (GRKs), thereby enhancing the association of the cellular protein, β-arrestin. Association of β-arrestin with the receptor not only uncoupled the receptor from the respective G protein that transduces the signal and thus blunted the receptor signaling (receptor desensitization); the β-arrestin also is involved in the agonist-induced, clathrin-coated vesicle-mediated receptor internalization. β-Arrestin itself also serves as an adapter molecule in the β_2-adrenergic receptor signaling such that a receptor-src kinase complex is formed through which activation of the MAP kinases ERK-1 and ERK-2 by the β_2-adrenergic receptor is accomplished [114]. Thus, the regulation of the arrestin activities by covalent modification, such as ubiquitination [115], could control the cellular trafficking of the receptor.

No question the cellular control of the delta opioid receptor signaling follows such pathways. The ability of agonist to induce the rapid phosphorylation of the delta opioid receptor has been demonstrated [116]. Expression of the dominant negative mutant of GRK or over expression of GRK5 resulted in the attenuation or potentiation of agonist-dependent phosphorylation of the delta opioid receptor [116]. Deletion of the last 31 amino acids of the delta opioid receptor resulted in the abolition of both GRK- and PKC-mediated agonist-dependent phosphorylation of the receptor [117]. Truncation of the mouse delta opioid receptor after Thr^{344} also blocked the ability of DPDPE to induce phosphorylation of the receptor [118]. The agonist-induced receptor phosphorylation sites were identified to be the Thr^{358} and Ser^{363} residues at the carboxyl tail domains of the receptor [119,120]. The phosphorylation of the receptor appears to have a casual relationship with the delta opioid receptor desensitization in the SK-N-BE cells [121]. Studies with GRK-dominant negative mutants and overexpression of GRK5 have suggested that receptor phosphorylation is involved in the delta opioid receptor desensitization [116,122]. Mutation of the last four Thr and Ser residues at the C-terminus of the delta opioid receptor to Ala would block the GRK and arrestin-mediated desensitization measured in *Xenopus* oocytes [122]. However, the delta opioid receptor lacking the C-terminal 31 amino acids that include the sites for agonist-induced phosphorylation can be rapidly desensitized by pretreating the CHO cells with DPDPE for 10 min [123]. Rapid desensitization of the delta opioid receptor inhibition of adenylyl cyclase activity required both receptor phosphorylation and internalization [124]. Thus, the cellular control of the delta opioid receptor trafficking represents an important step in the control of the receptor signaling.

The G protein–coupled receptor (GPCR) trafficking is a dynamic process. As shown in Figure 1, the rapid phosphorylation of the receptor after agonist binding and the recruited β-arrestin molecule participate in the dynamin-dependent endocytosis of the receptor via the clathrin-coated pits, the trafficking of the vesicles and the delivery of the vesicular contents to the early endosomes [see reviews in 125,126]. The receptors are further trafficked to the late endosomes where the decision for recycling or degradation takes place [127]. Since receptor endocytosis basically is a process to remove active receptors from the cell surface, in the past, such receptor trafficking has been considered to be a step in which the receptor signals are terminated. However, recent data have suggested that receptor endocytosis has other functions. The dephosphorylation and resensitization of the β_2-adrenergic and A_2-adenosine receptor required the receptor internalization and trafficking to the endosomes [128,129]. As discussed earlier, activation of the MAP kinases by β_2-

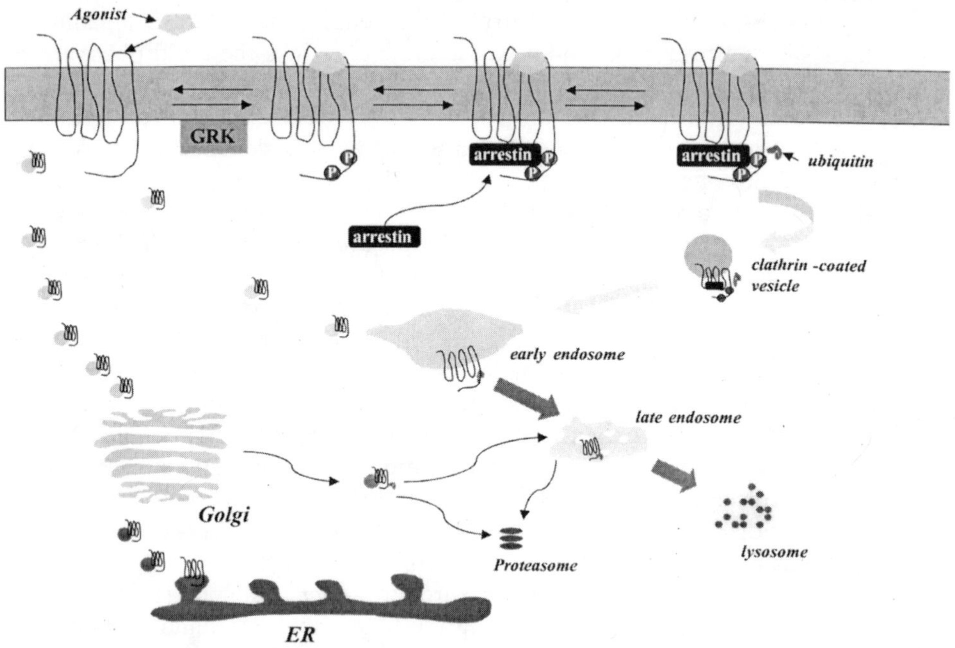

FIGURE 1 The cellular trafficking of GPCR.

adrenergic receptor could be blocked by inhibitors of receptor endocytosis, such as dominant negative arrestin [130]. Activation of the MAP kinases by α_2-adrenergic receptor was reported also to be dependent on the endocytosis of the receptor, but appears to be cell lines specific [131–133]. The fate of the activated MAP kinases was shown to be dependent on the internalized GPCR. This was demonstrated by the nonendocytosed mutant of PAR2 receptor-activated MAP kinases to translocate to nucleus, whereas the endocytosed PAR2 receptor-activated MAP kinases remained in the cytosol [134]. Thus, the GPCR signaling and the consequence of the signals are influenced by the receptor trafficking.

From the early studies reported by Chang et al. [135] and Law et al. [136], it is clear that opioid agonist would induce the delta opioid receptor internalization and subsequent downregulation in neuroblastoma cells. Similar downregulation of mu opioid receptor was observed in the neuroblastoma SHSY5Y cells [137]. Further, the internalized receptors were trafficked to the endosomes and subsequently to lysosomes as demonstrated with the intracellular accumulation of the ligand receptor complexes in the presence of chloroquine [138]. Such itinerary of the opioid receptor was substantiated after the cloning of the opioid receptors, with the development of receptor specific antibodies and the epitope-tagged receptor. Using the hemagglutinin (HA) epitope-tagged mu opioid receptor expressed in HEK293 cells, it could be demonstrated that agonists such as etorphine, DAMGO, could rapidly induce internalization of the receptor while morphine could not [105]. The HA-tagged delta opioid receptors expressed in neuroblastoma cells were demonstrated to colocalize with LAMP-2, a lysosomal marker, within 60 min after agonist addition [139]. In contrast to the mu opioid receptor, which could resensitize and recycle after internalization, the delta opioid receptor could not and is directly trafficked to the lysosome for degradation [140]. The agonist-induced endocytosis of the opioid receptor was demonstrated to involve the arrestin- and dynamin-dependent clathrin-coated pits pathway [141–145].

In addition to in vitro cell models, opioid agonists could induce the rapid endocytosis of the receptor in organo cultures or primary neuronal cultures, and also neurons in vivo. Treatment of longitudinal muscle-myenteric plexus preparation or the primary hippocampal neuron cultures with DAMGO resulted in internalization of the mu opioid receptor [146,147]. Similar observation was obtained with fluorescently labeled opioid peptides Fluo-dermorphin and Fluo-deltorphin [148]. Within 15 min of an intraperitoneal injection of etorphine, mu opioid receptor immunoreactivity was observed in the endosomal structures of the myenteric neurons of guinea pig ileum [149]. Again, rapid clustering of a spliced variant of mu opioid receptor MOR-1C was observed in the lateral septum of the mouse after intracerebroventricular injection of DAMGO [150]. Such studies extended the earlier studies in which in vivo administration of receptor-selective ligands such as morphiceptin [151], endormorphin-1 [152], or DADLE [153] resulted in the selective downregulation of the mu and delta opioid receptors. Thus, the in vivo trafficking of opioid receptors can be affected by agonist treatment.

3.1 Mechanism of Receptor Internalization

There are many mechanisms in which the GPCR trafficking could be regulated. As discussed earlier, the phosphorylation of the receptor and participation of β-arrestin in the agonist-induced receptor endocytosis have been firmly established [125–127]. Hence, the receptor's linear amino acid sequences that are involved in the phosphorylation and β-arrestin binding have been shown to be critical for the endocytosis process. The cytoplasmic tails of the GPCRs such as parathyroid hormone receptor [154], protease-activated receptor-1 and substance P receptor [155], thromboxane A2 receptor α and β [156],α_{1B}-adrenergic receptor [157], among others, have been reported to control the endocytosis, the trafficking to lysosome, and the recycling of these receptors. Other receptor domains, such as the intracellular loop III of the human gonadotropin-releasing hormone receptor [158], the intracellular loop II of cholecystokinin receptor [159], and transmembrane 6 and 7 of M2-muscarinic receptor [160], among others, could affect the internalization of these receptors. The reliance on primary sequence for directing GPCR trafficking is best illustrated by the identification of a kinase-regulated sequnce on β_2-adrenergic receptor that interacts with PDZ domain of the EPB50 [161]. Mutation of this sequence, DSLL at the carboxyl terminus of the receptor will disrupt the recycling of β_2-adrenergic receptor. Further, such sequence could be transplanted to GPCR such as the delta opioid receptor that normally traffics to lysosome and directs the rapid recycling of the resulting receptor chimera [162]. By delineating the primary sequence involved in the receptor trafficking, other cellular components participating in the regulation of this process can be identified.

Similar to other GPCRs, multiple sites within the opioid receptors participate in the regulation of receptor endocytosis. The importance of the carboxyl tail in the trafficking of the receptors has been established. Truncating the delta opioid receptor after Ser^{344} or mutation of Thr^{353} to Ala could block the agonist-induced receptor downregulation [163], while the mutation of the Ser/Thr residues between Ser^{344} and Ser^{363}, will retard the rate of receptor internalization [120,164]. After the identifying the agonist-induced phosphorylation sites of the delta opioid receptor by mutational analyses [119,120], it is apparent that these amino acid residues previously reported to participate in receptor trafficking are not phosphorylated in the presence of agonist. These amino acids could participate in the receptor interaction with β-arrestin as suggested by the pull-down assay and the BIACORE studies [165,166]. However, the exact amino acid sequence that is involved or whether the receptor has to be phosphorylated in the trafficking of delta opioid receptor remain to be determined. Whistler et al. [118] reported that phosphorylation was needed for agonist-induced delta opioid receptor endocytosis, while the data from our laboratory suggested otherwise [120,124]. The mutation of Ser^{363} to Ala completely blocked the agonist-induced phosphorylation of the delta opioid receptor, but could not eliminate the agonist-induced receptor internalization [124]. There is also apparent cell line dependence. For the same Ser^{344} truncated delta opioid receptor could be internalized in HEK293 cells but not in CHO cells [118,164], and the mutation of Ser/Thr residues in the carboxyl tail of the delta opioid receptor did not significantly alter the agonist-induced receptor internalization rate in HEK 293 cells [120]. This is not surprising since the delta opioid receptor internalization is β-arrestin dependent [118,143]. The affinity of β-arrestin to the agonist-receptor complex, and the cellular β-arrestin content should determine the rate and magnitude of delta opioid receptor being internalized. Nevertheless, the carboxyl tails of the opioid receptors have critical roles in directing the traffickings of these proteins. This was clearly established by the observations that trafficking of the internalized delta opioid receptor to the lysosomal compartments in the absence of agonist [167], that the distinct difference between mu opioid receptor and its carboxyl tail spliced variants to recycle and resensitize [168–171], and that the mu/delta opioid receptor chimeras could be downregulated more rapidly than the wild type [172] and the chimeras could be internalized by morphine while wild type could not [140]. In addition to serving as the recognition motif for cellular proteins within the endocytic pathways, the carboxyl tail as suggested by Whistler et al. [118] served as a brake for endocytosis which could be released upon receptor phosphorylation.

Motifs other than the carboxyl tail of the receptors might be involved also in the agonist-induced receptor internalization. Pak et al. [173] suggested that the mu opioid receptor was a substrate for the tyrosine-kinase, and that mutation of the four intracellular Tyr residues to Phe could attenuate the agonist-induced down-regulation of the receptor. One of these Tyr residues mutated was Tyr^{336} within the highly conserved $NP(X)_{2\text{-}3}Y$ motif of GPCR, a motif that has been identified as consensus binding sequence for adenosine diphosphate-ribosylation factor (ARF) [174]. ARF_6 has a regulatory role in the endocytosis and recycling of the transferrin receptor [175], and it targets the recycling vesicles to the plasma membrane [176]. Thus, it is not surprising that mutation of the Tyr within the $NP(X)_{2\text{-}3}Y$ motif could affect agonist-induced receptor trafficking, as in the case of β_2-adrenergic receptor mutation

[177–179]. Hence, the various motifs on the opioid receptor that are involved in the receptor interaction with other cellular proteins could in turn regulate the trafficking of the receptor.

One possible candidate for the directing of intracellular trafficking of delta opioid receptor is the recently identified G protein–coupled receptor-associated sorting protein (GASP) [180]. Using the yeast two-hybrid system, Whistler et al. [180] were able to identify a 1395-residue predicted brain enriched protein that preferentially interact with the carboxyl tail of the delta opioid receptor. The overexpression of GASP increased the rate and magnitude of delta opioid receptor being downregulated. The dominant negative mutant of GASP appeared to block the agonist-induced receptor down-regulation. Peripheral screening of various receptors' carboxyl tails-GST fusion proteins affinities for GASP appeared to correlate with the abilities of these receptors to recycle or not. All these data suggested that GASP could be one of the cellular proteins that direct the intracellular trafficking of the delta opioid receptor. The exact nature of this interaction and the exact role this protein might have in the intracellular trafficking of the delta opioid receptor toward lysosomes for degradation remain to be elucidated.

In addition to the linear primary sequence, the trafficking of GPCR can be regulated by three-dimensional signals. Covalent modification of the GPCR by conjugating polypeptides such as ubiquitin could establish such signals by the subsequent recruitment of other proteins within the endocytosis pathway. Ubiquitin molecule, a 76 amino acid polypeptide, is expressed in all eukaryotic cells. The conjugation of this polypeptide to the target proteins by the multienzyme cascade involving the E1s, E2s, and E3s enzymes has long been known to direct the degradation of cytosolic and nuclear proteins by proteasomes [see review in 181]. Normally, this involves the addition of multiubiquitin chains, i.e., the carboxy-termini glycine of ubiquitin is linked to the Lys^{48} of the preceding ubiquitin, to the ε-amino group of the lysine residue of the target protein. However, there is accumulating evidence to suggest a role of monoubiquitination in the endocytosis of plasma membrane proteins and their trafficking to the lysosomes [182,183]. In *Saccharomyces cerevisiae*, many of the plasma membrane proteins require ubiquitination in their cytoplasmic domains for their internalizations [184]. The yeast pheromone receptor Ste2p is ubiquitinated in the presence of α factor, and the internalization of Ste2p was impeded in E2 enzymes deficient yeast strain [185]. The region of Ste2p that was identified to be crucial for internalization was also required for the ubiquination of the receptor. In the case of growth hormone receptors (GHRs), polyubiquitination occurred prior to their recruitment to the clathrin-coated pit [186–188]. The agonist-induced ubiquitination has been reported with the opioid receptor [189], CXCR4 receptor [86], and β_2-adrenergic receptor [190]. Inclusion of proteasome inhibitors during chronic agonist treatment could prevent the downregulation of these receptors. In most of the receptors studied, the monoubiquitination process appears to participate in the endosomal sorting of the receptor, preventing the recycling of the proteins and shuttling the molecules to the multivesicular bodies of the late endosomes and subsequent degradation in the lysosome. This is supported by the observations that the ubiquitination of GHR regulates the lysosomal degradation [186,187] but not its internalization [188]. The mutation of the lysine residues within the degradative motif of CXCR4 [190] or the mutation of all 16 cytosolic lysine residues in the β_2-adrenergic receptor [115] did not affect the agonist-induced internalization of the receptor, but instead inhibited the degradation of these receptors. For receptors such as the epidermal growth factor receptor (EGFR), the overexpression of the Cbl proto-oncogene that encodes the ubiquitin ligase did not affect the EGFR internalization but stimulated significantly the lysosomal degradation of this receptor [191]. In addition to directing the lysosomal trafficking, ubiquitination of trans-acting endocytic protein(s) could also affect the agonist-induced receptor internalization. In the case of β_2-adrenergic receptor, the ubiquitination of β-arrestin, which also serves as the adaptor molecule for the E3 ligase, is essential for the endocytosis of the receptor [115]. The dependence on ubiquitin-conjugating enzymes and the Rsp5 ubiquitin ligase for the internalization of yeast receptor-ubiquitin chimera suggests also the ubiquitination of trans-acting endocytic protein(s) [192]. A possible trans-acting endocytic protein is the Eps15, a core component of clathrin-based endocytosis machinery that is ubiquitinated after activation of the EGFR [193,194]. Such ubiquitinated endocytic proteins could form multimers with other ubiquitinated proteins such as the receptor in the assembling of the complex that is needed for the budding of endocytic vesicles [182]. With the presence of ubiquitinlike proteins such as SUMO-1 and RUB-1 that could conjugate to proteins and control the proteins' activities (182), the monoubiquitination process could represent a dynamic control of the cellular trafficking of the receptor.

Ubiquitination of the opioid receptor has been reported. Petaja-Repo et al. [195,196] reported that >50% of the newly synthesized delta opioid receptors in HEK293S cells were retained within the endoplasmic reticulum, and these receptors, probably incorrectly folded, were deglycosylated and ubiquitinated for proteasome degradation. The same authors demonstrated subsequently that these receptors could be rescued with lipophilic opioid ligands that serve as chaperone for the receptor trafficking to the plasma membrane [197]. Chaturvedi et al. [189] reported that agonist-induced mu and delta opioid receptor downregulation was not affected by inhibitors of lysosomal proteolytic enzymes, but was attenuated by the inhibitors of proteasome inhibitors. Though these studies might be in disagreement with the confocal microscopy studies indicating the colocalization of the opioid receptor with lysosomal markers [139,162], the ability of proteasome inhibitors to affect the downregulation of the receptor suggests the opioid receptor trafficking is regulated similarly with other GPCRs [115]. The specific organelle that the delta opioid receptor is being targeted postendocytically, whether it is lysosomes or proteasomes, is under investigation.

3.2 Consequences for Receptor Internalization

The significance of the agonist-induced receptor endocytosis in the delta opioid receptor function has not been established. As discussed previously, there are conflicting data on the requirement of opioid receptor endocytosis in MAP kinase activation. Ignatova et al. [102] and Bohn et al. [198] have suggested that opioid receptor mediated modulation of MAP kinase activity required the endocytosis of the receptor, while Whistler and von Zastrow [103], Li et al. [199], and Trapaidze et al. [104] presented data that did not support the requirement of receptor internalization in the opioid agonist activation of MAP kinase activities. Whether the differences in the results were caused by different cell models used remained to be elucidated. Probably, the internalization and subsequent downregulation of the receptor have minimal role in the development of in vivo tolerance. This is best exemplified by the ability of both chronic etorphine and morphine treatment to elicit tolerance development, while only etorphine could downregulate and meanwhile morphine upregulate the mu opioid receptor [200]. The noncorrelation between degree of receptor downregulation and tolerance was observed also with chronic fentenyl or clocinnamox treatment [201]. Similarly, morphine and etorphine could desensitize the delta opioid receptor, while only etorphine could induce the downregulation of the receptor [202]. How morphine could induce delta opioid receptor to desensitize without stimulating the phosphorylation of the receptor or the internalization of the receptor remains an intriguing question to address.

However, it is clear that opioid receptor endocytosis is critical for the receptor to resensitize. Wolf et al. [170] reported the mutation of Thr^{394} of mu opioid receptor to Ala resulted in the rapid internalization and resensitization of receptor. Similar observations were reported with various spliced variants of mu opioid receptor, in which the rate of desensitization appears to correlate inversely with the resensitization properties of receptor [169]. Such observations and others have led to a hypothesis proposed by Whistler et al. [140] that the ability of various opioid agonists to produce tolerance is dependent on their "RAVE" values. In their hypothesis, agonist that induces rapid receptor internalization, e.g., etorphine, would develop less tolerance in animals than agonist such as morphine, which does not produce receptor internalization. Their hypothesis is based on the observation that in the receptor chimera construct in which the mu opioid receptor carboxyl tail domain was replaced by the delta opioid receptor carboxyl tail, morphine could produce receptor internalization, and that the drug would now induce receptor desensitization [140,203]. Ability of agonist to internalize the receptor also was proposed by the same group to be related to opioid dependence [204]. Agonist such as morphine, which does not induce receptor internalization, has a greater degree of "dependence" as measured by the increase in adenylyl cyclase activities. Meanwhile, agonist such as etorphine, which induces rapid receptor internalization and subsequent resensitization, has a lower degree of "dependence." Though this is an attractive model, this is not applicable to the delta opioid receptor tolerance and dependence. The magnitude of the increase in the adenylyl cyclase activities in cells expressing the delta opioid receptor after chronic opioid treatment does not appear to depend on the agonist used to treat the system chronically, but rather depends on the initial receptor desensitivity [205]. Morphine and other partial agonists in the delta opioid receptor system, e.g., levallorphan, could elicit similar level of adenylyl cyclase activity increase as that of agonists, such as DADLE. Hence, the endocytosis of the delta opioid receptor would have minimal effect on the increase in the adenylyl cyclase activity during chronic drug treatment, or the "dependence" response.

4 PERSPECTIVES

From the very beginning, it is abundantly clear that the complexity in the signal transduction of the delta opioid receptor goes beyond the simple involvement of receptor-G_i/G_o proteins and the effectors. With the ability of the activated G-protein heterotrimers generating two separate messengers, $G\alpha$ and $G\beta\gamma$ subunits, the myriad of effectors activated by the agonists have expanded continuously. The possibilities of coincident signaling and modulation of the signals activities are limitless. This situation is further complicated by the overall response to the opioid activation of G_i/G_o that could depend on the composition of the neurons expressing the receptor. This is best illustrated by the differential responses exhibited by the multiple adenylyl cyclase subtypes to the activation of the receptor. Further, the possibility for the delta opioid receptor to compartmentalize exists. In NG108-15 cells, the transient activation and inactivation of $PLC\beta_3$ by delta opioid receptors is correlated to the phosphorylation of the enzyme itself [206]. Under the conditions that delta opioid agonist could not activate the $PLC\beta_3$, the activation and phosphorylation of the same enzyme by other GPCR agonist such as LPA was observed. Such compartmentization of the delta opioid receptor-effector complexes provides an alterative model for the generally described homologous desensitization of the delta opioid receptor. The signaling of delta opioid receptor via the recruitment and scaffolding of cellular proteins will create membrane microdomains within the proximity of the receptor that will greatly affect the delta opioid receptor signaling. Though GPCR such as the delta opioid receptor does not contain motifs as in the case of tyrosine kinase receptor that could be recognized easily by proteins such as Grb2 that contains SH2 and SH3 domains, the receptor association with molecules such as β-arrestin and Gβγ could recruit subsequent cellular proteins. The ability of β-arrestin to function as adaptor molecule for the various kinases such as c-src has been documented [114]. The Gβγ subunits have been shown to interact with the PH domain of the PLCβ among other proteins [84]. The recruitment of molecules such as $PLC\beta_3$ and protein kinases to the receptor vicinity would provide a rapid control mechanism for the opioid receptor signaling. Hence, the immediate emphasis for the understanding of the delta opioid receptor signaling should be in the identification of cellular proteins that participate in the receptor signaling. Whether such proteins could be the homodimers or heterodimers of the receptors or protein kinases, the scaffolding of these proteins with the delta opioid receptor could be the basis for the reported delta opioid receptor subtypes that have yet been identified via the molecular cloning of the receptors.

The complexity in the delta opioid receptor signaling also obstructs the eventual understanding of cellular control and adaptation to the receptor activation. The current model of receptor phosphorylation and β-arrestin recruitment in turning off the signals is applicable to the delta opioid receptor regulation. However, it is increasingly clear that the trafficking of the delta opioid receptor also has a greater role in the regulation of receptor signaling. Not only does the internalization of the receptor serve as a means to reduce the amount of active receptor from the cell surface, but also the internalized receptor could continue their signaling processes inside the cells. An excellent example is the recruitment of the MAP kinase modules by the internalized β_2-adrenergic receptor-arrestin complexes [114]. Though contrasting data have been reported on the dependency of delta opioid receptor internalization and activation of the MAP kinases, the possibility that the internalized receptor could continue to signal is supported by the intriguing observation of the localization of the receptor with the nuclei fraction [207]. Thus, it is reasonable to hypothesize that the internalized receptor could contribute to the subsequent chronic responses to the drug, as suggested by Whistler and coworkers [203,204]. However, whether the receptor endocytosis is the dominant factor or a contributor among the many already identified to be involved in opioid tolerance development, e.g., the NMDA receptor [208,209], remains to be demonstrated. Regardless, the ability of the drug to induce tolerance could not be a direct correlation between ability to internalize or not as suggested by the proposed RAVE values [140]. Chronic morphine and etorphine treatment could produce tolerance and dependence in rodent, while chronic morphine treatment would upregulate and chronic etorphine treatment downregulate the opioid receptor content in various brain areas [200].

Currently, the working model suggests the delta opioid receptor is sorted from the early endosomes to the lysosomal degradative pathway. The trafficking of the receptor down this degradative pathway could be facilitated by cellular proteins such as the recently identified GASP [180]. The ability of protein such as GASP to distinguish the delta opioid receptor from the mu opioid receptor is critical because this will allow the correct sorting of the receptors into different trafficking pathway. However, from their studies, Whistler et al. reported some binding of the GASP protein with the mu opioid receptor carboxyl tail-GST fusion protein [180]. The exact difference in GASP's affinities for mu and delta opioid receptor is unknown. Since studies with chimera receptors suggested that carboxyl tails of the mu and delta opioid receptor contributed and were not sufficient in directing the receptors' traffic [172] and that proteasome inhibitors could attenuate the downregulation of the receptor [189], other receptor domains must participate in the sorting of the delta opioid receptor. The participation of the cytoskeleton, the small GTP-binding proteins, e.g., rab, in the trafficking of the delta opioid receptor should not be ignored.

In conclusion, the generation of the two immediate second messengers, $G\alpha$ and $G\beta\gamma$, provides the opportunities for delta opioid agonist to activate multiple effortor systems within a single cell. The scaffolding of cellular proteins with the receptor via adaptor proteins such as β-arrestin enables the rapid modification of the delta opioid receptor signaling. The trafficking of the signaling complexes will determine not only the frequency and amplitude of the delta opioid receptor signals, but also the content of the signals. By delineating the composition of the complexes, and the itinerary of the receptor, a better understanding of the delta opioid receptor signaling can be accomplished.

ACKNOWLEDGMENT

This research is supported in parts by NIH grants DA07339, DA11806, and K05 DA00513.

REFERENCES

1. Martin WR, Eades CG, Thompson JA, Huppler RE, Gilbert PE. J Pharmacol Exp Ther 1976; 197:517–532.

2. Hughes J, Smith TW, Kolsterlitz HW, Fothergill LA, Morgan BA, Morris HR. Nature 1975; 258:577–579.
3. Li CH, Chung D. Proc Natl Acad Sci USA 1976; 73:1145–1148.
4. Goldstein A, Tachibana S, Lowney LI, Hunkapiller M, Hood L. Proc Natl Acad Sci USA 1979; 76:6666–6669.
5. Sharma SK, Klee WA, Nirenberg M. Proc Natl Acad Sci USA 1977; 74:3365–3369.
6. Chang KJ, Miller RJ, Cuatrecasas P. Mol Pharmacol 1978; 14:961–970.
7. Evans CJ, Keith DE, Morrison H, Magendzo K, Edwards RH. Science 1992; 258: 1952–1955.
8. Kieffer BL, Befort K, Gaveriaux-Ruff C, Hirth CG. Proc Natl Acad Sci USA 1992; 89:12048–12052.
9. Chen Y, Mestek A, Liu J, Hurley JA, Yu L. Mol Pharmacol 1993; 44:8–12.
10. Fukuda K, Kato S, Mori K, Hishi M, Takeshima H. FEBS Lett 1993; 327:311–314.
11. Yasuda K, Raynor K, Kong H, Breder CD, Takeda J, Reisine T, Bell GI. Proc Natl Acad Sci USA 1993; 90:6736–6740.
12. Meng F, Xie GX, Thompson RC, Mansour A, Goldstein A, Watson SJ, Akil H. Proc Natl Acad Sci USA 1993; 90:9954–9958.
13. Li S, Zhu J, Chen C, Chen YW, Deriel JK, Ashby B, Liu-Chen LY. Biochem J 1993; 295:629–633.
14. Chen Y, Mestek A, Liu J, Yu L. Biochem J 1993; 295:625–628.
15. Zhu Y, King MA, Schuller AG, Nitsche JF, Reidl M, Elde RP, Unterwald E, Pasternak GW, Pintar JE. Neuron 1999; 24:243–252.
16. Hescheler J, Rosenthal W, Trautwein W, Schultz G. Nature 1987; 325:445–446.
17. Surprenant A, Shen KZ, North RA, Tatsumi H. J Physiol Lond 1990; 431:585–608.
18. North RA, Williams JT, Surprenant A, Christie MJ. Proc Natl Acad Sci USA 1987; 84:5487–5491.
19. Jin W, Lee NM, Loh HH, Thayer SA. Mol Pharmacol 1992; 42:1083–1089.
20. Fukuda K, Kato S, Morikawa H, Shoda T, Mori K. J Neurochem 1996; 67: 1309–1316.
21. Li LY, Chang KJ. Mol Pharmacol 1996; 50:599–602.
22. Prather PL, Loh HH, Law PY. Mol Pharmacol 1994; 45:997–1003.
23. Prather PL, McGinn TM, Erickson LJ, Evans CJ, Loh HH, Law PY. J Biol Chem 1994; 269:21293–21302.
24. Roerig SC, Loh HH, Law PY. Mol Pharmacol 1992; 41:822–831.
25. Wong YH, Conklin BR, Bourne HR. Science 1992; 255:339–342.
26. Law PY, Wu J, Koehler JE, Loh HH. J Neurochem 1981; 36:1834–1846.
27. Law PY, Loh HH. Mol Pharmacol 1993; 43:684–693.
28. McKenzie F, Milligan G. Biochem J 1990; 267:391–398.
29. Carter BD, Medzihradsky F. Proc Natl Acad Sci USA 1993; 90:4062–4066.
30. Tsu RC, Chan JSC, Wong YH. J Neurochem 1995; 64:2700–2707.
31. Puri SK, Cochin J, Volicer L. Life Sci 1975; 16:759–768.

32. Cruciani RA, Dvorkin B, Morris SA, Crain SM, Makman MH. Proc Natl Acad Sci USA 1993; 90:3019–3023.
33. Olianas MC, Onali P. J Pharmacol Exp Ther 1995; 275:1560–1567.
34. Makman MH, Dvorkin B, Crain SM. Brain Res 1988; 445:303–313.
35. Federman AD, Conklin BR, Schrader KA, Reed RR, Bourne HR. Nature 1992; 356:159–161.
36. Tang WJ, Gilman AG. Science 1991; 254:1500–1503.
37. Taussig R, Tang WJ, Hepler JR, Gilman AG. J Biol Chem 1994; 269: 6093–6100.
38. Tsu RC, Allen RA, Wong YH. Mol Pharmacol 1995; 47:835–841.
39. Toyoshi T, Ukai M, Kameyama T. Eur J Pharmacol 1992; 213:25–30.
40. Avidor-Reiss T, Nevo I, Saya D, Bayerwitch M, Vogel Z. J Biol Chem 1997; 272:5040–5047.
41. Crain SM, Shen KF. Neurochem Res 1996; 21:1347–1351.
42. Wu G, Lu ZH, Alfinito P, Ledeen RW. Neurochem Res 1997; 22:1281–1289.
43. Wu G, Lu ZH, Ledeen RW. Mol Brain Res 1997; 44:341–346.
44. Wang D, Quillan JM, Winans K, Lucas JL, Sadee W. J Biol Chem 2001; 276:34624–34630.
45. Wang D, Sadee W, Quillan JM. J Biol Chem 1999; 274:22081–22088.
46. Sarne Y, Rubovitch V, Fields A, Gafni M. Biochem Biophys Res Commun 1998; 246:128–131.
47. Motin LG, Bennett MR, Christie MJ. Neurosci Lett 1995; 193:21–24.
48. Stefani A, Surmeier DJ, Bernardi G. Brain Res 1994; 642:339–343.
49. Acosta CG, Lopez HS. J Neurosci 1999; 19:8337–8348.
50. Morikawa H, Mima H, Uga H, Shoda T, Fukuda K. Pflugers Archiv–Europ J Physiol 1999; 438:423–426.
51. Toselli M, Tosetti P, Taglietti V. Biophys J 1999; 76:2560–2574.
52. Nah SY, Unteutsch A, Bunzow JR, Cook SP, Beacham DW, Grandy DK. Brain Res 1997; 766:66–71.
53. Piros ET, Prather PL, Law PY, Evans CJ, Hales TG. Mol Pharmacol 1996; 50:947–956.
54. Prather PL, Song L, Piros ET, Law PY, Hales TG. J Pharmacol Exp Ther 2000; 295:552–562.
55. Catterall WA. Annu Rev Cell Dev Biol 2000; 16:521–555.
56. Angelotti T, Hofmann F. FEBS Lett 1996; 397:331–337.
57. Bourinet E, Soong TW, Sutton K, Slaymaker S, Mathews E, Monteil A, Zamponi GW, Nargeot J, Snutch TP. Nat Neurosci 1999; 2:407–415.
58. Kollmar R, Fak J, Montgomery LG, Hudspeth AJ. Proc Natl Acad Sci USA 1997; 94:14889–14893.
59. Lin Z, Lin Y, Schorge S, Pan JQ, Beierlein M, Lipscombe D. J Neurosci 1999; 19:5322–5331.
60. Perez-Reyes E. J Bioenerg Biomembr 1998; 30:313–318.
61. Moises HC, Rusin KI, Macdonald RL. J Neurosci 1994; 14:3842–3851.
62. Tang T, Kiang JG, Cote T, Cox BM. J Neurochem 1995; 65:1612–1621.
63. Ikeda SF. Nature 1996; 380:255–258.

64. Herlitze S, Garcia DE, Mackie K, Hille B, Scheuer T, Catterall WA. Nature 1996; 380:258–262.
65. Bean BP. Nature 1989; 340:153–156.
66. De Waard M, Liu H, Walker D, Scott VE, Gurnett CA, Campbell KP. Nature 1997; 385:446–450.
67. Qin N, Platano D, Olcese R, Stefani E, Birnbaumer L. Proc Natl Acad Sci USA 1997; 94:8866–8871.
68. Zamponi GW, Snutch TP. Curr Opin Neurobiol 1998; 8:351–356.
69. Safa P, Boulter J, Hales TG. J Biol Chem 2001; 276:38727–38737.
70. Garcia DE, Li B, Garcia-Ferreiro RE, Hernandez-Ochoa EO, Yan K, Gautam N, Catterall WA, Mackie K, Hille B. J Neurosci 1998; 18:9163–9170.
71. Jiang M, Gold MS, Boulay G, Spicher K, Peyton M, Brabet P, Srinivasan Y, Rudolph U, Ellison G, Birnbaumer L. Proc Natl Acad Sci USA 1998; 95:3269–3274.
72. Dascal N. Cell Signal 1997; 9:551–573.
73. Karschin C, Dissmann E, Stuhmer W, Karschin A. J Neurosci 1996; 16: 3559–3570.
74. Chen SC, Ehrhard P, Goldowitz D, Smeyne RJ. Brain Res 1997; 778:251–264.
75. Ikeda K, Kobayashi T, Ichikawa T, Usui H, Kumanishi T. Biochem Biophys Res Commun 1995; 208:302–308.
76. Kovoor A, Nappey V, Kieffer BL, Chavkin C. J Biol Chem 1997; 272: 27605–27611.
77. Torrecilla M, Marker CL, Cintora SC, Stoffel M, Williams JT, Wickman K. J Neurosci 2002; 22:4328–4334.
78. Huang CL, Jan YN, Jan LY. FEBS Lett 1997; 405:291–298.
79. Huang CL, Slesinger PA, Casey PJ, Jan NY, Jan LY. Neuron 1995; 15:1133–1143.
80. Chen J, Devivo M, Dingus J, Harry A, Li J, Sui J, Carty DJ, Blank JL, Exton JH, Stofel RH, Inglese J, Lefkowitz RJ, Logothetis DE, Hidebrandt JD, Iyengar R. Science 1995; 268:1166–1169.
81. Yan K, Gautam N. J Biol Chem 1996; 271:17597–17600.
82. Lim NF, Dascal N, Labarca C, Davidson N, Lester HA. J Gen Physiol 1995; 105:421–439.
83. Huang CL, Feng S, Hilgemann DW. Nature 1998; 391:803–806.
84. Rhee SG. Annu Rev Biochem 2001; 70:281–312.
85. Smart D, Lambert DG. J Neurochem 1996; 66:1462–1467.
86. Sanchez-Blazquez P, Rodriguez-Diaz M, Frejo MT, Garzon J. Eur J Neurosci 1999; 11:2059–2064.
87. Harrison C, Rowbotham DJ, Devi LA, Lambert DG. Eur J Pharmacol 1999; 379:237–242.
88. Sanchez-Blazquez P, Garzon J. J Pharmacol Exp Ther 1998; 285:820–827.
89. Xie W, Samoriski GM, McLaughlin JP, Romoser VA, Smrcka A, Hinkle PM, Bidlack JM, Gross RA, Jiang H, Wu D. Proc Natl Acad Sci USA 1999; 96:10385–10390.
90. Yoon SH, Lo TM, Loh HH, Thayer SA. Mol Pharmacol 1999; 56:902–908.

91. Murthy KS, Makhlouf GM. Mol Pharmacol 1996; 50:870–877.
92. Chan JS, Lee JW, Ho MK, Wong YH. Mol Pharmacol 2000; 57:700–708.
93. Yeo A, Samways DS, Fowler CE, Gunn-Moore F, Henderson G. J Neurochem 2001; 76:1688–1700.
94. Allouche S, Polastron J, Jauzac P. J Neurochem 1996; 67:2461–2470.
95. Lee JWM, Joshi S, Chan JSC, Wong YH. J Neurochem 1998; 70:2203–2211.
96. Law PY, McGinn TM, Campbell KM, Erickson LE, Loh HH. Mol Pharmacol 1997; 51:152–160.
97. Gupta K, Kshirsagar S, Chang L, Schwartz R, Law PY, Yee D, Hebbel RP. Cancer Res 2002; 62:4491–4498.
98. Burt AR, Carr IC, Mullaney I, Anderson NG, Milligan G. Biochem J 1996; 320:227–235.
99. Schultz S, Hollt V. Eur J Neurosci 1998; 10:1196–1201.
100. Loh HH, Liu HC, Cavalli A, Yang W, Chen YF, Wei LN. Mol Brain Res 1998; 54:321–326.
101. Belcheva MM, Vogel Z, Ignatova E, Avidor-Reiss T, Zippel R, Levy R, Young EC, Barg J, Coscia CJ. J Neurochem 1998; 70:635–645.
102. Ignatova EG, Belcheva MM, Bohn LM, Neuman MG, Coscia CJ. J Neurosci 1999; 19:56–63.
103. Whistler JL, von Zastrow M. J Biol Chem 1999; 274:24575–24578.
104. Trapaidze N, Gomes I, Cvejic S, Bansinath M, Devi LA. Mol Brain Res 2000; 76:220–228.
105. Keith DE, Murray SR, Zaki PA, Chu PC, Lissin DV, Kang L, Evans CJ, von Zastrow M. J Biol Chem 1996; 271:19021–19024.
106. Belcheva MM, Szùcs M, Wang D, Sadee W, Coscia CJ. J Biol Chem 2001; 276:33847–33853.
107. Pierce KL, Luttrell LM, Lefkowitz RJ. Oncogene 2001; 20:1532–1539.
108. Hawes BE, Luttrell LM, van Biesen T, Lefkowitz RJ. J Biol Chem 1996; 271:12133–12136.
109. Polakiewicz RD, Schieferl SM, Gingras AC, Sonenberg N, Comb MJ. J Biol Chem 1998; 273:23534–23541.
110. Mullaney I, Carr IC, Burt AR, Wilson M, Anderson NG, Milligan G. Cell Signal 1997; 9:423–429.
111. Wilson MA, Burt AR, Milligan G, Anderson NG. Biochem J 1997; 325: 217–222.
112. Hedin KE, Bell MP, Kalli KR, Huntoon CJ, Sharp BM, McKean DJ. J Immunol 1997; 159:5431–5440.
113. Lefkowitz RJ. J Biol Chem 1998; 273:18677–18680.
114. Luttrell LM, Ferguson SSG, Daaka Y, Miller WE, Maudsley S, Della Rocca GJ, Lin FT, Kawakatsu H, Owada K, Luttrell DK, Caron MG, Lefkowitz RJ. Science 1999; 283:655–661.
115. Shenoy SK, McDonald PH, Kohout TA, Lefkowitz RJ. Science 2001; 294:1307–1313.
116. Pei G, Kieffer BL, Lefkowitz RJ, Freedman NJ. Mol Pharmacol 1995; 48:173–177.

117. Zhao J, Pei G, Huang YL, Zhong FM, Ma L. Biochem Biophys Res Commun 1997; 238:71–76.
118. Whistler JL, Tsao P, Von Zastrow M. J Biol Chem 2001; 276:34331–34338.
119. Guo J, Wu Y, Zhang W, Zhao J, Devi LA, Pei G, Ma L. Mol Pharmacol 2000; 58:1050–1056.
120. Maestri-El Kouhen O, Wang G, Solberg J, Erickson LJ, Law PY, Loh HH. J Biol Chem 2000; 275:36659–36664.
121. Hasbi A, Polastron J, Allouche S, Stanasila L, Massotte D, Jauzac P. J Neurochem 1998; 70:2129–2138.
122. Koover A, Nappey V, Kieffer BL, Chavkin C. J Biol Chem 1997; 272: 27605–27611.
123. Wang CD, Zhou Z, Cheng Q, Wei J, Chen G, Li G, Pei G, Chi Z. Biochem Biophys Res Commun 1998; 249:321–324.
124. Law PY, Maestri-El Kouhen O, Solberg J, Wang W, Erickson LJ, Loh HH. J Biol Chem 2000; 275:32057–32065.
125. Krupnick JG, Benovic JL. Ann Rev Pharmacol Toxicol 1998; 38:289–319.
126. Roth BL, Willins DL, Kroeze WK. Drug Alcohol Depend 1998; 51:73–85.
127. Moore RH, Tuffaha A, Millman EE, Dai W, Hall HS, Dickey FF, Knoll FJ. J Cell Sci 1999; 112:329–338.
128. Zhang J, Barak LS, Winkler KE, Caron MG, Ferguson SSG. J Biol Chem 1997; 272:27005–27014.
129. Mundell SJ, Kelly E. Br J Pharmacol 1998; 125:1594–1600.
130. Daaka Y, Luttrell LM, Ahn S, Della Rocca GJ, Ferguson SSG, Caron MG, Lefkowitz RJ. J Biol Chem 1998; 273:685–688.
131. DeGraff JL, Gagnon AW, Benovic JL, Orsini MJ. J Biol Chem 1999; 274: 11253–11259.
132. Schramm NL, Limbird LE. J Biol Chem 1999; 274:24935–24940.
133. Pierce KL, Maudsley S, Daaka Y, Luttrell LM, Lefkowitz RJ. Proc Natl Acad Sci USA 2000; 97:1489–1494.
134. DeFea KA, Zalevssky J, Thoma MS, Dery O, Mulins RD, Bunnett NW. J Cell Biol 2000; 148:1267–1281.
135. Chang KJ, Eckel RW, Blanchard SG. Nature 1982; 296:446–448.
136. Law PY, Hom DS, Loh HH. Mol Pharmacol 1982; 22:1–4.
137. Zadina JE, Chang SL, Ge LJ, Kastin AJ. J Pharmacol Exp Ther 1993; 265:254–262.
138. Law PY, Hom DS, Loh HH. J Biol Chem 1984; 259:4096–4104.
139. Ko JL, Arvidsson U, Williams FG, Law PY, Elde R, Loh HH. Mol Brain Res 1999; 69:171–185.
140. Whistler JL, Chuang HH, Chu P, Jan LY, Von Zastrow M. Neuron 1999; 23:737–746.
141. Li JG, Luo LY, Krupnick JG, Benovic JL, Liu-Chen LY. J Biol Chem 1999; 274:12087–12094.
142. Zhang J, Ferguson SSG, Barak LS, Bodduluri S, Laporte S, Law PY, Caron MG. Proc Natl Acad Sci USA 1998; 95:7157–7162.

143. Zhang J, Ferguson SSG, Law PY, Barak LS, Caron MG. J Recept Signal Transduct Res 1999; 19:301–313.
144. Murray SR, Evans CJ, von Zastrow M. J Biol Chem 1998; 273:24987–24991.
145. Whistler JL, von Zastrow M. Proc Natl Acad Sci USA 1998; 95:9914–9919.
146. Sternini C, Brecha NC, Minnis J, D'Agostina G, Balestra B, Fiori E, Tonini M. Neuroscience 2000; 98:233–241.
147. Bushell T, Endoh T, Simen AA, Ren D, Bindokas VP, Miller RJ. Mol Pharmacol 2002; 61:55–64.
148. Lee MC, Cahill CM, Vincent JP, Beaudet A. Synapse 2002; 43:102–111.
149. Sternini C, Span M, Anton B, Keith DE, Bunnett NW, Von Zastrow M, Evans C, Brecha NC. Proc Natl Acad Sci USA 1996; 93:9241–9246.
150. Abbadie C, Pasternak GW. NeuroReport 2001; 12:3069–3072.
151. Tao PL, Lee HY, Chang LR, Loh HH. Brain Res 1990; 526:270–275.
152. Harrison C, Rowbotham DJ, Grandy DK, Lambert DG. Br J Pharmacol 2000; 131:1220–1226.
153. Tao PL, Chang LR, Law PY, Loh HH. Brain Res 1988; 462:313–320.
154. Huang Z, Chen Y, Nissenson RA. J Biol Chem 1995; 270:151–156.
155. Trejo J, Coughlin SR. J Biol Chem 1999; 274:2216–2224.
156. Parent JL, Labrecque P, Orsini MJ, Benovic JL. J Biol Chem 1999; 274: 8941–8948.
157. Wang J, Wang L, Zheng J, Anderson JL, Toews ML. Mol Pharmacol 2000; 57:687–694.
158. Myburgh DB, Millar RP, Hapgood JP. Biochem J 1998; 331:893–896.
159. Ding XQ, Rao RV, Kuntz SM, Holicky EL, Miller LJ. Mol Pharmacol 2000; 58:1424–1433.
160. Schlador ML, Grubbs RD, Nathanson NM. J Biol Chem 2000; 275:23295–23302.
161. Cao TT, Deacon HW, Reczek D, Bretscher A, Von Zastrow M. Nature 1999; 401:286–290.
162. Gage RM, Kim KA, Cao TT, Von Zastrow M. J Biol Chem 2001; 276:44712–44720.
163. Cvejic S, Trapaidze N, Syr C, Devi LA. J Biol Chem 1996; 271:4073–4076.
164. Trapaidze N, Keith DE, Cvejic S, Evans CJ, Devi LA. J Biol Chem 1996; 271:29279–29285.
165. Cen B, Yu Q, Guo J, Wu Y, Ling K, Cheng Z, Ma L, Pei G. J Neurochem 2001; 76:1887–1894.
166. Cen B, Xiong Y, Ma L, Pei G. Mol Pharmacol 2001; 59:758–764.
167. Tsao PI, von Zastrow M. J Biol Chem 2000; 275:11130–11140.
168. Law PY, Erickson LJ, El-Kouhen R, Dicker L, Solberg J, Wang W, Miller E, Burd AL, Loh HH. Mol Pharmacol 2000; 58:388–398.
169. Koch T, Schulz S, Schroder H, Wolf R, Raulf E, Hollt V. J Biol Chem 1998; 273:13652–13657.
170. Wolf R, Koch T, Schulz S, Klutzny M, Schroder H, Raulf E, Buhling F, Hollt V. Mol Pharmacol 1999; 55:263–268.

171. Koch T, Schulz S, Pfeiffer M, Klutzny M, Schroder H, Kahl E, Hollt V. J Biol Chem 2001; 276:31408–31414.
172. Afify EA, Law PY, Riedl M, Elde R, Loh HH. Mol Brain Res 1998; 54: 24–34.
173. Pak Y, O'Dowd BF, Wang JB, George SR. J Biol Chem 1999; 274:27610–27616.
174. Michell R, McCullock D, Lutz E, Johnson M, MacKenzie C, Fennell M, Fink G, Zhou W, Sealfon SC. Nature 1998; 392:411–414.
175. D'Souza-Schorey C, Li G, Colombo MI, Stahl PD. Science 1995; 267:1175–1178.
176. D'Souza-Schorey C, van Donselaar E, Hsu VW, Yang C, Stahl PD, Peters PJ. J Cell Biol 1998; 140:603–616.
177. Gabilondo AM, Krasel C, Lohse MJ. Eur J Pharmacol 1996; 307:243–250.
178. Barak LS, Menard L, Ferguson SS, Colapietro AM, Caron MG. Biochemistry 1995; 34:15407–15414.
179. Barak LS, Tiberi M, Freedman NJ, Kwatra MM, Lefkowitz RJ, Caron MG. J Biol Chem 1994; 269:2790–2795.
180. Whistler JL, Enquist J, Marley A, Fong J, Gladher F, Tsuruda P, Murray SR, von Zastrow M. Science 2002; 297:615–620.
181. Bonifacino JS, Weissman AM. Annu Rev Cell Dev Biol 1998; 14:19–57.
182. Hicke L. Nature Revievs/Mol Cell Biol 2001; 2:195–201.
183. Hicke L. Cell 2001; 106:527–530.
184. Rotin D, Staub O, Haguenauer-Tsapis R. J Membr Biol 2000; 176:1–17.
185. Hicke L, Riezman H. Cell 1996; 84:277–287.
186. van Kerkhof P, Sachse M, Klumperman J, Strous GJ. J Biol Chem 2001; 276: 3778–3784.
187. van Kerkhof P, Smeets M, Stours GJ. Endocrinology 2002; 143:1243–1252.
188. Govers R, ten Broeker T, van Kerkhof P, Schwartz AL, Strous GJ. EMBO J 1999; 18:28–36.
189. Chaturvedi K, Bandari P, Chinen N, Howells RD. J Biol Chem 2001; 276: 12345–12355.
190. Marchese A, Benovic JL. J Boil Chem 2001; 276:45509–45512.
191. Levkowitz G, Waterman H, Zamir E, Kam Z, Oved S, Langdon WY, Beguinot L, Geiger B, Yarden Y. Genes Dev 1998; 12:3663–3674.
192. Dunn R, Hicke L. J Biol Chem 2001; 276:25974–25981.
193. van Delft S, Govers R, Strous GJ, Verkleij AJ, Van Vergen en Henegouwen PM. J Biol Chem 1997; 272:14013–14016.
194. Polo S, Sigismund S, Faretta M, Guldi M, Capua MR, Bossi G, Chen H, De Camilli P, Di Flore PP. Nature 2002; 416:451–455.
195. Petaja-Repo UE, Hogue M, Laperriere A, Walker P, Bouvier M. J Biol Chem 2000; 275:13727–13736.
196. Petaja-Repo UE, Hogue M, Laperriere A, Bhalla S, Walker P, Bouvier M. J Biol Chem 2001; 276:4416–4423.
197. Petaja-Repo UE, Hogue M, Bhalla S, Laperriere A, Morello JP, Bouvier M. EMBO J 2002; 21:1628–1637.

198. Bohn LM, Belcheva MM, Coscia CJ. J Neurochem 2000; 74:574–581.
199. Li JG, Luo LY, Krupnick JG, Benovic JL, Liu-chen LY. J Biol Chem 1999; 274:12087–12094.
200. Tao PL, Law PY, Loh HH. J Pharmacol Exp Ther 1987; 240:809–816.
201. Chan KW, Duttory A, Yoburn BC. Eur J Pharmacol 1997; 319:225–228.
202. Law PY, Hom DS, Loh HH. Mol Pharmacol 1983; 24:413–424.
203. He L, Fong J, von Zastrow M, Whistler JL. Cell 2002; 108:271–282.
204. Finn AK, Whistler JL. Neuron 2001; 32:829–839.
205. Law PY, McGinn TM, Wick MJ, Erickson LJ, Evans CJ, Loh HH. J Pharmacol Exp Ther 1994; 271:1689–1694.
206. Strassheim D, Law PY, Loh HH. Mol Pharmacol 1998; 53:1047–1053.
207. Belcheva MM, Ignatova EG, Young EC, Coscia CJ. Biochemistry 1996; 35:14818–14824.
208. Trujillo KA, Akil HA. Science 1991; 251:85–87.
209. Elliot K, Kest B, Man A, Kao B, Inturrisi CE. Neuropharmacol 1995; 13:347–356.

6

Delta Opioid Receptors and G Proteins

Mary J. Clark and John R. Traynor
University of Michigan Medical School, Ann Arbor, Michigan, U.S.A.

1 INTRODUCTION

The delta opioid receptor is a member of the large family of seven trans-membrane-spanning G protein–coupled receptors (GPCRs), as discussed extensively in Chapter 2. Delta opioid receptors modulate many intracellular effectors through their activation of GTP-binding proteins (G proteins), including adenylyl cyclase, K^+ channels, Ca^{2+} channels, the MAP kinase cascade, phospholipase C, and intracellular Ca^{2+} release [1] (see Chap. 5).

2 G PROTEINS AND THE G PROTEIN CYCLE

Agonist occupancy of GPCRs, such as the delta opioid receptor, leads to physiological effects through interactions with heterotrimeric G proteins. Such G proteins consist of a Gα subunit and its Gβγ dimeric partner. There are four major families of Gα proteins with different profiles of effector interaction: 1) Gαs, which activate adenylyl cyclase; 2) Gαi/o, so-called inhibitory G proteins named for their ability to inhibit adenylyl cyclase, but interact with many effectors; 3) Gαq/11, which activate phospholipase C-β (PLC-β); and 4) Gα12/13, which may regulate small GTP-binding proteins. Delta opioid receptors, like mu and kappa opioid receptors, couple to members of the $G\alpha i/o$ family which comprises the subtypes $G\alpha i_1$, $G\alpha i_2$, $G\alpha i_3$, $G\alpha o$ (both A and B, also known as 1 and 2, isoforms), $G\alpha t$, and $G\alpha z$.

There are several steps from agonist binding to effector response via G protein activation:

1. Agonist binding to receptor results in conformational change of the receptor or stabilization of the active state of the receptor.
2. Activated receptor interacts with G protein causing GDP, which is bound to the $G\alpha$ subunit in the resting state, to dissociate and be replaced with GTP. This results in dissociation of $G\alpha$-GTP from $G\beta\gamma$ subunit complex.
3. Activated $G\alpha$-GTP or free $G\beta\gamma$ may stimulate or inhibit effectors.
4. Hydrolysis of GTP to GDP by the intrinsic GTPase activity of the $G\alpha$ subunit results in inactivation and reassociation of the $G\alpha$ and $G\beta\gamma$ subunits and termination of the signal.

Agonist activation increases the rate of guanine nucleotide exchange and therefore the amount of active $G\alpha$-GTP and $G\beta\gamma$. While this process follows agonist occupation of GPCRs, it can also occur in the absence of agonist since receptors, including the delta opioid receptor [2,3], may assume active conformations and so constitutively activate G protein. The dissociation of GDP is the rate-limiting step; the rate of hydrolysis of GTP to GDP by the intrinsic GTPase activity of the $G\alpha$ subunit is increased by RGS (regulator of G protein signaling) proteins, thereby reducing the lifetime of the active GTP bound form of $G\alpha$ [for reviews see 4,5]. This RGS protein induced increased rate of deactivation has been shown to reduce effector responses following agonist occupancy of several GPCRs, including the mu opioid receptor [6,7], although to date no information is available in delta receptor expressing systems.

Structurally, each $G\alpha$ subunit consists of two domains—a GTPase domain, and a α-helical domain. In between these two domains is a cleft where guanine nucleotide binds. Lipid modification of a Cys residue near the amino-terminus of the $G\alpha$ subunit allows for binding to membrane [8], and the carboxyl terminus of the protein appears important for interaction with receptor. Indeed, the last five residues of $G\alpha$ are believed to contribute to specificity of interaction [reviewed in 9]. However Ho and Wong [10] have demonstrated that the amino terminus of $G\alpha z$ is also a critical determinant of its coupling to the delta opioid receptor.

Synthetic peptides derived from the third intracellular loop of the delta receptor inhibit high-affinity binding of the delta peptide [^{3}H] DSLET and G-protein activation by DSLET by competing with the activated receptor for recognition sites on $G\alpha$ [11,12]. This is in agreement with studies on a variety of GPCRs that have demonstrated the third intracellular loop of the receptors to mediate much of the coupling between receptor and G protein [13,14].

Mutations in transmembrane domains III and VII of the delta opioid receptor enhance activation of G protein in the absence of agonist [15,16]. It is suggested that specific helix-helix interactions in the receptor contribute to maintaining the receptor in an inactive conformation that does not interact well with G proteins [15].

Gα subunits of the Gi/o type of G proteins can be ADP-ribosylated in the presence of pertussis toxin at Cys^{351}, four amino acids from the C-terminus. Petussis toxin sensitivity is the major method of identifying a role for Gαi/o proteins in GPCR-mediated signaling. This treatment prevents receptor-mediated G-protein activation and thus exchange of GTP for GDP and so blocks signaling by Gα and Gβγ. There are numerous examples of the use of this technique to identify coupling of the delta opioid receptor [e.g., 2,3,41,42,73,77]. One Gα protein in this class, Gαz, lacks the Cys residue that is the site for pertussis toxin action and so is insensitive to pertussis toxin treatment [see 17 for review].

The Gβγ dimer is an absolute requirement for the binding of the Gα subunit to receptor, for the formation of high affinity agonist binding and for receptor catalyzed activation of G protein [18,19]. Gβγ is tethered to the membrane by a lipid modification of the γ subunit [20] and acts as both a scaffolding protein [21,22] and a signaling molecule in its own right. Gβγ has been demonstrated to modulate adenylyl cyclases [23], PLC-β [24], voltage-gated Ca^{2+} channels [25], and extracellular signal–regulated kinases (ERK); [26], in addition to other signaling molecules.

Within the four families of Gα proteins 20 different subunits have been identified. In addition, five different β subunits and 12 γ subunits have been described. Although not all of these are able to form βγ dimers, there is obviously a potential for numerous combinations of Gα, Gβ, and Gγ subunits that could interact with the delta opioid receptor. The particular combinations in a cell could be important in governing which signaling pathways are activated.

3 DELTA OPIOID RECEPTOR ACTIVATION OF G PROTEIN SUBTYPES

It is pertinent to ask whether interaction of the delta receptor with particular G protein subunit combinations leads to specificity for particular signaling pathways. For example, in rat dorsal root ganglia delta opioid receptors inhibit adenylyl cyclase [27] but do not couple to Ca^{2+} channels [28]. A variety of experimental approaches have shown that the delta receptor generally appears to be promiscuous with regard to its interaction with members of the Gαi/o family. However, those experiments do confirm that this receptor does not normally couple to other Gα families, although this can be seen in transfected systems (Table 1).

TABLE 1 Gα Subunits Demonstrated to Couple to Agonist-Occupied Delta Receptors

Cell or tissue	Agonist	Gα subunits	Measure	Method	Ref.
In vitro					
NG108-15	DADLE	αi_2, αi_3, αo_2	G protein	CTX catalysed ADP-ribosylation	29
	DADLE	$\alpha i_3 > \alpha i_2 = \alpha o_1 = \alpha o_2$	G protein	[α-^{32}P]GTP-azidoanilide	31
	DADLE	αi_2	Adenylyl cyclase	Antisera	44
	DADLE	αi_2, αo	G protein	[α-^{32}P]GTP-azidoanilide	30
	DADLE	$\alpha o > \alpha i_1$	Ca^{2+} current	PTX catalyzed ADP-ribosylation	41
NS20Y	DADLE	αi_2, αi_3, αo_1, αo_2	G protein	[α-^{32}P]GTP-azidoanilide	31
N1E115	DADLE	αi_2, αi_3, αo_1, αo_2	G protein	[α-^{32}P]GTP-azidoanilide	31
ND8-47	DSLET	αi_2	Intracellular Ca^{2+}	Antisense	54
SH-SY5Y	DPDPE	$\alpha i_1 > \alpha o > \alpha i_2 = \alpha i_3$	G protein	[α-^{32}P]GTP-azidoanilide	33
	DPDPE	$\alpha i_{1/2} > \alpha o > \alpha i_3$	Adenylyl cyclase	Antisera	43
SK-N-BE	DPDPE/delt I	αi_2, αo_2	G protein	[α-^{32}P]GTP-azidoanilide	65
	Etorphine	αi_1, αi_2, αi_3			
CHO	DADLE	αi_2, αi_3, αo_2	G protein	[α-^{32}P]GTP-azidoanilide	32
	DPDPE	αi_1, αi_2, αi_3, αo, αz, αq	Receptor binding	Immunoprecipitation	35
	SNC80	αt_1	G protein	Coexpression of δOR and αt_1	37
COS-7	DPDPE	α_{14}	Inositol phosphates	Coexpression of δOR and α_{14}	39
HEK293	DPDPE	αz	Adenylyl cyclase/ MAPK	Coexpression of δOR and Gαz	55
Mouse PAG	[D-Ala[2]]Delt II	$\alpha i_2 > \alpha z$	G protein	Immunoelectrophoresis	40
Smooth muscle (guinea pig)	DPDPE	αi_2, αo	Adenylyl cyclase	Antisera	46
Striatum (mouse)	DADLE	αi_2	Adenylyl cyclase	Antisense	45
Cortex (rat)	DADLE	αo	Receptor binding	Immunoprecipitation	35
In vivo (mouse)					
Supraspinal	DPDPE/ [D-Ala[2]]-delt II	αi_3, αi_2	Antinociception	Antisera	58,59
Supraspinal	DPDPE/ [D-Ala[2]]-delt II	αi_3, αi_2	Antinociception	Antisense	60
Spinal	DPDPE	αi_1, αi_2, αi_3, αo, αs, αq, αz	Antinociception	Antisense	59
Spinal	DPDPE	αz	Antinociception	Antisense	60

Unlike pertussis toxin that ADP-riboslyates Gαi/o, cholera toxin specifically ADP-ribosylates Gαs. However, cholera toxin can ADP-ribosylate Gαi/o if complexed with agonist-bound receptor in the absence of exogenous GTP, i.e., a nucleotide-free Gαi/o associated with receptor. Using this methodology Roerig and colleagues [29] found that delta agonist–induced cholera toxin–mediated ADP-ribosylation of $G\alpha i_2$, $G\alpha i_3$, and

$G\alpha o$ in membranes of NG108-15 cells in a dose-dependent manner with similar potency, indicating a lack of discrimination by agonist-bound delta receptor for coupling to these inhibitory $G\alpha$ protein subunits. An alternative approach has used the delta agonist–stimulated photoaffinity labeling of $G\alpha$ subunits with the photoreactive GTP analogue, azidoanilido[α-^{32}P]GTP. In NG108-15, NS20Y, and N1E115 cells the delta peptide agonist DADLE stimulated [α-^{32}P]GTP-azidoanilide incorporation into $G\alpha i_2$, $G\alpha i_3$, $G\alpha o_1$, and $G\alpha o_2$ with little specificity and similar potency [30,31]. Using CHO cells, the magnitude, but not the selectivity, of labeling was shown to be dependent on the density of delta opioid receptors expressed [32]. In contrast, the delta agonist DPDPE applied to membranes of SH-SY5Y cells led to a maximal incorporation of the [α-^{32}P]GTP-azidoanilide into $G\alpha i_1$ and $G\alpha o$ that was three times greater than the labeling of $G\alpha i_3$ and $G\alpha i_2$ [33], suggesting specificity of activation. To eliminate the effects of varying receptor and G-protein density on determinations of delta receptor G-protein selectivity measurements, Moon et al. [34] expressed fusion proteins between the C-terminus of the delta receptor and $G\alpha i_1$ or $G\alpha o_1$ in HEK393 cells. In these cells DADLE activated $G\alpha i_1$ three times more efficiently than $G\alpha o_1$ as measured by agonist stimulated GTP turnover number. However, the agonist-occupied mu opioid receptor stimulated $G\alpha i_1$ as efficiently as agonist-occupied delta receptor, suggesting a lack of selectivity between receptors.

Rat brain delta opioid receptors are precipitated by antisera to $G\alpha o$ [35]. Coimmunoprecipitation of delta receptor binding activity from CHO cell membranes expressing delta opioid receptors, using antisera selective for various G proteins, confirm, however, that the delta opioid receptor couples to multiple G proteins. Unoccupied delta receptors were seen to associate with $G\alpha i_1$, $G\alpha i_3$, $G\alpha o$, $G\alpha z$, $G\alpha q$, and $G\beta_1$ and $G\beta_2$ but not $G\alpha i_2$. In contrast, when agonist (DPDPE) was introduced, the delta receptor was now shown to associate with $G\alpha i_2$, but no longer with $G\alpha i_1$ [36]. These findings are consistent with agonist-induced changes in receptor–G protein coupling.

Delta opioid receptors can also couple to transducin ($G\alpha t$) [37], and the nonselectivity of delta opioid receptor–G protein coupling extends to pertussis toxin–insensitive G proteins. Delta receptors activate $G\alpha z$ to mediate effects on adenylyl cyclase and the MAP kinase pathway in HEK293 cells [38] and have been found to activate $G\alpha 14$ (a type of Gq) to increase inositol phosphates when both proteins are expressed in COS-7 cells [39]. Garzon and colleagues [40] used a nonisotopic, immunoelectrophoretic technique in which membranes were treated with agonist (D-Ala^2-deltorphin II) in the presence of GTPγS and then solubilized and subjected to electrophoresis on agarose gels containing specific G protein antisera, to demonstrate coupling of the delta opioid receptor to $G\alpha i_2$ and $G\alpha z$ in membranes from mouse periaqueductal gray matter. D-Ala^2-deltorphin II was 10 times more potent in activating $G\alpha i_2$ than $G\alpha z$. DPDPE acted as a partial agonist in activating $G\alpha i_2$ and did not stimulate $G\alpha z$.

4 DELTA OPIOID RECEPTORS, G PROTEIN SUBTYPES, AND DOWNSTREAM EFFECTORS

There have been several studies to determine if delta receptors activate particular $G\alpha$ subtypes for specific effector responses. Delta agonist stimulation of voltage-dependent Ca^{2+} channels was restored in pertussis toxin treated neuroblastoma x glioma hybrid cells by patch pipette addition of either $G\alpha i$ or $G\alpha o$ protein subunits, with $G\alpha o$ being more effective [41] or by expression of pertussis toxin insensitive $G\alpha o$ [42]. Treatment of SH-SY5Y cells with antisera for specific $G\alpha$ subtypes prevented delta receptor–mediated adenylyl cyclase inhibition through each of the subtypes tested in the order $G\alpha i_1/G\alpha i_2 > G\alpha o > G\alpha i_3$ [43]. In contrast, studies of the same type in NG108-15 cells concluded that the delta receptor interacted specifically with $G\alpha i_2$ to cause inhibition of adenylyl cyclase [44]. Similarly, there is a reduced efficacy of the delta agonist DADLE to inhibit adenylyl cyclase in mouse striatal tissue, with no change in the ED_{50} following intracerebroventricular antisense to $G\alpha i_2$ [45], and in intestinal smooth muscle of the guinea pig, antibodies for $G\alpha i_2$ and $G\alpha o$ blocked adenylyl cyclase inhibition by DPDPE [46]. PKC-mediated phosphorylation of $G\alpha i_2$ attenuates delta opioid inhibition of adenylyl cyclase in NG108-15 cells [47] and smooth muscle cells [48], supporting the importance of $G\alpha i_2$ for signaling to adenylyl cyclase. Rat pituitary GH_3 cells expressing high levels (2.5 pmol receptor/mg protein) of delta opioid receptor inhibited an L-type Ca^{2+} current and adenlylyl cyclase, while cells expressing lower levels of delta opioid receptor (0.6 pmol receptor/mg) inhibited adenylyl cyclase but not the L-type Ca^{2+} current [49]. The cells expressing the low number of receptors activated fewer of the same G proteins as the higher-expressing cells with the same pattern of preference. These results indicate that different threshold densities of the delta opioid receptor are required to activate critical amounts of the same $G\alpha$ proteins necessary for efficiently coupling to adenylyl cyclase and L-type Ca^{2+} channels.

The role of $G\beta\gamma$, released from pertussis toxin–sensitive G proteins, in coupling activated receptor to effectors further complicates specificity. Delta opioids activate G protein–coupled inwardly rectifying K^+ channels [50]; activation of such channels occurs by direct binding of $G\beta\gamma$ to various regions of the channel [51]. Delta agonist–mediated increase in the release of Ca^{2+} from intracellular stores in NG108-15 cells is mediated by $G\beta\gamma$ subunits [52], and $G\beta$ antibodies inhibited DPDPE stimulated PLC-β activation and, therefore, Ins(1,4,5)P_3-dependent Ca^{2+} release and smooth muscle contraction in intestinal smooth muscle cells of the guinea pig [46]. However, the $G\alpha$

partner for the G$\beta\gamma$ appears to vary. Functional coupling of the delta opioid receptor and phospholipase C occurs through activation of Gαi_1 but not Gαo or Gαq when Gα subunits and receptor are expressed in *Xenopus* oocytes [53]. Treatment of ND8-47 cells with antisense oligodeoxynucleotide to Gαi_2 subunit messenger RNA inhibited DSLET-mediated stimulation of intracellular Ca^{2+}, while antisense oligodeoxynucleotide to Gαi_3 or Gαs message had no effect [54].

In HEK cells coexpressing the delta opioid receptor and a constitutively active mutant of Gαs and adenylyl cyclase type II, delta agonists cause G$\beta\gamma$-mediated stimulation of adenylyl cyclase type II [55]. Agonist action at delta receptors expressed in COS-7 cells or HEK 293 cells has been shown to stimulate the MAP kinase pathway, possibly mediated by G$\beta\gamma$ subunits [56,57].

5 IN VIVO STUDIES OF DELTA OPIOID RECEPTOR G-PROTEIN INTERACTIONS

The relatively nonselective coupling of delta opioid receptors to the various G-protein subtypes has also been observed in vivo. Intracerebroventricular injections of mice with antibodies specific for Gαi_3, but not Gαi_1 or Gαz, inhibited supraspinally mediated delta analgesia [58,59]. In support of their work, Sanchez-Blazquez and colleagues [60] demonstrated that injections of antisense oligodeoxynucleotides specific for Gαi_3 were able to reduce supraspinally-mediated delta analgesia, while antisense oligodeoxynucleotides specific for Gαi_1 and Gαx/z had no effect. In contrast, Standifier et al. [61] reported that intrathecal injections of antisense oligodeoxynucleotides specific for Gαi_1, Gαi_2, Gαi_3, Gαo, Gαs, Gαq, or Gαz were each able to reduce spinally mediated delta-mediated analgesia. Karim and Roerig [62] have reported that intrathecal treatment with antisense oligodeoxynucleotides specific for Gαz reduces intrathecal DPDPE-mediated antinociceptin in mice. Comparable studies with G$\beta\gamma$ are scarce, but knockdown of the Gγ2 subunit has been demonstrated to cause a significant reduction in DPDPE-induced antinociception in the mouse [63].

6 AGONIST-SPECIFIC ACTIVATION OF G PROTEIN

Mutation of Trp^{284} at the border of transmembrane domain IV and the third extracellular loop of the human delta receptor reduced the relative efficacy of the delta agonist TAN-67 to stimulate [^{35}S]GTPγS binding, increased the efficacy of SNC80, and did not change the efficacy of DPDPE, when expressed in CHO cells [64]. These results suggest agonist specific conformations of the delta opioid receptor for G-protein activation. Allouche and colleagues [65] have demonstrated agonist specific selectivity of delta receptor coupling to G proteins in SK-N-BE cells endogenously expressing the delta opioid receptor. Peptide agonists (DPDPE and deltorphin I) activated $G\alpha i_2$ and $G\alpha o_2$ as measured by [α-^{32}P]azidoanilide-GTP incorporation, with deltorphin I having a higher efficacy. In contrast, the alkaloid etorphine stimulated [α-^{32}P] azidoanilide-GTP incorporation into $G\alpha i_1$, $G\alpha i_2$, $G\alpha i_3$, and a pertussis toxin–insensitive $G\alpha$. To determine which of the delta agonist–activated $G\alpha$ subunits were able to mediate inhibition of adenylyl cyclase, SK-N-BE cells were treated with anti-sense oligodeoxynucleotide to the various $G\alpha$ subunit mRNAs. Maximal peptide mediated inhibition of adenylyl cyclase was reduced by $G\alpha o_2$, but not by $G\alpha i_2$, antisense oligodeoxynucleotide treatment. In contrast, maximal etorphine-mediated inhibition of adenylyl cyclase was not reduced by antisense oligodeoxynucleotide directed against various $G\alpha$ subunits, but a reduction in $G\alpha i_2$ levels did shift the dose response curve for etorphine > 10-fold to the right, while antisense to $G\alpha i_1$ or $G\alpha i_3$ had no effect. These results demonstrate specificity of agonist mediated receptor G protein coupling that may have implications for physiological actions. These findings are in line with suggestions that peptide and nonpeptide ligands bind at different sites on the delta opioid receptor [66,67] and support the hypothesis that different agonists induce different receptor conformations that may activate different G proteins with different efficiencies [68].

7 DELTA AGONIST EFFICACY AND G PROTEINS

The relative ability of compounds to activate G proteins can be determined by measuring the binding of [^{35}S]GTPγS, a nonhydrolyzable analog of GTP [69]. The relative efficacy of compounds that bind to the delta receptor, as determined by their ability to stimulate the incorporation of [^{35}S]GTPγS, is as follows in decreasing order of efficacy: BW373U86 = SNC80 > DSLET > deltorphin II = DPDPE = etorphine > levallorphan = diprenorphine > nalorphine = buprenorphine > naltrindole [3,70,71]. The peptidic delta ligand ICI 174864 is an inverse agonist in this assay, confirming that the delta receptor constitutively activates G protein [2,3]. However, agonist efficacy can vary among cell types and assay conditions owing to variations in receptor number, type, and concentration of G protein expressed. It may also vary depending on the effector measured due to sensitivity of the pathway and intracellular modulating proteins, such as RGS. At the α_2-adrenergic receptor expressed in NIH-3T3 cells, relative efficacy of agonists to stimulate the binding of [^{35}S]GTPγS is dependent upon the type of $G\alpha$ subunit available for coupling [72]. The role of different G protein subtypes in governing full and partial agonist efficacy at the delta receptor has not been determined, yet may be important. For example, the delta antagonists TIPP and TIPPψ have been found to inhibit adenylyl cyclase in GH_3 cells expressing the delta opioid receptor and in NG108-15 or N1E115 cells endogenously expressing the delta opioid receptor in an antagonist and pertussis toxin–sensitive manner [73]. The authors suggest that TIPP and TIPPψ may activate only a single G-protein subtype that mediates inhibition of adenylyl cyclase without being detected by assays measuring G-protein activation directly.

8 ROLE OF G PROTEINS IN CHRONIC EFFECTS OF DELTA OPIOIDS

Continual exposure of cells expressing delta opioid receptors with delta agonist decreases activation of G protein, as measured by [^{35}S]GTPγS binding [74–76] and inhibition of adenylyl cyclase [75,77,78]; that is, a tolerance develops. These changes are due to an uncoupling of receptor and G protein (desensitization) followed by receptor downregulation (see Chap. 5). Chronic treatment of delta receptor–expressing cells does not appear to cause a change in the number of G proteins [79]. Agonist-mediated downregulation of the delta receptor is seen in pertussis toxin–treated cells [77,80,81], as is receptor upregulation following exposure to inverse agonists [81]. Taken together, the results suggest that functional coupling of receptor to G proteins is not necessary for controlling cell surface receptor levels following chronic drug exposure, although a tight physical association may be required [80]. In contrast, activation of the delta receptor expressed in HEK 293 cells triggers a Gβγ-mediated translocation of the cytosolic G protein receptor kinases GRK2 or GRK3 to the membrane to phosphorylate agonist–occupied receptors, followed by cointernalization of the receptor with GRK2 or 3 [82].

Chronic agonist treatment at the delta receptor causes a pertussis toxin–sensitive supersensitization of adenylyl cyclase that is a cellular correlate of withdrawal and may define dependence at the cellular level [83]. Gαs appears to be required for this supersensitization, although the effect is not mediated by Gαs [84]. In CHO cells expressing the human delta opioid receptor, supersensitization following washout of chronic SNC80 is blocked by expression of α-transducin to scavenge free Gβγ subunits. This suggests a role for Gβγ in adenylyl cyclase supersensitization in these cells, though Gβγ does not mediate acute inhibition of adenylyl cyclase [85].

9 G PROTEINS AND CROSSTALK INVOLVING THE DELTA RECEPTOR

There are several indications that crosstalk between delta and other opioid systems may alter the pharmacology of delta agonists. For example, delta analgesia and respiratory depression are reduced in mu receptor knockout mice [86], and there is a decrease in delta-mediated [^{35}S]GTPγS binding detected by autoradiography in mu receptor knockout mice [87]. In addition, delta receptor knockout mice do not develop analgesic tolerance to morphine [88]. Although several mechanisms could account for these interactions, crosstalk occurring at the cellular level could contribute. Certainly compartmentalization of mu and delta receptors has been suggested [89]. Indeed, delta and mu receptors share a common pool of G proteins in cotransfected COS-7 cells [90] and also in SH-SY5Y cells endogenously expressing both mu and delta receptors [76]. However, this effect may be cell specific since it is not observed in SK-N-SH cells, the parent line of SH-SY5Y cells [90].

In addition to intracellular crosstalk, delta receptors have been shown to dimerize and form heterodimers with other receptors [91,92]. Putative delta-kappa receptor dimers have the properties of the previously described putative $kappa_2$ receptor and are activated synergistically by selective delta and kappa ligands, leading to enhanced G protein–mediated inhibition of adenylyl cyclase or stimulation of the MAP kinase pathway [93]. Cotransfection of mu and delta opioid receptors in GH_3 cells similarly leads to a synergistic inhibition of adenylyl cyclase when cells are treated simultaneously with mu and delta agonists, possibly owing to an interaction between the receptors that results in an enhancement of binding [94]. Expression of mu and delta receptors in COS cells that individually couple to pertussis toxin sensitive G protein is shown to form a receptor species that can inhibit adenylyl cyclase, albeit weakly, through a pertussis toxin–insensitive G mechanism [95]. Treatment of CHO cells coexpressing mu and delta opioid receptors with the mu agonist DAMGO or the delta agonist deltorphin II activates the MAP kinase pathway leading to phosphorylation of Erk1/2. Coadministration of the mu antagonist CTOP potentiates the signaling of deltorphin II; conversely, the signal generated by DAMGO is potentiated by deltorphin II or TIPPψ. Taken together, these results are all suggestive of delta receptor containing functional heterodimers that signal through G proteins [96].

10 CONCLUDING REMARKS

In many systems delta receptors appear to be promiscuous with regard to their ability to interact with a variety of G proteins of the Gi/o family. The actual combinations of Gα and Gβγ with which the receptor interacts are likely to be governed by the types and amounts of G protein expressed in particular cells, and the physiological and pharmacological outcome by the ability of particular G proteins to modulate various downstream effectors.

Importantly, there are additional controls on the ability of delta agonist–occupied receptor to activate specific G protein and downstream effectors that are thus far poorly understood. These include agonist-specific activated states of the receptor, the capacity of the delta receptor to form homocomplexes or heterocomplexes with other GPCRs, and accessory proteins that may scaffold and/or act in a functional capacity.

ACKNOWLEDGMENT

The authors thank NIDA for support (DA 00254 and DA 4087).

REFERENCES

1. Quock RM, Burley TH, Varga E, Hosohata Y, Hosohata K, Cowell SM, Slate CA, Ehlert FJ, Roeske WR, Yamamura HI. Pharmacol Rev 1999; 51:503–532.
2. Costa T, Herz A. Proc Natl Acad Sci USA 1989; 86:7321–7325.
3. Szekeres PG, Traynor JR. J Pharmacol Exp Ther 1997; 283:1276–1284.
4. Ross EM, Wilkie TM. Annu Rev Biochem 2000; 69:795–827.
5. De Vries L, Zheng B, Fischer T, Elenko E, Farquhar MG. Annu Rev Pharmacol Toxicol 2000; 40:235–271.
6. Potenza MN, Gold SJ, Roby-Shemkowitz A, Lerner MR, Nestler EJ. J Pharmacol Exp Ther 1999; 291:482–491.
7. Harrison C, Clark MC, Neubig RR, Traynor JR. Br J Pharmacol 2002; 135: 179P.
8. Linder ME, Middleton P, Hepler JR, Taussig R, Gilman AG, Mumby SM. Proc Natl Acad Sci USA 1993; 90:3675–3679.
9. Hamm HE. J Biol Chem 1998; 273:669–672.
10. Ho MKO, Wong YH. Mol Pharmacol 2000; 58:993–1000.
11. Merkouris M, Dragatsis I, Megaritis G, Konidakis G, Zioudrou C, Milligan G, Georgoussi Z. Mol Pharmacol 1996; 50:985–993.
12. Megaritis G, Merkouris M, Georgoussi Z. Receptors Channels 2000; 7:199–212.
13. Cotecchia S, Ostrowski J, Kjelsberg MA, Caron MG, Lefkowitz RJ. J Biol Chem 1992; 267:1633–1639.
14. Kjelsberg MA, Cotecchia S, Ostrowski J, Caron MG, Lefkowitz RJ. J Biol Chem 1992; 267:1430–1433.
15. Befort K, Zilliox C, Filliol D, Yue S, Kieffer BL. J Biol Chem 1999; 274:18574–18581.
16. Cavalli A, Babey A-M, Loh HH. Neuroscience 1999; 93:1025–1031.
17. Fields TA, Casey PJ. Biochem J 1997; 321:561–571.
18. Fung BK. J Biol Chem 1983; 258:10495–10502.
19. Blumer KJ, Thorner J. Proc Natl Acad Sci USA 1990; 87:4363–4367.
20. Iniguez-Lluhi J, Simon MI, Robishaw JD, Gilman AG. J Biol Chem 1992; 267:23409–23417.
21. Sternweis PC. J Biol Chem 1989; 261:631–637.

22. Pitcher JA, Inglese J, Higgins JB, Arriza JL, Casey PJ, Kim C, Benovic JL, Kwatra MM, Caron MG, Lefkowitz RJ. Science 1992; 257:1264–1267.
23. Tang WJ, Gilman AG. Science 1991; 254:1500–1503.
24. Katz A, Wu D, Simon MI. Nature 1992; 360:686–689.
25. De Waard M, Liu H, Walker D, Scott VE, Garnett CA, Campbell KP. Nature 1997; 385:446–450.
26. Crespo P, Xu N, Simmonds WF, Gutkind JS. Nature 1994; 369:418–420.
27. Makman MH, Dvorkin B, Crain SM. Brain Res 1988; 445:303–313.
28. Moises HC, Rusin KI, Macdonald RL. J Neurosci 1994; 14:5903–5916.
29. Roerig SC, Loh HH, Law PY. Mol Pharmacol 1992; 41:822–831.
30. Offermans S, Schultz G, Rosenthal W. J Biol Chem 1991; 266:3365–3368.
31. Prather PL, Loh HH, Law PY. Mol Pharmacol 1994; 45:997–1003.
32. Prather PL, McGinn TM, Erickson LJ, Evans CJ, Loh HH, Law PY. J Biol Chem 1994; 269:21293–21302.
33. Laugwitz K-L, Offermanns S, Spicher K, Schultz G. Neuron 1993; 10:233–242.
34. Moon H-E, Cavalli A, Bahia DS, Hoffmann M, Massotte D, Milligan G. J Neurochem 2001; 76:1805–1813.
35. Georgoussi Z, Milligan G, Zioudrou C. Biochem J 1995; 306:71–75.
36. Law SF, Reisine T. J Pharmacol Exp Ther 1997; 281:1476–1486.
37. Varge EV, Stropova D, Wang TKM, Roeske WR, Yamamura HI. J Pharmacol Exp Ther 2000; 292:209–214.
38. Tso PH, Yung LY, Wong YH. J Neurochem 2000; 74:1685–1693.
39. Ho MKC, Yung LY, Chan JSC, Chan JHP, Wong CSS, Wong YH. Br J Pharmacol 2001; 132:1431–1440.
40. Garzón J, García-España A, Sánchez-Blázquez P. J Pharmacol Exp Ther 1997; 281:549–557.
41. Hescheler J, Rosenthal W, Trautwein W, Schultz G. Nature 1987; 325:445–447.
42. Taussig R, Sanchez S, Rifo M, Gilman AG, Belardetti F. Neuron 1992; 8:799–809.
43. Carter B, Medzihradsky F. Proc Natl Acad Sci USA 1993; 90:4062–4066.
44. McKenzie FR, Milligan G. Biochem J 1990; 267:391–398.
45. Shen J, Shah L, Hsu H, Yoburn BC. Mol Brain Res 1998; 59:247–255.
46. Murthy KS, Makhlouf GM. Mol Pharmacol 1996; 50:870–877.
47. Strassheim D, Malbon CC. J Biol Chem 1994; 269:14307–14313.
48. Murthy KS, Grider JR, Makhlouf GM. Am J Physiol Cell Physiol 2000; 279:C925–934.
49. Prather PL, Song L, Piros ET, Law PY, Hales TG. J Pharmacol Exp Ther 2000; 295:552–562.
50. North RA, Williams JT, Suprenant A, Christie MJ. Proc Natl Acad Sci USA 1987; 84:5487–5491.
51. Huang C-L, Jan YN, Jan LY. FEBS Lett 1997; 405:291–298.
52. Miyamae T, Fukushima, Misu Y, Ueda H. FEBS Lett 1993; 333:311–314.
53. Tang T, Kiang JG, Côté TE, Cox BM. Mol Pharmacol 1995; 48:189–193.
54. Yoon SH, Lo T-M, Loh HH, Thayer SA. Mol Pharmacol 1999; 56:902–908.
55. Tsu RC, Chan JSC, Wong YH. J Neurochem 1995; 64:2700–2707.

56. Gutstein HB, Rubie EA, Mansour A, Akil H, Woodgett JR. Anesthesiology 1997; 87:1118–1126.
57. Tso PH, Wong YH, Wong YH. J Neurochem 2000; 74:1685–1693.
58. Sánchez-Blázquez P, Juarros JL, Martínez-Peña Y, Castro MA, Garzón J. Life Sci 1993; 53:PL381–386.
59. Sánchez-Blázquez P, Garzón J. Life Sci 1993; 53:PL129–PL134.
60. Sánchez-Blázquez P, García-España A, Garzón J. J Pharmacol Exp Ther 1995; 275:1590–1596.
61. Standifer KM, Rossi GC, Pasternak GW. Mol Pharmacol 1996; 50:292–298.
62. Karim F, Roerig SC. Pain 2000; 87:181–191.
63. Hosohata K, Logan JK, Varga E, Burley TH, Vanderah TW, Porreca F, Hruby VJ, Roeske WR, Yamamura HI. Eur J Pharmacol 2000; 392:R9–R11.
64. Hosohata Y, Varga EV, Stropova D, Li X, Knapp RJ, Hruby VJ, Rice KC, Nagase H, Roeske WR, Yamamura HI. Life Sci 2001; 68:2233–2242.
65. Allouche S, Polastron J, Hasbi A, Homburger V, Jauzac P. Biochem J 1999; 342:71–78.
66. Meng F, Ueda Y, Hoversten MT, Thompson RC, Taylor L, Watson SJ, Akil H. Eur J Pharmacol 1996; 311:285–292.
67. Von Zastrow M, Keith DE, Evans CJ. Mol Pharmacol 1993; 44:166–172.
68. Kenakin T. Nature Reviews. Drug Discov 2002; 1:103–110.
69. Traynor JR, Nahorski SR. Mol Pharmacol 1995; 47:848–854.
70. Clark MJ, Emmerson PJ, Mansour A, Akil H, Woods JH, Portoghese PS, Remmers AE, Medzihradsky F. J Pharmacol Exp Ther 1997; 283:501–510.
71. Lee KO, Akil H, Woods JH, Traynor JR. Eur J Pharmacol 1999; 378:323–330.
72. Yang Q, Lanier SM. Mol Pharmacol 1999; 56:651–656.
73. Martin NA, Terruso MT, Prather PL. J Pharmacol Exp Ther 2001; 298:240–248.
74. Breivogel CS, Selley DE, Childers SR. J Neurochem 1997; 68:1462–1472.
75. Remmers AE, Clark MJ, Liu XY, Medzihradsky F. J Pharmacol Exp Ther 1998; 287:625–632.
76. Alt A, Clark MJ, Woods JH, Traynor JR. Br J Pharmacol 2002; 135:217–225.
77. Law PY, McGinn TM, Wick MJ, Erikson LJ, Evans C, Loh HH. J Pharmacol Exp Ther 1994; 271:1686–1694.
78. Louie AK, Law PY, Loh HH. J Neurochem 1986; 47:733–777.
79. Lang J, Costa T. J Neurochem 1989; 53:1500–1506.
80. Chakrabarti S, Yang W, Law P-Y, Loh HH. Mol Pharmacol 1997; 52:105–113.
81. Zaki PA, Keith DE, Thomas JB, Carroll FI, Evans CJ. J Pharmacol Exp Ther 2001; 298:1015–1020.
82. Schultz R, Wehmeyer A, Schultz K. J Pharmacol Exp Ther 2002; 300:376–384.
83. Sharma SK, Klee WA, Nirenberg M. Proc Natl Acad Sci USA 1975; 72:3092–3096.
84. Ammer H, Schulz R. J Pharmacol Exp Ther 1998; 286:855–862.
85. Rubenzik M, Varga E, Stropova D, Roeske WR, Yamamura HI. Mol Pharmacol 2001; 60:1076–1082.
86. Matthes HW, Smadja C, Valverde O, Vonesch JL, Foutz AS, Boudinot E,

Denavit-Saubie M, Severini C, Negri L, Roques BP, Maldonado R, Keiffer BL. J Neurosci 1998; 18:7285–7295.
87. Park Y, Ma T, Tanaka S, Jang C-G, Loh HH, Ko KH, Ho IK. Brain Res Bull 2000; 52:297–302.
88. Zhu Y, King MA, Schuller AGP, Nitsche JF, Reidl M, Elde RP, Unterwald E, Pasternak GW, Pintar JE. Neuron 1999; 24:243–252.
89. Remmers AE, Clark MJ, Alt A, Medzihradsky F, Woods JH, Traynor JR. Eur J Pharmacol 2000; 396:67–75.
90. Shapiro M, Vogel ZZ, Sarne Y. Cell Mol Neurobiol 2000; 20:291–304.
91. Cvejic S, Devi LA. J Biol Chem 1997; 272:26959–26964.
92. Ramsay D, Kellett E, McVey M, Rees S, Milligan G. Biochem J 2002; 365:429–440.
93. Jordan BA, Devi LA. Nature 1999; 399:697–700.
94. Martin NA, Prather PL. Mol Pharmacol 2001; 59:774–783.
95. George SR, Fan T, Xie Z, Tse R, Tam V, Varghese G, O'Dowd BF. J Biol Chem 2000; 275:26128–26135.
96. Gomes I, Jordan BA, Gupta A, Trapaidze N, Nagy V, Devi LA. J Neuroscі 2000; 20:1–5.

7

Transcriptional Regulation of Delta Opioid Receptor Gene

Ping Sun and Horace Loh
University of Minnesota Medical School, Minneapolis, Minnesota

1 INTRODUCTION

The activities of the opioid receptors appear to be dependent on the receptor density on the cell surface. Though the general principle of "spare" receptor applies in the opioid receptor regulation of the second-messenger system, the agonist potency, and the effector it regulates are receptor density dependent [1,2]. The importance of opioid receptor density in the pharmacological actions of this drug class is supported by the established relationship between tissue (cell type)-specific expression patterns of a receptor type and the response to opioids, which is ultimately dependent upon the spatial and temporal regulation of gene expression. In addition, opioid receptor density on the cell surface can affect the rate of receptor desensitization [3–5], which is key to the development of drug tolerance and dependence [6]. Thus, the control of the expression of the opioid receptor will determine the opioid agonist activities and may modulate tolerance development.

In general, the localization of the opioid receptors coincides with the pharmacological action sites of corresponding opioids [7]. In the case of delta opioid receptor (DOR), there is a strong correlation between the presence of

DOR mRNA and delta agonist binding sites [8]. Compelling evidence from numerous studies, together with the heterogeneity of the cell type–specific distribution of DOR, indicates that the DOR expression is under strict spatiotemporal control. For example, DOR is distributed in various densities at different regions of the brain [9] and appears later than mu and kappa opioid receptors in development [10]. In addition, levels of DOR mRNA as well as delta-agonist-binding sites can be regulated by various agents in some neuronal cell lines. For example, nerve growth factor [11], ethanol [12], or retinoic acid [13] can upregulate DOR mRNA and delta-agonist-binding sites. On the other hand, activation of the protein kinase A pathway by cyclic AMP analogues results in downregulation of DOR mRNA and delta-agonist-binding sites [14]. All these studies suggest that the DOR expression can be regulated by certain extracellular signals at the transcription level. As the DOR expression level on the cell surface will eventually determine the delta-agonist activities and may affect tolerance development, the study of the transcriptional regulation of the DOR gene will not only promote the understanding of how the DOR gene is regulated in physiological and pharmacological contexts, but also may raise the possibility of maximizing the pharmacological benefits of delta opioids by manipulation of the DOR expression levels.

2 GENOMIC STRUCTURE OF δ-OPIOID RECEPTOR GENE

To study the transcriptional regulation of the DOR gene, mouse DOR genomic clones were isolated [15]. Subsequent analysis revealed that the mouse

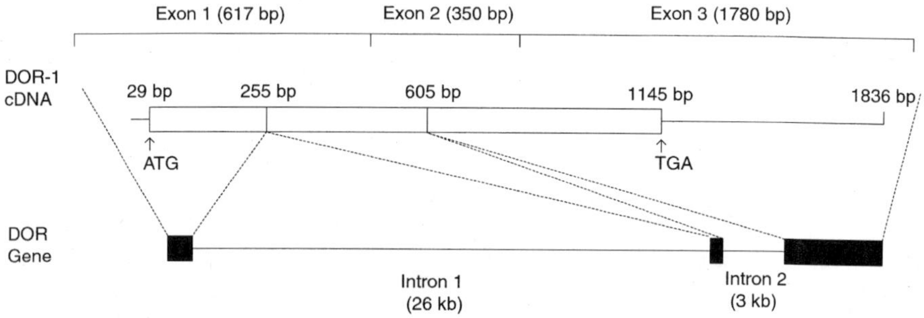

FIGURE 1 Structure of the mouse DOR gene. The bottom line represents mouse genomic DNA encompassing the DOR gene from 390 bp upstream of the ATG translation codon in exon 1 to 1.24 kb downstream of the TGA translation stop codon in exon 3. The 1.8-kb DOR-1 cDNA molecule is drawn above the DOR gene, relative to a fully spliced, undegraded DOR message (top line). (From Ref. 16.)

DOR gene spans 32 kb from transcription initiation sites located between 140 bp and 390 bp upstream of the ATG translation start codon to a polyadenylation site located 1.2 kb downstream of the TGA stop codon. RNase protection analysis of the 5′ ends of mouse brain poly(A)$^+$RNA and NG108-15 total RNA resulted in identical patterns of multiple protected fragments, suggesting that the DOR gene is transcribed from multiple initiation sites in the TATA-less, 80% G + C–rich sequence between 140 and 390

```
-1300  GAGCTCCCTGCTGTGACCTGGCTGTCTTCAGTCTTTTTGACTTGTAGGACCTTTATATATGTTTT
-1234  TCTTTTTTCTAGAAGGCTCTTCCTATTTCTTCCCTAAATTAGTTCCCCCACCTCTAAATCCACACA
-1168  ACACCTCAAGTGTCATTTCTTTTAGGAATATTCTCCTGACCCTACCATGCCCCGACATTCCTTGAA
-1102  TAGGACTTCATTGTGTTTTTTTTAGAATCGATGCATCTCTTCTGGTCTGAGCTGTAAGCTCTTGGG
-1036  AGCCGGGCCTTGGTCTGATTATGCACACCGATTACCTGGCTCTTTTGTTAGGCGCCCTCAGAGTG
-970   AATGTGGAGCTAGTGAACTGGGCGCAGGAAAGTGTGGCTGATAGGATGTTTTGTGGTTTTATTTA
-904   TCGCTCAAGCAGCCTGCTAGCTCAGAGCATAGAGCCAGAAGGAAAGTTGTGTCTCTAATAACATA
-838   CACACATCTATCCAGATAGGTGTGACCTTTTTCAGCTTGATGTAGTGAGATGGTTAGTGACCTTGA
-772   TTGCGGGGCTTATGTCTGCCATGCCGTGCCAAGCTGTACAGACCCAGTGGGCTTATTTCCAGTGT
-706   CTCCCCTTGAGGTTGTAGCTGTCATATTACTAGTCTTCATCCGGAGCCTTTGGGGGAACTGGACG
-640   GTTTTGTTTTTGTTTCTTGGTCAGGAATCACTTGGCAGTTCTAAAGAAGCGGCAAGGAGCAGAGC
-574   GCACACGTACACGGGGGTGGTGGTGGTGGTGGTGGTGGAAATCTGAGTCTTTACTCCGTTG
-508   TCTTCATCTCCCCAAAACCCTCTGAAGGCGGCAGGAGTTTTCTTCATCGTGCAGTCACAGAGAAT
-442   GAACTTGAGTGCTTAAGCGGCTCGCTCAAGGTTACTCCCATCGTCCGTGAAGAGGCGGGAGGCG
                                                         ^
       *Ikaros              AP-1
-376   GGATTCGCATCCAGGTTCTTCTGACTCCGAAGTGGCACGGGGGTGGGCGGGAGCGGGCGGGC
                ^                                              *
-310   TGGGGCCTGGAGTCAGCGGAGCCCAGGCAGGGAGCGCGCGCGAGGCCTCTCTGGCCCAGCTCG
                                        ^
                    Sp1/Sp3                                     Ets-1    USF
-244   GAGCCGGTGCAGCGCCCGCCCCCGATCGCGGGCGCAGACTGCGGCCGCTCCGGACCACGTGG
                           ^                                  ^          ^
                                 AP-2
-178   TGCGCGCAGCGGCTCGGGGCTCCCCGGGCTTGGGCGAGCACTGCGGGCGGTCAGGCGGAGGA
              ^       ^                          *   ^
-112   GGCGGGACGAGCGCTGCAGCCTGCGCGTCGGGGCCGTGGCCTCCGTTTCCCGCCACTCGCTG
-46    GGTCCCTGCGCCCAGGGCGCACGGTGGAGACGGACACGGCGGCGCCATG
```

FIGURE 2 Nucleotide sequence of DOR gene 5′ flanking region. Sequence is numbered relative to + 1 representing the A nucleotide in the ATG translation start codon, indicated by bold type. Positions of minor transcription initiation sites determined from estimates of RNase protected fragment sizes are indicated with bullets (ˆ). The two strongest transcription start sites are indicated by asterisks (*) at positions −142 and −324. The *cis* elements and corresponding *trans* factors that have been identified are underlined and labeled above the sequence, respectively. (From Ref. 16.)

nucleotides upstream from the translation start codon of mouse DOR gene [15] (Fig. 1). The amino acid coding sequence is divided into three exons and introns of 26 kb and 3 kb (Fig. 2). The first spliced site occurs at the N-terminal end of the first intracellular loop (Arg-73), and the second at the N-terminal end of the second extracellular loop (Asp-193). Interestingly, the locations of these splice sites are the same as those reported for the mouse μ-opioid receptor [16] and the human κ- and ì-opioid receptors [17,18]. The Asp-193 spice junction also coincides with a differential splice site in a rat orphan opioid receptor gene [18]. This suggests that an ancestral opioid receptor gene may have acquired these two introns prior to, or simultaneous with, its divergence into multiple genes.

3 TRANSCRIPTIONAL REGULATION OF DELTA OPIOID RECEPTOR GENE IN NEURONAL CELL LINES

To investigate the transcriptional regulation of the DOR gene, the cis elements and trans factors of the DOR promoter must be defined. As the mouse DOR gene shares >90% homology with its human counterpart, the study was carried out using a 1.3-kb DNA fragment upstream of the mouse DOR gene translation start site (−1300 to +1, with the translation start site designated as +1). The DOR gene shows the features of a typical housekeeping gene, containing no classical TATA box and no CCAAT box or a consensus initiator in the promoter region [19–21]. The minimal DNA sequence that is sufficient to provide the basal promoter activity in mouse neuronal cell lines is located between −262 and −141. A GC box (−226/−221) and an E box (−185/−180) contribute ~60% and 90% to the basal DOR promoter activity, respectively [22]. In vitro protein-DNA binding assays and in vivo transient transfection assays demonstrated that the upstream stimulatory factors (USF-1/USF-2) and Sp family factors (Sp1/Sp3) bind to the E box and GC box, respectively, and trans-activate the DOR promoter. Physical interactions are present between USF and Sp factors. Apparently, the E box–bound USF determines the extent of the physical interactions between USF and Sp factors and plays a decisive role in the DOR promoter activation. Interestingly, transcription factor Ets-1 can bind to an Ets-1-binding site (−192/−183) overlapping the E box, and confers ~50% of the DOR promoter activity by synergizing with USF in specific DNA binding [23]. As Ets-1 is only expressed at the developing stages of the mouse brain when the mouse brain DOR system is markedly developed, the identification of Ets-1 as a DOR promoter trans-activator implicates that Ets-1 may contribute to the maturation of mouse brain DOR system.

In addition, the E box cooperates with the −174/−152 region in the DOR promoter to confer cell type–specific promoter activities [24]. Using the yeast one-hybrid system, the AP-4 and MZF-1 transcription factors were identified interacting with the −174/−152 region. AP-4 acts as a trans-activator while MZF-1 as a repressor for the DOR promoter in NS20Y cells, a mouse neuronal cell line that constitutively expresses DOR but not H2.35 cells, a hepatic cell line. Collectively, the GC box, the composite Ets-1-binding site/E box and the −174/−152 region, together with their corresponding transcription factors, such as Sp1/Sp3, USF-1/USF-2, and AP-4/MZF-1, that may vary in concentrations as well as in isoforms or partners combinations in different spatiotemporal settings, may account at least partially for the constitutive spatiotemporal expression of the DOR gene in the nervous system.

Moreover, transcription factor AP-1 and an AP-1 binding site at −355/−349 in the DOR promoter are reportedly responsible for the induced DOR promoter activities in NG108-15 neuronal cells treated with phorbol ester *O*-tetradecanoylphorbol 13-acetate, while AP-2 binding of an AP-2 binding site at −157/−150 accounts for the upregulation of DOR promoter activities 48 h after treatment with forskolin [25]. This study shows that AP-2 mediates the upregulation of DOR mRNA levels by cAMP/PKA activation, while AP-1 mediates the enhancement of the DOR promoter activities by PKC activation.

4 TRANSCRIPTIONAL REGULATION OF DELTA OPIOID RECEPTOR GENE IN T-CELL LINES

Although DOR is mainly confined to the nervous system, it is also found in immune cells such as T and B cells. Accumulating evidence shows that endogenous and synthetic delta opioids can modulate T-cell proliferation, cytokine production, and calcium mobilization, through the DOR on T cells [26–29]. DOR transcripts expression has been detected in human and murine T cells [30], and can be markedly increased by treatment with concanavalin A or anti-CD3-ε, through a transcriptional mechanism [31]. In addition, the enhanced expression of DOR transcripts is correlated with greater capacity of delta opioids to affect the T cell's functions [32,33]. Thus, investigation into the transcriptional regulation of the DOR gene in T cells may not only provide insights to the tissue (cell type)-specific expression of DOR, but also may raise the possibility of regulating the immunomodulatory effects of delta opioids on T cells through manipulation of the inducible expression of DOR. USF binding to the E box in the mouse DOR promoter was found to be fundamental for both the constitutive and the enhanced promoter activities in the resting or phytohemagglutinin (PHA)-activated EL-4 cells, a mouse T-cell line that constitutively expresses DOR transcripts [34]. In addition, both in vivo and in vitro experiments demonstrated that increased binding activity of Ikaros at an Ikaros-binding site (−378/−374) in the DOR promoter is required for the stimulated transcription of the DOR gene in PHA-activated EL-4 cells [35]. Nuclear Ikaros isoforms (Ikaros-1 and Ikaros-2) exhibit increased expression in PHA-activated El-4 cells [35], which changes the ratio of different homo- and heterodimers of Ikaros [36]. Subsequent studies revealed that the augmented formation of Ikaros-2 homodimers results in the increased binding activity at the putative Ikaros binding site in PHA-activated EL-4 cells. Ikaros-2 homodimers specifically bind to the −378/−374 Ik binding site and exert position-dependent trans-activation effect on the DOR promoter via functional synergy with the E box–bound USF [34].

T lymphocytes are exposed to endogenous opioid peptides in vivo. For example, circulating β-endorphin originating from the pituitary and enkephalin peptides originating from the adrenal medulla continuously bathe T lymphocytes. In addition, T lymphocytes may produce and release their own opioids [37,38]. As Ikaros has been reported to set threshold for T-cell activation [39] and plays an important role in the T-cell homeostasis [40], the link between Ikaros and the stimulated transcription of the DOR gene in activated T cells implicates an active role for endogenous opioids in modulating the functions and homeostasis of activated T cells.

5 FUTURE DIRECTIONS

The direct goal of the study on the transcriptional regulation of the DOR gene is to better understand how the DOR gene is regulated in physiological and pharmacological contexts, by delineating the complete network of cis elements, trans factors, and epigenetic means, as well as the underlying mechanisms that are responsible for the spatiotemporal expression of DOR. Hopefully, this information could be applied for modulation of the DOR expression for either scientific research or clinical therapies. Moreover, although it remains an open question, a better understanding of opioid receptor gene regulatory mechanisms might be useful in the ultimate goal of solving some fundamental problems in opioid pharmacology such as tolerance and dependence. A number of positive or negative cis elements and trans factors have so far been identified to contribute to the transcriptional regulation of the DOR promoter, as summarized in Figure 3. In the future, studies could be interesting in the following aspects. First, since the 80% G + C–rich sequence encompassing transcription initiation sites in the DOR promoter contains an abundance of CpG dinucleotides, this promoter may well be subjected to regulation by developmental and/or tissue-specific methylation [41]. Actually, there are convincing preliminary data showing that the DOR gene is under the regulation of tissue-specific methylation [42]. Thus, future research targeted on epigenetic regulatory means like methylation and the resultant histone modification and chromatin structure change [43] will undoubtedly provide a

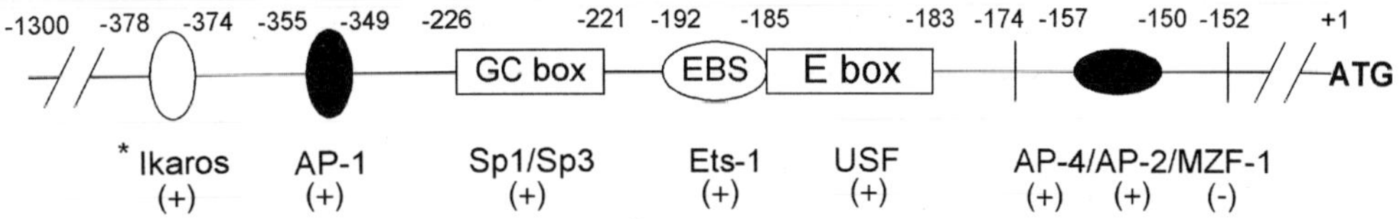

FIGURE 3 A schematic representation of the *cis* elements and corresponding *trans* factors that have been identified in the mouse DOR promoter. The translation start site (ATG) is designated as +1. The asterisk (*) marked transcription factor and its cognate binding site are identified in T cells, with the rest in neuronal cells. The (+) and (−) indicate positive and negative effects, respectively. EBS stands for Ets-1-binding site.

new angle of view for understanding the transcriptional regulation of the DOR gene. Second, different members of the USF, Sp, or Ikaros families may vary in concentrations and protein interaction combinations in different spatiotemporal settings, which may lead to significant differences in the DOR promoter activity and thereby the DOR transcripts expression. So, studies focused on the functional and physical interactions of transcription factors as well as the resultant chromatin modification/remodeling effects, if any, will be helpful for a better understanding of the spatiotemporal expression patterns of the DOR gene. Third, the traditional opioid receptor gene structure has been challenged. For example, the μ-opioid receptor gene was reported to span ~53 kb with four exons and three introns [44–46]. However, a number of recent studies reported that the mu opioid receptor gene spans >250 kb and consists of 14 exons, through which different splice variants of the μ-opioid receptor gene are generated [47–51]. Though no conclusive evidence is available for a contiguous endogenous mRNA generated from these variants, these studies provide new concepts on the structures and functions of the opioid receptor genes.

In view of the strong similarity among mu, kappa, and delta opioid receptors, it is tempting to explore the genomic structure and function of the DOR gene for potential functional elements, including additional promoter region [52,53]. Genomic studies on the DOR gene will hopefully give rise to a whole new perspective for understanding and modulating the spatial and temporal expression of the DOR gene. Finally, an important consideration will be to learn whether and how ligand binding can trigger the transcriptional regulation of the DOR gene. It has been reported that treatment of NG108-15 cells with etorphine [54] and cortical astrocyte primary cultures with DPDPE [55] resulted in downregulation and upregulation of the DOR mRNA, respectively. Questions arise as to what and how signal transduction pathways downstream of the ligand–receptor interaction turn on the transcriptional regulation of the DOR gene, and what functional significance the transcriptional regulation may have on the delta opioids' effects on cells. It will be intriguing to seek the answers to these questions, which may provide potential targets for therapeutic interventions aimed at modulating the DOR expression and consequently the drug activities and tolerance development. In addition, many putative transcription factor–binding sites in the present DOR promoter remain to be evaluated, particularly those responding to the action of growth factors and cytokines.

Besides transcription regulation, regulatory events at the posttranscription level such as alternative transcripts splicing, mRNA stability, and translation efficiency may also play important roles in the spatiotemporal expression of opioid receptor genes. Interestingly, it is recently reported that cell-permeant delta opioids can interact directly with DOR in the endoplasmic reticulum and rescue them from the degradative pathway as pharmacological chaperones, thus modulating the DOR expression on the cell surface [56]. It is foreseeable that future studies in all these areas will be geared toward understanding how the DOR expression is regulated in physiological and pharmacological contexts.

ACKNOWLEDGMENTS

This research is supported in part by National Institutes of Health grants DA-00564, DA-01583, DA-11806, and KO5-DA-70554, and by the A. and F. Stark Fund of the Minnesota Medical Foundation.

REFERENCES

1. Burt AR, Carr IC, Mullaney I, Anderson NG, Milligan G. Biochem J 1996; 320:227–235.
2. Law PY, Mcginn TM, Wick MJ, Erickson LJ, Evans CJ, Loh HH. J Pharmacol Exp Ther 1994; 271:1689–1694.
3. Pak Y, Kouvelas A, Scheideler MA, Rasmussen J, O'Dowd BF, George SR. Mol Pharmacol 1996; 50:1214–1222.
4. Law PY, Erickson LJ, El-Kouhen R, Dicker L, Solberg J, Wang W, Miller E, Burd AL, Loh HH. Mol Pharmacol 2000; 58:388–398.
5. Law PY, Maestri-El Kouhen O, Solberg J, Wang W, Erickson LJ, Loh HH. J Biol Chem 2000; 275:32057–32065.
6. Law PY, Loh HH. Studies on the Molecular Mechanism of Opioid Tolerance: Receptor Desensitization and Phosphorylation. Submitted paper. Minneapolis, MN: University of Minnesota, 2002.
7. Goldstein A, Naidu A. Mol Pharmacol 1989; 36:265–272.

8. Monsour A, Fox CA, Burke S, Meng F, Thompson RC, Akil H, Watson SJ. J Comp Neurol 1994; 350:412–438.
9. Monsour A, Khachaturian H, Lewis ME, Akil H, Watson SJ. Trends Neurosci 1988; 11:308–314.
10. Zhu Y, Hsu M, Pintar JE. J Neurosci 1998; 18:2538–2549.
11. Abood ME, Tao Q. J Pharmacol Exp Ther 1998; 274:1566–1573.
12. Jenab S, Inturrisi CE. Brain Res Mol Brain Res 1997; 47:44–48.
13. Beczkowska IW, Buck J, Inturrisi CE. Brain Res Bull 1996; 39:193–199.
14. Gylys KH, Tran N, Magendzo K, Zaki P, Evans CJ. Neuroreport 1997; 8: 2369–2372.
15. Augustin LB, Felsheim RF, Min BH, Fuchs SM, Fuchs JA, Loh HH. Biochem Biophys Res Commun 1995; 207:111–119.
16. Min BH, Augustin LB, Felsheim RF, Fuchs JA, Loh HH. Proc Natl Acad Sci USA 1994; 91:9081–985.
17. Minami M, Toya T, Katao Y, Maekawa K, Nakamura S, Onogi T, Kaneka S, Satoh M. FEBS Lett 1993; 329:291–295.
18. Wang JB, Johnson PS, Perico AM, Hawkins AL, Griffin CA, Uhl GR. FEBS Lett 1994; 338:217–222.
19. Gill G, Tjian R. Curr Opin Genet Dev 1992; 2:236–242.
20. Smale ST, Baltimore D. Cell 1989; 57:103–113.
21. Javahery R, Khachk A, Lo K, Zenzie-Gregory B, Smale ST. Mol Cell Biol 1994; 14:116–127.
22. Liu HC, Shen JT, Augustin LB, Ko JL, Loh HH. J Biol Chem 1999; 274: 23617–23626.
23. Sun P, Loh HH. J Biol Chem 2001; 276:45462–45469.
24. Liu HC, Loh HH. Transcriptional Regulation of Mouse δ-Opioid Receptor Gene: AP4 and MZF1 Play Divergent Roles in DOR Transcriptional Regulation. Submitted paper. Minneapolis, MN: University of Minnesota, 2002.
25. Woltje M, Kraus J, Hollt V. J Neurochem 2000; 74:1355–1362.
26. Linner KM, Quist HE, Sharp BM. J Immunol 1995; 154:5049–5060.
27. Sharp BM, McKean DJ, McAllen K, Shahabi NA. Ann NY Acad Sci 1998; 840:420–424.
28. Singh VK, Bajpai K, Narayan P, Yadav VS, Dhawan VC, Haq W, Mathur KB, Agarwal SS. Neuroimmunomodulation 1999; 6:355–360.
29. Sharp BM, Gekker G, Li MD, Chao CC, Peterson PK. Biochem Pharmacol 1998; 56:289–292.
30. Gaveriaux C, Peluso J, Simonin F, Laforet J, Kieffer B. FEBS Lett 1995; 369: 272–276.
31. Li MD, McAllen K, Sharp BM. J Leukoc Biol 1999; 65:707–714.
32. Sharp BM, Shahabi NA, Heagy W, McAllen K, Bell M, Huntoon C, McKean DJ. Proc Natl Acad Sci USA 1996; 93:8294–8299.
33. Shahabi NA, McAllen K, Matta SG, Sharp BM. Cell Immunol 2000; 205:84–93.
34. Sun P, Loh HH. Transcriptional Regulation of Mouse δ-Opioid Receptor Gene: Ikaros-2 and USF Synergize in *Trans*-Activating Mouse δ-Opioid

Receptor Gene in T Cells. Submitted paper. Minneapolis, MN: University of Minnesota, 2002.
35. Sun P, Loh HH. J Biol Chem 2002; 277:12854–12860.
36. Georgopoulos K, Winandy S, Avitahl N. Annu Rev Immunol 1997; 15:155–176.
37. Blalock JE. Prog Neuroendocrinimmunol 1988; 1:9–12.
38. Wesley HJ, Kleiss KJ, Kelley KW, Wong PKY, Yuen PH. J Exp Med 1986; 163:1589–1594.
39. Avitahl N, Winandy S, Friedrich C, Jones B, Ge Y, Georgopoulos K. Immunity 1999; 10:333–343.
40. Nichogiannopoulou A, Trevisan M, Friedrich C, Georgopoulos K. Semin Immunol 1998; 10:119–125.
41. Christophe D, Pichon B. Mol Cell Endocrinol 1994; 100:155–158.
42. Wang GL, Loh HH. Role of Methylation in the Transcriptional Regulation of Mouse δ-Opioid Receptor Gene. Unsubmitted paper. Minneapolis, MN: University of Minnesota, 2002.
43. Kadonaga JT. Cell 1998; 92:307–313.
44. Wang JB, Imai Y, Eppler CM, Gregor P, Spivak CE, Uhl GR. Proc Natl Acad Sci USA 1993; 90:10230–10234.
45. Thompson RC, Mansour A, Akil H, Watson SJ. Neuron 1993; 11:903–913.
46. Chen Y, Mestek A, Liu J, Hurley JA, Yu L. Mol Pharmacol 1993; 44:8–12.
47. Bare LA, Mansson E, Yang D. FEBS Lett 1995; 354:213–216.
48. Zimprich A, Simon T, Hollt V. FEBS Lett 1995; 359:142–146.
49. Pan YX, Xu J, Bolan EA, Chang A, Mahurter L, Rossi GC, Pasternak GW. FEBS Lett 2000; 466:337–340.
50. Pan YX, Xu J, Bolan EA, Abbadie C, Chang A, Zuckerman A, Rossi GC, Pasternak GW. Mol Pharmacol 1999; 56:396–403.
51. Pan YX, Xu J, Mahurter L, Bolan EA, Xu MM, Pasternak GW. Proc Natl Acad Sci USA 2001; 98:14084–14089.
52. Ko JL, Minnerath SR, Loh HH. Biochem Biophys Res Commun 1997; 234: 351–357.
53. Lu S, Loh HH, Wei LN. Mol Pharmacol 1997; 52:415–420.
54. Kim DS, Chin H, Klee WA. FEBS Lett 1995; 376:11–14.
55. Trorlin T, Eriksson PS, Hansson E, Ronnback L. Neurosci Lett 1997; 232:67–70.
56. Petaja-Repo UE, Hogue M, Bhalla S, Laperriere A, Morello JP, Bouvier M. EMBO J 2002; 21:1628–1637.

8

Benzhydrylpiperazines as Nonpeptidic Delta Opioid Receptor Ligands

Michael J. Bishop
GlaxoSmithKline Research and Development, Research Triangle Park, North Carolina, U.S.A.

Robert W. McNutt
Ardent Pharmaceuticals, Inc., Durham, North Carolina, U.S.A.

1 INTRODUCTION

The clinical treatment of moderate to severe pain relies on traditional opioid analgesics, such as morphine (**1**) [1a,b] and fentanyl (**2**) (Fig. 1) [1]. Morphine is one of several known related alkaloids isolated or derived from the opium poppy (*Papaver somniferum*). The pharmacological properties of opium extracts have been recognized for thousands of years, and their ability to dampen the sensation of pain has been used clinically for centuries. Fentanyl and related 4-substituted piperidine opioids were discovered over the last 75 years and have found great therapeutic utility in pain management [2]. While these two classes of opioid ligands produce profound analgesia, their use is also associated with some deleterious physiological effects, such as respiratory depression, muscle rigidity, emesis, constipation, and physical dependence [3]. The clinical need for a potent analgesic lacking these harmful and sometimes fatal additional effects continues today, and the search for ligands that meet

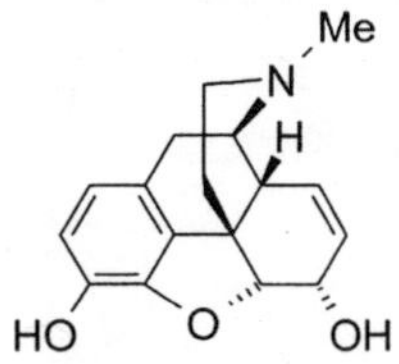

Morphine, **1**

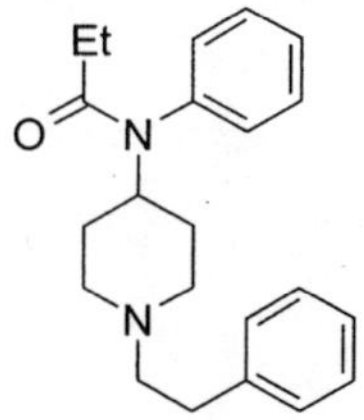

Fentanyl, **2**

FIGURE 1 Structures of morphine and fentanyl.

this need has been a research focus for pharmacologists and medicinal chemists for decades.

These powerful analgesics relieve pain primarily through agonism of mu opioid receptors. A realistic hope that scientists might be able to separate detrimental opioid physiological effects from analgesia and other desirable effects can be traced back to the mid-1960s, when structure–activity relationship data on opiates led Portoghese and Martin to suggest the existence of multiple opioid receptors [4,5a,b]. By the middle to late 1970s, extensive research had led to the characterization of three opioid receptors with distinct pharmacological profiles. These three opioid receptors (named mu, delta, and kappa) have now been cloned, expressed, and sequenced [6]. Since the 1970s considerable research has focused on delta and kappa opioid receptors, as agonism, partial agonism, or even antagonism at one of these receptors or subtypes could lead to a therapeutically useful agent for pain, drug addiction, urological or gastrointestinal disorders, possibly lacking the side effects associated with mu opioid receptor agonism. The identification of enkephalins as endogenous delta opioid receptor ligands [7] coupled with early evidence that delta selective ligands could produce analgesia without some of the side effects of mu agonism [8] encouraged delta opioid research.

To facilitate a thorough evaluation of the physiological effects of delta opioid receptor activation, significant effort has been expended over the past two decades to identify selective agonists and antagonists [9a–c]. Small peptides provided early examples of delta opioid receptor–selective ligands [10]. While peptides are often excellent pharmacological tools, nonpeptidic small molecules with better metabolic stability and pharmacokinetic properties were sought for more robust in vivo experimentation and the clinical development of therapeutic agents. An exciting breakthrough in delta opioid research was Portoghese's discovery of the delta opioid receptor antagonist naltrindole, the first reported delta-selective nonpeptide opioid ligand [11]. Over the past decade, nonpeptidic delta opioid receptor–selective agonists have also been discovered. Selective delta opioid receptor agonists have been

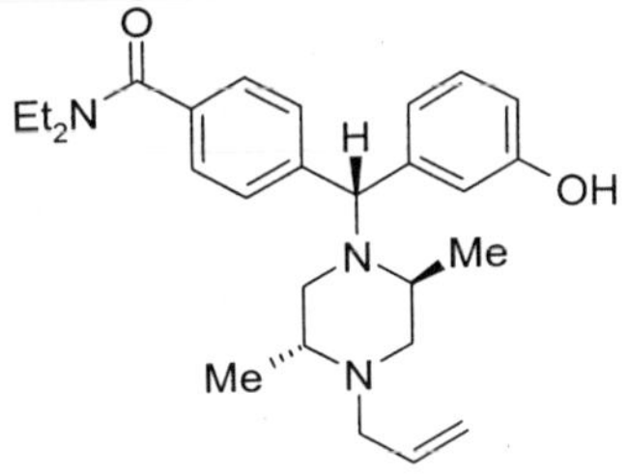

FIGURE 2 Structure of BW373U86 (compound **3**).

disclosed in several structural classes, including morphinans [12a,b], octahydroisoquinolines [13a,b], phenoxyethylpiperidines [14], and benzhydrylpiperazines [15]. In 1992, BW373U86 (**3**) (Fig. 2), a benzhydrylpiperazine, was the first potent nonpeptidic delta opioid receptor full agonist disclosed [16]. Considerable opioid research at Burroughs Wellcome in the 1980s and early 1990s focused on the benzhydrylpiperazines, and this general class of compounds has been explored by other laboratories since the BW373U86 disclosure. This chapter will provide a medicinal chemistry viewpoint of the benzhydrylpiperazine opioids, covering the delta opioid receptor structure–activity relationship in this series, the synthetic routes, and a glimpse of related delta opioid receptor ligand series that have been reported and were likely inspired by the benzhydrylpiperazines.

2 DISCOVERY OF BENZHYDRYLPIPERAZINE OPIOID RECEPTOR LIGANDS

In the late 1970s, the molecular biology and pharmacology of opioid receptors were under investigation in the laboratories of Kwen-Jen Chang at Burroughs Wellcome. Early research into delta receptor biology relied on peptidic ligands, such as the enkephalins, as molecular tools. The metabolic instability, formulation difficulties, and in vivo absorption and distribution characteristics made these suboptimal tools for in-depth in vivo studies. The availability of rat brain membranes expressing the delta receptor to evaluate ligand binding, and the use of mouse vas deferens tissue to identify the functional response of delta-receptor ligands made it possible to search for structurally novel opioids. However, the ability to screen large numbers of compounds against a receptor did not exist as it does today. Careful selection of compounds for in vitro evaluation was necessary.

The Burroughs Wellcome compound collection was carefully studied in search of structures that might have opioid activity. Medicinal chemists considered the structures of the morphinoids as well as the enkephalins in choosing novel compounds to test for delta receptor activity. Among the small set of molecules selected for testing were benzhydrylpiperazines. Structures reminiscent of morphinoids would include a phenol and a basic amine (a tyramine-like pharmacophore) with significant structural constraint. The initial benzhydrylpiperazines sent for testing preserved this motif. Additionally, Leu- and Met-enkephalins (Tyr-Gly-Gly-Phe-Leu and Tyr-Gly-Gly-Phe-Met, respectively), delta receptor agonists that were known at that time, have a second phenyl ring (part of a phenylalanine residue), as do the benzhydrylpiperazines. Compounds were evaluated for opioid receptor affinity via displacement of radiolabeled ligands from rat brain membranes (RBM) [17a,b]. Function activity was evaluated via inhibition of electrically stimulated twitch in mouse vas deferens (MVD) and guinea pig ileum (GPI) tissue preparations for the delta opioid and mu opioid receptors, respectively [18]. This initial testing identified a relatively simple benzhydrylpiperazine, compound **4a** (Fig. 3), as a selective delta opioid receptor antagonist (mu receptor IC_{50} = 7000 nM, delta receptor IC_{50} = 200 nM, pA_2 from mouse vas deferens = 5.7). Simple manipulation of the substitution on the piperazine nitrogen distal to the 4′-chloro-3-benzhydrol provided interesting structure–activity data. Removal of the methyl group as well as replacement with an ethyl group (compounds **4b** and **4c**, respectively) provided similar delta opioid receptor antagonists. Lengthening the chain to *n*-propyl and *n*-butyl incrementally reduced delta opioid receptor affinity (**4d** and **4e**, respectively), with the *n*-butyl compound having a delta receptor IC_{50} of 1.5 μM. The *N*-allylpiperazine analogue, compound **4f**, provided an exciting result. Delta opioid receptor affinity had increased (IC_{50} = 50 nM), and the mouse vas deferens assay revealed this compound to be a relatively weak but effective delta receptor agonist (MVD ED_{50} = 2 μM). The *N*-allylpiperazine present in this first agonist would prove to be a key structural feature, present in many of the potent benzhydrylpiperazine opioid agonists produced at Burroughs Wellcome. After the identification of this first

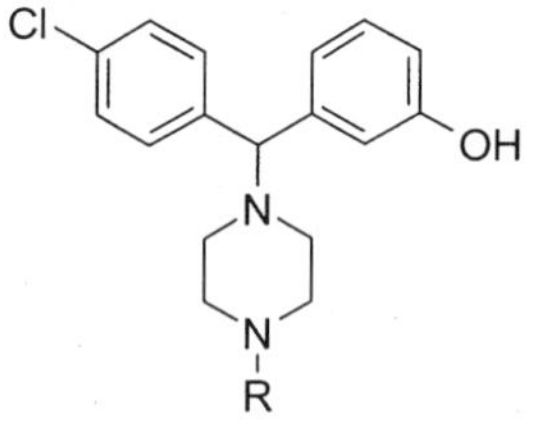

Compound	R
4a	**Me**
4b	**H**
4c	**Et**
4d	**Pr**
4e	**Bu**
4f	**Allyl**

Compounds **4a-f**

FIGURE 3 Early benzhydrylpiperazine opioids.

benzhydrylpiperazine delta receptor agonist, medicinal chemistry research focused on developing the SAR to discover highly potent, selective delta receptor agonists.

2.1 Early SAR of Benzhydrylpiperazine Opioids

Early medicinal chemistry efforts in the benzhydrylpiperazine series focused on structural modifications to explore opioid receptor potency and selectivity, with a goal of identifying potent, selective delta opioid receptor agonists for potential therapeutic use in the treatment of severe pain. Potent antagonists were also desirable as valuable pharmacological tools. With compound **4f** in hand and initial SAR indicating the value of the allyl group for agonist activity, much of the early structural exploration focused on two areas: 1) additional small alkyl substitution on the piperazine ring; 2) substitution on and heterocyclic replacement of the nonphenolic aryl ring. Based on the precedent of the morphinoids and enkephalins, the phenolic —OH was hypothesized to be valuable for optimal opioid potency and was therefore a common feature of most of the early benzhydrylpiperazine analogues. While the most potent compounds to date retain the phenolic —OH, we now know that it is not critical for obtaining potent and selective delta-opioid receptor agonists.

With the allyl group on the piperazine providing delta opioid receptor agonism, substitution on the carbons of the piperazine ring was explored. For example, N-allyldimethylpiperazine analogues were prepared, and the placement and relative stereochemistry (early in the program, mixtures of stereoisomers were often tested) of the two methyl groups affected delta opioid receptor potency and selectivity (see Table 1). Interesting analogues include a direct analog of **4f** containing a *cis*-3,5-dimethyl-*N*-allylpiperazine substituent that was a modestly potent, selective delta receptor antagonist (**5a**). The analogue featuring a *cis*-2,5-dimethyl substitution pattern (**5b**) had slightly weaker delta opioid receptor binding affinity and less delta selectivity, and was not evaluated for functional activity. The *cis*-2,3-dimethyl analogue (**5c**) exhibited potent, selective affinity for the delta opioid receptor (IC_{50} = 10 nM, 300-fold selective vs. the mu receptor), but displayed weak agonist activity in the mouse vas deferens assay (ED_{50} = 20 μM). The *trans*-2,3-dimethyl analogue (**5d**) possessed much more delta opioid agonism, with an ED_{50} of ~0.4 μM. A breakthrough was realized with the *trans*-2,5-dimethyl analogue, which provided a more potent, selective delta opioid agonist. The two pairs of enantiomers were separated for this compound, and the more potent pair (**5e**) displayed delta opioid agonist activity with an ED_{50} of 40 nM in the MVD assay, with 20-fold tissue selectivity. While preferred absolute stereochemistry was addressed in a later series of analogs, these results sug-

TABLE 1 Delta and Mu Opioid Activity of Dimethylpiperazine Analogues[a]

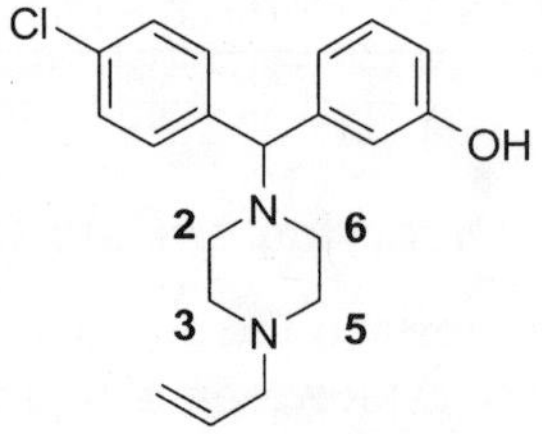

Compound (piperazine substitution)	Delta agonism (MVD) ED_{50} (nM)	Mu agonism (GPI) ED_{50} (nM)	Delta binding (RBM) IC_{50} (nM)	Mu binding (RBM) IC_{50} (nM)
4f (no methyls)	2000	—	50	3500
5a (*cis*-3,5-dimethyl)	(pA_2 = 6.6)	—	40	2000
5b (*cis*-2,5-dimethyl)	—	—	70	500
5c (*cis*-2,3-dimethyl)	20,000	—	10	3000
5d (*trans*-2,3-dimethyl)	429	—	30	3000
5e (*trans*-2,5-dimethyl)	40	800	15	90

[a] Each value is an average of at least three runs. See Refs. 17, 18 for a description of the assays.

gested that *cis*-3,5-dimethyl-*N*-allylpiperazine analogues would lead to useful delta receptor antagonists, while *trans*-2,5-dimethyl-*N*-allylpiperazines might be useful for the development of potent agonists.

The SAR of the nonphenolic ring was extensively explored. Removal of the *para*-chloride of the initial hit **4a** to give an analogue with no substitution on the phenyl ring (compound **6**, Fig. 4) provided a compound with similar affinity for the delta receptor but with much less delta receptor selectivity (only fivefold delta selectivity for **6** vs. 35-fold binding selectivity for **4a**). This supplied the initial implication that *para* substitution on the phenyl ring, even with a group as small as a chloride, was useful for delta receptor selectivity. This proved to be true for the agonists also, as the des-chloro analogue of *trans*-2,5-dimethyl agonist **5e** (compound **7**) showed no agonist selectivity for the delta receptor (MVD ED_{50} = 78 nM, GPI ED_{50} = 19 nM).

Numerous substituents were explored at the *para* position of this phenyl ring, in search of potent, selective delta opioid receptor agonists. It is important to note at this point that in the 1980s, when much of this work took place, it was often more efficient to test mixtures of stereoisomers to quickly look for potent delta opioid receptor agonists than to separate out the individual isomers (stereoselective routes were eventually developed to reduce the need for difficult separations). For this reason, selectivity data (especially

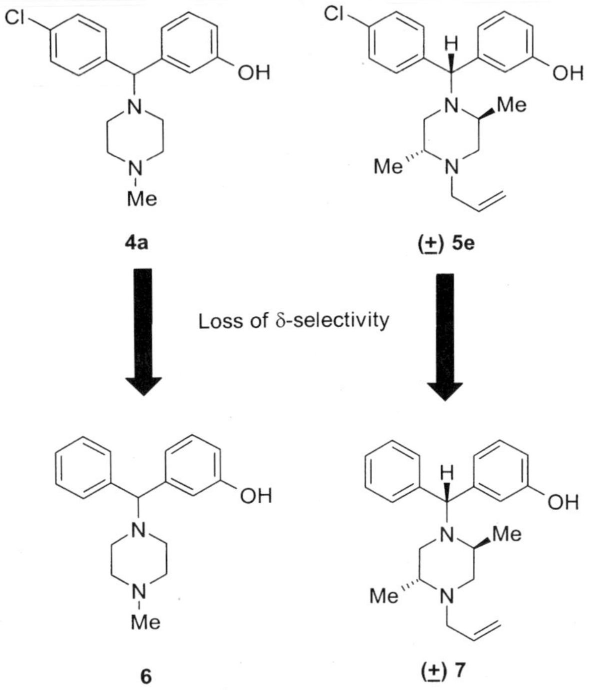

FIGURE 4 Effect of *para* substitution on opioid selectivity.

lack of selectivity) should not be overinterpreted, as individual stereoisomers could possess much greater delta opioid receptor selectivity than the mixture of stereoisomers.

A quick screen of functional groups at the *para* position indicated that several groups were tolerated by the delta receptor. Table 2 highlights a subset of compounds made to explore this region of the molecule, a set to compare various *para* substituents in a series while keeping the *N*-allyl-*trans*-dimethylpiperazine and the *meta*-phenol on the other ring of the benzhydryl group constant. Some functional agonism potency at the delta opioid receptor was retained by most small functional groups tested at the *para* position. Compounds with chloro- (**9**), cyano- (**11**), and methylsulfonyl (**15**) groups provided more potent delta opioid receptor functional agonism than the compound with only a proton at the *para* position (**8**). However, the two standout motifs were sulfonamides and carboxamides. A large number of sulfonamide and carboxamide analogues were synthesized and tested in an effort to identify potent, selective delta opioid receptor agonists, and a few of these are included in Table 2 to illustrate the basic SAR. Small alkyl

TABLE 2 *Para* Substitution SAR[a]

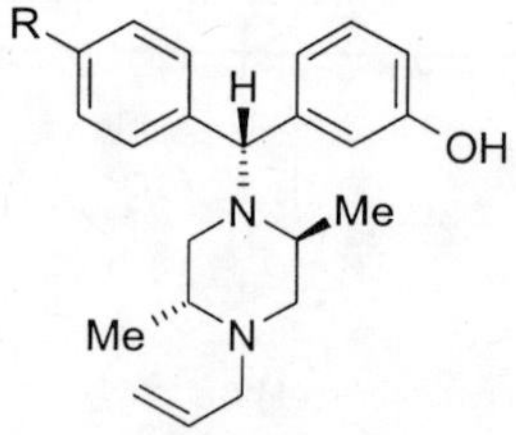

R	Compound	Delta agonism (MVD) ED_{50} (nM)	Mu agonism (GPI) ED_{50} (nM)	Delta binding (RBM) IC_{50} (nM)	Mu binding (RBM) IC_{50} (nM)
8	H	78	19	1.3	25
9	Cl	40	800	15	90
10	Br	100	—	80	200
11	CN	10	88	3.8	14
12	Me	230	—	10	40
13[b]	$-CH_2OH$	230	—	7	150
14	$-CO_2H$	95	37	14	8.4
15	$S(O)_2CH_3$	42	5600	1.3	40
16	$S(O)_2N(Me)_2$	37	1000	2.6	500
17	$S(O)_2N(Et)_2$	2.6	1800	20	100
18	$S(O)_2N(iPr)_2$	6.6	460	20	200
19	$S(O)_2N(nBu)_2$	450	—	80	200
20	$S(O)_2$pyrrolidine	5	640	1	40
21	$S(O)_2NPh(Me)$	10	170	2	20
22	$C(O)NH_2$	44	167	2	14
(±)-3	$C(O)N(Et)_2$	0.2	143	3	20
23	C(O)N(Me)(Et)	0.4	—	4	11
24	C(O)NH(Et)	20	118	3	9
25	C(O)N(Me)(cPr)	0.45	—	2	15
26	C(O)N(Me)(nPr)	1.2	—	3.7	17
27	C(O)pyrrolidine	1.7	—	2.3	130
28	C(O)N(Me)(Ph)	3.2	—	9.5	24

[a] Relative stereochemistry—compounds in this table are mixtures of both enantiomers, unless otherwise indicated. See Refs. 17, 18 for a description of the assays.
[b] Mixture of four compound—stereochemical mixture at the benzhydryl position. Each value is an average of at least three runs.
Source: Refs. 17, 18.

groups were tolerated on both, with potency dropping off as the straight alkyl groups lengthened beyond propyl. Two small alkyl groups gave more potent compounds than the corresponding monoalkyl compounds (compare (±)-**3** vs. **24**). While some loss in potency was observed, phenyl rings were also tolerated as one of the substituents. In both the carboxamides and sulfonamides, diethyl seemed to provide the best delta opioid receptor potency with good selectivity in the functional assays. The carboxamide (±)-**3**, (±) BW373U86, showed subnanomolar agonist potency in the MVD assay and excellent functional selectivity (MVD vs. GPI).

Heterocycles were explored as replacements for the nonphenolic ring (Table 3) [19]. Unsubstituted thiazoles, pyridines, and thiophenes gave active agonists, but with little or no selectivity. While an *N*-methylimidazole had little if any opioid agonist activity in this series, often additional substituents on the heterocycles provided potent compounds with delta opioid receptor selectivity. For instance, compound **34**, a thiophene with a diethyl carboxamide, displayed selective delta opioid receptor agonism (although no binding selectivity was observed).

TABLE 3 Heterocycle SAR[a]

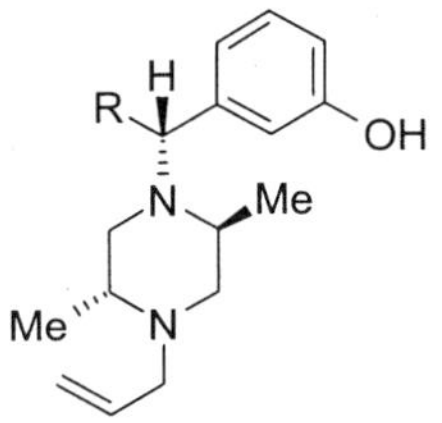

R	Compound	Delta agonism (MVD) ED_{50} (nM)	Mu agonism (MVD) ED_{50} (nM)	Delta binding (RBM) IC_{50} (nM)	Mu binding (RBM) IC_{50} (nM)
8	Phenyl	78	19	1.3	25
29	2-Thiazole	49	7.2	5	2.5
30	3-Pyridine	30	30	3.2	2.1
31	3-Thiophene	25	2.3	24	0.8
32	2-Thiophene	16	27	3.5	1.5
33	*N*-Me-2-imidazole	>1000	>1000	900	170
34	2-(4-Diethyl carbamoyl)thiophene	0.8	210	0.8	1.0

[a] Relative stereochemistry—compounds in this table are mixtures of both enantiomers. Each value is an average of at least three runs. See Refs. 17, 18 for a description of the assays.

While many early compounds were tested as mixtures of enantiomers, when these compounds were made stereoselectively or separated, one enantiomer was usually more potent. For instance, the (+)-enantiomer of BW373U86 (containing the *R* stereochemistry at the benzhydryl position, *S* stereochemistry at the 2 position of the piperazine, and *R* stereochemistry at the 5 position) was significantly more potent at both the delta and mu opioid receptors (Table 4).

Within the series of compounds containing the (2*S*,5*R*)-dimethylpiperazine, there is not always a preferred benzhydryl epimer. The epimer with *R* stereochemistry at this position was certainly preferred for BW373U86 with respect to delta opioid receptor agonism, but for other compounds there was little difference in the activity of the epimers. For example, pyridine containing compounds **35** and **36** have similar potencies at the delta and mu opioid receptors (Table 4). As demonstrated by the activity of related compounds (see Sec. 2.3), the stereochemical features of (+)-BW373U86 are not essential to the delta opioid receptor pharmacophore.

The discovery of BW373U86 from this early effort enabled a significant amount of pharmacological investigation into the effects of a selective delta opioid agonist. Studies on BW373U86, which is a potent, selective, delta

TABLE 4 Stereochemical Comparisons[a]

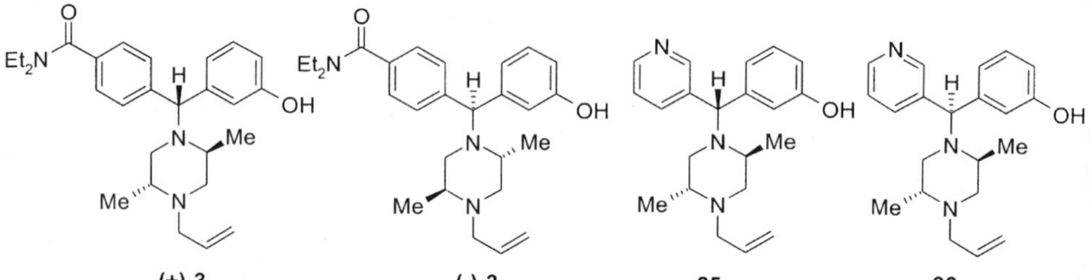

(+)-3 **(-)-3** **35** **36**

Compound		δ-Agonism (MVD) ED_{50} (nM)	μ-Agonism (GPI) ED_{50} (nM)	δ-Binding (RBM) IC_{50} (nM)	μ-Binding (RBM) IC_{50} (nM)
(±)-**3**	(±)-BW373U86	0.2	143	3	20
(+)-**3**	(+)-BW373U86	0.17	85	1.2	5.0
(−)-**3**	(−)-BW373U86	50	2900	24	540
35		30	12	0.6	0.85
36		30	4.3	6.0	1.3

[a] Each value is an average of at least three runs.

opioid receptor agonist, revealed that the in vivo analgesic efficacy and potency of this compound did not meet our expectations for treatment of severe pain, and proconvulsant activity precluded further development [20]. However, BW373U86 did not cause respiratory depression in laboratory animals and may in fact reverse or block typical mu opioid agonist induced respiratory depression [21]. Much of the research efforts in the benzhydrylpiperazine series at Burroughs Wellcome then shifted from the development of delta selective compounds to identifying nonpeptidic mixed delta/mu agonists, with the hypothesis of incorporating strong analgesic properties due to mu agonism, with a reduced side effect profile resulting from delta agonism [22]. This work is described briefly in Section 3.

After the disclosure of the structure and opioid activity of BW373U86, scientists outside of Burroughs Wellcome began exploring benzhydrylpiperazine opioids, and made significant contributions to our understanding of the delta opioid receptor SAR.

2.2 SNC80 and Other SAR Insights After BW 373U86

The first public disclosure of BW373U86 took place in 1992 at an INRC meeting [16]. This novel, nonpeptidic opioid agonist with delta opioid receptor selectivity was of interest to many researchers studying opioid receptor pharmacology. Soon after this disclosure, scientists at the National Institutes for Drug Addiction (NIDA) and the National Institutes of Health (NIH), led by Kenner Rice, began to make significant contributions to the benzhydrylpiperazine opioid field. In fact, Rice has been the leader in defining the SAR around BW373U86 and related compounds in the literature. In 1994, Rice's group published a paper describing the phenolic methyl ether of BW373U86 [23]. This ether, SNC80 (compound **37**, Fig. 5), had much greater affinity for

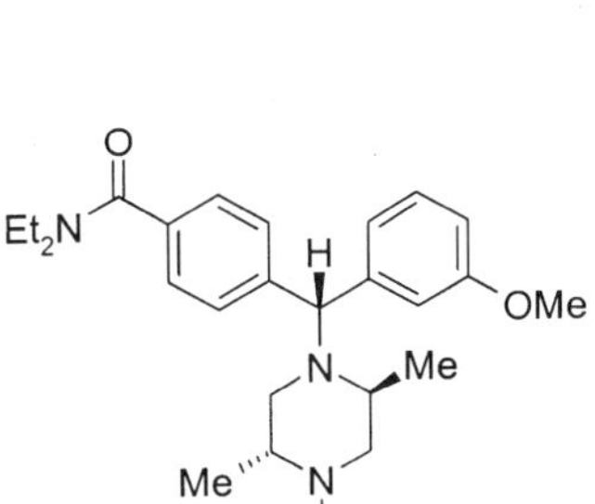

SNC80, **37**

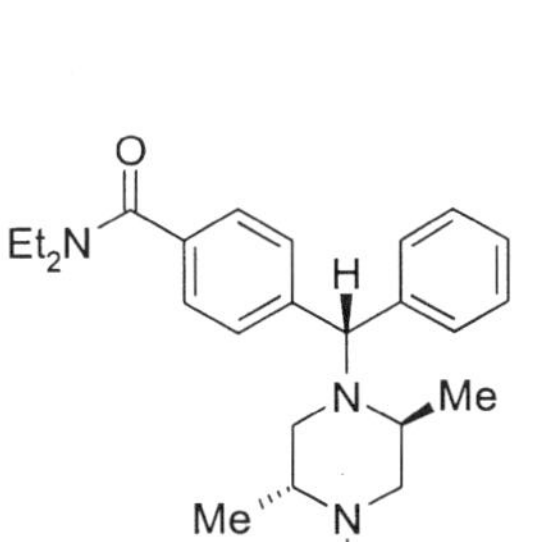

38

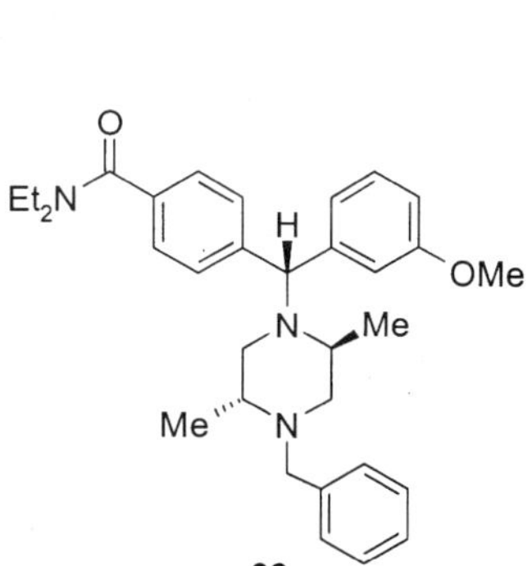

39

FIGURE 5 Structure of SNC 80 and related compounds.

the delta opioid receptor than the mu opioid receptor. While not as potent an agonist as BW373U86, SNC80 is much more selective with respect to receptor affinity (2327-fold vs. 31-fold) and also more selective with respect to agonist activity (1996-fold vs. 715-fold). Rice not only demonstrated that the phenol —OH is not critical to potent binding, but an H-bond acceptor is not needed either, as evidenced by a compound unsubstituted at this position (**38**, Fig. 5) which retained significant potency and delta opioid receptor selectivity (delta binding K_i = 0.5 nM, MVD IC_{50} = 10.5 nM, agonist mu/ delta ratio = 813) [24]. In fact, considerable structural variations on this phenyl ring are tolerated.

Rice's group also demonstrated an important contribution to the favored substituents on the piperazine nitrogen that is distal to the benzhydryl system. In addition to the small alkyls and highly favored allyl-type analogues, benzyl groups are well tolerated. An SNC80 analogue with an *N*-benzyl (compound **39**) proved to be a potent full agonist that was only fourfold weaker than SNC80 with respect to delta receptor binding [25].

Research groups at other pharmaceutical companies have also presented significant contributions to the SAR around BW373U86. An exciting example is the work of Pfizer scientists in replacing the preferred amide of BW373U86 with a tetrazole. Appending a carboxylic acid–containing chain off of the tetrazole (compound **40**, Fig. 6) provided a potent, delta selective compound tailored to a specific therapeutic application (delta receptor pIC50 = 9.8, >5000-fold selective) [26]. Compound **40** is highly bioavailable after oral administration, but does not penetrate into the brain, making it suitable for peripheral indications such as irritable bowel syndrome, and avoiding any centrally mediated undesirable effects.

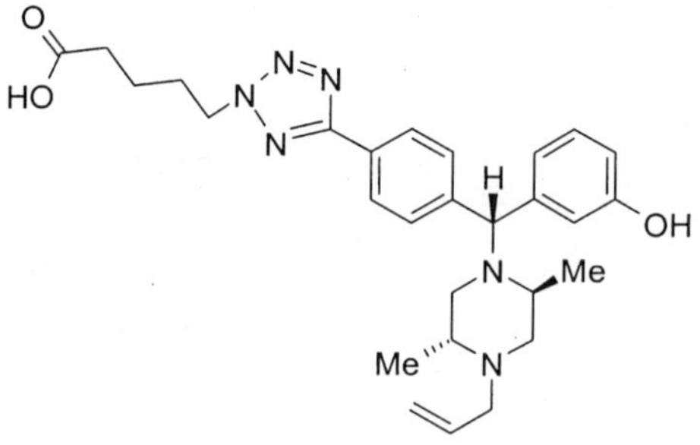

40

FIGURE 6 Tetrazole replacement of the carboxamide.

2.3 New Delta-Selective Ligands Related to Benzhydrylpiperazines

In addition to the disclosure of new benzhydrylpiperazine opioids, several research groups have published or filed patent applications on structurally similar delta opioid receptor agonists. These publications and patent applications have added to our understanding of the minimum pharmacophore. The most exciting revelations have dealt with the role of the benzhydryl stereocenter and the piperazine ring nitrogens.

Rice and coworkers reported that the piperazine nitrogen proximal to the benzhydryl group is not critical to opioid activity by replacing the piperazine ring with a piperidine ring (Table 5) [27]. For example, an *N*-allylpiperidine analogue (compound **41**) of SNC80 was sevenfold less potent than SNC80 with respect to delta-opioid receptor binding, and was equipotent to the direct analogue (SNC80 without the two methyl groups on the piperazine, compound **42**). These compounds lacked mu opioid activity. In this series, the allyl group on the nitrogen did not add to the binding potency, and the des-allyl compound (compound **43**) was at least as potent as its parent, compound **41**. A boost in potency was achieved by removing the stereocenter via a carbon-carbon double bond between the diarylmethyl and the piperidine ring (compound **44**). Wei and coworkers at AstraZeneca have also explored compounds in this series and reported a number of potent, selective delta opioid receptor agonists [28]. An impressive addition from a drug discovery standpoint is the simple piperidine compound **45** (Fig. 7),

TABLE 5 Piperidine SAR

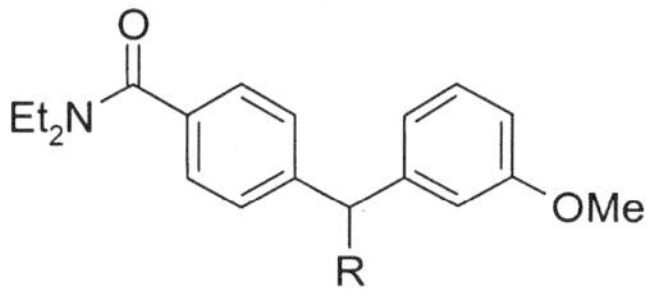

R	Compound	Delta binding (RBM) K_i (nM±SEM)	Mu binding (RBM) K_i (nM±SEM)
37	(SNC80)	4±0.18	3970±170
41	N-allyl-4-piperidinyl	27±2	>6300
42	N-allyl-1-piperazinyl	27±2	>6300
43	4-Piperidinyl	20±2	>6300
44	4-Piperidinylidene	5±0.3	>6300

Source: Ref. 27.

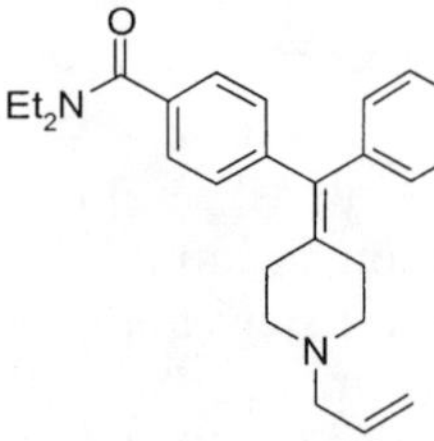

44

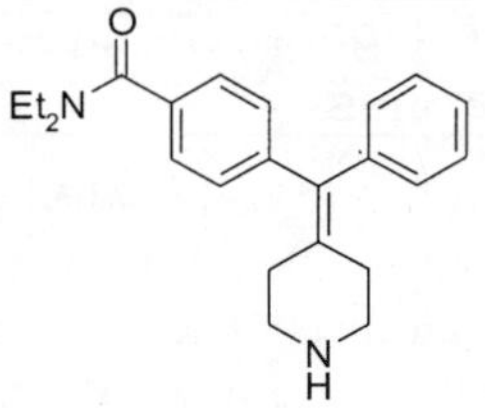

45

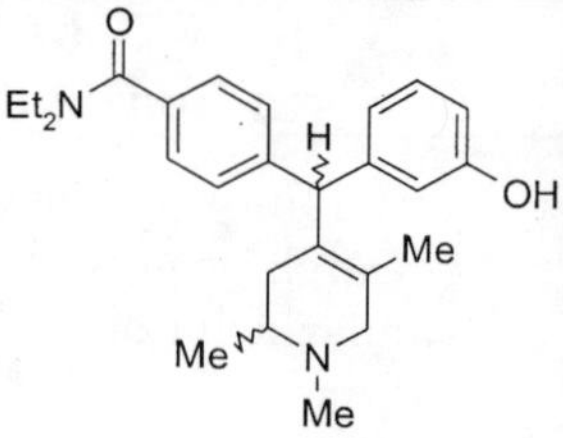

46

FIGURE 7 Examples of piperazine replacements.

which is as potent and selective as SNC80, but which is considerably more metabolically stable and displayed excellent oral bioavailability in a rat model ($F = 90$–100%). Related compounds with a tetrahydropyridine ring were discovered to be delta opioid receptor agonists by Burroughs Wellcome scientists (e.g., compound **46**, with a MVD $ED_{50} = 52$ nM and a delta receptor binding $IC_{50} = 2.5$ nM) [29]. These tetrahydropyridine compounds demonstrated that the proximal nitrogen was not crucial, but did not eliminate the benzhydryl stereocenter.

Another successful approach to replacing the benzhydryl stereocenter has been to synthesize and evaluate diarylamines with pendant basic amines. A simple transposition of the nitrogen and carbon atoms of the benzhydryl-piperazine leads to diarylaminopiperidine delta opioid agonists, as Carroll and coworkers at the Research Triangle Institute (RTI) demonstrated [30]. Scientists at AstraZeneca [31] and the R.W. Johnson Research Institute [32] have also disclosed delta opioid agonists possessing the carbon-nitrogen transposition. Boyd and coworkers at R.W. Johnson were the first to report on compounds where the aminopiperidine was further constrained as an aminotropane. Compound **47** has a $K_i = 0.4$ nM at the delta opioid receptor, with 14000-fold selectivity versus the mu opioid receptor (Fig. 8).

Patent applications have been filed covering compounds without the distal nitrogen constrained in a ring directly attached to the benzhydryl carbon. SmithKline Beecham scientists filed a patent application on diaryldiamines typified by compound **48** [33], and Meiji Seika Kaisha researchers filed a patent application covering diphenylalkylpiperidines as mu opioid agonists—for instance, compound **49** [34].

All of these additional reports have helped define the minimum pharmacophore within the benzhydrylpiperazine-related structural class. Compound **45**, described above, clearly shows that the *N*-allyl group, the phenolic —OH, and the nitrogen proximal to the benzhydryl system are not needed for delta opioid receptor binding, agonism, or selectivity, and that the benzhydryl

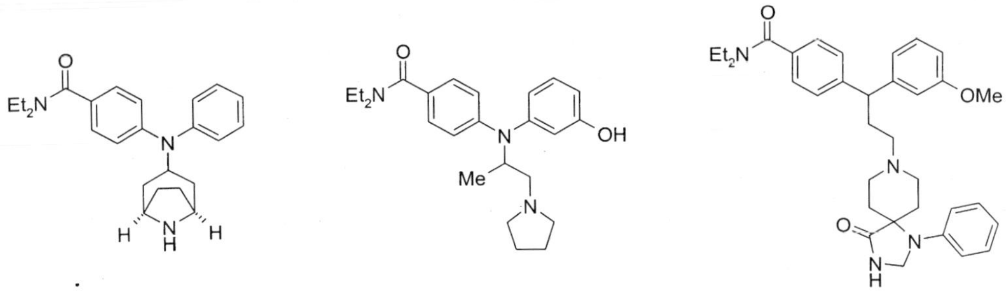

47 48 49

FIGURE 8 Disclosed opioids related to benzhydrylpiperazines.

stereocenter can be eliminated. The *para*-diethylamide or a related substituent (sulfonamide, tetrazole) at that position on the phenyl ring appears to be important for delta opioid receptor selectivity and strong potency. Several of the newer structural classes reported also appear to have better pharmacokinetic properties compared to BW373U86 and SNC80, and may improve the chances for successful development of an orally administered delta opioid receptor agonist as a therapeutic agent.

2.4 Benzhydrylpiperazine Opioids and the Message-Address Concept

In 1977, Schwyzer described a "message-address" concept for the binding selectivity of seven-transmembrane receptor ligands [35a,b], and Portoghese refined the concept for opioid receptors [36]. This theory generally states that there is a region of the receptor that recognizes the opioid pharmacophore (the "message"), and another region that is unique for the delta receptor type and confers selectivity (the "address"). If one wishes to extend the message-address concept that Portoghese developed for delta opioids to the benzhydrylpiperazine series, the phenol/piperazine region of BW373U86 could correspond to the tyrosine residue found in enkephalins, the putative opioid "message" domain essential for receptor recognition. The diethylamide is appropriately positioned as the delta opioid "address" region, the part of the molecule that confers opioid receptor-type selectivity. Dondio and coworkers have previously stated this hypothesis, and have published an elegant application of the diethylamide as a nonaromatic delta opioid address (Fig. 9) [37].

However, a recent paper by Coop and Jacobsen uses results from a four-point recognition model of the delta opioid receptor to question whether the oxymorphindoles and benzhydrylpiperazines could be binding in similar

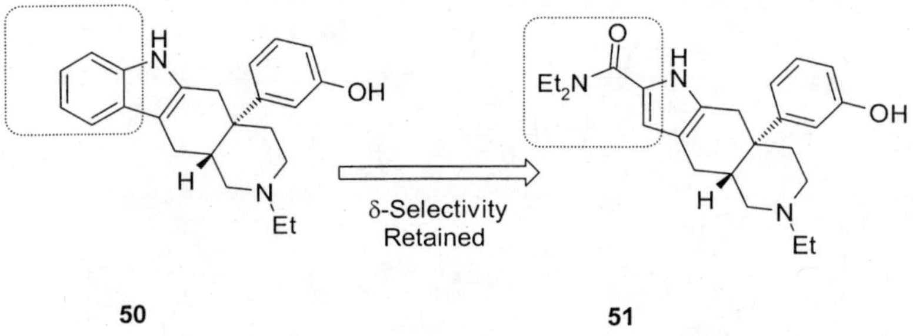

FIGURE 9 Nonaromatic delta opioid "address" model (Dondio et al. [37])

orientations [38]. They concluded that the amide and indole regions do not interact with the same regions of the receptor, as suggested by Dondio's results. Even if these classes do bind differently, the Burroughs Wellcome in vitro results in the benzhydrylpiperazine series certainly support Dondio's hypotheses that the diethylamide in the *para* position is a delta opioid "address" region, as the location of the amide on BW373U86 and related structures has a profound influence on delta/mu opioid receptor selectivity (Fig. 10). Movement of the diethylamide from the *para* position of BW373U86 to the *meta* position of the phenyl ring (**53**) produced a 30-fold increase in mu activity and a 30-fold decrease in delta activity. Also, complete removal of the amide resulted in a nonselective (and much weaker) opioid agonist (**52**). Changing the *meta*-diethylamide to a *meta*-*N*-methylanilide resulted in both a 12-fold increase in delta agonism and a threefold increase in mu agonist activity [39]. This result led to the identification of a series of potent, mixed mu/delta opioid agonists.

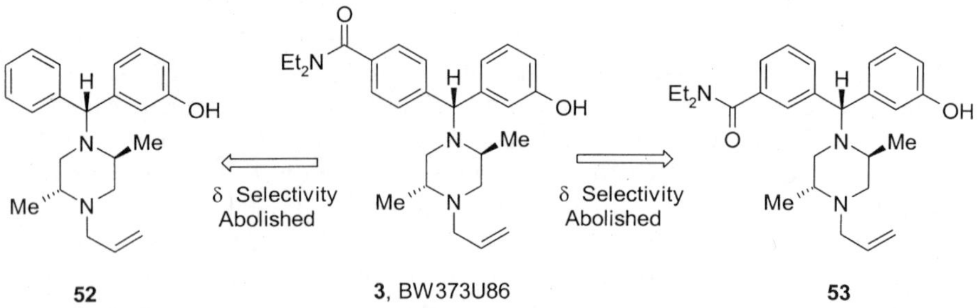

FIGURE 10 Delta opioid selectivity and the carboxamide.

3 MIXED MU/DELTA OPIOID AGONISTS IN THE BENZHYDRYLPIPERAZINE SERIES

The discovery of delta-selective agonists such as BW373U86 has enabled significant pharmacological evaluation of the physiology related to the delta opioid receptor. The evaluation of BW373U86, which is a potent, selective, delta opioid receptor agonist in the benzhydrylpiperazine series, led to the conclusion that the analgesic efficacy and potency of this compound did not meet expectations for the treatment of severe pain, and proconvulsant activity precluded further development [20]. However, BW373U86 did not cause respiratory depression in laboratory animals and may in fact reverse or block typical mu opioid agonist induced respiratory depression [21]. This exciting result added to the literature evidence that delta opioid receptor agonists modulate and/or potentiate mu opioid effects [22].

To exploit the exciting intermodulatory effects of delta and mu receptor ligands, research efforts in the benzhydrylpiperazine series at Burroughs Wellcome shifted from the development of delta-selective compounds to identifying mixed delta/mu agonists that could show strong analgesic properties due to mu agonism, with a reduced side effect profile resulting from delta agonism. Mixed delta/mu agonists were discovered in the benzhydrylpiperazine series by moving the *para*-diethylamide portion of the molecule to the *meta* position of the phenyl where it no longer conferred delta selectivity. Changing the *meta*-diethylamide to a *meta*-*N*-methylanilide resulted in a potency increase at both the delta and mu receptors, and a series of 3-((αR)-α-((2S,5R)-4-allyl-2,5-dimethyl-1-piperazinyl)-3-hydroxybenzyl)-*N*-alkyl-*N*-arylbenzamides were synthesized to explore the mu and delta opioid receptor agonism SAR [40]. Many small functional groups are tolerated on the aryl ring, and *N*-methyl through *N*-propyl substituents can be placed on the amide nitrogen and retain activity. Larger groups on the nitrogen lead to delta

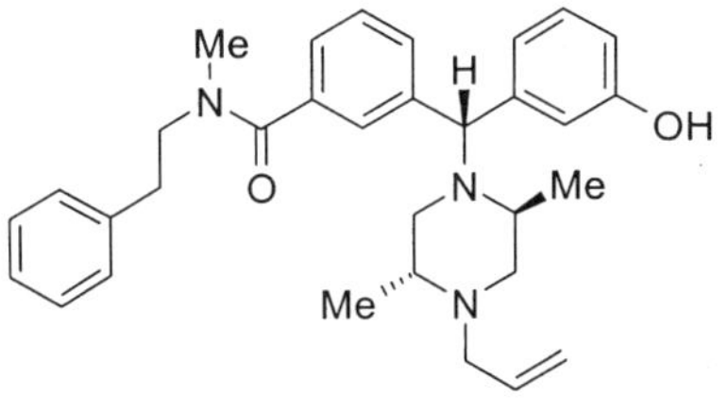

54

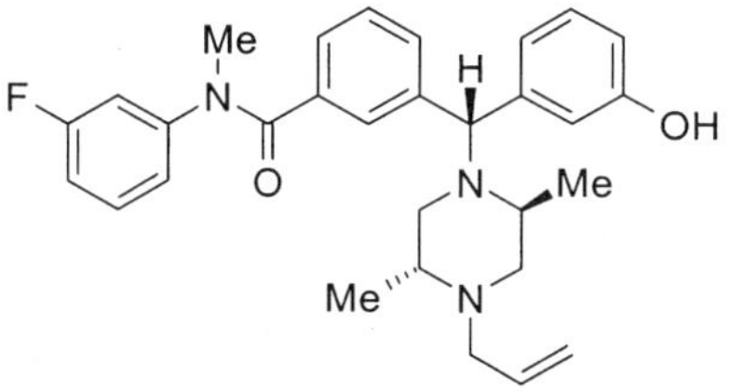

55

FIGURE 11 Mixed delta/mu opioid agonists.

selectivity; these compounds may be putting a lipophilic group back into part of the delta address region that was occupied by the *para*-diethylamide.

The identification of several compounds that have *in vitro* potencies in the 1–5 nM range at both the mu and delta receptors, such as compounds **54** and **55** (Fig. 11), has provided useful tools to test the pharmacological effects of mixed delta/mu opioid agonists in human clinical trials.

4 SYNTHETIC ROUTES TO BENZHYDRYLPIPERAZINES

Several synthetic routes to benzhydrylpiperazine opioids have been published or presented, including stereoselective routes. The most common route to building the benzhydrylpiperazine core involves displacement of a halogen from the benzhydryl central carbon by a piperazine (Scheme 1).

Displacement of a chloride, conveniently derived from the benzhydrol, has often been utilized. However, running the reaction with sodium iodide in acetonitrile improved yields of the displacement reaction, presumably via a Finklestein reaction (Scheme 2).

There are a number of synthetic routes available to access the benzhydrols needed for this approach. The first published route to SNC80 used a Friedel-Crafts acylation of toluene with *m*-anisoyl chloride to yield a benzophenone, followed by sodium borohydride reduction of the ketone (after the three-step elaboration of the *para*-methyl group into the diethylcarboxamide, Scheme 3) [23,24].

An approach that has allowed for considerable structural variety and was used most frequently at Burroughs Wellcome to construct the appropriate benzhydrols is the addition of an aryllithium or an aryl Grignard reagent to an appropriate benzaldehyde. This can be an extremely effective approach, as demonstrated by the construction of the complete benzhydrol needed for SNC80 in 91% yield (Scheme 4) [27,29].

Many of the most potent benzhydrylpiperazine opioids that have been reported contain an *N*-allyl-*trans*-2,5-dimethylpiperazine. The racemic *N*-

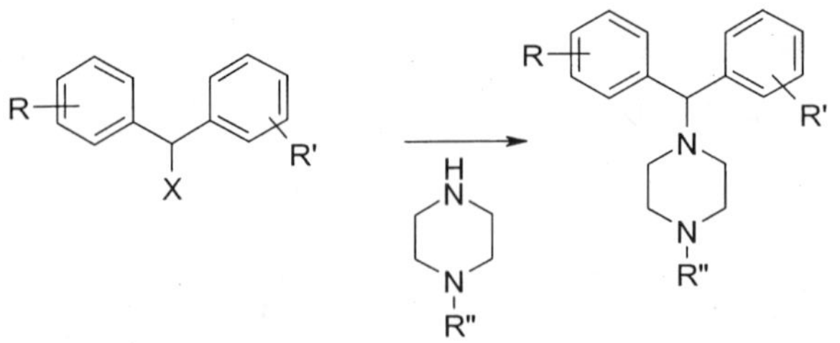

SCHEME 1 General displacement route to benzhydrylpiperazines.

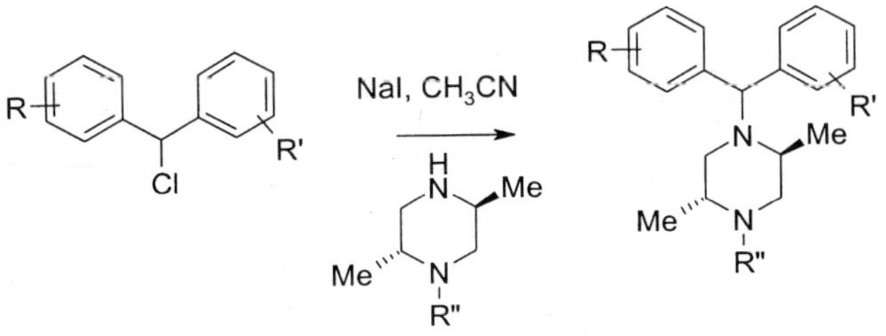

SCHEME 2 Improved displacement route.

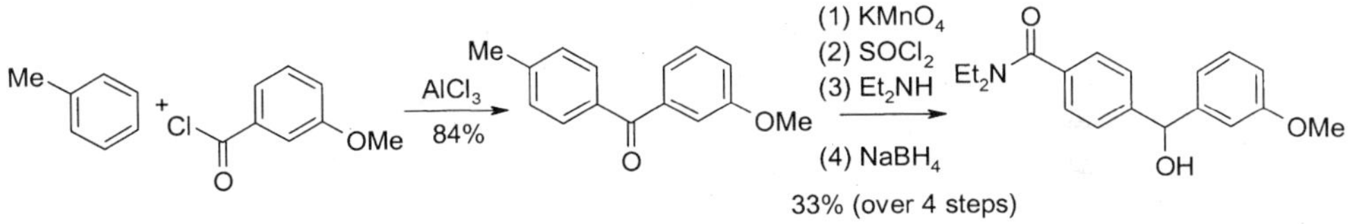

SCHEME 3 Friedel-Crafts route to benzhydrol intermediates.

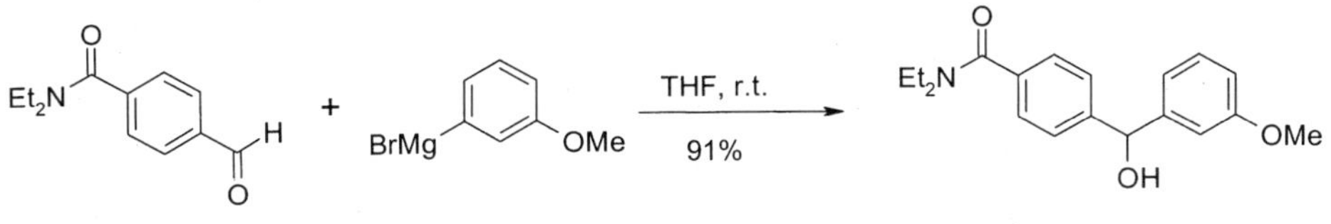

SCHEME 4 Grignard addition route to benzhydrol intermediates.

allyl-*trans*-2,5-dimethylpiperazine can be prepared on large scale using the three-step approach illustrated in Scheme 5 [19]. However, the preferred stereoisomer of the final product usually contains the 1-benzhydryl-(2*S*,5*R*)-2,5-dimethylpiperazine. While classical resolution of the final products has been used to separate stereoisomers [19], it is usually more efficient to synthesize benzhydrylpiperazines using the enantiopure (−)-(2*R*,5*S*)-1-allyl-2,5-dimethylpiperazine.

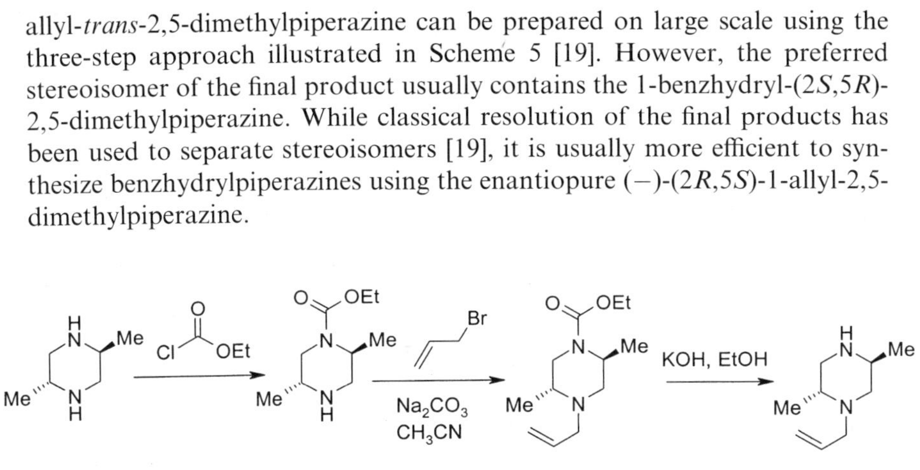

SCHEME 5 Three-step route to N-allyl-*trans*-2,5-dimethylpiperazine.

(−)-(2*R*,5*S*)-1-Allyl-2,5-dimethylpiperazine has been prepared by direct enantiospecific synthesis [29,39] and via classical resolution of the racemic piperazine [23,29]. Kilo-scale batches of (−)-(2*R*,5*S*)-1-allyl-2,5-dimethylpiperazine have been prepared from *trans*-2,5-dimethylpiperazine by the three-step monoallylation shown in Scheme 5, followed by a resolution using di-*p*-toluoyl-D-tartaric acid. This resolution has also been achieved in a two-stage process using (−)-camphoric acid followed by di-*p*-toluoyl-D-tartaric acid, giving (−)-(2*R*,5*S*)-1-allyl-2,5-dimethylpiperazine in >99% optical purity.

Enantiospecific syntheses have utilized the chirality available in D-alanine and L-alanine. For instance, coupling and cyclization (after the necessary deprotection) of N-allyl-N-BOC-D-alanine with L-alanine methyl ester, followed by lithium aluminum hydride reduction of the diketopiperazine provided (−)-(2R,5S)-1-allyl-2,5-dimethylpiperazine (Scheme 6) [27,39]. Racemization was not observed during the synthesis.

Carrying out the benzhydrylpiperazine formation via nucleophilic displacement as depicted in Scheme 2 using the enantiopure (−)-(2R,5S)-1-allyl-2,5-dimethylpiperazine provides two stereoisomers, epimeric at the benzhydryl position. These epimers are separable by common chromatographic techniques (although these are sometimes difficult separations), and the absolute stereochemistry of the products has been proven via x-ray crystallography [19,23,40,41]. For large-scale preparation, a stereoselective approach to the final products is preferred to minimize waste and reduce costs. Two distinct stereoselective routes to benzhydrylpiperazines have been described.

One approach utilizes Katritzky's route to tertiary amines to construct the benzhydrylpiperazine (Scheme 7) [42]. This method involves addition of an aryl Grignard reagent to a "masked iminium," an adduct formed from the piperazine, an appropriate benzaldehyde, and benzotriazole. In solution, this adduct is in equilibrium with an iminium ion formed by elimination of the benzotriazole, and it is likely that the Grignard reagent adds to this species. Assuming that the iminium ion is the reactive entity, the stereoselectivity appears to rely on two separate events: 1) preferential formation of one of the

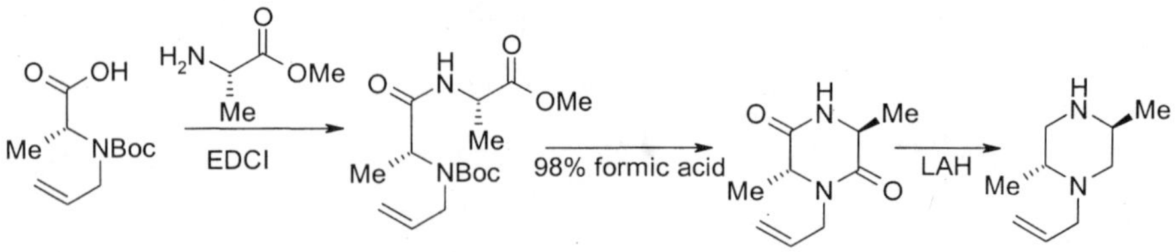

SCHEME 6 A stereoselective synthesis of N-allyl-*trans*-2,5-dimethylpiperazine.

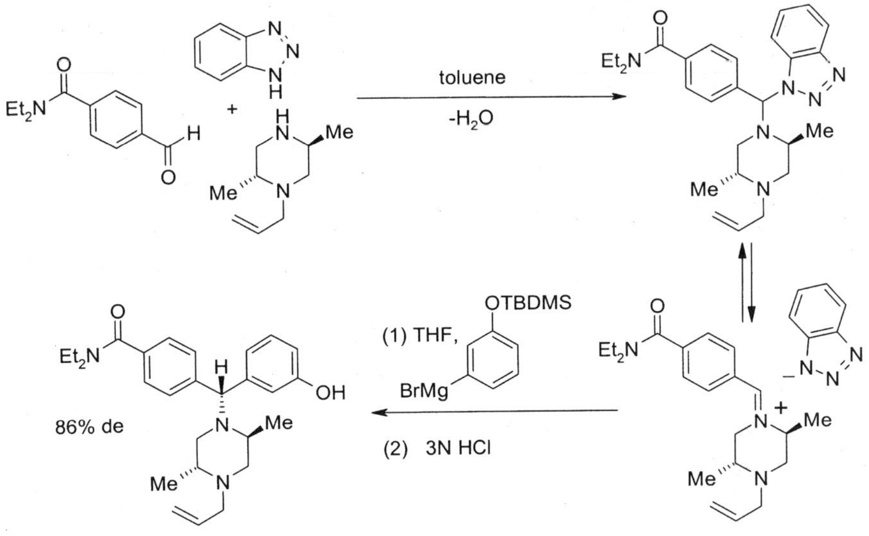

SCHEME 7 A "masked iminium" approach to benzhydrylpiperazines.

two possible iminium ions (**56a** vs. **56b**) (Fig. 12); and 2) facial selectivity of the nucleophilic attack on the iminium ion. It is reasonable to assume that a greater proportion of adducts will dissociate to give the sterically less demanding iminium ion, **56a**, with the methyl-bearing carbon of the piperazine *trans* to the aromatic ring. Next, the nucleophile must approach **56a** from the *Si* face, away from the methyl group, to give the observed diastereoselectivity. The steric effect of the methyl group on the piperazine ring is apparently responsible for this substantial stereoselectivity. Products

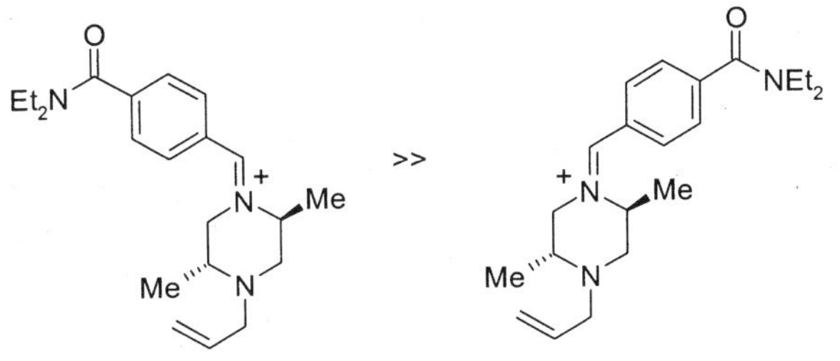

56a **56b**

FIGURE 12 Structures of proposed iminium ion intermediates.

are typically obtained in a 9:1 diastereomer ratio that can be purified to >99% isomeric purity by recrystallization or chromatography. Enantiopurity is dictated by the purity of the precursor piperazine. One limitation of this method is that piperazines without a substituent in the directing position will not provide stereocontrol of the benzhydryl position.

Delorme and coworkers have published a stereoselective route that is effective with a wide range of amines, including those without a stereocenter on the amine (Scheme 8) [43]. Chiral reduction of the appropriate benzophenone (as a chromium tricarbonyl complex) using Corey's oxazaborolidine approach afforded the benzhydrol with 91% ee. Treatment with tetrafluoroboric acid followed by the piperazine gave the desired benzhydryl piperazine without any erosion of stereochemical purity after decomplexation. In addition to simplifying analogue synthesis, these two complementary routes provide a useful base for the future development of stereoselective manufacturing routes.

For the rapid production of analogues, parallel synthesis techniques are often desirable. Morphy and coworkers have published two solid-supported syntheses of benzhydrylpiperazines and related compounds [44,45]. Both of these approaches utilize a displacement of the benzhydryl bromide by an amine (piperazine, e.g.) to build the molecules. However, these two routes are complementary, "traceless linker" approaches. In one, the piperazine is attached to the solid support while the molecule is built, allowing significant variation on other parts of the molecule. Cleavage is effected via quaterniza-

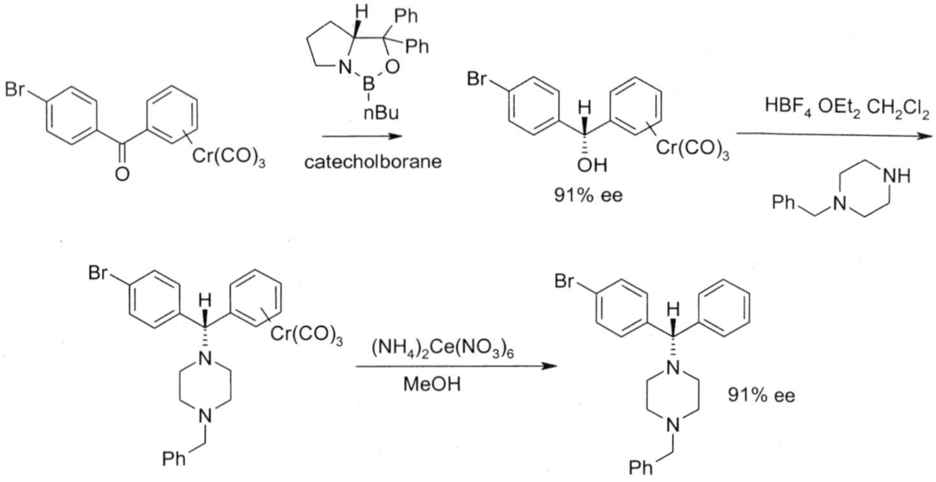

SCHEME 8 A stereoselective reduction approach to benzhydrylpiperazines.

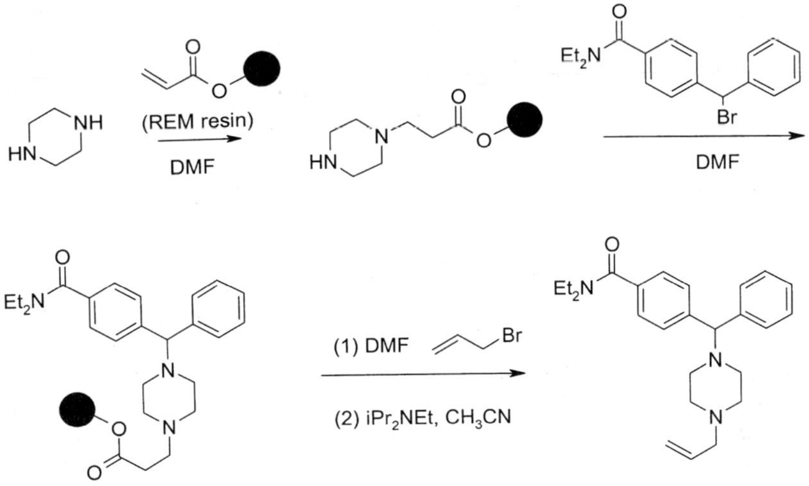

SCHEME 9 A solid-supported synthesis via piperazine attachment.

tion of the piperazine linker nitrogen, followed by Hofmann elimination (Scheme 9).

In the second route, a carboxylic acid at the *para* position of a benzhydryl bromide is attached to the resin as an ester (Scheme 10). After constructing various molecules with different amines attached at the benzhydryl position, a facile aminolysis of the linker ester with $AlCl_3$ and diethylamine provided products with a *para*-diethylamide at the point of previous resin attachment.

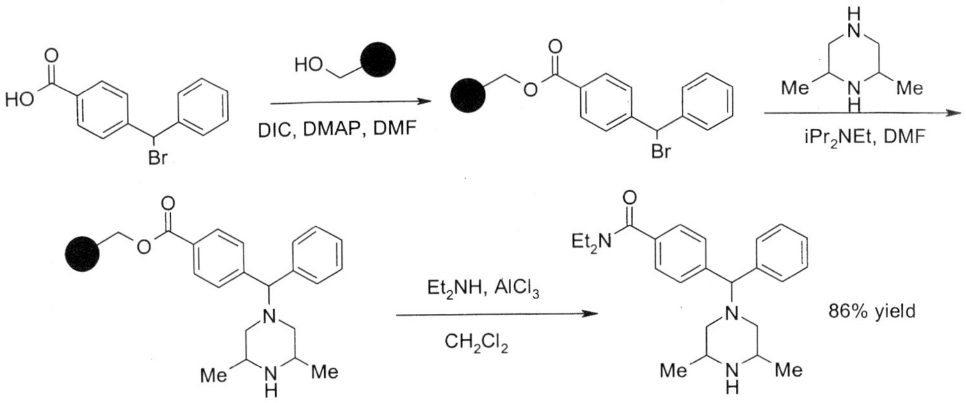

SCHEME 10 A solid-supported synthesis via carboxylic acid attachment.

The diverse synthetic routes developed to make benzhydrylpiperazines have provided complementary tools to aid in the construction of a variety of analogues. The current routes have also provided high quality tools for the development of efficient manufacturing routes to benzhydrylpiperazines.

5 CONCLUSIONS

Benzhydrylpiperazines represent an exciting and useful class of opioid agonists, both delta receptor selective opioid agonists and mixed delta/mu opioid receptor agonists. These ligands have been valuable tools for exploring opioid receptor pharmacology; BW373U86 or its methyl ether, SNC80, has been used in studies published in over 50 journal articles. The benzhydrylpiperazines have also provided the inspiration for other exciting delta opioid agonist structures, such as the piperidinylidenemethyl and diarylamines typified by compounds (**45**) and (**47**), respectively.

The clinical evaluation of benzhydrylpiperazines is underway at Ardent Pharmaceuticals, with an injectable mixed delta/mu opioid receptor agonist in trials for the treatment of severe pain. With the development of delta receptor selective opioid agonists with optimized pharmacokinetic properties, other benzhydrylpiperazines (or related agents) are expected to undergo clinical evaluation for the treatment of a variety of indications. Synthetic routes are available to construct these compounds on a reasonable scale and to allow the further development of manufacturing processes.

REFERENCES

1a. Parrot T. J Am Board Fam Pract 1999; 12:293–306.
1b. Jadad AR, Browman GP. JAMA 1995; 274:1870–1873.
2. Casy AF, Parfitt RL. Opioid Analgesics and Receptors. New York: Plenum Press, 1986:Chaps. 6–8.
3. Duthie DJR, Nimmo WS. Br J Anaesth 1987; 59:61–77.
4. Portoghese PS. J Med Chem 1965; 8:609–616.
5a. Martin WR. Pharmacol Rev 1967; 19:463–521.
5b. Martin WR, Eades CG, Thompson JA, Huppler RE, Gilbert PE. J Pharmacol Exp Ther 1976; 197:517–532.
6. Simon EJ, Gioannini TL. In: Herz A, ed. Opioids I. Berlin: Springer-Verlag, 1993:3–21.
7. Hughes J, Smith TN, Kosterlitz HW, Fothergill LA, Morgen BA, Morris HR. Nature 1976; 258:577–579.
8. Vaught JL, Takemori AE. J Pharmacol Exp Ther 1979; 208:86–90.
9a. Dondio G, Ronzoni S, Petrillo P. Expert Opin Ther Pat 1999; 9:353–374.
9b. Dondio G, Ronzoni S, Petrillo P. Expert Opin Ther Pat 1997; 7:1075–1098.

9c. Maw GN, Middleton DS. Annu Rep Med Chem 2002; 37:159–166.
10. Zimmerman DM, Leander JD. J Med Chem 1990; 33:895–902.
11. Portoghese PS, Sultana M, Takemori AE. Eur J Pharmacol 1988; 146:185–186.
12a. Dappen MS, Pitzele BS, Rafferty MF. U.S. Patent 5223507. Chem Abstr 1993; 119:152105.
12b. Schmidhammer H. PCT application: WO9531464. Chem Abstr 1995; 124: 176606.
13a. Dondio G, Clarke GD, Giardina G, Petrillo P, Petrone G, Ronzoni S, Visentin L, Vecchietti V. Analgesia 1995; 1:394–399.
13b. Nagase H, Wakita H, Kawai K, Endo T, Matsumoto O. JP 04275288 A2. Chem Abstr 1992; 119:49376.
14. Tsushima M, Tadauchi K, Asai K, Miike N, Imai M, Kudo T. PCT Int. Appl. WO 0160796 A1. Chem Abstr 2001; 135:195564.
15. Chang K-J, Rigdon GC, Howard JL, McNutt RW. J Pharmacol Exp Ther 1993; 267:852–857.
16. Lee PHK, McNutt RW, Chang K-J. A non-peptide delta-opioid receptor agonist BW 373U86 suppresses naloxone-precipitated morphine abstinence. Meeting of the International Narcotics Research Conference, Keystone, CO, June 23–27, 1992, Abstract 34.
17a. Chang K-J, Cuatrecasas P. J Biol Chem 1979; 254:2610–2618.
17b. Chang K-J, Wei ET, Killian A, Chang JK. J Pharmacol Exp Ther 1983; 227:403–408.
18. Kramer TH, Davis P, Hruby VJ, Burks TF, Porreca F. J Pharmacol Exp Ther 1993; 266:577–584.
19. Boswell GE, McNutt RW, Bubacz DG, Davis AO, Chang K-J. J Heterocyclic Chem 1995; 32:1801–1818.
20. Comer SD, Hoenicke EM, Sable AI, McNutt RW, Chang K-J, deCosta BR, Mosberg HI, Woods JH. J Pharmacol Exp Ther 1993; 267:888–895.
21. O'Neill SJ, Collins MA, Pettit HO, McNutt RW, Chang K-J. J Pharmacol Exp Ther 1997; 282:271–277.
22. Su YF, McNutt RW, Chang KJ. J Pharmacol Exp Ther 1998; 287:815–823.
23. Calderon SN, Rothman RB, Porreca F, Flippen-Anderson JL, McNutt RW, Xu H, Smith LE, Bilsky EJ, Davis P, Rice KC. J Med Chem 1994; 37:2125–2128.
24. Calderon SN, Rice KC, Rothman RB, Porreca F, Flippen-Anderson JL, Kayakiri H, Xu H, Becketts K, Smith LE, Bilsky EJ, Davis P, Horvath R. J Med Chem 1997; 40:695–704.
25. Furness FM, Zhang X, Coop A, Jacobson AE, Rothman RB, Dersch CM, Xu H, Porreca F, Rice KC. J Med Chem 2000; 43:3139–3196.
26. Maw GN. Discovery of non-peptide δ-opioid receptor agonists. Book of Abstracts, 221st ACS National Meeting, San Diego, 2001:MEDI-168.
27. Zhang X, Rice KC, Calderon SN, Kayakiri H, Smith L, Coop A, Jacobson AE, Rothman RB, Davis P, Dersch CM, Porreca F. J Med Chem 1999; 42: 5455–5463.

28. Wei ZY, Brown W, Takasaki B, Plobeck N, Delorme D, Zhou F, Yang H, Jones P, Gawell L, Gagnon H, Schmidt R, Yue SY, Walpole C, Payza K, St-Onge S, Labarre M, Godbout C, Jakob A, Butterworth J, Kamassah A, Morin PE, Projean D, Ducharme J, Roberts E. J Med Chem 2000; 43:3895–3905.
29. Chang K-J, Boswell GE, Bubacz DG, Collins MA, Davis AO, McNutt RW. PCT Int. Appl. WO9315062. Chem Abstr 1993; 121:83367.
30. Thomas JB, Herault XM, Rothman RB, Burgess JP, Mascarella SW, Xu H, Horel RB, Dersch CM, Carroll FI. Bioorg Med Chem Lett 1999; 9:3053–3056.
31. Pelcman B, Roberts E. PCT Int. Appl. WO9828270. Chem Abs 1998; 129: 108996.
32. Boyd RE, Carson JR, Codd EE, Gauthier AD, Neilson LA, Zhang S-P. Bioorg Med Chem Lett 2000; 10:1109–1111.
33. Dondio G, Ronzoni S. PCT Int. Appl. WO9636620. Chem Abs 1996; 126: 74601.
34. Tsushima M, Tadauchi K, Asai K, Miike N, Kudo T. PCT Int. Appl. WO0170689. Chem Abs 2001; 135:272955.
35a. Schwyzer R. NY Acad Sci 1977; 297:3–26.
35b. Schwyzer R. Biochemistry 1986; 25:6335–6342.
36. Portoghese PS, Sultana M, Nagase H, Takemori AE. J Med Chem 1988; 31:281–282.
37. Dondio G, Ronzoni S, Eggleston DS, Artico M, Petrillo P, Petrone G, Visentin L, Farina C, Vecchietti V, Clarke GD. J Med Chem 1997; 40:3192–3198.
38. Coop A, Jacobsen AE. Bioorg Med Chem Lett 1999; 9:357–362.
39. Bubacz DG, Davis AO, Dickerson SH, Harris PA, McNutt RW, Collins MA, Chang K-J. Synthesis of novel mixed delta-mu opioid agonists. Book of Abstracts, 213th ACS National Meeting, San Francisco 1997:MEDI-050.
40. Bishop MJ, Garrido DM, Boswell GE, Collins MA, Harris PA, McNutt RW, O'Neill SJ, Wei K, Chang KJ. J Med Chem 2003; 46:623–633.
41. Katsura Y, Zhang X, Homma K, Rice KC, Calderon SN, Rothman RB, Yamamura HI, Davis P, Flippen-Anderson JL, Xu H, Becketts K, Foltz EJ, Porreca F. J Med Chem 1997; 40:2936–2947.
42. Bishop MJ, McNutt RW. Bioorg Med Chem Lett 1995; 5:1311–1314.
43. Delorme D, Berthelette C, Lavoie R, Roberts E. Tetrahedron Asym 1998; 9:3963–3966.
44. Cottney J, Rankovic Z, Morphy JR. Bioorg Med Chem Lett 1999; 9:1323–1328.
45. Barn DR, Cottney J, Caulfield WL, Morphy JR. Bioorg Med Chem Lett 1999; 9:1329–1334.

9

Delta-Selective Ligands Related to Naltrindole

D.J. Daniels and P.S. Portoghese
University of Minnesota, Minneapolis, Minnesota, U.S.A.

1 INTRODUCTION

Opioid antagonists have been crucial as pharmacological tools in opioid research [1]. Historically, the ability of naloxone or naltrexone to reversibly antagonize an opioid agonist effect in an apparently competitive fashion was an important criterion for establishing the involvement of an opioid receptor–mediated effect. Naloxone and naltrexone are useful in this regard because they are universal antagonists; that is, they are able to antagonize the agonist effects mediated through multiple opioid receptors. However, because these ligands possess low selectivity, they are not useful for investigating the pharmacology mediated through specific opioid receptor types. Consequently, an armamentarium of highly selective opioid antagonists is now available for this purpose. Such antagonists have been invaluable for determining the selectivity of opioid ligands and opioid receptor mechanisms. In this chapter we focus on the rationale for the design of nonpeptide, delta-selective opioid antagonists related to naltrindole and their utility as pharmacological tools.

2 DESIGN RATIONALE FOR DELTA OPIOID ANTAGONISTS

The rationale for the design of the first nonpeptide, delta-selective opioid antagonists was based on the message-address concept put forth by Schwyzer [2] to analyze the structure-activity relationship of ACTH-related peptide hormones. Accordingly, "sychnologically" organized peptide hormones contain a "message" sequence and "address" sequence of amino acids residues contiguous with one another in the peptide chain. The message component is the feature common to the family of peptides and is proposed to be involved with signal transduction, while the address component provides additional affinity by binding to a unique subsite that is not necessary for the transduction process. For a group of peptides associated with a family of closely related receptors, the message component is similar or identical, while the address component is variable and determines the selectivity for a particular type of receptor. This concept is illustrated for a family of receptor types where each receptor has two major subsites (Fig. 1): the message subsite, which is largely conserved for all of the receptor types, and a nonconserved address subsite that is responsible for selectivity.

Chavkin and Goldstein [3] pointed out that endogenous opioid peptides conform to a message-address motif, and it was suggested that the invariant tetrapeptide sequence Tyr-Gly-Gly-Phe can be viewed as the message while subsequent amino acid residues constitute the address. A modification of this

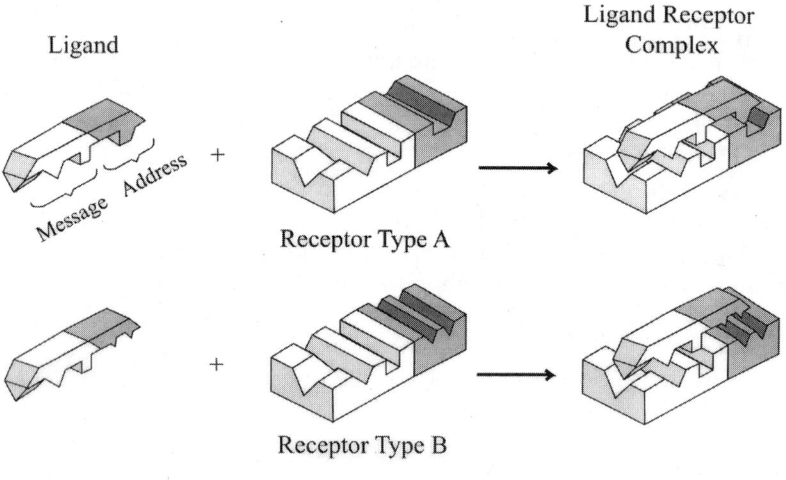

FIGURE 1 A cartoon of the message-address concept as a basis for the selectivity of a family of sychnologically organized peptides.

concept, as applied to the opioid peptides is that the Tyr^1 residue contains the message component, while the address sequence starts with Phe^4, with the intervening Gly-Gly sequence serving as a spacer that connects the message and address components. This interpretation is consistent with the structure-activity relationships of nonpeptide opioid ligands such as morphine, which contain only one aromatic ring that apparently mimics the Tyr^1 residue.

This concept was tested by attaching a peptide address to an opiate pharmacophore [4]. When the address segment of leucine-enkephalin (delta-selective) or dynorphin A (kappa-selective) was linked to oxymorphone through a spacer, a change in selectivity was observed. The mu-selective opiate, oxymorphone, was transformed to a delta-selective ligand by the attachment of the delta address (Phe-Leu) of leucine-enkephalin. Similarly, a kappa-selective ligand was obtained upon attachment of the kappa address (Phe-Leu-Arg-Arg-TLe-OMe) of dynorphin A (Fig. 2).

The determination of opioid receptor selectivity by the address component in the above hybrid ligands suggested that it was feasible to develop nonpeptide, delta-selective antagonists by attaching a nonpeptide element to an opiate in order to mimic a key recognition element in the address. The concept was modified to a more general approach to include antagonists. Thus, the opioid receptor family was viewed as having a homologous recognition site (message recognition site) that accommodates the pharmacophore and a neighboring variable recognition site (address) that confers selectivity.

In light of the preceding study, the design of nonpeptide delta opioid receptor antagonists was explored by utilizing an antagonist pharmacophore derived from naltrexone **1** for the message component joined to an aromatic moiety that was believed to be a key delta address component that mimics Phe^4 of enkephalin. The message and address components of leucine-enkephalin and the proposed nonpeptide antagonist are compared in Figure 3. Naltrexone was selected for two major reasons. First, as a universal opioid

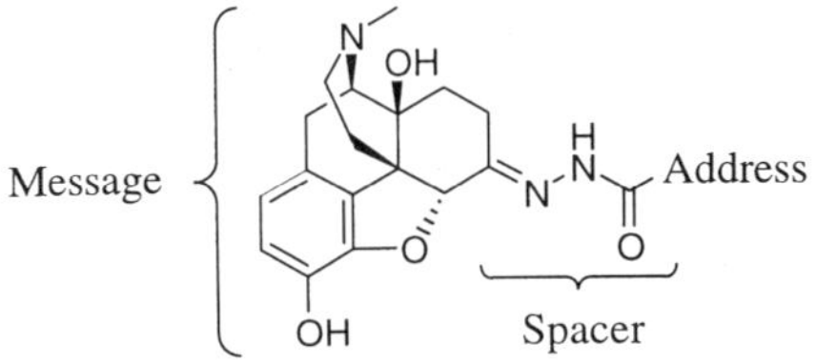

Address	Selectivity
Phe-Leu-OH	Delta
Phe-Leu-Arg-Arg-Ile-OMe	Kappa

FIGURE 2 Hybrid structures consisting of an opiate (message) and a peptide (address).

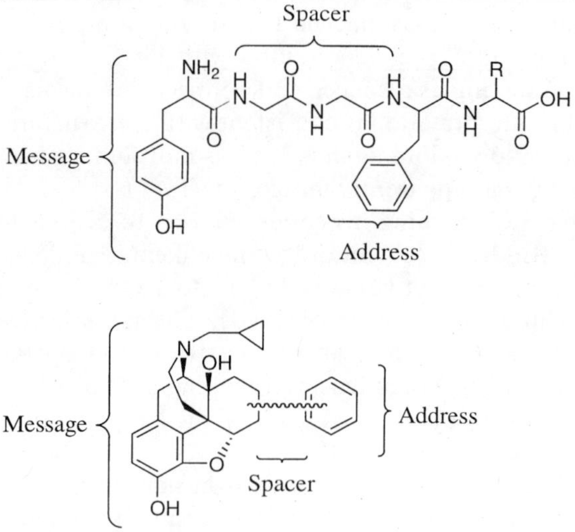

FIGURE 3 An approach to the design of a nonpeptide, delta-selective opioid antagonist based on the message-address concept. The message and address components of the delta-selective peptide enkephalin (upper) are compared to those in an opiate (lower).

receptor antagonist, it contains a message component that will be accommodated by all three types of opioid receptors; second, it offered the opportunity to synthesize ligands that contain a spacer fused to the C ring of the morphinan structure, thereby restricting the conformation of the address. Also, the availability of naltrexone together with its 6-keto functionality permitted easy access to a variety of analogues.

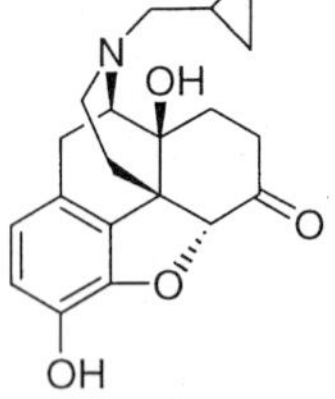

1

Since the orientation of the Phe^4 phenyl group in the delta receptor-bound conformation of enkephalins was not known, the obvious strategy was to obtain series of compounds that were altered in their orientation of the aromatic ring. In addition, conformational rigidity was important because a properly oriented, rigid address should confer greater selectivity for the target receptor by excluding conformations that could bind to other opioid receptor types.

3 NALTRINDOLE (NTI), A HIGHLY POTENT AND SELECTIVE DELTA OPIOID ANTAGONIST

3.1 NTI

The first series of compounds that possessed selective delta receptor antagonism contained the indole moiety [5,6]. The prototypical nonpeptide delta antagonist, naltrindole **2** (NTI), and other indoles in this series were accessible in a single step through the Fischer indole synthesis. By way of comparison, the enkephalin analogue ICI-174864, **3** [7], in use as a delta antagonist, had a potency that was 1/530 that of NTI (Ke = 0.13 nM, mouse vas deferens (MVD) vs. [D-Ala^2, D-Leu^5]enkephalin (DADLE)) [8]. In terms of binding, NTI (K_i, delta = 0.031 nM) has a >1000-fold greater affinity than ICI74864 [8]. The large difference between the selectivity profiles of NTI and naltrexone illustrates the dramatic effect exerted by the indole moiety. Thus, NTI is 240 times more potent than naltrexone as an antagonist at delta receptors [8]. Figure 4 outlines many of the analogs that are based on NTI.

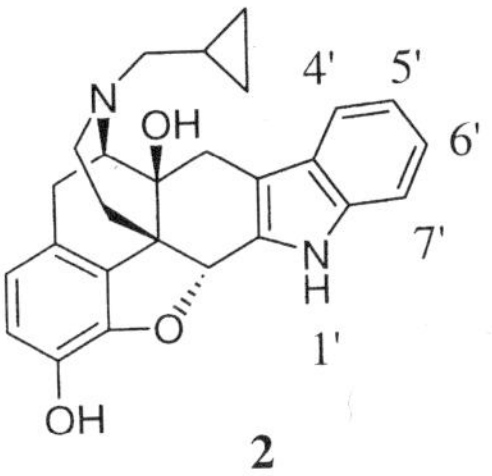

2

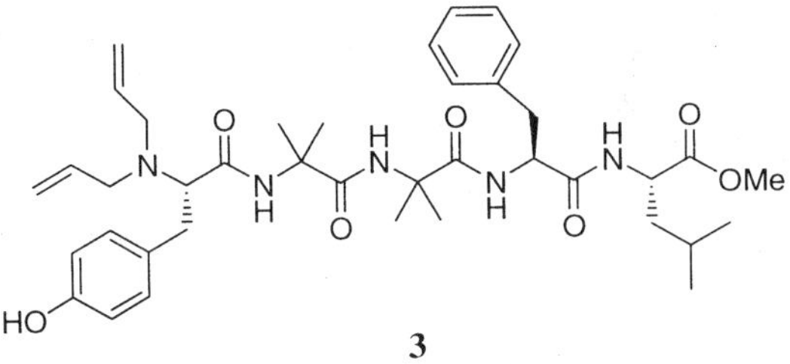

3

3.2 Structure-Activity Relationship of NTI

Substitution of the indole moiety of NTI generally results in reduced potency [8,9]. Substitution at the 5′ or 6′ position leads to a decrease of delta opioid antagonist potency by at least 1 order of magnitude. This is accompanied by a decrease in selectivity, as there is generally no substantial diminution of antagonist potency at mu or kappa receptors. At the 5′ position, the decreased potency change appears to be correlated with an increase in the size and polarity of the substituent. Substitution at the 1′ or 7′ position has a less detrimental effect on delta antagonist potency. In this regard, 1′-benzyl ana-

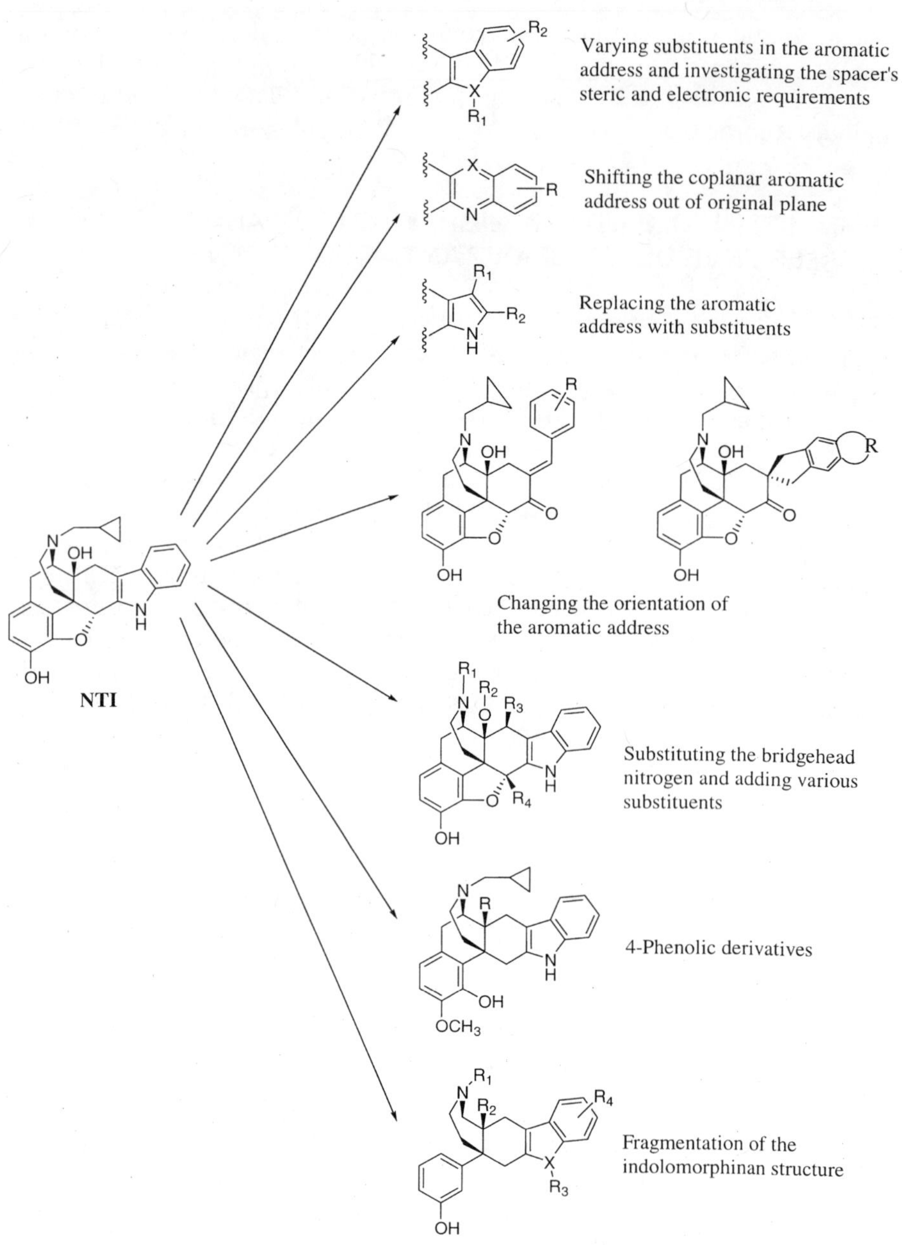

FIGURE 4 Summary of analogues based on naltrindole (NTI).

logues have been described as potent, long-lasting delta opioid receptor antagonists [10].

The phenolic group was found to be necessary for maintaining high delta antagonist activity, which suggests that the message component of NTI and naltrexone bind to a similar message recognition locus. Other modifications such as acetylation of the 14-hydroxy group or replacement of the cyclopropylmethyl with allyl afforded relatively smaller reductions of delta antagonist potency.

It has been shown that replacement of the N-cyclopropylmethyl (CPM) or N-allyl group with methyl at the basic nitrogen in opiate structures generally results in a change from antagonist to agonist activity [11]. This possibility was explored with the oxymorphone-derived analogue oxymorphindole **4** (OMI). Since oxymorphone, **5**, is a potent mu-selective agonist, it was anticipated that its indole derivative **4** might be a delta-selective agonist if the conformational requirements of delta agonists and antagonists are similar [8]. On the MVD, OMI **4** acted as a partial agonist (65% maximum response), and it was virtually inactive on the guinea pig ileum preparation (GPI) [8]. When tested in the mouse abdominal stretch assay, OMI was found to be a highly potent delta antagonist when administered ICV [12]. Surprisingly, **4** was found to be a kappa agonist at a 20-fold higher dose. Similarly, NTI also displayed kappa agonist activity in the higher dose range. As these ligands possess high delta binding selectivity, a possible explanation for these observations is that the agonist effect is mediated through the release of dynorphin. Further studies that addressed the molecular modification of the N-substituent have led to full agonists with moderate activities in vitro. The phenethyl derivative **6** (IC_{50} = 171 nM, MVD) is such an example [13]. However, in vivo studies revealed a pharmacological profile for **6** similar to that of NTI and OMI. Thus, the phenethyl group does not afford the potency-enhancing effect that is generally observed in mu-selective opiates.

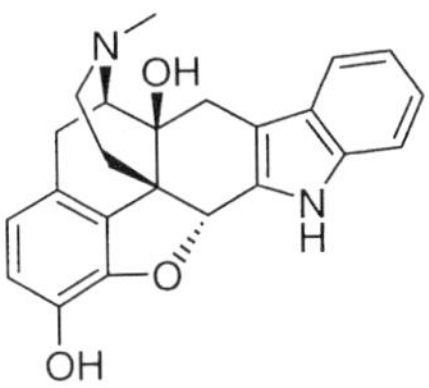

4

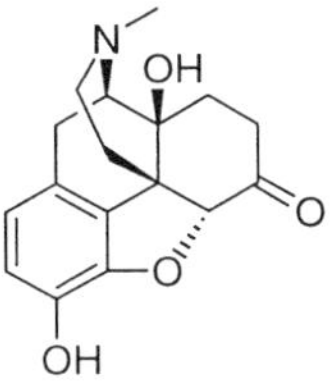

5

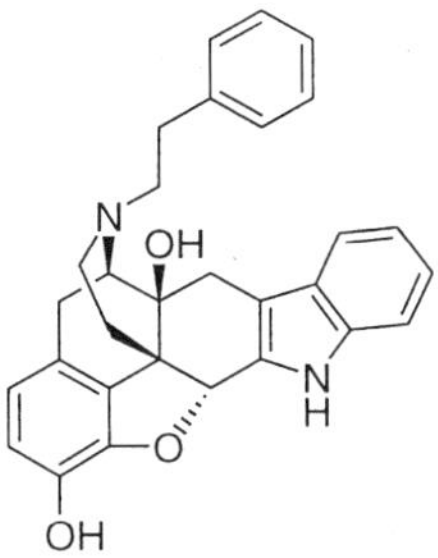

6

3.3 Spacer Modifications to NTI

The pyrrole component of the indole group functions mainly as a rigid spacer to hold the benzene moiety, which was considered to be the relevant nondelta address component that mimics the Phe^4 phenyl group of enkephalin. This was suggested by the observation that pyrrolomorphinan **7** is a mu-selective ligand with substantially decreased delta opioid receptor activity relative to NTI [14]. Since the pyrrole moiety of NTI functions as a spacer, other heterocycles can play a similar role [15]. Replacement of the pyrrole moiety of NTI with furan led to the benzofuran analogue, naltriben **8** (NTB) [15], which is as potent as NTI and highly delta-selective (K_e = 0.27 nM vs. DADLE; K_i, delta= 0.013 nM). Analogues with a quinoline or quinoxaline system **9** are substantially less potent and less selective than NTI, which possibly reflects the differential orientation of the delta address mimic (benzene moiety) between five-membered and six-membered spacers [15]. These results further illustrate the regio-requirements of the delta opioid receptor address.

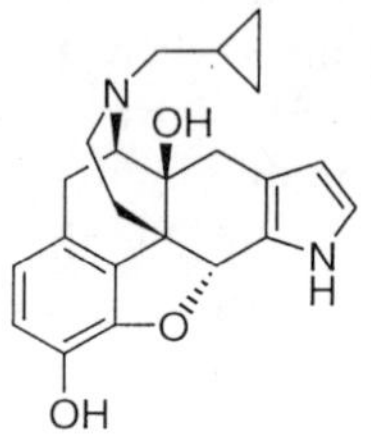

7

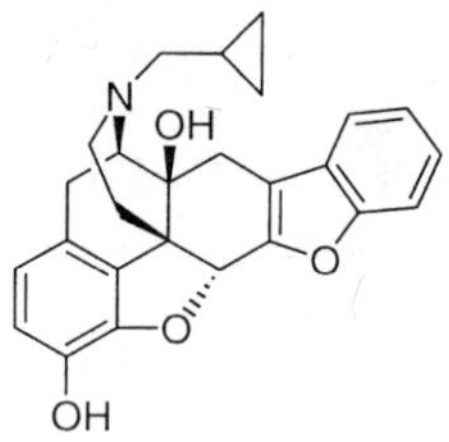

8

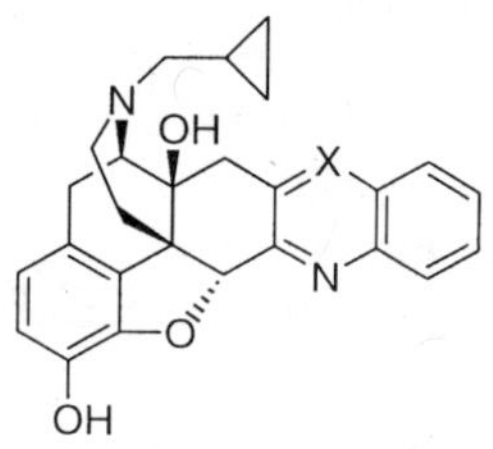

9, X = CH_2 or O

3.4 Investigations and Modifications to the Delta Address of NTI

The aromatic address of NTI confers delta selectivity by enhancing the affinity for delta receptors with a concomitant reduction in the affinity for nondelta opioid receptors. This is likely accomplished through two separate mechanisms. First, affinity for the delta opioid receptor may be enhanced by interaction between the aromatic address of NTI and the nonconserved Trp-284 residue of the delta opioid receptor, as it has been shown that site-directed mutagenesis of Trp-284 located at the top of transmembrane 6 (TM6) leads to a decrease in selectivity of delta selective ligands [16]. Second, the aromatic address of NTI may sterically interfere with binding to mu and kappa receptors. In this regard, it has been shown that replacement of the benzene moiety of the indole ring with cyclohexane **10** [15] or with alkyl groups substantially decreases delta antagonist potency. This is exemplified in a series of 2′,3′-disubstituted pyrrolomorphinans (**11**) [17]. Both **10** and members of **11** are less potent at delta receptors than NTI; however, they still are delta-selective, presumably because the alkyl groups attached to the pyrrole moiety sterically interfere with binding to mu and kappa receptors [17].

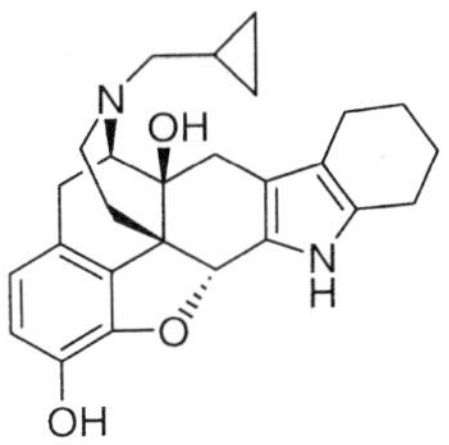

10

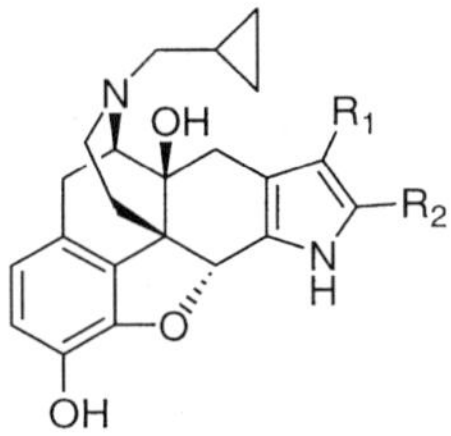

11, R_1 and R_2 = alkyl

Additional evidence in support of the idea that steric hinderance contributes to the lower potency of delta antagonists at wild-type mu and kappa receptors was obtained from site-directed mutagenesis studies [18]. Single point mutations were made in all three types of opioid receptors with the focus on two positions at the extracellular end of TM6 and TM7. It was found that NTI could bind both mutant mu and kappa receptors with greatly enhanced affinity when a bulky aromatic residue at the top of TM7 was replaced with the smaller alanine residue (mu-Trp318Ala and kappa-Tyr312-Ala) [18]. These results suggest that the reduction in the steric bulk of these residues increases access of NTI into the central binding cavity that recognizes the antagonist pharmacophore and are consistent with the finding that the address or substituents on the pyrrole moiety confer selectivity by interfering with binding at wild-type receptors.

Molecular dynamic simulations of enkephalin and NTI were consistent with the idea that Phe^4 of enkephalin and the indolic benzene moiety of NTI both bind to a common delta address subsite [15]. The results of these simulations show that the phenyl group of Phe^4 is restricted to the same conformational space as the indolic benzene moiety of NTI, suggesting that they both may bind to the same locus of the delta address. However, it is unlikely that the phenyl group of Phe^4 and the indolic benzene moiety superpose, since conformations leading to complete superposition of both rings were not observed during the simulations. One possible explanation in this regard is that NTI stabilizes the delta receptor in an antagonist conformational state that is different from that of an agonist state [19].

The benzene moiety of NTI is conformationally restricted and coplanar to ring C of the morphinan structure as a consequence of ring fusion with the indole system. The effect of noncoplanar delta address mimics in opiates having orthogonally orientated aromatic groups was investigated with ligands such as benzylidenenaltrexone **12** (BNTX) [20], the 7-spiroindanyl derivatives **13** [21] and **14** [22], and 7-phenylnaltrexone **15** [23]. All of these ligands possess good delta antagonist activity, but are less potent compared to NTI. The lack of coplanarity between ring C and the address moiety is most likely the principal reason for the lower delta antagonist potencies.

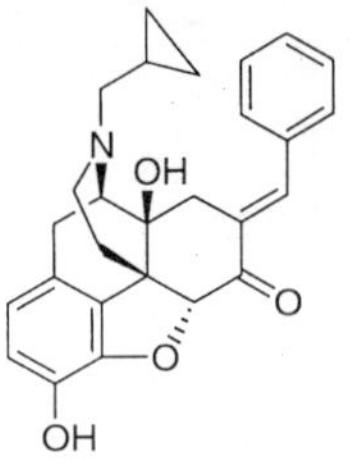

12

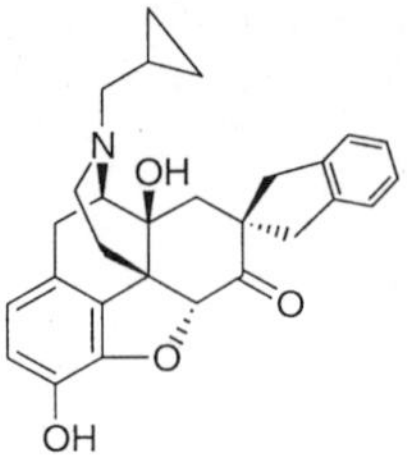

13

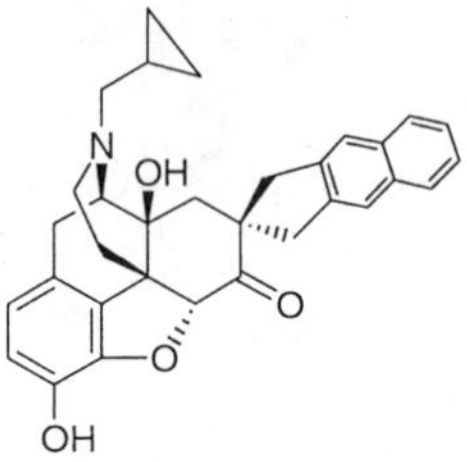

14

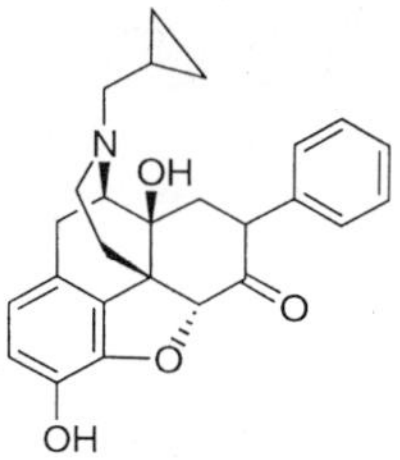

15

The existence of delta receptor subtypes, $delta_1$ and $delta_2$, has been suggested based on in vivo pharmacological studies [24,25]. BNTX 12 has been classified as a $delta_1$ antagonist because it selectively antagonizes the agonist effect of [D-Pen^2,D-Pen^5]enkephalin (DPDPE) [26]. Since NTB **8** selectively blocks [D-Ser^2,D- Leu^5, Thr^6]enkephalin (DSLET) [27] and deltorphin II [28], it is considered to be a $delta_2$ antagonist. It is not known whether these putative subtypes represent different affinity states for the same receptor, different gene products/splice variants, different receptor aggregation states (i.e., monomer, homodimer, heterodimer, oligomer, etc.), or differential access to receptors. While in vivo pharmacological selectivity has been observed, studies in vitro show little or no differences between putative $delta_1$ and $delta_2$ ligands [29,30]. In a recent study, it was shown using CHO cells expressing the human delta opioid receptor that BNTX **12** and NTB **8** were not selective for either the putative $delta_1$ or $delta_2$ opioid receptors in binding or functional assays [31].

The observation that the preferred conformation of the benzylidene aromatic group bears an orthogonallike relationship to ring C of BNTX originally led to the proposal that this orientation favors $delta_1$ activity [32]. In contrast, the coplanar aromatic group of NTB may help confer $delta_2$ selectivity. The rationale for the design of the 7-spiroindanyl derivatives (**13** and **14**) was to rigidly hold the benzene moiety of the indanyl substituent in an orthogonal position relative to ring C of the morphinan structure, similar to the aromatic group conformation in BNTX. Indeed, BSINTX **14** was found to be a potent, selective $delta_1$ antagonist in vivo [22]. This may reflect a difference between the topography of the address subsite in putative $delta_1$ and $delta_2$ receptors.

Replacement of the N-cyclopropylmethyl group of **13** with N-methyl led to SIOM **16**, with full delta agonist activity in the MVD assay (IC_{50} = 23 nM). However, in the mouse tail flick assay, SIOM at low doses acted as a $delta_1$ antagonist, but at higher doses it displayed $delta_1$ agonist activity [21,33]. The $delta_1$ agonist potency of **16** is consistent with molecular dynamic simulations which showed that the conformation of the Phe^4 phenyl group of the $delta_1$ agonist DPDPE [34,35] more closely matches the orientation of **14** and **16** when compared to the $delta_2$ antagonist NTB **8**.

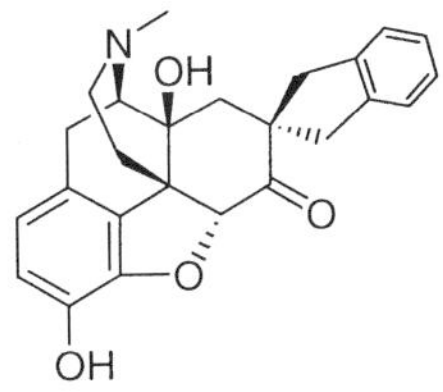

16

4 NALTRINDOLE AS A LEAD FOR THE DEVELOPMENT OF NONPEPTIDE DELTA LIGANDS

4.1 Analogues of NTI Utilizing the Indolomorphinan Framework

Following the initial report [5] on NTI, a variety of indolomorphinan analogues were synthesized based upon its structure (Fig. 4). The Toray group has disclosed analogues, **17**, whose indole moiety contains a ring that is attached through the 1′ and 6′ position [36,37]. In this series the most potent antagonists (pA_2 = 9–10 in the MVD bioassay) were those where X is oxygen or methylene [36]. Schmidhammer and the Astra group have described various 14-alkoxy derivatives of NTI [38–43] typified by **18**. Reports from the National Institutes of Health have investigated various indolomorphinan derivatives including phenyl, phenoxy or benzyloxy additions to the indole moiety [44], 4-phenolic substitutions [45], N-alkyl substitutions [46,47], and 8β substitutions [48]. Of interest are the 4-

phenolic-substituted NTI analogues, **19** and **20**, created by reductive epoxy ring opening [45]. Compound **19** was found to be a selective delta antagonist with lower potency than NTI.

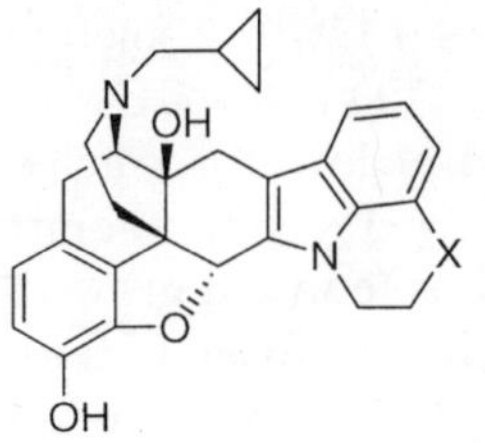

17

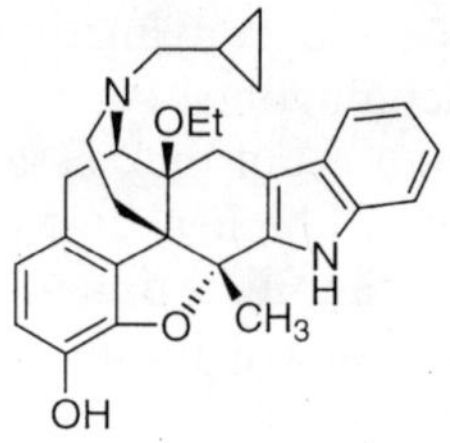

18

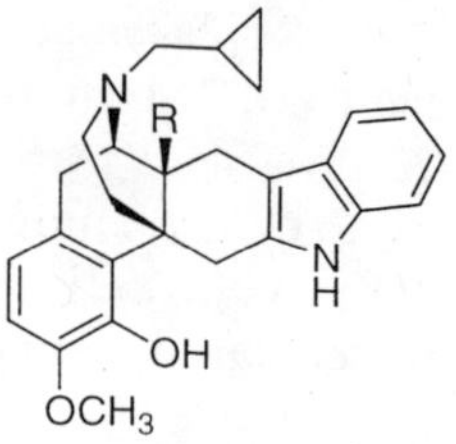

19, R = H
20, R = OH

4.2 Isoquinoline Analogues Based on NTI

The Toray group was the first to disclose mimics of NTI based on modification of the indolomorphinan framework to afford indolo[2,3-g]octahydroisoquinolines typified by structure **21** [49]. The major difference between NTI and **21** is that ring A is constrained by a methylene bridge in the former (Fig. 5). Using an analogous approach, the SmithKline Beecham group in Milan developed delta-selective antagonists (e.g., **22**) without the bridgehead hydroxyl group [50,51]. Additional compounds in this series featured diverse heterocyclic spacers such as furan and [3,2-g]pyrrole [50]. In general, derivatives based on this scaffold are less potent than NTI with varying delta selectivity. This suggests that retention of selectivity was accomplished mainly through reduced affinity for mu and kappa receptors rather than enhanced affinity for delta receptors.

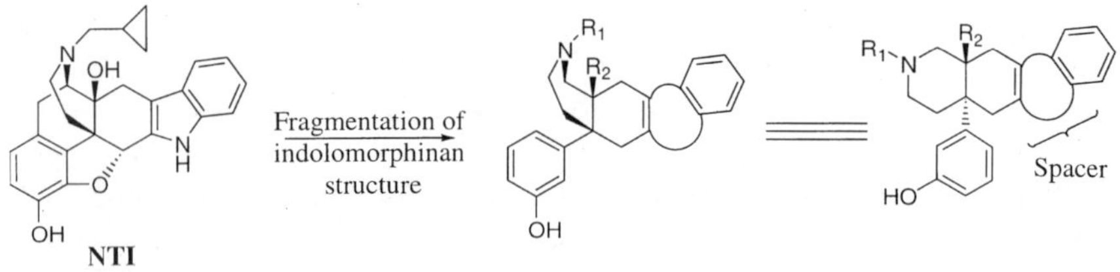

FIGURE 5 Fragmentation of the indolomorphinan framework.

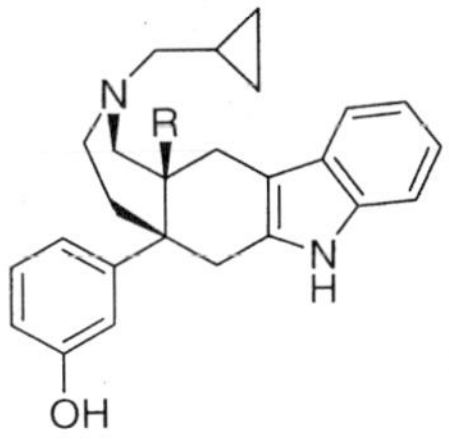

21, R = OH
22, R = H

4.3 Conversion of Delta Antagonists to Agonists

Since substitution of the N-cyclopropylmethyl group for methyl at the basic nitrogen of opiates generally results in a change from opioid antagonist to agonist activity, this approach was employed for the design of delta agonists. This led to the synthesis of N-methyl analogues typified by compound **23** [49–51]. Introduction of a six-membered ring in the octahydroisoquinoline series resulted in **24** as a racemate (SB 213698 [50] or TAN-67 [52]). Although **24** is a potent (IC_{50} 6.6 nM, MVD) and selective delta agonist in vitro, it afforded only weak antinociceptive acivity in the mouse tail flick assay [53–55]. It was discovered that (–)-**24** (having the same absolute configuration as natural (–)-morphine) produced more potent antinociception [54–56], whereas its (+)-enantiomer displayed nociceptive properties that were blocked by intrathecal treatment with NTI or (–)-**24** [55]. Thus, the reduced effect of the racemate is apparently due to antagonism between enantiomers.

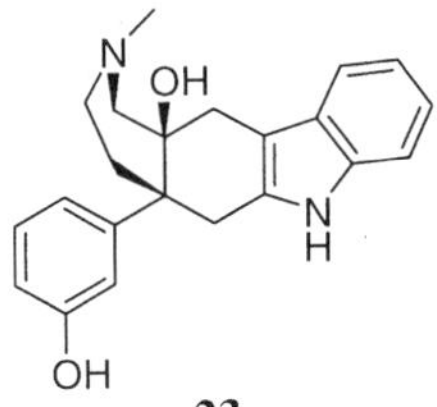

23

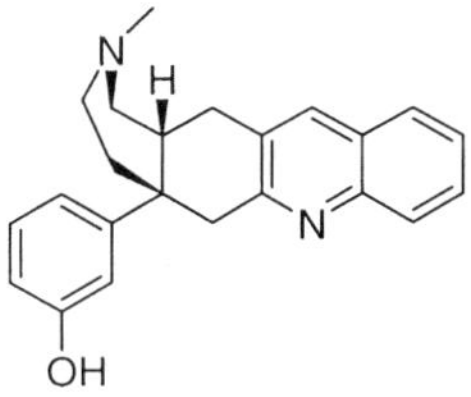

24

Subsequent studies revealed that N-methyl/ethyl analogues of pyrrolo-octahydroisoquinoline, e.g., **25**, exhibit good delta receptor affinity (K_i 0.9 nM) and selectivity, and behaved as full agonists in the MVD assay (IC_{50} 25 nM) [57]. The corresponding morphinan analogue **26** also displayed a similar profile [37,50]. It is likely that the delta selectivity of these ligands originates from an exclusion mechanism at mu and kappa opioid receptors, as has been discussed for **11** and **12** [17]. Delta-selective isoquinoline and pyrrole derivatives have been reviewed [37,58].

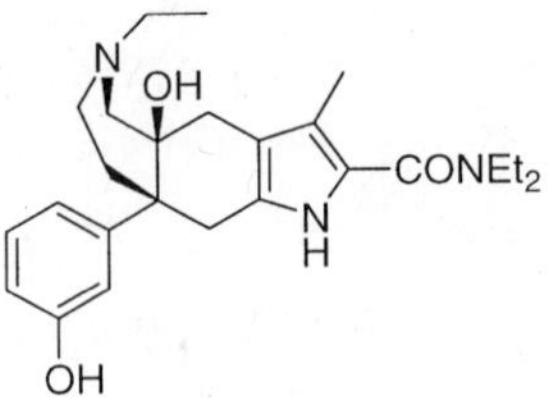

25

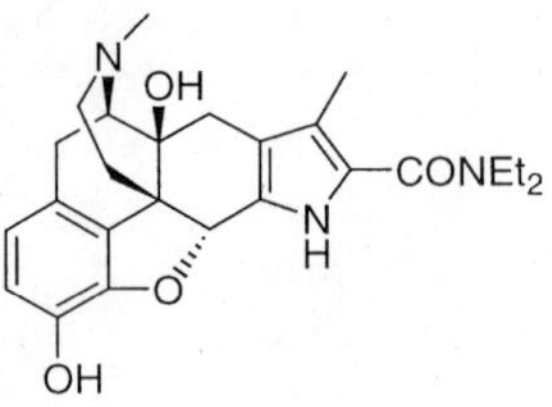

26

4.4 Affinity Labels Derived from Naltrindole

An effort to design delta opioid receptor antagonist affinity labels led to the synthesis of naltrindole isothiocyanate regioisomers **27** (NTII) [59]. The 5′ regioisomer (5′-NTII) is a highly selective and potent nonequilibrium delta receptor antagonist [60]. It produces a time-dependent blockage of delta receptors in the MVD assay and affords insurmountable antagonism of delta agonists in vivo without attenuating the antinociceptive effect of mu agonists [61]. It appears that 5′-NTII is pharmacologically selective for putative $delta_2$ opioid receptors, as it more potently antagonizes the antinociceptive effect of deltorphin II and DSLET relative to that of DPDPE [25]. Other delta affinity labels include a series of N-benzylnaltrindoles (BNTI) that contain electrophilic moieties on the benzyl group [62].

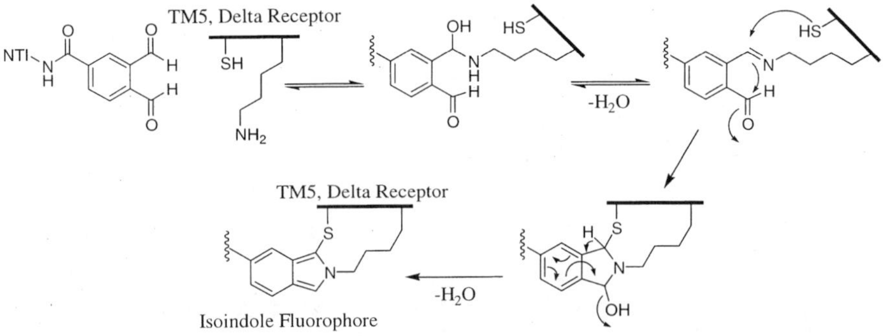

FIGURE 6 PNTI irreversibly binding to neighboring cysteine and lysine residues at the top of transmembrane 5 of the delta opioid receptor.

A new approach to affinity labels recently has been applied to the design of the delta opioid receptor affinity label, **28** (PNTI). This type of affinity label has been named a "reporter affinity label" because it reports the crosslinking of neighboring cysteine and lysine residues through the generation of an isoindole fluorophore [63]. PNTI selectively covalently binds delta receptors, as indicated by the generation of specific fluorescence. The unexpected finding that PNTI functions as an irreversible delta agonist rather then an irreversible delta antagonist in the MVD preparation has led to the proposal that crosslinking of neighboring Cys and Lys residues at the top of TM5 causes axial rotation of this helix which leads to activation of the delta receptor (Fig. 6) [64].

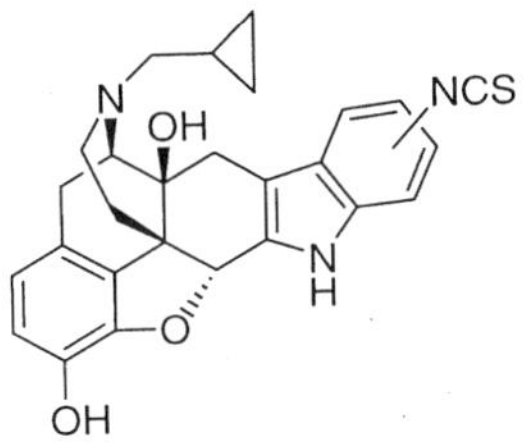

27

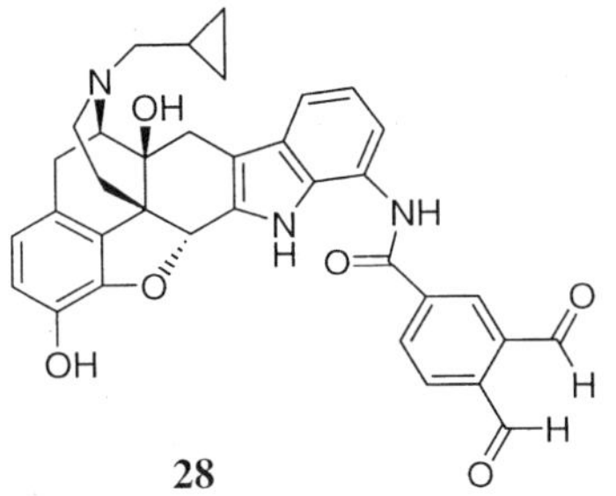

28

5 MODULATION OF THE MU OPIOID RECEPTOR BY DELTA-SELECTIVE ANTAGONISTS

Interaction between mu and delta opioid receptors were first reported by Vaught and Takemori, who observed potentiation of morphine-induced antinociception by a delta agonist [65]. Reports [reviewed in 66] that followed also suggested the interaction between mu and delta opioid receptors and its possible significance in morphine tolerance and physical dependence [67–70]. Studies with delta antagonists have demonstrated that the chronic effects of morphine are blocked without significantly diminishing its antinociceptive action [61,70,71]. Through use of selective delta antagonists such as NTI, NTB, and 5′-NTII, it was revealed that putative delta$_2$, but not delta$_1$, opioid receptors are involved in this interaction [70,71]. Subsequent studies using antisense oligodeoxynucleotides and delta receptor knockout mice have supported the evidence implicating a mu–delta interaction in the development of morphine tolerance and dependence [72,73].

The important role of delta receptors in the development of morphine tolerance and physical dependence has prompted the search for mixed mu agonist/delta antagonist ligands as an approach to analgesics devoid of these side effects [74,75]. DIPP-NH_2[Ψ] **29**, a peptide that has been reported [74] to be a mu agonist in the guinea pig ileum assay and delta antagonist in the MVD, produces potent antinociception with reduced tolerance and physical dependence when compared to morphine.

This approach also has been employed in the design of naltrexone-derived ligands with mu agonist/delta antagonist properties [75,76]. One such compound, **30** (SoRI 9409), was reported not to induce tolerance in mice and produced fewer withdrawal signs when challenged with naloxone in acute and chronic morphine dependence models [75]. However, discrepancy between the in vivo/in vitro data requires additional investigation in order to better define the mechanism of the improved in vivo profile [76,77].

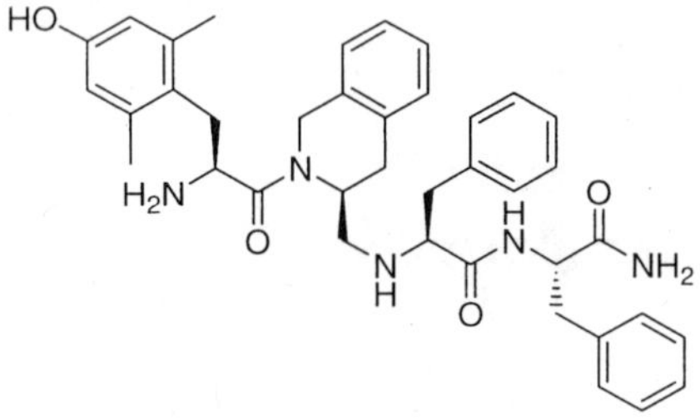

29

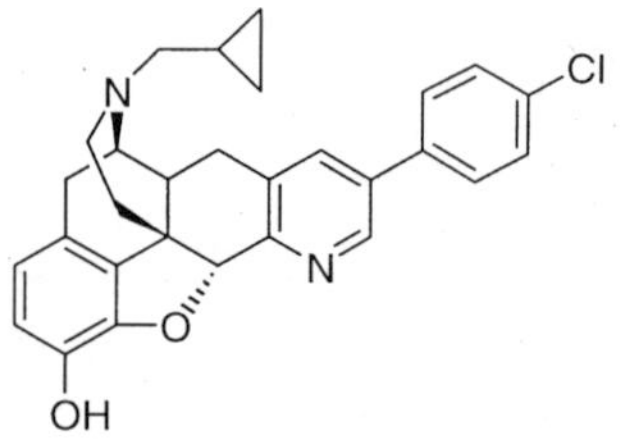

30

It has not yet been determined whether the synergy between mu and delta receptors is a consequence of direct association between receptors or due to functional modulation involving neuronal circuitry. In view of evidence for mu-delta opioid receptor heterodimers [78,79] in cultured cells, there is reason to believe that similar interactions occur in vivo. In this connection, it is noteworthy that morphine pellet implantation changes the receptor selectivity in the brain for heroin and its metabolite 6-monoacetylmorphine from a mu to delta opioid receptor-mediated response [80]. Since heroin appears to activate putative $delta_1$ receptors under these conditions, perhaps the transition from mu to delta selectivity may in tolerant mice reflect changes in the distribution of putative mu–delta heterodimers relative to homomers [81]. The development of selective ligands that target opioid receptor heterodimers should help to further delineate the pharmacology of these receptors. The design of such ligands is in progress [82,83].

6 CONCLUSION

The opioid peptides can be viewed to contain two elements: an essential message component that is recognized by a homologous receptor subsite common to the opioid receptor family, and an address element that is recognized by a unique nonconserved subsite that confers selectivity to the ligand. Application of the message-address concept has led to the design of highly potent nonpeptide delta opioid ligands. Combining the universal opioid antagonist, naltrexone, with a strategically located address mimic has led to the design of the prototypical delta receptor antagonist, naltrindole (NTI).

Structure–activity relationship studies of NTI reveal that the pyrrole moiety functions as a rigid spacer that directs the delta address mimic (benzene moiety) to the address subsite. Selectivity is conferred by a combination of increased affinity for delta receptors and decreased affinity for mu and kappa receptors through an exclusion mechanism. The structure of NTI has been used as a template to design congeners that have delta receptor–mediated pharmacological profiles. Molecular modification of the indolomorphinan framework of NTI has afforded hundreds of analogues with delta antagonist or agonist activity. In addition to the indolomorphinans, these include pyrrolomorphinans, pyrrolo-octahydroisoquinolines, 7-spiroindanylmorphinans, and 7-benzylidene derivatives of morphinan.

In addition to their important use as pharmacological tools, there are potential clinical applications for nonpeptide delta opioid ligands. Though they have not been extensively investigated, areas of interest for clinical potential are as immunosuppressive agents [84,85], antitussive [86], preventive agents for alcohol and drug abuse [87–90], and as analgesics devoid of tolerance and physical dependence [61,70,71].

REFERENCES

1. Zimmerman DM, Leander JD. J Med Chem 1990; 33:895–902.
2. Schwyzer R. Ann NY Acad Sci 1977; 297:3–26.
3. Chavkin C, Goldstein A. Proc Natl Acad Sci USA 1981; 78:6543–6547.
4. Lipkowski AW, Tam SW, Portoghese PS. J Med Chem 1986; 29:1222–1225.
5. Portoghese PS, Sultana M, Nagase H, Takemori AE. J Med Chem 1988; 31:281–282.
6. Portoghese PS, Sultana M, Takemori AE. Eur J Pharmacol 1988; 146:185–186.
7. Cotton R, Giles MG, Miller L, Shaw JS, Timms D. Eur J Pharmacol 1984; 97:331–332.
8. Portoghese PS, Sultana M, Takemori AE. J Med Chem 1990; 33:1714–1720.
9. Toray Ind., Inc., WO9107966, 1991; WO9407896, 1994.
10. Korlipara VL, Takemori AE, Portoghese PS. J Med Chem 1994; 37:1882–1885.
11. Casey AF, Parfitt RT. Opioid Analgesics. New York: Plenum Press, 1986.
12. Takemori AE, Sultana M, Nagase H, Portoghese PS. Life Sci 1992; 50:1491–1495.
13. Portoghese PS, Larson DL, Sultana M, Takemori AE. J Med Chem 1992; 35:4325–4329.

14. Portoghese PS, Nagase H, Lipkowski AW, Larson DL, Takemori AE. J Med Chem 1988; 31:836–841.
15. Portoghese PS, Nagase H, MalonyHuss KE, Lin CE, Takemori AE. J Med Chem 1991; 34:1715–1720.
16. Valiquette M, Vu HK, Yue SY, Wahlestedt C, Walker P. J Biol Chem 1996; 271: 18789–18796.
17. Farouz-Grant F, Portoghese PS. J Med Chem 1997; 40:1977–1981.
18. Metzger TG, Paterlini MG, Ferguson DM, Portoghese PS. J Med Chem 2001; 44:857–862.
19. Portoghese PS. J Med Chem 1991; 34:1757–1762.
20. Portoghese PS, Sultana M, Nagase H, Takemori AE. Eur J Pharmacol 1992; 218:195–196.
21. Portoghese PS, Sultana M, Moe T, Takemori AE. J Med Chem 1994; 37:579–585.
22. Ohkawa S, DiGiacomo B, Larson DL, Takemori AE, Portoghese PS. J Med Chem 1997; 40:1720–1725.
23. Gao P, Larson DL, Portoghese PS. J Med Chem 1998; 41:3091–3098.
24. Sofuoglu M, Portoghese PS, Takemori AE. J Pharmacol Exp Ther 1991; 257: 676–680.
25. Jiang Q, Takemori AE, Sultana M, Portoghese PS, Bowen WD, Mosberg HI, Porreca F. J Pharmacol Exp Ther 1991; 257:1069–1075.
26. Mosberg H, Hurst R, Hruby VJ, Gee K, Yamamura HI, Galligan JJ, Burks TF. Proc Natl Acad Sci USA 1983; 80:5871–5874.
27. Handa BK, Lane AC, Lord JA, Morgan BA, Rance MJ, Smith CFC. Eur J Pharmacol 1981; 70:531–540.
28. Erspamer V, Melchiorri P, Falconieri-Erspamer G, Negri L, Corsi R, Severini C, Barra D, Simmaco M, Kreil G. Proc Natl Acad Sci USA 1989; 86:5188–5192.
29. Connor MA, Keir MJ, Henderson G. Neuropharmacology 1997; 36:125–133.
30. Toll L, Polgar WE, Auh JS. Eur J Pharmacol 1997; 323:261–267.
31. Parkhill AL, Bidlack JM. Eur J Pharmacol 2002; 451:257–264.
32. Portoghese PS. Farmaco 1993; 48:243–251.
33. Portoghese PS, Moe ST, Takemori AE. J Med Chem 1993; 36:2572–2574.
34. Hruby VJ, Kao LF, Petti BM, Karplus M. J Am Chem Soc 1988; 110:3351–3359.
35. Schiller PW, Weltrowka G, Nguyen TMD, Lemieux C, Chung NN, Marsden BJ, Wilkes BC. J Med Chem 1991; 34:3125–3132.
36. Toray Ind., Inc., WO9711948, 1997.
37. Dondio G, Silvano R, Perrillo P. Exp Opin Ther Patents 1997; 7:1075–1098.
38. Astra AB, WO9531463, 1995.
39. Schmidhammer H, Daurer D, Wieser M, Monory K, Borsodi A, Elliott J, Traynor JR. Bioorganic Med Chem Lett 1997; 7:151–156.
40. Schmidhammer H, Krassnig R, Greiner E, Schutz J, White A, Berzetei-Gurske IP. Helv Chim Acta 1998; 81:1064–1069.
41. Schmidhammer H, Schwarz P, Wei ZY. Helv Chim Acta 1998; 81:1215–1222.
42. Biyashev D, Monory K, Benyhe S, Schuetz J, Koch M, Schmidhammer H, Borsodi A. Helv Chim Acta 2001; 84:2015–2021.

43. Schuetz J, Dersch CM, Horel R, Spetea M, Koch M, Meditz R, Greiner E, Rothman RB, Schmidhammer H. J Med Chem 2002; 45:5378–5383.
44. Ananthan S, Johnson CA, Carter RL, Clayton SD, Rice KC, Xu H, Davis P, Porreca F, Rothman RB. J Med Chem 1998; 41:2872–2881.
45. Coop A, Rothman RB, Dersch C, Partilla J, Porreca F, Davis P, Jacobson AE, Rice KC. J Med Chem 1999; 42:1673–1679.
46. McLamore S, Ullricj T, Rothman RB, Xu H, Dersch C, Coop A, Davis P, Porreca F, Jacobson AE, Rice KC. J Med Chem 2001; 44:1471–1474.
47. Ullrich T, Dersch CM, Rothman RB, Jacobson AE, Rice KC. Bioorganic Med Chem Lett 2001; 11:2883–2885.
48. Yu H, Prisinzano T, Dersch CM, Marcus J, Rothman RB, Jacobson AE, Rice KC. Bioorganic Med Chem Lett 2002; 12:165–168.
49. Toray Ind., Inc., EP-0485636-A, 1992.
50. L Zambeketti, SPA, WO9301186, 1993.
51. Dondio G, Clarke GD, Giardina G, Petrillo P, Rapalli L, Ronzoni S, Vecchietti V. Regul Pept 1994; 21:43–44.
52. Nagase H, Kawai K, Toray Ind., Inc., JP04297458, 1992.
53. Nagase H, Kawai K, Hayakawa J, Wakita H, Mizusuna A, Matsuura H, Tajima C, Takezawa Y, Endoh T. Chem Pharm Bull 1998; 46:1695–1702.
54. Dondio G, Clarke GD, Giardina G, Petrillo P, Petrone G, Ronzoni S, Visentin L, Vecchietti V. Analgesia 1995; 1:398–399.
55. Nagase H, Yajima Y, Fujii H, Kawamura K, Narita M, Kamei J, Suzuki T. Life Sci 2001; 68:2227–2231.
56. Kamei J, Kwawi K, Mizusuna A, Saitoh A, Morita K, Narita M, Tseng LF, Nagase H. Eur J Pharm 1997; 322:27–30.
57. Dondio G, Ronzoni S, Eggleston DS, Artico M, Petrillo P, Petrone G, Visentin L, Farina C, Vecchietti V, Clarke GD. J Med Chem 1997; 40:3192–3198.
58. Dondio G, Ronzoni S, Petrillo P. Exp Opin Ther Patents 1999; 9:353–374.
59. Portoghese PS, Sultana M, Nelson WL, Klein P, Takemori AE. J Med Chem 1992; 35:4086–4091.
60. Portoghese PS, Sultana M, Takemori AE. J Med Chem 33:1547–1548.
61. Abdelhamid EE, Sultana M, Portoghese PS, Takemori AE. J Pharmacol Exp Ther 1991; 258:299–303.
62. Korlipara VL, Takemori AE, Portoghese PS. J Med Chem 1995; 38:1337–1343.
63. Le Bourdonnec B, El Kouhen R, Poda G, Law PY, Loh HH, Ferguson DM, Portoghese PS. J Med Chem 2001; 44:1017–1020.
64. Le Bourdonnec B, El Kouhen R, Lunzer MM, Law PY, Loh HH, Portoghese PS. J Med Chem 2000; 43:2489–2492.
65. Vaught J, Takemori A. J Pharmacol Exp Ther 1979; 208:86–94.
66. Rothman RB, Holiday JW, Porreca F. In: Herz A, ed. Opioids I. Berlin: Springer-Verlag, 1993:217–237.
67. Rothman RB, Westfall TC. Fed Proc 1980; 39:385.
68. Vaught JL, Rothman RB, Westfall TC. Life Sci 1982; 30:1443–1455.
69. Heyman JS, Porreca F. Prog Opioid Res 1989; 75:467–470.
70. Miyamoto Y, Portoghese PS, Takemori AE. J Pharmacol Exp Ther 1993; 264:1141–1145.

71. Miyamoto Y, Brown WD, Portoghese PS, Takemori AE. J Pharmacol Exp Ther 1994; 270:37–39.
72. Sánchez-Blázquez P, García-España A, Garzón J. J Pharmacol Exp Ther 1997; 280:1423–1431.
73. Zhu Y, King MA, Schuller AG, Nitsche JF, Reidl M, Elde RP, Unterwald E, Pasternak GW, Pintar JE. Neuron 1999; 24:243–252.
74. Schiller PW, Fundytus ME, Merovitz L, Weltorski G, Nguyen TMD, Lemieux C, Chung NN, Coderre T. J Med Chem 1999; 42:3520–3526.
75. Wells JL, Bartlett JL, Ananthan S, Bilsky EJ. J Pharmacol Exp Ther 2001; 297:597–605.
76. Ananthan S, Kezar HS III, Carter RL, Saini SK, Rice KC, Wells JL, David P, Xu H, Dersch CM, Bilsky EJ, Porreca F, Rothman RB. J Med Chem 1999; 42:3527–3538.
77. Xu H, L YF, Rice KC, Ananthan S, Rothman RB. Brain Res Bull 2001; 55:507–511.
78. George SR, Fan T, Xie Z, Tse R, Tam V, Varghese G, O'Dowd BF. J Biol Chem 275:26128–26135.
79. Gomes I, Jordan BA, Gupta A, Trapaidze N, Nagy V, Devi LA. J Neurosci 2000; 20, 1–5.
80. Rady JJ, Holmes BB, Portoghese PS, Fujimoto JM. Proc Soc Exp Biol Med 2000; 334:93–101.
81. Rady JJ, Portoghese PS, Fujimoto JM. Jpn J Pharmacol 2002; 88:319–331.
82. Daniels DJ, Kulkarni A, Portoghese PS. Bivalent ligands designed to probe receptor heterodimers/hetero-oligomers: pharmacological evidence for interactions between opioid receptor types. International Narcotic Research Conference, Monterey, CA, 2002:S47.
83. Bhushan RG, Sharma SK, Portoghese PS. Design and synthesis of kappa–delta opioid bivalent ligands as probes to study opioid receptor dimerization. 224th ACS National Meeting, Boston, 2002:MEDI-242.
84. Arakawa K, Akami T, Okamoto M, Akioka K, Nakai I, Oka T, Nagase H. Transplant Proc 1993; 25:783–784.
85. House RV, Thomas PT, Kozak JT, Hargava HNB. Neurosci Lett 1995; 198:119–122.
86. Kamei J, Iwamoto Y, Suzuki T, Misawa M, Nagase H, Kasuya Y. Eur J Pharmacol 1993; 249:161–165.
87. Suzuki T, Mori T, Tsuji M, Misawa M, Nagase H. Life Sci 1994; 55:PL339–344.
88. Krishnan-Sarin S, Portoghese PS, Li TK, Froehlich JC. Pharmacol Biochem Behav 1995; 52:153–159.
89. Calcagnetti DJ, Keck BJ, Quatrella LA, Schechter MD. Life Sci 1995; 56:475–483.
90. Reid LD, Glick SD, Menkens KA, French ED, Bilsky EJ, Porreca F. Neuroreport 1995; 6:1409–1412.

10

Endogenous Peptides for Delta Opioid Receptors and Analogues

Victor J. Hruby
University of Arizona, Tucson, Arizona, U.S.A.

Henry I. Mosberg
University of Michigan, Ann Arbor, Michigan, U.S.A.

1 INTRODUCTION

The discovery of the endogenous opioid peptides leucine-enkephalin (H-Tyr-Gly-Gly-Phe-Leu-OH) and methionine-enkephalin (H-Tyr-Gly-Gly-Phe-Met-OH) [1] (Fig. 1) and the suggestion, first from pharmacological studies [2,3] and later from molecular cloning [4–7], that there are three opioid receptors (mu, delta, and kappa) have had an enormous impact on the search for better opioid analgesics that would not possess the severe toxicities and other undesirable biological activities of the alkaloid morphine and related compounds. The delta opioid receptor, and selective ligands for this receptor that also have minimal side effects, have been a major target in opioid research ever since.

Most of the early work, much of it done in industry, sought to modify the enkephalin structures to make them more stable in relation to proteolytic breakdown and improve their bioavailability. The conversion of enkephalin to [D-Ala^2]enkphalin and other D-amino acid–containing derivatives was a

H-Tyr-Gly-Gly-Phe-Leu(Met)-OH Leucine (Methionine)-Enkephalin	
H-Tyr-c[D-Pen-Gly-Gly-Phe-D-Pen]-OH DPDPE	
H-Tyr-c[D-Pen-Gly-DAla-Phe-DPen]-OH [DAla^{3}]-DPDPE	
H-Tyr-c[D-Cys-Phe-D-Pen]-OH JOM-13	
H-Tyr-D-Ala-Phe-Glu-Val-Val-Gly-NH2 Deltorphin II	

FIGURE 1 Structure of key compounds discussed.

major advance during this time. It was important for structure–activity relationship studies because it demonstrated that substitution of L-amino acids in the 2-position greatly decreased binding affinities and biological activities of enkephalin derivatives and analogues, whereas D-amino acid improved these activities [8]. Despite these and other initial improvements in enkephalins, most of the analogues obtained were only modestly selective for the delta opioid receptor, and none of the early compounds led to a useful drug for human medicine.

The importance of the D-amino acid in the 2-position of opioid peptides was strongly reinforced by the discovery in amphibian skins of the highly delta opioid receptor selective peptide ligands, the deltorphins [9,10]. This discovery stimulated much additional research on the potential of delta ligands for the treatment of pain and drug abuse, but again despite their high potency and relatively good stability against proteolytic hydrolysis, to date no deltorphin analogues or derivatives have been developed into drugs.

A major contribution to the design of opioid receptor selective ligands came with the realization by the research groups of Schiller [11] and Hruby [8,12–14] that conformational restriction of the flexible enkephalins and other linear opioid peptides could lead to new insights into the conformational structure–biological activity relationships of opioid ligands, and that these insights could then be exploited for further design of more potent, stable, selective, and bioavailable opioid ligands. This review, therefore, focuses on the exploitation of conformational considerations and how these have led to more potent and selective delta ligands, and to greater insights into the three-dimensional structural pharmacophore for agonists at the delta opioid receptor, and to testable hypotheses about the topographical nature of the delta opioid pharmacophore.

2 CONFORMATIONALLY CONSTRAINED DELTA OPIOID LIGAND

Several early linear analogues of the enkephalins such as [D-Ala^2,D-Leu^5]enkephalin and [D-Ser^2,Leu^5,Thr^6]enkephalin (DSLET) [15] were somewhat delta opioid receptor selective, and others such as H-Tyr-D-Ala-Gly-N-MePhe-NH-CH_2-CH_2-OH (DAMGO, [16] were somewhat mu opioid receptor selective (reviewed in [8]). However, it was the design and discovery of cyclic-constrained peptide opioid ligands that led to the first potent and selective delta opioid ligands.

Using D-diamino-α-amino acids in position 2 of enkephalin, DiMaio, Schiller, and co-workers [11,17] made a series of cyclic enkephalin analogues in which the ε-amino group was cyclized to the C-terminal group. Most of these analogues were slightly mu-opioid receptor selective. Mosberg, Hruby, and co-workers substituted the constrained amino acid D-penicillamine (D-β,β-dimethylcysteine, D-Pen) (Fig. 1) into the 2-position of enkephalin and/or a D-Cys (DPDCE) or D-Pen (DPDPE) in the 5-position of enkephalin and prepared 14-membered-ring compounds that were highly delta opioid receptor selective [12–14] especially as measured in the classical mouse vas deferens (delta) versus guinea pig ileum (mu) in vitro. Also in binding assays, using delta receptors from both rat brain and NG108-15 cells containing the delta opioid receptor and using association rate constants and dissociation rate constants to get true equilibrium binding constants [19], DPDPE showed high delta opioid receptor selectivity. DPDPE has been an extremely useful ligand from several perspectives. Not only is it a potent and highly delta opioid receptor selective ligand both in binding [18] and, especially, in in vitro functional assays such as the MVD (delta) vs. GPI (mu) assays [12,13], but it also maintains this high selectivity in in vivo functional analgesic assays. Furthermore, it does not have the gastrointestinal and other undesirable side effects of mu agonists [19]. Moreover, it is very stable to proteolytic degradation both to whole-brain homogenates and to all protease enzymes against which it has been tested [20,21]. It does cross the blood–brain-barrier (BBB) well, similar to morphine [20,22–26], but unfortunately is not potent when given peripherally, presumably because it readily is pumped out of the brain as well. Glycopeptide analogues related to DPDPE, on the other hand, do not cross the BBB any better than DPDPE, but nonetheless are much more efficacious analgesics when given peripherally [27–30] and appear to be excellent candidates as novel opioid analgesics without the undesirable side effects of current opioids. These compounds have mixed agonist activities at the delta and mu opioid receptors, much like the superpotent delta/mu ligand biphalin [31], which also has few of the toxic side effects associated with mu opioid ligands [31] and crosses the BBB as well [32,33].

Because of its unique biological properties and high stability, as well as its cyclic-constrained structure, DPDPE and its analogues, especially [L-Ala^3]- and [D-Ala^3]DPDPE, which have unique biological profiles [34], have been the subject of numerous structure–biological activity studies (see [35] for a recent review), as well as extensive biophysical studies including NMR in conjunction with computational chemistry [36–38], X-ray crystallography [39 40], molecular dynamics [41], and evaluation of its three-dimensional pharmacophore [42,43], all of which have been used for the design of new ligands as well as for de novo design of highly potent and delta opioid receptor selective nonpeptide ligands [42]. In addition, the [D-Cys^2,des-Gly^3] analogue of DPDPE, H-Tyr-c[D-Cys,Phe-D-Pen]-OH (JOM-13) [44], though slightly less selective than DPDPE also has received extensive examination using similar biophysical tools [45], reviewed in [46] and has been used to explore the bound structure to the delta opioid receptor. These aspects of delta ligands are discussed further below.

3 OTHER HIGHLY POTENT SELECTIVE DELTA OPIOID RECEPTOR ANALOGUES

We consider selectivity for delta over mu of 1000-fold in either binding affinity and/or biological activity (MVD vs. GPI) to indicate a highly selective ligand. Figure 2 lists a few such analogues with the corresponding binding affinity (IC_{50}) ratios and/or bioassay potency (EC_{50}) ratios. As can be seen in Figure 2, there are two major ways to increase selectivity of cyclic enkephalin analogues for the delta opioid receptor: 1) extend the cyclic structure of DPLPE at the C-terminal residues; and/or 2) substitute the Phe^4 residue with halogens in the para position. In most cases increased selectivity is a result of both enhanced affinity (or efficacy) at the delta opioid receptor relative to

Compound – IC_{50} (nM), EC_{50} (nM)	Ref.
H-Tyr-c[D-Pen-Gly-Phe-L-Cys]-Trp-OH IC_{50} ratio – 2,500; EC50 ratio – 670; MVD EC_{50} – 42 nM	51
H-Tyr-c[D-Pen-Gly-Phe(p-F)-L-Pen]-Phe-OH IC_{50} ratio – 3,800; EC_{50} ratio – 45,000; MVD EC_{50} – 16 pM	52
H-Tyr-c[D-Pen-Gly-Phe(p-Br)-L-Pen]-Phe-OH IC_{50} ratio – 21,000; EC_{50} ratio – 19,000; MVD EC_{50} – 180 pM	52
H-Tyr-c[D-Pen-Gly-Phe(pI)-D-Pen]-OH EC_{50} ratio – 17,400; MVD – 2.6 nM	53

FIGURE 2 Highly delta opioid receptor selective agonists.

DPDPE *and* decreased affinity/efficacy at the mu opioid receptor. In these cases perhaps it can be stated that the ligands are now specific for the delta opioid receptor as agonists.

4 DELTORPHINS AS DELTA OPIOID RECEPTOR LIGANDS

As mentioned previously, the most delta opioid receptor selective naturally occurring ligands are the deltorphins which are found in the skins of frogs. These compounds have D-amino acids in the 2-position (D-Ala,*D*Met) and the epimerization has been shown to be a post-translational modification of the peptide structure mediated by a specific enzyme. There have been extensive structure–activity studies that have been well reviewed [10]. The use of highly constrained β-Me-2′,6′-Me_2-Tyr (TMT) and β-iPr-Phe analogues of deltorphin have been examined in linear analogues [47–49] and in cyclic analogues to provide insight into the preferred side chain conformation for these linear delta opioid receptor ligands. One of the most interesting amino acids used for structure–activity studies has been the 2′,6′-dimethyl Tyr (DMT) analogues, which have led to highly potent delta receptor ligands. This area has also been reviewed [10] and will not be discussed here.

5 BIOACTIVE CONFORMATION OF CONFORMATIONALLY CONSTRAINED DELTA PEPTIDE LIGANDS

The development of cyclic, conformationally restricted opioid peptide ligands allowed significant advances in the elucidation of the bioactive conformation(s) of opioid peptides, since these more rigid analogues are less subject to the dynamic averaging that typifies flexible, linear peptides. As the first of the conformationally constrained, highly delta selective peptides, DPDPE was an especially attractive target for conformational analysis. Several groups, ourselves included, proposed binding conformations for DPDPE based upon experimental and/or computational results [50–57]. Typically these studies revealed small sets of possible conformational families, several of which were represented in more than one report. However, no consensus-proposed bioactive model was found and, indeed, considerable differences are apparent. It is likely that residual conformational lability in DPDPE underlies much of this disagreement, a view consistent with results from molecular dynamics simulations [58,59]. Three causes for this residual flexibility are likely: the flexibility of the central glycine residue; the disulfide bridge, which can be right-handed or left-handed (±~115°), which together result in several accessible ring conformations [60]; and the expected conformational lability of the exocyclic Tyr^1 residue (and to a lesser extent the Phe^4 side chain) [39].

Two approaches have been employed to reduce the residual flexibility of the Gly^3 residue of DPDPE. In the first of these, a series of cyclic disulfide containing tetrapeptides, representing des-Gly^3 DPDPE analogues, was explored [44]. The smaller (vs. DPDPE) constrained 11-membered tripeptide ring in this series results in a slight divergence in structure/conformation–activity relations compared to DPDPE in that replacement of the D-Pen^2 residue of DPDPE by D-Cys^2 is necessary for optimal delta binding affinity. The resulting JOM-13 has, as noted previously, somewhat improved delta binding affinity compared to DPDPE, but is not quite as selective. Interestingly, this tetrapeptide series yielded a structurally related analogue, JOM-6, differing from JOM-13 only in having a C-terminal carboxamide (vs. the carboxylate of JOM-13) and in being cyclized via an ethylene dithioether (vs. the JOM-13 disulfide), yet displaying a shift to a high affinity, fairly selective mu ligand [44,60].

The second approach employed to reduce the flexibility attributable to the Gly^3 of DPDPE was to replace this glycine with other amino acids. In the first such example, Gly^3 was replaced by Aib (aminoisobutyric acid, α,α-dimethylglycine) [61]. The observation that Aib^3-DPDPE displayed in vitro binding and bioactivity behavior similar to DPDPE refuted the earlier dogma that substitution of Gly^3 in enkephalin analogues invariably leads to drastic activity losses and suggested that conformational analysis of this more constrained analogue might help resolve the uncertainty of the bioactive conformation of DPDPE and related structures. Subsequently, it was shown that small L-amino acid substitutions for Gly^3 in DPDPE were well tolerated and that [L-Ala^3]DPDPE, in particular, displayed similar delta binding affinity and improved delta selectivity compared with DPDPE [62].

Conformational analyses of JOM-13 and [L-Ala^3]DPDPE have proven to be critical for the determination of the bioactive conformation of enkephalin-like peptides at the delta receptor. ^{1}H-NMR studies of JOM-13 in aqueous solution revealed that this tetrapeptide exists in two distinct conformations on the NMR time scale as evidenced by two sets of resonances [63]. Large differences in the observed chemical shifts and coupling constants for the D-Cys^2 residue in the two conformers suggested that the major differences between the two NMR conformers reside in the disulfide portion of the molecule; however, a paucity of conformationally informative nuclear Overhauser enhancement (NOE) interactions precluded the development of a detailed structural model from the NMR studies. In order to develop such a model a thorough conformational analysis of JOM-13 was undertaken, in which the NMR data were complemented by x-ray diffraction results and by molecular mechanics calculations [64]. The results indicate that the 11-

membered cyclic structure in JOM-13 is quite well defined, but exists in two low-energy conformers that differ in orientation of the disulfide (similar to observations noted above for DPDPE), as originally predicted from the earlier NMR study [63]. Both conformers are structurally very similar, leading to the conclusion that the 11-membered ring forms a reasonably rigid structural scaffold that holds the pharmacophore elements (Tyr amine and phenolic groups and Phe aromatic ring) in an appropriate arrangement for interaction with the delta binding site.

Parallel studies probing the conformation of [L-Ala^3]DPDPE provided key insights into the bioactive conformation of DPDPE. As noted above, [L-Ala^3]DPDPE displays binding and in vitro bioassay properties similar to those of DPDPE, while [D-Ala^3]DPDPE is a somewhat poorer ligand. Subsequent elucidation of the x-ray structures of [L-Ala^3]DPDPE and [D-Ala^3]DPDPE [39] and their comparison with the x-ray structure of DPDPE [40] revealed that while all three structures exhibit differences in the disulfide regions, the backbone structures of the latter two are virtually identical, while that of [L-Ala^3]DPDPE differs in the region of residue 3 [39]. These results were originally suggested to represent distinct agonist (DPDPE and [D-Ala^3]DPDPE) and antagonist ([L-Ala^3]DPDPE) conformations; however, subsequently Shenderovich et al. [43] demonstrated that DPDPE could adopt a [L-Ala^3]DPDPE-like (i.e., like the x-ray structure of [L-Ala^3]DPDPE) conformation at little energy cost, while the DPDPE x-ray structure represented a high-energy conformation of [L-Ala^3]DPDPE. The authors concluded, then, that since both compounds display similar activity, the correct bioactive conformation resembles the x-ray structure of [L-Ala^3]DPDPE, not that of DPDPE. This conclusion was also arrived at by Mosberg and co-workers, who, likewise, reasoned that a similar binding conformation must exist for DPDPE, [L-Ala^3]DPDPE, and JOM-13 and proposed that this conformation resembled the [L-Ala^3]DPDPE x-ray structure [65,66].

While the studies noted above were crucial for defining the bioactive conformation of the tri- or tetrapeptide cycles of JOM-13 and DPDPE, elucidating the bioactive conformation of the highly flexible Tyr^1 residue required alternate approaches. To determine which side chain conformations are essential for binding affinity and selective biological activities of DPDPE and related peptides, Hruby and colleagues prepared the four diastereoisomeric analogues of DPDPE containing a modestly χ space constrained (i.e., with limited conformational possibilities about the C^{α}–C^{β} and C^{β}–C^{γ} bonds) Tyr analogue, β-Me-Tyr (four isomers) [67], and a highly biased χ space constrained Tyr analogue, β-Me-2′,6′-Me_2-Tyr (TMT, four isomers) [47] and examined their binding affinities and functional bioactivities at the MVD (delta) and GPI (mu) (see [68] for a detailed discussion of χ-constrained amino acids and their use in peptide and peptide mimetic design). As can be

TABLE 1 Binding Affinities and In Vitro Bioactivities of Tyr^1 Topographically Constrained Analogues of DPDPE in χ Space

Compound	Binding Affinities (nM)		Potencies	
	δ	μ	MVD (δ)	GPI (μ)
DPDPE	1.2	720	3.9	11,300
1 H-(2S,3S)-[β-$MeTyr^1$]DPDPE[a]	440	19,000	240	7,200
2 H-(2S,3R)-[β-$MeTyr^1$]DPDPE[a]	85	>20,000	15	34,000
3 H-(2R,3S)-[β-$MeTyr^1$]DPDPE[a]	~20,000	>40,000	700	49,000
4 H-(2R,3R)-[β-$MeTyr^1$]DPDPE[a]	430	23,000	21,000	>100,000
5 H-(2S,3S)-[β-Me-2′,6′-Me_2Tyr^1] DPDPE[b]	210	720	170	290
6 H-(2S,3R)-[β-Me-2′,6′-Me_2Tyr^1] DPDPE[b]	5.0	4,300	1.8	0% at 60 μM
7 H-(2R,3S)-[β-Me-2′,6′-Me_2Tyr^1] DPDPE[b]	0% at 10 μM	8% at 10 μM	28% at 10 μM	7% at 82 μM
8 H-(2R,3R)-[β-Me-2′,6′-Me_2Tyr^1] DPDPE[b]	3,500	77,000	2,200	50,000

[a] Data from [67].
[b] Data from [47].

seen in Table 1, substitution of the four isomers of β-Me-Tyr^1 into DPDPE reduces binding affinity and bioactivity at both the delta and mu receptor for all four isomers. The most potent and delta opioid receptor selective analogue is **2**, which shows about 70-fold loss in binding affinity relative to DPDPE, but is five times more potent than **1** and **4** in binding affinity and is more selective. The χ^1 bias from both NMR and computational studies [68] for the (2S,3R)β-Me-Tyr^1 residue is for the trans (±180°) χ^1 conformer. This conformational requirement is even more apparent from the MVD vs. GPI results, in which the selectivity for **2** is now >2000, and the potency at the MVD is 15 nM, similar to that observed for DPDPE, itself. All other isomers have binding affinities and bioactivities in the high nanomolar to micromolar range, and all other isomers (**1,3,4**) are somewhat delta opioid receptor selective. Interestingly, when the more bulky and constrained amino acid TMT is placed in the 1-position a generally more potent and selective series of analogues is obtained (Table 1) [47] than for the β-Me-Tyr series, and all four isomers retain the same conformation as DPDPE in the cyclic 14-membered ring. Now, however, a much greater differentiation in potency and selectivity is obtained, with the (2S,3R) analogue **6** having very similar binding affinities and biological potencies as DPDPE except for the very interesting antagonist activity at the mu opioid receptor [47] in the in vitro GPI assay. Furthermore, now due to the large energy difference between *g*(−), *g*(+) and trans for the χ^1 torsional angle for various TMT isomers [see 47,68], and the fact that the (2S,3R)TMT isomer is biased toward the trans conformation (±180°), it can be concluded that for the delta opioid receptor, the Tyr in the trans conformation is greatly preferred by 2 to 3 kcal/mole, the same energy difference as the energy penalty the 2S,3S isomer would have to pay if it assumed a trans χ^1 conformation. Finally, it is interesting to note that the (2S,3S)-TMT-containing analogue **5**, has the same binding affinity for the mu receptor as DPDPE (Table 1), and as a balanced delta/mu ligand has the most potent antinociceptive activity of any of the analogues [48].

Mosberg and co-workers utilized a similar type of approach for constraining the flexibility of Tyr in JOM-13 [69]. In an effort to better determine the binding conformation of the Tyr^1 residue, a series of conformationally constrained analogues of Tyr (HO-Tic: 7-hydroxy-1,2,3,4-tetrahydroisoquinoline-3- carboxylic acid; Hai: 6-hydroxy-2-aminoindan-2-carboxylic acid; Hat: 6-hydroxy-2-aminotetralin-2-carboxylic acid, and c-Hpp and t-Hpp:

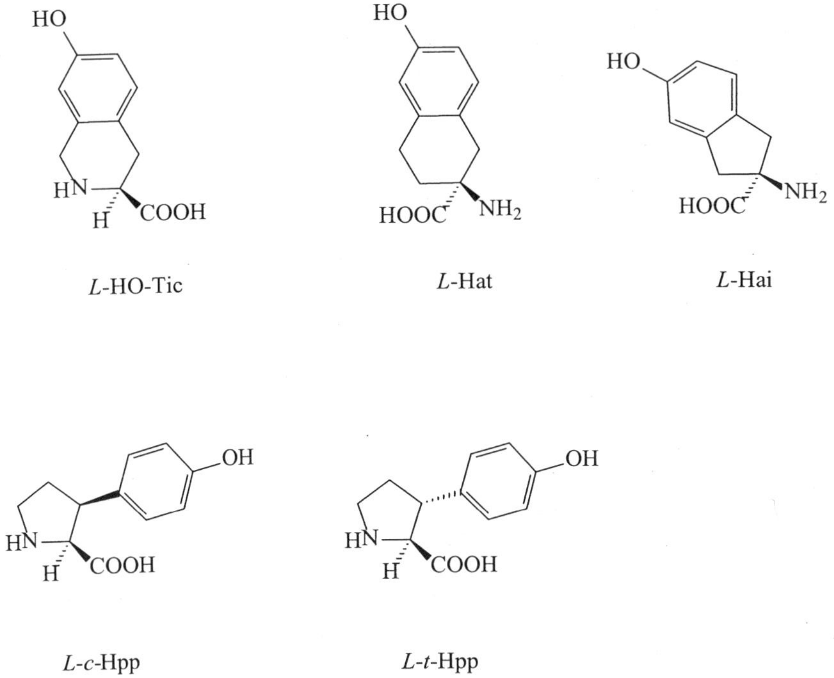

FIGURE 3 Constrained tyrosine replacements.

cis- and trans-3-(4′-hydroxy)-phenylproline, respectively, Fig. 3), were prepared and each was utilized as a Tyr[1] replacement in analogues of JOM-13. For each resulting analogue, the range of conformational space available to the Tyr replacement residue is somewhat different, while the conformation of the 11-membered ring is the same (and identical to that in JOM-13). Consequently, if more than one of the resulting analogues displayed good delta receptor binding affinity, then the binding conformation of the Tyr residue in JOM-13 would be confined to the region of energetically accessible conformational space common to the Tyr replacement in all the high affinity analogues. As shown in Table 2, both the *c*-Hpp and *t*-Hpp analogues exhibit binding affinities similar to those of JOM-13. Examining the intersection of energetically accessible conformational space for these two analogues reduced the likely bioactive conformations of Tyr[1] in the parent, JOM-13, to only two possibilities, differing only in the choices of φ^2 (the angle about the N^α–C^α bond of D-Cys[2]), which could be either ~160° or ~70°. In both conformers χ^1 for the Tyr side chain is ~180° and ψ^1 (the angle about the Tyr C^α–CO bond) is ~160°. Since a residue 1 conformation with $\varphi^2 \sim 160°$ allowed better superpositions of the high-affinity c-Hpp[1], t-Hpp[1], and Tyr[1] analogues with the moderate affinity Hat[1] and Hai[1] analogues, we proposed [69] that this was the more likely binding conformation, the underlying assumption being that the binding conformations and binding modes of structurally similar ligands would be similar. However, ligand–receptor docking of JOM-13 clearly shows that only the *alternate*, $\varphi^2 \sim 70°$ conformer can fit the binding pocket (see below).

To eliminate the remaining uncertainty regarding the bioactive conformations of DPDPE and JOM-13, the conformation of the Phe[3,4] side chain, Phe in these peptides was replaced by β-MePhe (all four isomers). The results [70,71] clearly suggested that both peptides bind to the delta receptor

TABLE 2 Binding Affinities of [X[1]]JOM-13 Analogues[a]

Structure	K_i (nM) [³H]DAMGO	K_i (nM) [³H]DPDPE	$K_i(\mu)/K_i(\delta)$
Tyr-c[D-Cys-Phe-D-Pen]OH (JOM-13)	52±4.4	0.74±0.08	70
Hat-c[D-Cys-Phe-D-Pen]OH	230±17	20±4.4	12
Hai-c[D-Cys-Phe-D-Pen]OH	840±40	13±1.2	65
HO-Tic-c[D-Cys-Phe-D-Pen]OH	>10,000	2400±440	>4.2
t-Hpp-c[D-Cys-Phe-D-Pen]OH	110±19	0.66±0.06	170
c-Hpp-c[D-Cys-Phe-D-Pen]OH	720±57	2.4±0.18	300

[a] Data from [69].

with $Phe^{3,4}$ in a gauche ($\chi^1 = -60°$) conformation. This was confirmed in both peptide series by replacing Phe with Δ^EPhe (in which the phenyl side chain is locked into a trans orientation) and with Δ^ZPhe (which approximates the proposed gauche orientation). In both JOM-13 [72] and DPDPE [73], the Δ^ZPhe-containing analogue displayed considerably higher binding affinity (20–400 fold) than the corresponding Δ^EPhe analogue.

6 OPIOID RECEPTORS AND LIGAND/RECEPTOR BINDING MODELS

Opioid receptors belong to the rhodopsin-like G protein–coupled receptor (GPCR) family, a large (>1000 sequences) group of structurally related transmembrane proteins that includes receptors for ligands of vastly varying size (biogenic amines to glycoproteins) [74]. With the publication in the mid-1990s of low-resolution electron microscopy structures of rhodopsin [75–78] it became possible to construct models of other GPCRs in the rhodopsin family. Mosberg and co-workers developed a distance geometry–based approach that allowed the refinement of such models and employed this approach to propose structural models of mu, delta, and kappa opioid receptors [66,78]. These receptor models, which were developed without ligand-related structural bias, allowed proposed bioactive conformations of DPDPE, JOM-13, and related ligands to be docked and evaluated. As described previously [66] the docking procedure employed key ligand–receptor interactions, known from extensive literature reports, and was done in a stepwise fashion, starting first with rigid alkaloid ligands and proceeding through known covalent ligands, to DPDPE and JOM-13. As alluded to above, from docking of JOM-13 to the delta receptor, it was immediately obvious that the originally proposed bioactive conformation for Tyr^1 (with $\varphi^2 \sim 160°$) was incompatible with receptor binding site and that the alternative conformation ($\varphi^2 \sim 70°$) was indeed correct.

The publication of the x-ray structure for rhodopsin [79] has made possible the construction of better receptor models. Using such homology models for the delta and mu receptor, Mosberg and co-workers have been able to identify specific structural features of the mu and delta receptor ligand binding sites that form the basis of the delta selectivity of JOM-13 and the mu selectivity of the structurally related JOM-6 (differing from JOM-13 only in having a C-terminal carboxamide and, in ring size, being cyclized as the ethylene dithioether) [60]. In particular, the presence of E229 in the mu receptor (in place of the corresponding D210 of the delta receptor) causes an adverse electrostatic interaction with C-terminal carboxylate-containing ligands, resulting in the observed preference of ligands with an uncharged C-terminus for the mu receptor. Unlike JOM-13 (and DPDPE) in which a gauche orientation of the Phe^3 side chain is required for optimal delta receptor binding, the Phe^3 side chain of JOM-6 must be in a trans orientation for high-affinity mu binding. This difference can be largely attributed to the steric effect of replacement of L300 of the delta receptor by W318 of the mu receptor. Consequently a gauche side chain orientation is energetically disfavored. These hypotheses were validated by preparing appropriate receptor mutants in which proposed key residues were mutated and examining the binding behavior of JOM-13, JOM-6, and analogues combining features of each [80].

7 FUTURE DIRECTIONS

The development and validation of models for the bioactive conformation of constrained delta receptor peptide ligands, when considered in combination with known SAR for these ligands and with consistent receptor structural models, open the door to exciting new prospects. Chief among these are the structure-based design of nonpeptide delta ligands with improved metabolic resistance and better bioavailability (e.g. [49]), the extrapolation of ligand–receptor interaction models to the design of selective mu, kappa, or orphanin receptor ligands, and the development of an understanding (and exploitation of this understanding) of the structural basis (ligand and receptor) of agonism versus antagonism. In this regard the application of the recently developed plasmon waveguide resonance (PWR) spectroscopy showing that delta agonist, antagonist, and inverse agonist ligand binding to the human delta opioid receptor leads to different conformational states of the ligand–receptor complex [81,82] offers exciting possibilities for new insights into the structural and dynamic bases for these biological effects in transduction. Advances in these areas will not only provide critical insights into the relationship of structure and function, but also have a major impact on health care.

ACKNOWLEDGMENTS

This work was supported by grants from NIDA to VJH and HIM. We also gratefully acknowledge the outstanding collaboration of our students, postdoctoral associates, and pharmacological and biophysical colleagues, without whose efforts this report would not be possible.

REFERENCES

1. Hughes J, Smith TW, Kosterlitz HW, Fothergill LA, Morgan BA, Morris HR. Nature 1975; 258:577–579.

2. Martin WR, Eades CD, Thompson JA, Huppler RA, Gilbert PE. J Pharmacol Exp Therap 1976; 197:517–532.
3. Iwamoto ET, Martin WR. Med Res Rev 1981; 1:411–440.
4. Evans CJ, Keith DE Jr, Morrison H, Magendzo K, Edwards RH. Science 1992; 258:1951–1955.
5. Kieffer BL, Befort K, Gaveriaux-Ruff C, Herth CG. Proc Natl Acad Sci USA 1992; 89:10252–12048.
6. Knapp RJ, Maltynska E, Fang L, Li X, Babin E, Nguyen M, Santoro G, Varga EV, Hruby VJ, Roeske WK, Yamamura HI. Life Sci (Pharm Lett) 1994; 54:PL 463–469.
7. Miotto K, Magendzo K, Evans CJ. In: Tseong LF, ed. The Pharmacology of Opioid Peptides. New Jersey: Harwood, 1995:57–71.
8. Hruby VJ, Gehrig C. Med Res Rev 1989; 9:343–401.
9. Ersparmer V, Melchoris P, Falconieri-Erspamer G, Negri L, Serereni C, Barr D, Simmaco M, Kreil G. Proc Natl Acad Sci USA 1989; 86:5188–5192.
10. Lazarus LH, Bryant SD, Cooper PS, Salvadori S. Prog Neurosci 1999; 57: 377–420.
11. DiMaio J, Schiller PW. Proc Natl Acad Sci USA 1980; 77:7262–7266.
12. Mosberg HI, Hurst R, Hruby VJ, Galligan JJ, Burks TF, Gee K, Yamamura HI. Life Sci 1983; 32:2565–2569.
13. Mosberg HI, Hurst R, Hruby VJ, Gee K, Yamamura HI, Galligan JJ, Burks TF. Proc Natl Acad Sci USA 1983; 80:5871–5874.
14. Mosberg HI, Hurst R, Hruby VJ, Gee K, Akiyama K, Yamamura HI, Galligan JJ, Burks TF. Life Sci 1984; 33(suppl 1):447–450.
15. Fournié-Zaluski MC, Belleney J, Gaeel G, Margret B, Roques BP. Mol Pharmacol 1981; 20:484–491.
16. Handa BK, Lane AC, Lord JAH, Morgan BA, Rance MJ, Smith CFC. Eur J Pharmacol 1981; 70:531–540.
17. Maio JD, Nguyen TM-D, Lemieux C, Schiller PW. J Med Chem 1982; 25: 1432–1438.
18. Akiyama K, Gee KW, Mosberg HI, Hruby VJ, Yamamura HI. Proc Natl Acad Sci USA 1985; 82:2543–2547.
19. Porreca F, Mosberg HI, Hurst R, Hruby VJ, Burks TF. Life Sci 1983; 33 (suppl 1):457–460.
20. Weber SJ, Greene DL, Hruby VJ, Yamamura HI, Porreca F, Davis TP. J Pharmacol Exp Therap 1992; 263:1308–1316.
21. Greene DL, Hau VS, Abbruscato TJ, Bartosz H, Misicka A, Lipkowski AW, Hom S, Gillespie TJ, Hruby VJ, Davis TP. J Pharmacol Exp Therap 1996; 277:1366–1375.
22. Weber SJ, Greene DL, Sharma SD, Yamamura HI, Kramer TH, Burks TF, Hruby VJ, Hersh LB, Davis TP. J Pharmacol Exp Therap 1991; 259:1109–1117.
23. Weber SJ, Abbruscato TJ, Brownson EA, Lipkowski AW, Polt R, Misicka A, Haaseth RC, Bartosz H, Hruby VJ, Davis TP. J Pharmacol Exp Therap 1993; 266:1649–1655.

24. Hruby VJ, Davis TP, Polt R, Porreca F, O'Brien D, Yamamura HI, Bartosz H, Szabo L, Gillespie TJ, Williams SA, Misicka A, Lipkowski AW, Qian X, Li G, Patel D, Bonner G. Analgesia 1995; 1:469–472.
25. Williams SA, Abbruscato TJ, Hruby VJ, Davis TP. J Neurochemistry 1996; 66:1289–1299.
26. Thomas SA, Abbruscato TJ, Hruby VJ, Davis TP. J Pharmacol Exp Therap 1997; 280:1235–1240.
27. Polt R, Porreca F, Szabo LZ, Bilsky EJ, Davis P, Abbruscato TJ, Davis TP, Horvath R, Yamamura HI, Hruby VJ. Proc Natl Acad Sci USA 1994; 91:7114–7118.
28. Polt R, Szabo L, Hruby VJ, Davis TP, Porreca F, Yamamura HI. In: Epton R, ed. Solid Phase Synthesis and Combinatorial Libraries. Kingswinford, England: Mayflower Scientific, 1996:277–280.
29. Egleton RD, Mitchell SA, Huber JD, Janders J, Stropova D, Polt R, Yamamura HI, Hruby VJ, Davis TP. Brain Res 2000; 881:37–46.
30. Bilsky EJ, Egleton RD, Mitchell SA, Palian MM, Davis P, Huber JD, Jones H, Yamamura HI, Janders J, Davis TP, Porreca F, Hruby VJ, Polt R. J Med Chem 2000; 43:2586–2590.
31. Horan PJ, Mattia A, Bilsky EJ, Weber SJ, Davis TP, Yamamura HI, Malatynska E, Applyard SM, Slaninová J, Misicka A, Lipkowski AW, Hruby VJ, Porreca F. J Pharmacol Exp Therap 1993; 265:1446–1454.
32. Abbruscato TJ, Williams SA, Misicka A, Lipkowski AW, Hruby VJ, Davis TP. J Pharmacol Exp Therap 1996; 276:1049–1057.
33. Abbruscato TJ, Thomas SA, Hruby VJ, Davis TP. J Neurochemistry 1997; 69:1236–1245.
34. Haaseth RC, Horan PH, Bilsky EJ, Davis P, Zalewska T, Slaninova J, Yamamura HI, Weber SJ, Davis TP, Porreca F, Hruby VJ. J Med Chem 1994; 37:1572–1577.
35. Hruby VJ, Agnes RS. Biopolymers (Peptide Sci) 2000; 51:391–410.
36. Hruby VJ, Kao L-F, Pettitt BM, Karplus M. J Am Chem Soc 1988; 110: 3351–3359.
37. Matsunaga TO, Collins N, Yamamura S, Ramaswami V, O'Brien DF, Hruby VJ. Biochemistry 1993; 32:13180–13189.
38. Shenderovich MD, Kövér KE, Nikiforovich GV, Jiao D, Hruby VJ. Biopolymers 1996; 38:141–156.
39. Collins N, Flippen-Anderson JL, Haaseth RC, Deschamps JR, George C, Kövér K, Hruby VJ. J Am Chem Soc 1996; 118:2143–2152.
40. Flippen-Anderson JL, Hruby VJ, Collins N, George C, Cudney B. J Am Chem Soc 1994; 116:7523–7531.
41. Pettitt BM, Matsunaga TO, Al-Obeidi F, Gehrig CA, Hruby VJ, Karplus M. Biophysical J 1991; 60:1540–1544.
42. Liao S, Alfaro-Lopez J, Shenderovich MD, Hosohata K, Lin J, Li X, Stropova D, Davis P, Jernigan KA, Porreca F, Yamamura HI, Hruby VJ. J Med Chem 1998; 41:4767–4776.
43. Shenderovich MD, Liao S, Qian X, Hruby VJ. Biopolymers 2000; 53:565–580.

44. Mosberg HI, Omnaas JR, Smith CB, Medzihradsky F. Life Sci 1988; 43: 1013–1020.
45. Lomize AL, Flippen-Anderson JL, George C, Mosberg HI. J Am Chem Soc 1994; 116:429–436.
46. Mosberg HI. Biopolymers (Peptide Sci) 1999; 51:426–439.
47. Qian X, Shenderovich MD, Kövér KE, Davis P, Horváth R, Zalewska T, Yamamura HI, Porreca F, Hruby VJ. J Am Chem Soc 1996; 118:7280–7290.
48. Bilsky EJ, Qian X, Hruby VJ, Porreca F. J Pharmacol Exp Therap 2000; 293: 151–158.
49. Liao S, Shenderovich M, Kövér KE, Zhang Z, Hosohata K, Davis P, Porreca F, Yamamura HI, Hruby VJ. J Peptide Res 2001; 57:257–276.
50. Mosberg HI, Sobczyk-Kojiro K, Subramanian P, Crippen GM, Ramalingam K, Woodard RW. J Am Chem Soc 1990; 112:822–829.
51. Nikiforovich GV, Balodis J. FEBS Lett 1988; 227:127–130.
52. Hruby VJ, Kao LF, Pettitt BM, Karplus M. J Am Chem Soc 1988; 110:3351–3359.
53. Keys C, Payne P, Amsterdam P, Toll L, Loew G. Mol Pharmacol 1988; 33:528–536.
54. Froimowitz M. Biopolymers 1990; 30:1011–1025.
55. Froimowitz M, Hruby VJ. Int J Peptide Prot Res 1989; 34:88–96.
56. Nikiforovich GV, Golbraikh AA, Shenderovich MD, Balodis J. Int J Peptide Prot Res 1990; 36:209–218.
57. Wilkes BC, Schiller PW. In: Rivier JE, Marshall GR, eds. Peptides: Chemistry, Structure, and Biology. Leiden, The Netherlands: ESCOM, 1990:341–343.
58. Smith PE, Dang LX, Pettitt BM. J Am Chem Soc 1991; 113:67–73.
59. Pettitt BM, Matsunaga TO, Al-Obeidi F, Gehrig CA, Hruby VJ, Karplus M. Biophys J 1991; 60:1540–1544.
60. Mcfadyen IJ, Ho JC, Mosberg HI, Traynor JR. J Peptide Res 2000; 55:255–261.
61. Haaseth RC, Sobczyk-Kojiro K, Medzihradsky F, Smith CB, Mosberg HI. Int J Peptide Prot Res 1990; 36:139–146.
62. Haaseth RC, Horan PJ, Bilsky EJ, Davis P, Zalewska T, Slaninova J, Yamamura HI, Weber SJ, Davis TP, Porreca F, Hruby VJ. J Med Chem 1994; 37:1572–1577.
63. Mosberg HI, Sobczyk-Kojiro K. In: Renugopalakrishnan V, Carey PR, Smith ICP, Huang SG, Storer AC, eds. Proteins: Structure, Dynamics and Design. Leiden, The Netherlands: ESCOM, 1991:105–109.
64. Lomize AL, Flippen-Anderson JL, George C, Mosberg HI. J Am Chem Soc 1994; 116:429–436.
65. Lomize AL, Pogozheva ID, Mosberg HI. Biopolymers 1996; 38:221–234.
66. Pogozheva ID, Lomize AL, Mosberg HI. Biophys J 1998; 75:612–634.
67. Toth G, Russell KC, Landis G, Kramer TH, Fang L, Knapp R, Davis P, Burks TF, Yamamura HI, Hruby VJ. J Med Chem 1992; 35:2384–2391.
68. Hruby VJ, Li G, Haskell-Luevano C, Shenderovich MD. Biopolymers (Peptide Sci) 1997; 43:219–266.

69. Mosberg HI, Lomize AL, Wang C, Kroona H, Heyl DL, Ma W, Sobczyk-Kojiro K, Mousigian C, Porreca F. J Med Chem 1994; 37:4371–4383.
70. Hruby VJ, Toth G, Gehrig C, Kao L, Knapp R, Lui G, Yamamura HI, Kramer T, Davis P, Burks TF. J Med Chem 1991; 34:1823–1830.
71. Mosberg HI, Omnaas JR, Lomize A, Heyl DL, Nordan I, Mousigian C, Davis P, Porreca F. J Med Chem 1994; 37:4384–4391.
72. Mosberg HI, Dua RK, Pogozheva ID, Lomize AL. Biopolymers 1996; 39:287–296.
73. Mosberg HI, unpublished observations.
74. Watson S, Arkinstall S. The G-Protein Linked Receptor Facts Book. San Diego: Academic Press, 1994.
75. Unger VM, Schertler GFX. Biophys J 1995; 68:1776–1786.
76. Unger VM, Hargrave PA, Baldwin JM, Schertler GFX. Nature 1997; 389: 203–211.
77. Davies A, Schertler GFX, Gowen BE, Saibil HR. J Structur Biol 1996; 117: 36–44.
78. Pogozheva ID, Lomize AL, Mosberg HI. Biophys J 1997; 72:1963–1985.
79. Palczewski K, Kumasaka T, Hori T, Behnke CA, Motoshima H, Fox BA, Le Trong I, Teller DC, Okada T, Stenkamp RE, Yamamoto M, Miyano M. Science 2000; 289:739–745.
80. Mosberg HI, Fowler CB. J Peptide Res 2002; 60:329–335.
81. Salamon Z, Cowell S, Varga E, Yamamura HI, Hruby VJ, Tollin G. Biophys J 2000; 79:2463–2474.
82. Salamon Z, Hruby VJ, Tollin G, Cowell S. J Peptide Res 2002; 60:322–328.

11

Deltorphins

Lucia Negri and Elisa Giannini
University La Sapienza, Rome, Italy

1 INTRODUCTION

> We started out our studies on amphibian opioid peptides in the wake of the explosion of research on mammalian endogenous ligands for opiate receptors. We asked ourselves whether the never disappointing amphibian skin contained related molecules. The question was reasonably posed in the light of the outcome of our previous, yearlong research showing that amphibian skin peptides often had counterparts in mammalian CNS and gastrointestinal tract [1].

With these words Vittorio Erspamer depicted the rational basis of the pharmacological research on peptides and proteins of the amphibian skin. This 10-year period of research has shown that amphibian skin, with its generally large peptide content, offers a rich source of secretory peptide compounds, analogues of mammalian neuropeptides and hormones.

The first peptide family of amphibian opiates was discovered in 1981 and named dermorphins [2,3]. Until the discovery of mammalian endomorphins by Zadina et al. [4], these peptides represented the most potent and selective mu opiate receptor agonists identified in living organisms. Nine years later, deltorphins were discovered in the amphibian skin. These peptides are still the most potent and selective delta opiate agonists available today [5].

The structure of the deltorphins provided the basis for synthetic derivatives and analogues that clarified the functional role of the delta opiate system.

A unique characteristic of amphibian opioid peptides is the presence in the second N-terminal position of a D-amino acid residue that confers to these compounds high resistance against enzyme degradation. Hence amphibian opioids, unique among naturally occurring opioid peptides, can act centrally after peripheral administration.

Amphibian opiate peptides have been found only in the skin of South American hylid frogs belonging to the subfamily Phyllomedusinae (*Phyllomedusa, Agalychnis, and Pachymedusa* spp.). Although pharmacologists discovered these opiates in Amazonian frogs comparatively recently, the Matses of the upper Amazonian basin unveiled the pharmacological properties of amphibian skin opiates long ago. For centuries they had habitually applied the dried skin secretions of *Phyllomedusa bicolor*, called "sapo" (the Spanish word for toad), to cuts in their skin during shamanistic hunting rituals. The abundance of deltorphins and dermorphins acting together with or probably synergistically with the other active peptides present in these secretions (caerulein, phyllokinin, phyllomedusin, sauvagine, adrenoregulin, and other, still unknown substances) might have caused the hunters' analgesia and behavioral excitation [6].

The D-amino acid–containing opioid peptides issue from precursors showing a common preproregion (22-residue signal peptide and an 18- to 25-residue acidic prosequence) with precursors of the peptide antibiotics dermaseptins (24- to 34-residue polycationic and α-helical amphipathic peptides). Of the three types of dermal glands (mucous, lipid, and serous) in the skin of *Phyllomedusa bicolor*, only the serous glands are specifically involved in the biosynthesis and secretion of dermaseptins and deltorphins. The serous glands are the largest glands in the *Phyllomedusa* skin, they lie deeper in the epidermis, they are lined by epithelium that is more a syncytium, and they are surrounded by a layer of myoepithelial cells involved in the holocrine rapid discharge of secretory products collected in roundish granules. The granules do not bud off from the membranes on the Golgi apparatus, but seem to be generated in the vacuoles of the vacuolated stage during gland development. Although mucous and serous glands were detected in skin of *Ph. bicolor* tadpoles, the serous glands gain access to the outer surface of the skin only after metamorphosis [7].

2 A BRIEF HISTORY OF THE DISCOVERY OF DELTORPHINS (TABLE 1)

The screening of a cDNA library prepared from the skin of *Ph. sauvagei* established the amino acid sequence of several dermorphin precursors. The

TABLE 1 Natural Occurring Deltorphins and Their Origin

Tyr-D-Met-Phe-His-Leu-Met-Asp-NH2	D-Met-deltorphin, Ph. sauvagei
Tyr-D-Ala-Phe-Asp-Val-Val-Gly-NH2	D-Ala-deltorphin-I, Ph. bicolor
Tyr-D-Ala-Phe-Glu-Val-Val-Gly-NH2	D-Ala-deltorphin-II, Ph. bicolor
Tyr-D-Leu-Phe-Ala-Asp-Val-Ala-Ser-Thr-Ile-Gly-Asp-Phe-Phe-His-Ser-Ile-NH2	D-Leu-deltorphin-17, Ph. burmaisteri
Tyr-D-Ile-Phe-His-Leu-Met-Asp-NH2	D-Ile2-deltorphin, Pachymedusa dacnicolor, Agalychnis annae

sequence of one of these cDNAs indicated the existence of another peptide that contained methionine as the second amino acid [8]. This peptide was subsequently isolated from the skin of *Ph. sauvagei* and proved to have higher affinity and selectivity for δ-opiate receptors than any other known natural compound [9–11]. It has been given various names: deltorphin, D-Met-deltorphin, dermenkephalin, and deltorphin A. Here we will use the original name, D-Met-deltorphin.

Subsequently, two additional peptides with even higher affinity for the delta opiate receptor were isolated from the skin of *Ph. bicolor* [12]. Like dermorphin, these peptides contain D-alanine as the second amino acid. They have been termed D-Ala-deltorphin-I and D-Ala-deltorphin-II.

Screening of cDNA libraries from the skin of *Ph. bicolor* revealed the sequence of four precursors for D-Ala-deltorphin-I and -II [13]. In the Brazilian frog *Ph. burmaisteri*, Barra et al. [14] identified another deltorphin-like peptide, Tyr-D-Leu-Phe-Ala-Asp-Val-Ala-Ser-Thr-Ile-Gly-Asp-Phe-Phe-His-Ser-Ile-NH2, which they termed D-Leu-deltorphin-17 due to the D-Leu at position 2 of its linear chain of 17 amino acid residues. During screening of cDNA libraries prepared from skin of two additional species of Phyllomedusinae, the *Pachymedusa dacnicolor* and *Agalychnis annae*, using sequence information from cDNAs encoding dermorphin and deltorphin precursors from *Ph. sauvagei* and *Ph. bicolor*, Wechselberger et al. [15] identified, in addition to four copies of dermorphin, a sequence that contains the genetic information for a novel peptide: Tyr-D-Ile-Phe-His-Leu-Met-Asp-NH2. This peptide, analogue to D-Met-deltorphin, contains a D-Ile at the second position.

All the natural deltorphins, like the mu agonist dermorphins, contain the N-terminal sequence Tyr-D-Xaa-Phe, where the aromatic residues of Tyrl and Phe3 are of L configuration and the D-Xaa in the second position of the molecule is a D amino acid (D-Ala or D-Met or D-Leu). The D-enantiomer is encoded, however, by the codon for the L isomer in the precursor cDNA [8,13,15]. Thus, L-Xaa2 must be converted to D-Xaa2 by an unusual posttranslational reaction that presumably takes place in the precursor itself.

Because L-Xaa2-containing peptides have never been found in amphibian skin extracts, the epimerization mechanism probably involves a quantitative inversion of the chirality of the α-carbon of the amino acid residue, rather than a racemization, which would yield an equimolar mixture of L and D isomers [16,17]. Enzymes catalyzing the formation of D amino acids are so far known only in yeast [18]. From Bombina skin secretions Kreil et al. [19] recently purified a 52-kDa glycoprotein which catalyzes the reaction Ile-Ile-Gly to Ile-D-allo-Ile-Gly. The partial conversion of Ile to D-allo-Ile in peptide linkage proceeds without the addition of cofactors.

Similar to mammalian prohormones, all opioid peptides in amphibian skin precursors are flanked by paired dibasic amino acids (Lys-Arg). Moreover, the precursor sequence contains an additional Gly residue at its carboxyl terminus; this extra residue is required for the carboxamidation of the mature heptapeptide [20].

Several groups have actively searched for deltorphins in mammalian tissues. By using polyclonal and monoclonal antibodies specific for the various parts of deltorphins or their precursors' proteins, investigators have succeeded in immunostaining structures in the mouse brain (accessory olfactory bulb and selected neurons in the mesencephalon, as well as in the striatum and nucleus accumbens) and in the rat brain (immunoreactive nerve fibers in the amygdala, lateral hypothalamus, hippocampus substantia nigra, periaqueductal gray, and locus ceruleus) as well as in the respiratory system of perinatal rats [21–23]. Although these results argue in favor of endogenous synthesis and processing of pro-deltorphins in mammalian tissues, the exact chemical nature of these immunoreactive substances has never been established.

3 STRUCTURE-ACTIVITY RELATIONSHIP

Despite a common N-terminal tripeptide (Tyr-D-Xaa-Phe), the two groups of opioid peptides, dermorphins and deltorphins, differ enormously in receptor selectivities but bind to their own receptors with similar affinities. The N-terminal domain contains the minimum sequence essential for binding to opioid receptors whereas the C-terminal domain contains the address requisites for receptor selectivity.

The N-terminal tetrapeptides of D-Met-deltorphin and D-Ala-deltorphins did not show preference for delta receptors over mu receptors. The common determinants concurring to the remarkably efficient targeting of deltorphins towards the delta receptors were identified through structure-activity relationship studies conducted on an extensive series of synthetic analogues. The following structural requirements explain why the deltorphins are such potent and selective delta agonists: a phenolic side chain (Tyr) and a protonated nitrogen at the N-terminus; a D-isomer in the second position, which restricts peptide conformation and confers biological stability; a second aromatic center (Phe), common to all opioid peptides; an anionic residue (Asp7 in D-Met-deltorphin; Asp4 and Glu4 in D-Ala-deltorphins) in the C-terminal tetrapeptide, which increases electrostatic ligand repulsion by the negatively charged mu receptor and electrostatic binding to the positively charged site of the delta receptor; and a hydrophobic region associated with residues in the C-terminal address domain [1,24].

Recent site-directed mutagenesis and chimeric receptor molecules indicated that specific δ-opiate receptor binding requires receptor amino acid sequences 291–300, namely the terminal portion of extracellular loop 3 and the initial sequence of transmembrane segment VII, where Arg292 could actively participate in electrostatic binding to ligand anionic residues [25]. The hydroxyl group of Tyr in the peptide could form a hydrogen bond with transmembrane II Asp95 [26] or transmembrane III Asp128 [27].

The tertiary structure of deltorphin assessed by combined use of nuclear magnetic resonance (2D NMR) and spectroscopy in DMSO, indicated a common S-shaped arrangement in the deltorphin N-terminal peptide. This structure contains a type II′ β-turn in which D-Xaa2 lies sandwiched between β Tyr1 and Phe3 in a trans configuration. The folded C-terminal tail comes into close contact with the tripeptide amino end (unlike the linear, flexible C-terminus of the mu agonist dermorphins) and places the Tyr1 and Phe3 aromatic rings in definite orientations that are best suited for the delta-receptor. Moreover, Bryant et al. indicated a similar extended tertiary architecture for D-Met-deltorphin and D-Ala-deltorphin-I but unique compact topographies for D-Ala-deltorphin-II [28–31].

In the D-Met-deltorphin molecule, substitution of the positively charged His4 by a variety of amino acids is generally detrimental. An aliphatic side chain and L-isomer at the fifth residue appear critical for activity, but the C-terminal residue in the sixth and seventh position can generally be replaced by other amino acids with only marginal effects [32].

While D-Ala-deltorphins have delta-binding affinity similar to D-Met-deltorphin (0.3–2.0 nM), they consistently have the highest delta-opiate selectivity. The rank order of selectivity (Kiδ/Kiμ) is D-Ala-deltorphin-I = D-Ala-deltorphin-II (3000–4000) > D-Met-deltorphin (700) >> D-Ile-deltorphin (100) >> D-Leu-deltorphin heptadecapeptide or its N-terminal decapeptide fragment.

The high delta-selectivity of D-Ala-deltorphins can be attributed to their C-terminal tetrapeptide sequence in which the anionic residue plays an important role. Elimination of the charge at the fourth position normally results in opiates that have similar delta and mu affinity and generally lack selectivity. Substitution of Gly in the fourth position would permit D-Ala-deltorphins to assume a more extended conformation, dramatically increasing their μ-affinity and potency. The hydrophobic qualities of the residues at the fifth and the sixth positions (Val5–Val6) are crucial in maintaining the affinity and selectivity of D-Ala-deltorphins, as evidenced in peptide analogues in which the aliphatic quality of the side chain was enhanced [33–36].

4 D-Ala-DELTORPHINS RECOGNIZE TWO DELTA OPIATE RECEPTOR SUBTYPES

When D-Ala-deltorphins became available, pharmacological and biochemical studies provided evidence of distinct $delta_1$ and $delta_2$ opiate receptor subtypes in the rodent CNS. A study comparing the binding properties of [^{3}H]deltorphin-I and DPDPE in rat brain synaptosomes provided the first biochemical evidence for two delta opiate receptor subtypes [37]. In mouse brain homogenates and in rat brain slices, D-Ala-deltorphin-II binds preferentially to a population of delta opiate receptors that develops after weaning and would correspond to $delta_2$ subtypes [38,39]. The accumulated evidence demonstrates that D-Ala-deltorphin-II exerted antinociceptive action at a $delta_2$ receptor subtype while D-Ala-deltorphin-I (and DPDPE) interacted with greater specificity at a $delta_1$ receptor subtype. Using selective delta opiate antagonists [40,41], tolerance development [42], and delta opiate receptor knockdown (antisense oligonucleotide-treated) mice and rats [43–46] numerous pharmacological studies also suggested that spinal and supraspinal antinociception produced by DPDPE/D-Ala-deltorphin-I and D-Ala-deltorphin-II in rats and mice are mediated by distinct delta opiate receptor subtypes.

Despite pharmacological evidence of two distinct delta opioid receptors, molecular biologists have not yet succeeded in cloning delta-receptor subtypes. The available evidence implies that the relatively small change in the longer side chain on Glu (1.5 Å), due to the methylene C–C bond, could influence the global conformation of D-Ala-deltorphin-II, exerting a role in selecting the postreceptor transduction pathway by differentially activating delta-receptors.

5 PHARMACOKINETICS OF AMPHIBIAN OPIATES

Both D-Met-deltorphin and D-Ala-deltorphins exhibit long half-lives in brain homogenates and plasma, ($t_{1/2} = 57$ and 130 min for deltorphin and $t_{1/2} = 4$ and 6 h for D-Ala-deltorphins). The metabolites, generated by metalloendoproteases are represented by the N-terminal penta-, tetra-, and tripeptides. Stability depends on the presence of D-isomer and C-terminal amidation, but also on the peptide conformation in solution because elongating the N-terminal of D-Ala-deltorphin-I and -II destabilizes the peptides and reduces their half-lives to a few minutes [47–50].

D-Ala-deltorphin-I and -II transverse the blood brain barrier in vivo and in vitro [51]. Recently, D-Ala-deltorphin-II was identified as a transport substrate of organic anion transporting polypeptides (Oatp/OATP), a family of polyspecific membrane transporters, strongly expressed at the rat and human blood brain barrier [52]. Modified analogues of these peptides were synthesized to improve their transit across the blood brain barrier [48,49,53]. Because they resist enzyme degradation and can cross endothelial barriers into the CNS, the deltorphins meet the criteria for peptides with potential for systemic administration.

In cultured cortical neurons, by confocal microscopy, Lee and colleagues [54] demonstrated a receptor-mediated internalization of fluorescently labeled D-Ala-deltorphin and a retrograde transport of the peptide within nerve cell bodies that might be involved in mediating some of the long-term transcriptional effects of opioids.

6 PHARMACOLOGICAL PROPERTIES

Among deltorphins, D-Ala-deltorphin-I is the most potent δ-opiate agonist. Deamidation and shortening of the D-Ala-deltorphin-I molecule both cause a sharp decay in potency less evident for D-Met-deltorphin [5,12,33]. The putative natural deltorphin with D-Ile2 displays potency and affinity for the δ-receptor significantly lower than that of deltorphins [15]. The unusually large molecule Leu-deltorphin-17 has a very low affinity for the δ-receptor, but its shortened homologue (1–10) displays excellent delta-opiate activity, which is reversed by the opiate antagonist naltrindole and comparable to that of the enkephalins [5,14] (Table 2).

6.1 Antinociception

D-Ala-deltorphin-II induces δ-opiate receptor–mediated analgesia in frogs [55] and also in the invertebrate land snail (*Cepaea nemoralis*) [56]. When administered by intrathecal injection in rats, D-Ala-deltorphin-II produces a dose-related inhibition of the tail-flick response (threshold 0.6 nmol/rat). Its inhibitory effect lasts 10–60 min, depending on the dose, and is naltrindole reversible [57]. Wang et al. [25] demonstrated that D-Ala-deltorphin-II inhibited Aδ and C fiber–evoked responses from nociceptive neurons in the superficial and deeper dorsal horn of the rat medulla.

Conversely, when injected ICV in rats, D-Ala-deltorphin-II was a weak partial agonist: only doses >30 nmol produced some degree of antinocicep-

TABLE 2 Inhibitory Potencies of Deltorphins and Some Analogues on Electrically Evoked Contractions of Mouse Vas Deferens (MVD) and Guinea Pig Ileum (GPI), and on the Specific Binding of 0.3 nM [^{3}H]D-Ala-Deltorphin-I and of 0.5 nM [^{3}H]DAGO at Delta and Mu Sites in Rat Brain Membranes

	IC50 (nM)		K_i (nM)		
Peptides	MVD	GPI	δ	μ	δ/μ
D-Ala-delt-I	0.18±0.02	1239±203	0.78±0.08	1985±224	3.9×10^{-4}
D-Ala-delt-II	0.37±0.03	2500±170	1.03±1.09	2222±233	4.6×10^{-4}
D-Met-delt	0.97±0.05	1476±185	1.18±0.21	693±37	1.7×10^{-3}
D-Leu-delt	2480±378	>5000	>10000	>10000	—
D-Leu-delt(1–10)	37.0±3.9	1648±403	—	—	—
D-Ile-delt	7.0±0.9	4200±475	24±3	1021±57	2.3×10^{-2}
D-aIle-delt	70±8.2	3200±307	54±6.5	452±68	1.2×10^{-1}
(D-Ala^2,Gly^4)delt	2.62±0.32	22±3	3.26±0.37	13.5±2.1	2.4×10^{-1}
(D-Ala^2,His^4)delt	0.83±0.09	143±27	0.25±0.02	83.0±5.7	3.0×10^{-3}
D-Ala-delt-I(1–4)	>2000	>3000	1254±103	195±31	6.4
D-Met-delt(1–4)	—	—	1289±111	8±0.4	161

tion in the tail flick test to radiant heat, none of the doses tested eliciting the maximum achievable response. This partial antinociception was accomplished with an in vivo occupancy of >97% of brain δ-opiate receptors and of 17% of mu-opiate receptors. Naloxone (0.1 mg/kg, SC) and naloxonazine (10 mg/kg, IV) antagonized the antinociception, but the selective δ-opiate antagonist naltrindole did not [58]. Experiments in rats showed that D-Ala-deltorphin-I behaved as a full antinociceptive agonist at doses between 6.5 and 52 nmol and that the analgesia was partially reduced by naltrindole (Negri, unpublished data). D-Met-deltorphin, in a comparable range of doses, induced a naloxone sensitive analgesia. D-Ile-deltorphin, which displays delta/mu selectivity about 2 orders of magnitude lower than that of D-Ala-deltorphins, induces a mu-mediated antinociception (AD50=4.1[1.9–8.6] μg/rat, ICV; = 4.4 nmol/rat) [59–61].

Studies in progress (personal unpublished data) show that high doses of D-Met-deltorphin, D-Ile-deltorphin, and the D-Ala-deltorphin analogue with a His residue in the fourth position [D-Ala2, His4]deltorphin (Tyr-D-Ala-Phe-His-Val-Val-Gly-NH2) invariably induce barrel rotations. Characteristic motor dysfunction such as hindlimb jerking, barrel rolling, circling, ataxia, and unusual contorted postures start within 2 min and last 20–60 min depending on the dose and the severity of the syndrome. In our experiments, none of these motor effects were antagonized by pre-administration of opiate antagonists, naloxone (3 mg/kg, SC), naltrexone (10 mg/kg, SC) and naltrindole (3 mg/kg, SC). However, barrel rotations and motor dysfunctions were completely blocked by the noncompetitive NMDA antagonist dextrorphan (5 nmol/rat, ICV) and by the σ1-receptor agonist (+)- SK&F 10047 (4 mg/kg, SC) [60] (Table 3).

In mice, D-Ala-deltorphin-II shows a moderate dose and time-related antinociceptive effect when administered ICV (EC50 = 2.1 nmol/mouse). The peptide is half as potent as morphine [62], and the analgesic effect is antagonized by naltrindole. Repeated injection of D-Ala-deltorphin-II induces tolerance to the antinociceptive effect. There is no cross tolerance between antinociception induced by D-Ala-deltorphin-II and that generated by either mu or $delta_1$ opiate receptor agonists [42,62]. The finding that intrathecal injection of D-Ala-deltorphin-II has a higher analgesic effect than DPDPE is probably related more to the predominance of δ2-receptors in the spinal cord than to a prevalence of supraspinal $delta_1$ receptors [63,64]. Isobolographic analysis revealed that supraspinal/spinal antinociceptive interactions for both the $delta_1$ agonist, DPDPE, and the $delta_2$ agonist, D-Ala-deltorphin-II, were synergistic in many nociceptive tests, suggesting that compounds acting through delta-opiate receptors may have sufficient potency for eventual clinical applications [65].

Data suggest that the delta-agonists play a predominantly modulatory role in antinociception rather than a primary role. In homozygote mice with a disrupted mu opiate receptor gene, Matthes et al. [66] and Sora et al. [67] demonstrated that delta agonist–induced analgesia is reduced. Previous findings showed that in mice [68,69] and in rats [58,70] the intensity of delta opiate analgesia depends on coactivation of mu opiate receptors by endogenous or exogenous opiates. Stress associated with the ICV injection may

TABLE 3 ICV Doses of Deltorphins Inducing Behavioral Effects

	Dose (nmol/rat, ICV)		
Peptides	Locomotion[a]	Analgesia[b]	Motor effects[b]
D-Ala-delt-I	0.026–0.26	6.5	—[c]
D-Ala-delt-II	0.06–3.8	>30	—[c]
(D-Ala^2, His^4)delt	1.2–25.5	3.2	38
D-Met-delt	1.04–20.8	10.4	31
D-Ile-delt	—	2.2	7.5
D-alle-delt	—	5.4	11

[a] Range of doses stimulating locomotion.
[b] Minimum doses producing analgesia or motor effects in 100% rats.
[c] Motor effects were absent up to the highest dose tested (130 nmol/rat).

activate mu opiate receptors through the release of endogenous opiates and thus potentiate the antinociceptive responses to delta opiate agonists. Moreover, delta opioid agonists can be regarded as potential drugs for the treatment of chronic pain: in rats, intrathecal administration of D-Ala-deltorphin-II dose-dependently antagonized the cold water allodynia which developed after sciatic nerve injury [71], and ICV administration of D-Ala-deltorphin-II significantly reversed the hyperalgesic response associated with peripheral inflammation [72].

6.2 Locomotor Behavior

Injections of D-Ala-deltorphin-II into the rat brain ventricles, ventral tegmental area, and nucleus accumbens invariably increase locomotor activity and induce stereotyped behavior [59,73]. The ambulatory activity is intermittent and usually intercalated by rearing events. The increase in rearing and locomotor activity is dose related over the range of 0.026–0.26 nmol/rat for D-Ala-deltorphin-I, of 0.06–3.8 nmol/rat for D-Ala-deltorphin-II, and 1.04–20.8 nmol/rat for D-Met-deltorphin. The motor activity is antagonized by the delta selective antagonist naltrindole and by high doses of the mu antagonist naloxone but is unaffected by the μ1-selective antagonist naloxonazine. ICV administration of 1.3 nmol/rat of D-Ala-deltorphin-II increases social contacts [59]. Local application of D-Ala-deltorphin-II to the nucleus accumbens but not to the nucleus caudatus increases extracellular dopamine concentrations (by up to 120%). It also stimulates locomotor activity and stereotypies [73]. Repeated ICV injection of D-Ala-deltorphin-II in naive rats induces tolerance to the stimulant effects, whereas repeated daily injections or continuous infusion of morphine result in sensitization to the behavioral activating effects of the delta opiate agonist [74].

6.3 Other Pharmacological Effects

Further pharmacological effects of deltorphins have been demonstrated under various experimental conditions. D-Ala-deltorphin improves memory consolidation in a passive avoidance apparatus in mice; this effect is abolished by naltrindole [75]. D-Ala-deltorphin-II caused hypothermia in cold-adapted animals [76]. In contrast to mu opiate agonists, D-Ala-deltorphin-I, at low doses, stimulates respiratory activity in fetal lambs, and this effect is blocked by simultaneous administration of naltrindole [77]. The peptide D-Ala-deltorphin-II inhibits diarrhea induced by castor oil and colonic bead expulsion, but it leaves the rate of transit through the small intestine unchanged [78,79]. By the SC route D-Ala-deltorphin-I inhibits acidified alcohol-induced gastric mucosal lesions [80], but by the ICV route, it fails to affect gastric secretion [81]. The peptide is involved also in the control of ingestive behavior. It stimulates the intake of food [82] and of sucrose [83], and when administered in conjunction with angiotensin II it increases water consumption in rats [84].

Data on bioactivity on immunocompetent cells provide evidence that D-Ala-deltorphin-I potently (10-9 to 10-11 M) and persistently (up to 4 days) enhances Con A–induced mouse spleen cell proliferation [85]. The peptide increases uptake of thymidine and production of interferon-γ in phytohemagglutinin-activated human lymphocytes [86] and is 100 times more potent than SNC80 in inhibiting the production of p24 antigen, an index of HIV-1 expression, in Jurkat cells stably transfected with a cDNA encoding for the delta-opiate receptor [87].

In human subjects, deltorphin inhibits the secretion of growth hormone and ACTH induced by insulin-induced hypoglycemia and modulates the secretion of pituitary luteinizing hormone in women [88–91].

7 CONCLUSIONS

The discovery of the amphibian opiate peptides, apart from the intriguing problem of the occurrence of their analogues in mammalian central and peripheral nervous system, has provided new insights into the functional role of the mu- and delta-opiate systems. In particular, the deltorphin peptides, which appear to be non addicting analgesic drugs [92,93], which circumvent the known effects on gastrointestinal transit and depression of respiration associated with treatment by alkaloid opiates, may offer an excellent means to counteract acute or chronic pain. In the broad sense, the discovery of these potent opiates confirms that the amphibian skin and its secretions offer an inexhaustible supply of biologically active peptides. Although the amphibian peptides identified to date were isolated from methanol extracts of amphibian skin and thus are of small or relatively small molecular mass (700–4600 daltons), during the last 2 years a new field of amphibian skin protein research has led to the discovery of proteins that are externally secreted by syncytial cells forming the wall of the integument glands upon electrical stimulation of the skin of the living frog. By repetitive electrical stimulation at weekly intervals, several milligrams of bioactive proteins can be collected from a single frog. This is a novel biotechnology that can compete with genetic engineering in the production of high quantities of biological active proteins for pharmacological research.

REFERENCES

1. Erspamer V. Int J Dev Neurosci 1992; 10:3–30.
2. Montecucchi PC, De Castiglione R, Piani S, Gozzini L, Erspamer V. Int J Peptide Prot Res 1981; 17:275–279.

3. Broccardo M, Erspamer V, Falconieri-Erspamer G, Improta G, Linari G, Melchiorri P, Montecucchi PC. Br J Pharmacol 1981; 73:625–631.
4. Zadina JE, Laszlo H, Ge L-J, Kastin AJ. Nature 1997; 386:499–502.
5. Erspamer V. Heatwole H, ed. Amphibian Biology. Surrey: Beatty & Sons Publ., 1994:178–350.
6. Erspamer V, Falconieri-Erspamer G, Severini C, Potenza R, Barra D, Mignogna G, Bianchi A. Toxicon 1993; 31:1089–1111.
7. Lacombe C, Cifuentes-Diaz C, Dunia I, Auber-Thomay M, Nicolas P, Amiche M. Eur J Cell Biol 2000; 79:631–641.
8. Richter K, Egger R, Kreil G. Science 1987; 238:200–202.
9. Kreil G, Barra D, Simmaco M, Erspamer V, Falconieri-Erspamer G, Negri L, Severini C, Corsi R, Melchiorri P. Eur J Pharmacol 1989; 162:123–128.
10. Mor A, Delfour A, Sagan S, Amiche M, Pedrelles P, Rossier J, Nicholas J. FEBS Lett 1989; 225:269–274.
11. Lazarus LH, Wilson WE, De Castiglione R, Guglietta A. J Biol Chem 1989; 264:3047–3050.
12. Erspamer V, Melchiorri P, Falconieri-Erspamer G, Negri L, Corsi R, Severini C, Barra C, Simmaco M, Kreil G. Proc Natl Acad Sci USA 1989; 86:5188–5192.
13. Richter K, Egger R, Negri L, Corsi R, Severini C, Kreil G. Proc Natl Acad Sci USA 1990; 87:4836–4839.
14. Barra D, Mignogna G, Simmaco M, Pucci P, Severini C, Falconieri-Erspamer G, Negri L, Erspamer V. Peptides 1994; 15:199–202.
15. Wechselberger C, Severini C, Kreil G, Negri L. FEBS Lett 1998; 429:41–43.
16. Heck SD, Faraci WS, Kelbaugh PR, Saccomano NA, Thadeio PF, Volkmann RA. Proc Natl Acad Sci USA 1996; 93:4036–4039.
17. Kreil G. A Rev Biochem 1997; 66:337–345.
18. Watanabe Y, Muro T, Sugihara A, Shimada Y, Nagao T, Takenishi S, Tominaga Y. Biochim Biophys Acta 1997; 1337:40–46.
19. Kreil G, Mollay C, Grassi J, Mignogna G, Barra D. A peptidyl-L,D-isomerase from skin secretions of Bombinae. 25th FEBS Meeting, Copenaghen, (abstract S19.2), 1998:36.
20. Mollay C, Wichta J, Kreil G. FEBS Lett 1986; 202:251–254.
21. Tooyama I, Abe H, Renda T, Erspamer V, Kimura H. Proc Natl Acad Sci USA 1993; 90:9635–9639.
22. Yu S, Zhao T, Fan M, Tooyama I, Kimura H, Renda TG. Peptides 2000; 21:1657–1662.
23. Sunday ME, Haley KJ, Emanuel RL, Torday JS, Tooyama I, Kimura H, Renda T, Erspamer V. Am J Respir Cell Mol Biol 2001; 25:447–456.
24. Hruby VJ, Gehrig C. Med Res Rev 1989; 9:343–401.
25. Wang X, Yan JQ, Zhang KM, Mokha S. Brain Res 1996; 739:235–243.
26. Kong H, Raynor K, Yasuda K, Moe ST, Portoghese PS, Bell GI, Reisine T. J Biol Chem 1993; 268:23055–23058.
27. Befort K, Tabbara L, Bausch S, Chavkin C, Evans C, Kieffer B. Mol Pharmacol 1996; 49:216–223.

28. Nikiforovich GV, Parakash O, Gehrig CA, Hruby VJ. J Am Chem Soc 1993; 115:3399–3406.
29. Amodeo P, Motta A, Tancredi T, Salvadori S, Tomatis R, Picone D, Saviano G, Temussi PA. Pept Res 1992; 5:48–55.
30. Bryant SD, Salvadori S, Attila M, Lazarus LH. J Am Chem Soc 1993; 115: 8503–8504.
31. Riand J, Baron D, Nicolas P, Benajiba A, Teng Y, Naim M. J Biomol Struct Dyn 1999; 17:445–460.
32. Salvadori S, Guerrini R, Forlani V, Bryant SD, Attila M, Lazarus LH. Amino Acids 1994; 7:291–304.
33. Melchiorri P, Negri L, Falconieri-Erspamer G, Severini C, Corsi R, Soaje M, Erspamer V, Barra D. Eur J Pharmacol 1991; 195:201–207.
34. Bryant SD, Attila M, Salvadori S, Guerrini R, Lazarus LH. Pept Res 1994; 7:175–184.
35. Benedetti E, Isernia C, Nastri F, Pedone C, Saviano M, Mierke DF, Melchiorri P, Negri L, Potenza RL, Severini C, Erspamer V. Eur J Org Chem 1998; 1:2279–2287.
36. Sasaki Y, Ambo A, Suzuki K. Biochem Biophys Res Commun 1991; 180:822–827.
37. Negri L, Potenza R, Corsi R, Melchiorri P. Eur J Pharmacol 1991; 196:335–336.
38. Kitchen I, Leslie FM, Kelly M, Barnes TJ, Crook RJ, Hill A, Borsodi A, Toth P, Melchiorri P, Negri L. J Pharmacol Exp Ther 1995; 275:1597–1607.
39. Negri L, Severini C, Lattanzi R, Potenza RL, Melchiorri P. Br J Pharmacol 1997; 120:989–994.
40. JIang Q, Takemori A, Sultana M, Portoghese P, Bowen W, Mosberg H, Porreca F. J Pharmacol Exp. Ther 1991; 257:069–1075.
41. Sofuoglu M, Portoghese P, Takemori A. J Pharmacol Exp Ther 1991; 257:676–680.
42. Mattia A, Vanderah T, Mosberg H, Porreca F. J Pharmacol Exp Ther 1991; 258:583–587.
43. Bilsky E, Bernstein R, Hruby V, Rothman R, Lai J, Porreca F. J Pharmacol Exp Ther 1996; 277:491–501.
44. Lai J, Bilsky E, Rothman R, Porreca F. NeuroReport 1994; 5:1049–1052.
45. Rossi GC, Su W, Leventhal L, Su H, Pasternak GW. Brain Res 1997; 753:176–179.
46. Sanchez-Blazquez P, Garcia-Espana A, Garzon J. J Pharmacol Exp Ther 1997; 280:1423–1431.
47. Marastoni M, Tomatis R, Balboni G, Salvadori S, Lazarus LH. Farmaco 1991; 46:1273–1279.
48. Negri L, Lattanzi R, Orrù L, Severini C, Scolaro B, Rocchi R. J Med Chem 1999; 42:400–404.
49. Thomas SA, Abbruscato TJ, Hau VS, Gillespie TJ, Zsigo J, Hruby VJ, Davis TP. J Pharmacol Exp Ther 1997; 281:817–825.
50. Sasaki Y, Chiba T, Amibo A, Suzuki K. Chem Pharm Bull 1994; 42:592–594.

51. Fiori A, Cardelli P, Negri L, Savi MR, Strom R, Erspamer V. Proc Natl Acad Sci USA 1997; 94:9469–9474.
52. Gao B, Hagenbuch B, Kullak-Hblick GA, Benke D, Aguzzi A, Meier PJ. J Pharmacol Exp Ther 2000; 294:73–79.
53. Tomatis R, Marastoni M, Balboni G, Guerrini R, Capasso A, Sorrentino L, Santagata V, Caliendo G, Lazarus LH, Salvadori S. J Med Chem 1997; 40:2948–2952.
54. Lee MC, Cahill CM, Vincent JP, Beaudet A. Synapse 2002; 43:102–111.
55. Stevens CW. J Pharmacol Exp Ther 1996; 276:440–448.
56. Thomas AW, Kavaliers M, Prato FS, Ossenkopp KP. Peptides 1997; 18:703–709.
57. Improta G, Broccardo M. Peptides 1992; 13:1123–1126.
58. Negri L, Improta G, Lattanzi R, Potenza RL, Luchetti F, Melchiorri P. Br J Pharmacol 1995; 116:2931–2938.
59. Negri L, Noviello V, Angelucci F. Eur J Pharmacol 1991; 209:163–168.
60. Negri L, Lattanzi R, Tabacco F, Orrù L, Melchiorri P. Pharmacol Res 1999; S39:50.
61. Negri L, Noviello L, Noviello V. Eur J Pharmacol 1996; 296:9–16.
62. Jiang Q, Mosberg H, Porreca F. Life Sci 1990; 47:PL43–PL47.
63. Raffa R, Martinez R, Porreca F. Eur J Pharmacol 1992; 216:453–456.
64. Portoghese PS, Takemori AE. Eur J Pharmacol 1993; 246:145–150.
65. Kovolevski CJ, Bian D, Ruby H, Lay JJ, Ossipov MH, Porreca F. Brain Res 1999; 843:12–17.
66. Matthes HW, Maldonado R, Simonin F, Valverde O, Slowe S, Kitchen I, Befort K, Dierich A, Le MM, Dolle P, Tzavara E, Hanoune J, Roques BP, Kieffer BL. Nature 1996; 383:819–823.
67. Sora I, Funada M, Uhl GR. Eur J Pharmacol 1997; 324:R1–R2.
68. Heyman JS, Vaught JL, Mosberg HI, Haaseth RC, Porreca F. Eur J Pharmacol 1989; 165:1–10.
69. Porreca F, Heyman JS, Mosberg HI, Omnaas JR, Vaught JL. J Pharmacol Exp Ther 1987; 241:393–398.
70. Adams JU, Tallarida RJ, Geller EB, Adler MW. J Pharmacol Exp Ther 1993; 266:1261–1267.
71. Mika J, Przewlocki R, Przewlocka B. Eur J Pharmacol 2001; 415:31–37.
72. Fraser GL, Gaudreau GA, Clarke PB, Menard DP, Perkins MN. BR J Pharmacol 2000; 129:1668–1672.
73. Longoni R, Spina L, Mulas A, Carboni E, Garrau L, Melchiorri P, Di Chiara G. J Neurosci 1991; 11:1565–1576.
74. Melchiorri P, Maritati M, Negri L, Erspamer V. Proc Natl Acad Sci USA 1992; 89:3696–3700.
75. Pavone F, Populin R, Castellano C, Kreil G, Melchiorri P. Peptides 1990; 11:591–594.
76. Broccardo M, Improta G. Neurosci Lett 1992; 139:209–212.
77. Cheng PY, Wu D, Decena J, Soong Y, McCabe S, Szeto HH. Eur J Pharmacol 1993; 203:85–88.

78. Broccardo M, Improta G. Pharmacol Res 1992; 25:5–6.
79. Broccardo M, Improta G. Eur J Pharmacol 1992; 218:69–73.
80. Gyires KAZR, Toth G, Darula Z, Furst S. Life Sci 1997; 60:1337–1347.
81. Improta G, Broccardo M. Neuropharmacology 1994; 33:977–981.
82. Yu WZ, Ruegg H, Bodnar RJ. Pharmacol Biochem Behav 1997; 56:353–361.
83. Ruegg H, Yu WZ, Bodnar RJ. Physiol Behav 1997; 62:121–128.
84. Yu WZ, Bodnar RJ. Peptides 1997; 18:241–245.
85. Stefano GB, Melchiorri P, Negri L, Hughes TK Jr, Scharrer B. Proc Natl Acad Sci USA 1992; 89:9316–9320.
86. Noviello L, Papadia S, Capobianchi MR, Negri L. Eur J Pharmacol 1997; 332:R1–R2.
87. Sharp BM, Gekker G, Li MD, Chao CC, Peterson PK. Biochem Pharmacol 1998; 56:289–292.
88. Bondanelli M, Ambrosio MR, Valentini A, degli Uberti EC. Hum Reprod 1998; 13:1159–1162.
89. Degli Uberti EC, Ambrosio MRLV, Portaluppi F, Bondanelli M, Trasforini G, Margutti G, Salvadori S. J Clin Endocrinol Metab 1993; 77:1490–1494.
90. Degli Uberti EC, Salvadori S, Trasforini G, Margutti A, Ambrosio MR, Portaluppi F, Vergnani L, Pansini R. Neuroendocrinology 1992; 56:907–912.
91. Degli Uberti EC, Salvadori S, Trasforini G, Margutti A, Ambrosio MR, Rossi R, Portaluppi F, Pansini R. J Clin Endocrinol Metab 1992; 75:370–374.
92. Rapaka RS, Porreca F. Pharmacol Res 1991; 8:1–8.
93. Hutcheson DM, Matthes HW, Valient E, Sanchez-Blazquez P, Rodiguez-Diaz M, Grazon J, Kieffer BL, Maldonado R. Eur J Neurosci 2001; 13:153–161.

12

Opioid Peptide-Derived Delta Antagonists, Inverse Delta Agonists, and Mixed Mu Agonist/Delta Antagonists

Peter W. Schiller
Clinical Research Institute of Montreal, Montreal, Quebec, Canada

1 INTRODUCTION

Delta (δ) opioid antagonists are of interest as both pharmacological tools and potential therapeutic agents. Obviously, they are required as essential tools in the elucidation of δ-opioid receptor mediated in vitro and in vivo effects. Recently, several compounds originally characterized as δ-opioid antagonists have been identified as inverse δ-agonists. High-affinity inverse δ-agonists with high δ-selectivity are needed to investigate the spontaneous activity of δ-opioid receptors in vitro and in vivo.

δ-Opioid antagonists may have therapeutic potential as immunosuppressants [1,2] and for the treatment of cocaine and alcohol addiction [3,4]. Furthermore, δ-opioid receptor blockade with δ-antagonists has been shown to reduce the development of morphine tolerance and dependence in mice and rats [5,6], suggesting the possibility of the combined use of a mu-type opioid analgesic and a δ-opioid antagonist in chronic pain treatment. Moreover, it has been shown that a δ-antagonist reversed μ-agonist-induced respiratory depression [7,8] and enhanced colonic propulsion [9]. These observations may also lead to clinical applications.

Both peptide and nonpeptide δ-opioid antagonists have been developed. The nonpeptide δ-antagonist naltrindole and its analogues are described in Chapter 9 of this volume. In the present chapter, the development of opioid peptide-derived δ-antagonists, inverse δ-agonists, and mixed μ-agonist/δ-antagonists is reviewed.

2 ENKEPHALIN-DERIVED δ-OPIOID ANTAGONISTS

Leu-enkephalin or its analogues have been converted into δ-opioid antagonists through two different kinds of structural modifications: N,N-dialkylation of the N-terminal amino group, and deletion of the positive charge of the N-terminal amino group.

2.1 N,N-Dialkylated Enkephalin Analogues with δ-Antagonist Activity

A number of δ-selective antagonists have been obtained through diallylation of the N-terminal amino group of enkephalin peptides. The design of these analogues was based on analogy with the well-known N-allyl substituted alkaloid opiate antagonists. Whereas the N-monoallylated derivative of leu-enkephalin turned out to be a weak partial agonist [10], N,N-diallylated leu-enkephalin was found to be a moderately potent, δ-selective antagonist in the mouse vas deferens (MVD) assay [11]. Replacement of the Gly^3-Phe^4 peptide bond in the latter derivative with a thiomethylene moiety resulted in a compound, N,N-diallyl-Tyr-Gly-GlyΨ[CH_2*S*]Phe-Leu-OH (ICI 154129), which also was a δ-selective antagonist with modest δ-receptor binding affinity (K_i^δ = 778 nM) [12,13]. To reduce the structural flexibility of N,N-diallylated leu-enkephalin, the -Gly^2-Gly^3-dipeptide unit was replaced by either an -Aib^2-Aib^3 unit (Aib = α-aminoisobutyric acid) [14] or by a rigid spacer such as *p*-aminobenzoic acid (-NH-Φ-CO-) [15]. One of these conformationally restricted analogues, N,N-diallyl-Tyr-Aib-Aib-Phe-Leu-OH (ICI 174864) was a moderately potent δ-antagonist in the MVD assay (K_e = 36.4 nM) and showed about four times higher δ-receptor affinity (K_i^δ = 193 nM) than ICI 154129 and quite high preference for δ-receptors over μ-receptors (K_i^μ/K_i^δ = 128) in the receptor binding assays (Table 1). ICI 174864 has been a useful tool in opioid research for many years and later on was identified as an inverse δ opioid agonist [16] (see Sec. 4). The analogue N,N-diallyl-Tyr-*p*-NH-Φ-CO-Phe-Leu-OH showed δ-antagonist potency similar to that of ICI 174864; however, no receptor-binding data were reported for this compound. Evaluation of the dimeric ligand (N,N-diallyl-Tyr-Gly-Gly-Phe-Leu-NH-CH_2-$)_2$

TABLE 1 Enkephalin-Derived δ Opioid Antagonists: K_e Values and Receptor-Binding Affinities

Compound	MVD K_e [nM][b]	Receptor binding assays[a]			
		K_i^{δ} [nM]	K_i^{μ} [nM]	K_i^{μ}/K_i^{δ}	Ref.
$(allyl)_2$Tyr-Aib-Aib-Phe-Leu-OH (ICI 174864)	36.4[c]	193[d]	24,700[e]	128	13
Boc-Tyr-Pro-Gly-Phe-Leu-Thr (O^tBu)-OH	38.7	297–945[f]	31,150	33–105	21
(2S)-Mdp-D-Ala-Gly-Phe-Leu-NH_2	28.1	11.7	192	16.4	22
(2S)-Mdp-c[D-Pen-Gly-Phe(pF)-Pen]-Phe-OH	0.785	2.32[e]	406	175	24

[a] Displacement of $[^3H]$DAMGO (μ-selective) and $[^3H]$DSLET (δ-selective) from rat brain membrane binding sites.
[b] Determined against DPDPE.
[c] Determined against DSLET.
[d] Displacement of $[^3H]$DADLE from guinea pig brain membrane binding sites.
[e] Displacement of $[^3H]$DAMGO from guinea pig brain membrane binding sites.
[f] Displacement of three different δ-receptor radioligands from rat brain membrane-binding sites.

in the MVD assay revealed that this compound also was a fairly potent and selective δ-antagonist [17]. However, studies with related dimers that were truncated on one side indicated that this bivalent ligand did not simultaneously bind to two distinct δ-receptor-binding sites.

A number of N,N-dialkylated analogues of leu-enkephalin were prepared and tested in the MVD and guinea pig ileum (GPI) assays, but not in receptor-binding assays [18]. Among these, the N,N-dibenzylated analogue was a δ-antagonist with modest potency (K_e = 210 nM), whereas the N,N-di-2-phenylethyl and N,N-dioctyl analogues showed significant δ-agonist activity. N,N-dialkylated analogues of leu-enkephalin containing melphalan (Mel) in place of Phe^4 were prepared with the goal of obtaining irreversible δ-antagonists [19]. At high concentrations (10 μM), both $(benzyl)_2$Tyr-Gly-Gly-Mel-Leu-OH and $(allyl)_2$Tyr-Aib-Aib-Mel-Leu-OH showed weak irreversible δ-antagonism in the MVD assay.

2.2 Enkephalin-Derived δ-Opioid Antagonists Lacking a Positively Charged N-Terminal Amino Group

In 1992, it was reported that Boc-Tyr-Pro-Gly-Phe-Leu-Thr-OH, an opioid peptide lacking a positive charge, showed weak δ-antagonist potency (K_e = 560 nM) in the MVD assay [20]. Subsequently, it was shown that a derivative of this peptide containing an O-t-butyl-protected Thr^6 residue, Boc-Tyr-Pro-

Gly-Phe-Leu-Thr(O^{t}Bu), was a somewhat more potent δ-antagonist ($K_e \sim 30$ nM) [21] (Table 1). However, the δ-receptor affinity of this compound determined in the rat brain membrane-binding assay was very weak (K_i^δ = 300–1000 nM), diminishing the impact of this report.

More recently, Lu et al. [22] synthesized an analogue of the potent enkephalin analogue H-Dmt-D-Ala-Gly-Phe-Leu-NH_2 (Dmt = 2′,6′-dimethyltyrosine), in which the N-terminal amino group was replaced with the neutral and almost isosteric methyl group. This was achieved by replacement of Dmt with (2*S*)-2-methyl-3-(2,6-dimethyl-4-hydroxyphenyl)propanoic acid ((2*S*)-Mdp), for which a stereospecific synthesis based on Evans chiral enol chemistry was developed [23]. The resulting peptide, (2*S*)-Mdp-D-Ala-Gly-Phe-Leu-NH_2, turned out to be a quite potent δ-antagonist (K_e = 28 nM) in the MVD assay, with quite high δ-receptor binding affinity (K_i^δ = 11.7 nM) and marked δ-receptor selectivity (K_i^μ/K_i^δ = 16.4), as determined in the rat brain membrane-binding assays (Table 1). In agreement with the receptor-binding data, this compound showed relatively weaker μ-antagonism (K_e = 154 nM) in the GPI assay. Subsequently, the (2*S*)-Mdp^1 analogue of the nonselective cyclic agonist peptide H-Tyr-c[D-Cys-Gly-Phe(*p*NO_2)-D-Cys]-NH_2 was shown to be a potent μ-, δ-, and κ-antagonist, indicating that elimination of the positively charged amino group in combination with 2′,6′-dimethylation of the Tyr^1 aromatic ring may represent a generally applicable structural modification to convert opioid agonist peptides into antagonists at all three opioid receptors [24]. Indeed, replacement of Tyr^1 with (2*S*)-Mdp permitted the conversion of almost any opioid peptide agonist into an antagonist, whereby receptor selectivity was often maintained or even improved [24]. An example is the (2*S*)-Mdp^1-analogue of the potent and highly δ-selective cyclic enkephalin analogue H-Tyr-c[D-Pen-Gly-Phe(*p*F)-Pen]-Phe-OH [25], which showed high δ-antagonist activity (K_e = 0.785 nM) in the MVD assay and retained high δ-receptor binding affinity (K_i^δ = 2.32 nM), high δ- versus μ-selectivity (K_i^μ/K_i^δ = 175) and high δ- versus κ-selectivity (K_i^μ/K_i^δ = 6600) in the receptor-binding assays [24]. (2*S*-Mdp^1)-c[D-Pen-Gly-Phe(*p*F)-Pen]-Phe-OH represents the first selective δ-opioid antagonist with a cyclic enkephalin-derived peptide structure.

3 TIP(P) PEPTIDES AND PEPTIDOMIMETICS: HIGHLY POTENT AND SELECTIVE δ-OPIOID ANTAGONISTS

The results of structure-activity studies on opioid peptides revealed that analogues consisting entirely of aromatic amino acids, such as H-Tyr-D-Phe-Phe-NH_2 [26] and H-Tyr-D-Phe-Phe-Phe-NH_2 [27] were quite potent and selective μ-agonists. Systematic replacement of the amino acids in these two peptides with conformationally constrained, cyclic L- and D-aromatic amino acids led to the discovery of a new class of δ-opioid antagonists, the so-called TIP(P) peptides, which contain an L-1,2,3,4-tetrahydroisoquinoline-3-carboxylic acid (Tic) residue in the 2 position of the peptide sequence [27]. The two prototype peptides, TIPP (H-Tyr-Tic-Phe-Phe-OH) and TIP (H-Tyr-Tic-Phe-OH), showed high δ-antagonist activity and δ-receptor selectivity (Table 2). As a δ-antagonist TIPP is about eight times more potent than ICI 174864 in vitro and eight times less potent than naltrindole, and it is much more δ-selective than either one of the latter two antagonists. In an effort to further improve the δ-antagonist potency and δ-receptor selectivity of TIPP, numerous analogues were prepared and pharmacologically characterized in vitro. The results of these structure-activity studies are reviewed in the following section.

3.1 TIP(P) Analogues

Whereas TIPP and TIP are stable in aqueous buffer solution (pH 7.7), both peptides were shown to undergo slow spontaneous diketopiperazine formation with concomitant cleavage of the Tic-Phe peptide bond in DMSO and MeOH [28]. This degradation can be prevented by replacement of the Tic-Phe peptide bond with a reduced peptide bond. Indeed, the pseudopeptide analogues H-Tyr-TicΨ[CH_2NH]Phe-Phe-OH (TIPP[Ψ]) and H-Tyr-TicΨ[CH_2NH]Phe-OH (TIP[Ψ]) were found to be highly stable against chemical and enzymatic degradation and, furthermore, showed slightly higher δ-antagonist activity and much higher δ-receptor selectivity than their respective parent peptides [29] (Table 2). In particular, TIPP[Ψ] displayed subnanomolar δ-receptor-binding affinity and extraordinary δ-receptor selectivity ($K_i^{\mu}/K_i^{\delta} = 10{,}500$), being > 500 times more δ-selective than naltrindole.

TIPP and TIPP[Ψ] are both stable in aqueous solution at physiological pH and show similar conformational behavior [29]. Therefore, the further design of analogues was based on the use of both of them as parent peptides. In the following review of structure-activity relationships, each position in the tetrapeptide sequence (Tyr^1, Tic^2, Phe^3, Phe^4) is considered in turn.

In comparison with the TIPP parent peptide, an analogue containing an N-methylated Tyr^1 residue showed 10-fold increased δ-antagonist potency in the MVD assay and seven fold higher δ-receptor selectivity in the receptor binding assays [30] (Table 2). Replacement of Tyr^1 in opioid peptides with Dmt was first performed with the δ-agonist peptide DPDPE (H-Tyr-c[D-Pen-Gly-Phe-D-Pen]OH), and was shown to result in a substantial increase in δ receptor binding affinity, an even more pronounced increase in μ-receptor affinity, and, consequently, a decrease in δ-selectivity [31]. Similarly, the Dmt^1-analogue of TIPP (DIPP) displayed 25-fold enhanced δ-antagonist

TABLE 2 Antagonist Activities (K_e Values) and Opioid Receptor Binding Affinities of δ-Antagonists of the TIP(P) Class

Compound	MVD K_e [nM][b]	Receptor binding assays[a] K_i^{δ} [nM]	K_i^{μ} [nM]	K_i^{μ}/K_i^{δ}	Ref.
H-Tyr-Tic-Phe-Phe-OH (TIPP)	4.80	1.22	1,720	1,410	27
H-Tyr-Tic-Phe-OH (TIP)	16.1	9.07	1,280	141	27
H-Tyr-TicΨ[CH_2NH]Phe-Phe-OH (TIPP[Ψ])	2.89	0.308	3,230	10,500	29
H-Tyr-TicΨ[CH_2NH]Phe-OH (TIP[Ψ])	9.06	1.94	10,800	5,570	29
Tyr(NMe)-Tic-Phe-Phe-OH	0.436[c]	1.29	13,400	10,400	30
H-Dmt-Tic-Phe-Phe-OH (DIPP)	0.196	0.248	141	569	30
H-Hmt-Tic-Phe-Phe-OH	0.473	0.132	1,540	11,700	32
H-Hmt-Tic-Phe-OH	0.961	0.458	2,110	4,610	32
H-Tyr-(3′-I)-Tic-Phe-Phe-OH	97.4 (IC50)[d]	24.8	5,230	211	30,33
H-Tyr-(3′-Br)-Tic-Phe-Phe-OH	PA (e = 0.16)[e]	3.62	18,500	5,110	30,37
H-Tyr(3′-Cl)-Tic-Phe-Phe-OH	PA (e = 0.12)[e]	3.00	1,550	517	30,37
H-Tyr(3′-F)-Tic-Phe-Phe-OH	13.0	1.62	3,070	1,900	30,37
H-Tyr(3′-I)-TicΨ[CH_2NH]Phe-Phe-OH	19.2	2.08	2,660	1,280	34
H-Tyr(3′-I)-Tic-Phe-OH	141	60	12,100	202	34
H-Tyr-(2*S*,3*S*)-βMeTic-Phe-Phe-OH	2.09	6.81[f]	>10,000[g]	>1,470	35
H-Tyr-(2*S*,3*R*)-βMeTic-Phe-Phe-OH	1.61	0.53[f]	>10,000[g]	18,900	35
H-Tyr-Tic-D-Phe-Phe-OH	PA (e = 0.45)[e]	6.76	4,000	592	35
H-Tyr-Tic-Trp-Phe-OH	2.56	0.301	1,790	5,950	36
H-Tyr-Tic-Trp-OH	6.23	7.55	5,000	662	36
H-Tyr(3′-I)-Tic-Trp-Phe-OH	19.8 (IC40)[e]	2.20	1,630	741	36
H-Tyr-Tic-Hfe-Phe-OH	0.408[c]	0.277	1,990	7,180	30
H-Tyr-Tic-Phe(*p*F)-Phe-OH	1.62	1.65	5,480	3,320	38
H-Tyr-Tic-Phe(*p*Cl)-Phe-OH	1.60	1.26	8,130	6,450	38
H-Tyr-Tic-Phe(*p*Br)-Phe-OH	3.35	0.382	1,200	3,140	38
H-Tyr-Tic-Phe(*p*I)-Phe-OH	2.88	0.570	1,190	2,090	38
H-Tyr-Tic-2-Nal-Phe-OH	3.07[c]	1.31	6,330	4,830	30
H-Tyr-Tic-(2*S*,3*R*)βMePhe-Phe-OH	0.192	0.38[f]	>10,000[g]	>26,300	35
H-Tyr-Tic-Leu-Phe-OH	8.59	2.84	904	318	36
H-Tyr-Tic-Ile-Phe-OH	13.1	4.37	6,460	1,480	36
H-Tyr-Tic-Nva-Phe-OH	8.17	2.62	6,900	2,630	36
H-Tyr-Tic-Cha-Phe-OH	0.438	0.611	3,600	5,890	36
H-Tyr-TicΨ[CH_2NH]Cha-Phe-OH (TICP[Ψ])	0.219	0.259	1,050	4,050	36
H-Tyr-Tic-Phe-Phe(*p*NO_2)-OH	3.30	0.703	2,890	4,110	41
H-Tyr-Tic-Trp-Phe(*p*NO_2)-OH	4.40	0.330	1,520	3,340	41
H-Tyr-Tic-Phe-Phe(*p*F)-OH	2.97	1.26	3,210	2,550	38
H-Tyr-Tic-Phe-Phe(*p*Cl)-OH	2.52	1.02	2,070	2,030	38
H-Tyr-Tic-Phe-Phe(*p*Br)-OH	2.93	1.42	2,870	2,020	38
H-Tyr-Tic-Phe-Phe(*p*I)-OH	2.37	0.509	2,690	5,280	38
Tyr($N(CH_2)_5CH_3$)-Tic-Phe-Phe-OH	4.28	1.10	1,080	982	38
Tyr(NCH_2-cyclopropyl)-Tic-Phe-Phe-OH	28.2[c]	4.84	6,910	1,430	38
Tyr($N(CH_2CH_3)_2$)-Tic-Phe-Phe-OH	3.29	1.22	7,940	6,510	38

TABLE 2 Continued

Compound	MVD K_e [nM][b]	Receptor binding assays[a] K_i^{δ} [nM]	K_i^{δ} [nM]	K_i^{μ}/K_i^{δ}	Ref.
H-Tyr-TicΨ[CH_2NCH_3]Phe-Phe-OH	2.89[c]	0.842	13,400	15,900	38,42
H-Tyr-TicΨ[$CH_2NCH_2CH_3$]Phe-Phe-OH	14.4	6.31	4,530	718	38
H-Tyr-TicΨ[$CH_2N(CH_2)_5CH_3$]Phe-Phe-OH	13.1	66.6	18,800	282	38
H-Tyr-TicΨ[$CH_2NCH_2CH_2Ph$]Phe-Phe-OH	13.8[c]	4.51	1,130	251	38
H-Tyr-Tic-Phe-Asp-Val-Val-Gly-NH_2	22.8 (IC_{25})[e]	6.49	9,230	1,420	43,44
H-Tyr-Tic-Phe-Phe-Val-Val-Gly-NH_2	6.36	1.44	3,280	2,280	43
H-Tyr-Tic-Phe-Gly-Tyr-Pro-Ser-NH_2	19.9	1.26	865	685	43
H-Tyr-Tic-Phe-Phe-Tyr-Pro-Ser-NH_2	2.90	0.578	1,250	2,160	43
H-Tyr-Tic-Phe-Phe-Leu-Arg-Arg-Ile-Arg-Pro-Lys-NH_2	11.5	13.3	49.1	3.69	43
H-Dmt-Tic-OH	6.55	1.84	1,360	739	37
N,N-Me_2Dmt-Tic-OH	5.01	5.93	5,720	965	37
H-TyrΨ[CH_2NH]Tic-Phe-Phe-OH	180	24.1	103	4.27	58
H-TyrΨ[CH_2NH]MeTic-Phe-Phe-OH	160	29.4	49.9	1.70	58
ICI 174684	36.4[h]	193[i]	24,700[j]	128	13
Naltrindole	0.636	0.687	12.2	17.8	27,37

[a] Displacement of [^{3}H]DAMGO (μ-selective) and [^{3}H]DSLET (δ-selective) from rat brain membrane binding sites.
[b] Determined against DPDPE.
[c] Determined against deltorphin I.
[d] Agonist, value indicates IC_{50} [nM].
[e] Partial agonist (e = intrinsic efficacy).
[f] Displacement of [^{3}H]DPDPE from NxG108CC15 cell membrane binding sites.
[g] Displacement of [^{3}H]sufentanil from rat brain membrane binding sites.
[h] Determined against DSLET.
[i] Displacement of [^{3}H]DADLE from guinea pig brain membrane binding sites.
[j] Displacement of [^{3}H]DAMGO from guinea pig brain membrane binding sites.

potency ($K_e = 0.196$ nM) and slightly lower but still very significant δ-receptor selectivity ($K_i^{\mu}/K_i^{\delta} = 569$), as compared to the TIPP parent [30]. Substitution of 2′-hydroxy,6′-methyltyrosine (Hmt) for Tyr[1] in TIPP also increased δ-receptor binding affinity about 10-fold but did not alter μ-receptor affinity [32]. Thus, H-Hmt-Tic-Phe-Phe-OH is a δ-antagonist with subnanomolar potency ($K_e = 0.473$ nM) and very high δ-receptor selectivity ($K_i^{\mu}/K_i^{\delta} = 11{,}700$). The corresponding tripeptide analogue H-Hmt-Tic-Phe-OH showed only slightly lower δ-antagonist potency and slightly lower δ-selectivity.

Substitution of a halogen at the 3′ position of the Tyr[1] aromatic ring of TIPP had interesting effects on the intrinsic efficacy of the peptide. The Tyr(3′-

I)-analogue showed a displacement curve with an IC_{50} value of 46.0 nM in a binding competition assay based on displacement of [^{3}H]diprenorphine from N4TG1 cell membrane δ-binding sites [33]. Surprisingly, this displacement curve was shifted to the right (IC_{50} = 126 nM) in the presence of 100 mM Na^+ and 0.1 mM 5′-guanylylimidodiphosphate [Gpp(NH)p], indicating that this compound behaved as a δ-opioid agonist. Examination of [Tyr(3′-I)]TIPP in the MVD assay revealed that this compound was indeed a full δ-agonist (IC_{50} = 97 nM, Table 2), the effect of which was antagonized by the TIPP parent peptide (K_e = 11 nM). This result indicated that monoiodination at the 3′ position of the Tyr^1 aromatic ring of the δ-antagonist TIPP converted it to a full δ-agonist. Bromination or chlorination at the 3′ position of Tyr^1 in TIPP resulted in partial δ-agonists with respective intrinsic efficacies (e) of 0.16 and 0.12, whereas the Tyr(3′-F) analogue was again a pure δ-antagonist [30]. Thus, systematic substitution of halogen atoms beginning with fluorine and in the order of the periodic table produced a progressive increase in the intrinsic efficacy and, as evident from the receptor-binding assay data, a concomitant decrease in δ-receptor affinity. Interestingly, the Tyr(3′-I)-analogues of TIPP[Ψ] and TIP retained δ-antagonist properties [34] (Table 2), indicating that the conversion to an agonist caused by iodination of TIPP may not be due to a direct local effect of the iodine substituent, but may be the result of an overall conformational change of the peptide.

Methylation of the β-carbon of the Tic^2 residue of TIPP did not have much of an effect on the activity profile [35]. Both the (2*S*,3*S*)-$\beta MeTic^2$ and the (2*S*,3*R*)-$\beta MeTic^2$ analogues were slightly more potent δ-antagonists than the TIPP parent peptide, and both of them retained high δ-receptor selectivity. An NMR study revealed that the βMeTic residue in both analogues was in the *gauche*$^+$ configuration [35], as is the case with the Tic residue in the TIPP parent peptide.

A number of structural modifications were performed at the 3 position of the TIPP peptide sequence. The D-Phe^3 analogue of TIPP was found to be a partial δ-agonist [35]. The Trp^3 analogues of both TIP and TIPP displayed two- to threefold higher δ-antagonist potency and about fivefold higher δ-receptor selectivity than their respective parent peptides [36]. Replacement of Tyr^1 in [Trp^3]TIPP with Tyr(3′-I) resulted in an analogue with nearly full δ-agonist properties [36], in parallel to the δ-agonist behavior shown by its iodinated parent H-Tyr(3′-I)-Tic-Phe-Phe-OH (Table 2). The results of tryptophan fluorescence decay measurements carried out in aqueous buffer (pH 7.5, 20°C) revealed that the Trp^3 side chain in H-Tyr-Tic-Trp-Phe-OH and H-Tyr-Tic-Trp-OH was exposed to the aqueous environment, whereas in a significant proportion of the conformers of H-Tyr(3′-I)-Tic-Trp-Phe-OH that same side chain was somewhat shielded from the solvent and located in a more hydrophobic environment provided by the other aromatic residues in the molecule [36,37]. It thus appears that iodination at the 3′ position of Tyr resulted in a conformational change, which may be related to the observed conversion of the antagonist into an agonist.

Replacement of Phe^3 in TIPP with homophenylalanine (Hfe) resulted in a compound with subnanomolar δ-antagonist potency and very high δ-selectivity [30]. Substitution of a halogen atom (F,Cl,Br,I) in the *para* position of Phe^3 led to compounds with slightly higher δ-antagonist potency and higher δ receptor selectivity than the TIPP parent peptide [38]. A TIPP analogue containing 3-(2′-naphthyl)alanine (2-Nal) in place of Phe^3 showed δ-antagonist potency comparable to that of the parent peptide and slightly increased δ-receptor selectivity [30]. The (2*S*,3*R*)-β$MePhe^3$ analogue of TIPP turned out to be a highly potent δ-antagonist with extraordinary δ-selectivity [35]. In (2*S*,3*R*)-βMePhe the *gauche(-)* and *trans* side chain conformations are allowed, whereas the *gauche(+)* conformation is strongly disfavored. The result of a molecular mechanics study of TIPP indicated that in the receptor-bound conformation the Phe^3 conformation may be *gauche(-)* [39]. Interestingly, the Ala^3 analogue of TIP was found to retain moderate δ-antagonist potency [40]. This finding indicated that an aromatic residue in the 3 position of the TIP peptide sequence is not a *conditio sine qua non* for δ-antagonist activity. Subsequently, TIPP analogues containing a Leu, Ile, or Nva (norvaline) residue in place of Phe^3 were shown to have only slightly lower δ-antagonist potency than the parent peptide and similar δ-selectivity [36] (Table 2). Substitution of Phe^3 in TIPP with cyclohexylalanine (Cha) resulted in a compound, H-Tyr-Tic-Cha-Phe-OH (TICP), with substantially increased δ-antagonist potency and higher δ-selectivity than the parent peptide. The corresponding pseudopeptide analogue H-Tyr-TicΨ[CH_2NH]Cha-Phe-OH (TICP[Ψ]) showed even higher δ-antagonist activity (K_e = 0.219 nM) and equally high δ-selectivity ($K_i^{\mu}/K_i^{\delta}=4050$) [36]. TICP[Ψ] has about the same extraordinarily high δ-antagonist potency as H-Dmt-Tic-Phe-Phe-OH, but is seven times more δ-selective. It is a 13 times more potent δ-antagonist than TIPP[Ψ], and equally stable to chemical degradation owing to the presence of the reduced peptide bond. A direct comparison under identical assay conditions revealed that TICP[Ψ] was about three times more potent and 230 times more δ-selective than naltrindole [37].

Structural modifications made at the Phe^4 residue involved the introduction of various substituents in the *para* position of the aromatic ring. The Phe(pNO_2)^4 (*para*-nitrophenylalanine) analogues of both TIPP and [Trp^3]TIPP were potent δ-antagonists showing slightly higher δ-receptor binding affinity and δ-selectivity than TIPP [41] (Table 2). Similarly, halogeneration (F, Cl, Br, I) at the para position of Phe^4 of TIPP resulted in four compounds that all were slightly more potent and more selective δ-antagonists than TIPP [38]. These halogenated analogues are significantly more lipophilic than the TIPP parent peptide, as indicated by their hydrophobicity parameters k′ determined by HPLC [38].

The lipophilicity of TIPP and TIPP[Ψ] was further enhanced by introduction of various alkyl-or arylalkyl substituents either at the N-terminal amino group of TIPP or at the secondary amino group of TIPP[Ψ] [38]. TIPP analogues containing an n-hexyl or two ethyl groups attached to the N-terminal amino function retained δ-antagonist potency comparable to that of the parent peptide as well as high δ-selectivity, whereas the analogue carrying a cycloproylmethyl group at the N-terminus was a somewhat less potent δ-antagonist. Introduction of a methyl substituent at the secondary amino group of the reduced peptide bond in TIPP[Ψ] produced a compound with unchanged δ-antagonist potency, subnanomolar δ-receptor binding affinity, and extraordinary δ-selectivity ($K_i^{\mu}/K_i^{\delta} = 15{,}900$) [38,42]. TIPP[Ψ] analogues with an ethyl-, *n*-hexyl-, or phenylethyl substituent attached to the secondary amino group showed only about fourfold lower δ-antagonist potency than TIPP[Ψ] and still quite high δ-receptor selectivity. HPLC determination of the k′ values indicated that, like the Phe^4-halogenated TIPP analogues described above, all alkyl- or arylalkyl-substituted TIPP and TIPP[Ψ] analogues had substantially increased lipophilic character. Thus, it can be expected that these various substitutions by themselves or in combination may improve the ability of these δ-antagonists to cross the blood-brain barrier (BBB).

The effect of substituting a Tic residue in the 2 position of larger δ-, μ-, and κ-selective opioid peptides was examined (Table 2). [Tic^2]Deltorphin-I turned out to be a partial δ-agonist in the MVD assay and was eight times more δ-selective than the deltorphin-I parent in the receptor-binding assays [43,44]. A deltorphin-I analogue in which the entire N-terminal tetrapeptide segment had been replaced with the TIPP sequence was a pure δ-antagonist with an in vitro activity profile similar to that of TIPP [43]. The Tic^2- and Tic^2,Phe^4-analogues of dermorphin also were found to be potent and selective δ-antagonist. An analogue of [D-Pro^{10}]dynorphin A(1-11)-NH_2 containing the TIP tripeptide segment at the N-terminus, H-Tyr-Tic-Phe-Phe-Leu-Arg-Arg-Ile-Arg-Pro-Lys-NH_2, was a potent δ-antagonist as well ($K_e = 11.5$ nM). Interestingly, this analogue was also a moderately potent κ-antagonist against dynorphin A(1-13) in the GPI assay, ($K_e = 279$ nM) with moderate κ-receptor-binding affinity ($K_i^{\kappa} = 24.5$ nM) [43]. Subsequently, Tic analogues of dynorphin A were also shown to have quite high δ-antagonist potency and κ-antagonist activity [45]. None of all the other TIP(P) analogues listed in Table 2 displayed significant κ-receptor binding affinity.

The dipeptide H-Dmt-Tic-OH has been reported to be an "ultraselective" δ-opioid antagonist ($K_i^{\delta} = 0.022$ nM, $K_i^{\mu}/K_i^{\delta} = 150{,}773$) [46]. However, subsequent evaluations of this compound indicated that it had much lower δ-receptor binding affinity and much lowe δ-receptor selectivity than originally reported: $K_i^{\delta} = 1.84$ nM, $K_i^{\mu}/K_i^{\delta} = 739$ (Table 2) [37]; $IC_{50}(\delta) = 1.6$ nM, $IC_{50}(\mu)/IC_{50}(\delta) = 558$ [47]; and $K_i^{\delta} = 1.6$ nM [48]. In a direct comparison under identical assay conditions, H-Dmt-Tic-OH showed ~30 times lower δ-antagonist potency and six times lower δ-selectivity than TICP[Ψ] (Table 2) [37]. Furthermore, H-Dmt-Tic-OH was found to be unstable in organic solvents due to diketopiperazine formation (Nguyen and Schiller, unpublished results). The occurrence of diketopiperazine formation was eliminated by preparing the N,N-dimethylated analogue, N,N-Me_2Dmt-Tic-OH [49]. This stable compound turned out to be a quite potent and very selective δ antagonist. In a direct comparison under identical assay conditions, N,N-Me_2Dmt-Tic-OH showed ~25 times lower δ-antagonist potency than TICP[Ψ] and four times lower δ-receptor selectivity (Table 2) [37]. Pagé et al. [47] prepared H-Dmt-Tic dipeptide analogues with substitution of large hydrophobic groups at position 6 or 7 of Tic. Among these, compounds containing a 4-F-Ph or 4-OMe-Ph group at the 7 position of Tic turned out to be fairly potent and quite selective δ-antagonists.

TIPP, TIPP[Ψ], and TICP[Ψ] have become valuable pharmacological tools. Highly selective δ-opioid receptor radioligands were obtained by preparing these three peptides in tritiated form [50–52]. [^{125}I]TIPP[Ψ] was prepared and shown to be a stable δ-receptor radioligand suitable for in vivo studies [53]. Visconti et al. [54] showed that TIPP[Ψ] had very high δ-receptor binding selectivity ratios against all μ- and κ-receptor subtypes (μ_1, μ_2, κ_1, κ_2, κ_3) and used this compound in an improved μ_1-receptor-binding assay.

3.2 The Receptor-Bound Conformation of TIP(P)-Related Peptides

Proposed models of the receptor-bound conformation of TIP(P)-related δ-opioid antagonists were described in detail in a recent review article [37]. These models were based on molecular mechanics studies of TIP(P)-related antagonists and comparison of the resulting low energy conformers with the structurally rigid nonpeptide δ-antagonist naltrindole. One model was obtained through identification of a low-energy conformer of TIP showing good spatial overlap of the centroids of its Tyr^1 and Tic^2 aromatic rings and its N-terminal amino group with the corresponding aromatic rings and the nitrogen atom in naltrindole [55]. This model is characterized by a *trans* peptide bond between Tyr^1 and Tic^2 (Fig. 1). An alternative model resulted from theoretical conformational analysis of the weak δ-opioid dipeptide antagonist H-Tyr-Tic-NH_2 and was also based on spatial overlap of the Tyr^1 and Tic^2 aromatic rings and N-terminal amino group with the corresponding moieties in naltrindole [56] (Fig. 1). In this model the peptide bond between the Tyr^1 and Tic^2 residues has the *cis* conformation. A subsequently

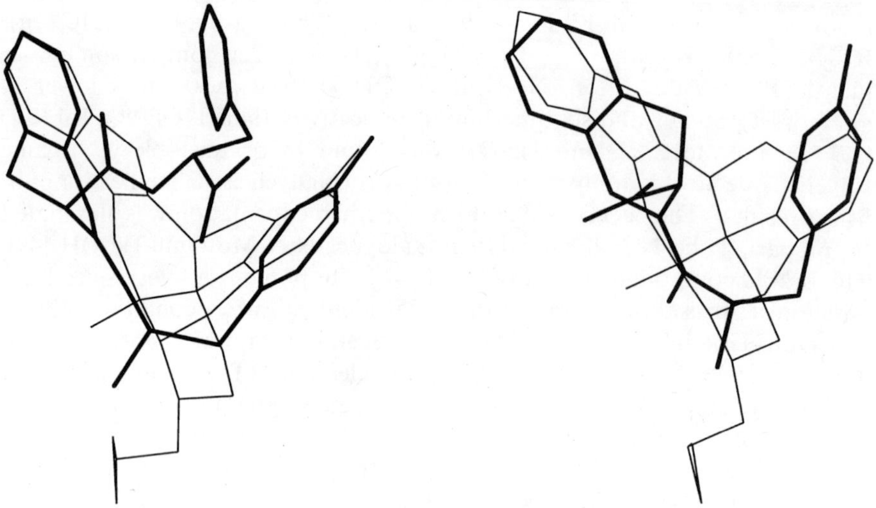

FIGURE 1 (Left) Model of the receptor-bound conformation of TIP containing all-*trans* peptide bonds (heavy lines) in spatial overlap with naltrindole (light lines). (Right) Model of the receptor bound conformation of H-Tyr-Tic-NH_2 containing a cis peptide bond (heavy lines) in spatial overlap with naltrindole (light lines). In both cases the N-terminal amino group and the Tyr^1 and Tic^2 aromatic rings of the peptide are superimposed on the corresponding pharmacophoric moieties in the alkaloid structure.

performed comparative assessment of these two models was based on theoretical conformational analysis of six δ-opioid peptide antagonists of this class: H-Tyr-Tic-NH_2, H-Tyr-Tic-Ala-OH, H-Tyr-Tic-Phe-OH (TIP), H-Tyr-TicΨ[CH_2NH]Phe-OH (TIP[Ψ]), H-Tyr-Tic-Phe-Phe-OH (TIPP), and H-Tyr-TicΨ[CH_2NH]Phe-Phe-OH (TIPP[Ψ]) [39]. Low energy conformers consistent with both models were identified for all six compounds. However, conformers corresponding to the all-*trans* peptide bond model generally were lower in energy than conformers corresponding to the *cis* peptide bond–containing model. Moreover, better coplanarity of the peptide aromatic rings with the corresponding aromatic rings in naltrindole was observed with conformers corresponding to the all-*trans* peptide bond model than with conformers corresponding to the *cis* peptide bond containing model in the spatial overlap studies. Nevertheless, both models remained plausible candidate structures for the receptor-bound conformation of δ-antagonists of the TIPP class. Both models differ from the crystal structure of TIPP [57]. However, the structure in the crystalline state is affected by numerous intermolecular hydrophobic interactions between TIPP molecules and does not represent the receptor-bound conformation.

A final assessment of the two models was based on the synthesis and pharmacological characterization of two peptides, H-TyrΨ[CH_2NH]Tic-Phe-Phe-OH and H-TyrΨ[CH_2NH]MeTic-Phe-Phe-OH (MeTic = L-3-methyl-1,2,3,4-tetrahydroisoquinoline-3-carboxylic acid), in which a *cis* peptide bond between the Tyr^1 and Tic^2 (or $MeTic^2$) residues is sterically forbidden [58]. Both compounds turned out to be δ-opioid antagonists with moderate potency against DPDPE in the MVD assay. They showed ~20–25 times lower δ-receptor binding affinity than TIPP, but still about sevenfold higher δ-receptor affinity than the δ-antagonist ICI 174864. The results of molecular mechanics calculations indicated that both analogues assumed low-energy conformations that are consistent with the all-*trans* peptide bond model of the receptor-bound conformation of TIPP. No low-energy conformers with a *cis* or *cis*-like peptide bond ("ω" = 0°) between Tyr^1 and Tic^2 were found for either peptide. This result led to the conclusion that δ-opioid peptide antagonists containing an N-terminal H-Tyr-Tic dipeptide segment assume a receptor-bound conformation in which the peptide bond between the Tyr^1 and Tic^2 residues must be in the *trans* conformation.

4 INVERSE δ-OPIOID AGONISTS

In a pioneering study, Costa and Herz showed that the δ-opioid receptor–selective enkaphalin analogue ICI 174864 was able to inhibit GTPase activity in NG108-15 neuroblastoma-glioma cells [16]. This compound, previously shown to be a δ-antagonist in the MVD assay (see Sec. 2.1), was thus identified as the first example of a compound with inverse agonist properties at a G protein–coupled receptor. However, ICI 174864 is an inverse δ-opioid agonist with relatively weak δ-receptor-binding affinity (Table 2).

Several of the Tic^2-containing peptides characterized as δ-opioid antagonists in the MVD assay were subsequently shown to be inverse δ-agonists. In the adenylate cyclase assay using HEK293s cells stably transfected with hDORs, the pseudopeptide TICP[Ψ] [36,37] behaved as an inverse δ-agonist that, in comparison with ICI 174864, had about the same high negative efficacy but 110-fold higher δ-receptor binding affinity [59]. TICP[Ψ] is the most potent and most selective full inverse δ-opioid agonist reported to date. [^{3}H]TICP[Ψ] was used in a study to demonstrate that short-term inverse agonist treatment increases the δ-receptor-binding capacity for this inverse δ-agonist in HEK cells stably expressing the human δ-receptor [59]. The dipeptide H-(2*S*,3*R*)-Tmt-Tic-OH (Tmt = β-methyl-2′,6′-dimethyltyrosine) has also been shown to have about the same negative efficacy as ICI 174864, as determined in the [^{35}S]GTPγS binding assay using Chinese hamster ovary

(CHO) cells transfected with hDOR [60]. In comparison with TICP[Ψ], this dipeptide showed about the same high δ- versus μ-receptor selectivity but 35 times lower δ-receptor-binding affinity [61]. Finally, the dipeptide derivative N,N-Me_2Dmt-Tic-OH also turned out to be an inverse δ-agonist in the [^{35}S]GTPγS binding assay (HEK cells stably transfected with hDOR) with a negative efficacy similar to that of ICI 174864 [62]. However, in a direct comparison with TICP[Ψ], N,N-Me_2Dmt-Tic-OH displayed ~25 times lower δ-receptor affinity and fourfold lower δ- versus μ-receptor selectivity [37] (see Sec. 3.1.).

TIPP and TIPP[Ψ] were confirmed as neutral antagonists in assays using a number of cellular models expressing endogenous or transfected δ-opioid receptors [59,63,64]. Recently, it was reported that TIPP exhibited properties of δ-agonist, neutral δ-antagonist, or inverse δ-agonist, depending on the cellular assay systems used [65]. The significance of this finding in relation to the neutral δ-antagonist behavior of TIPP demonstrated in other in vitro assays needs to be further examined.

5 MIXED μ-AGONIST/δ-ANTAGONISTS

Two studies indicated that selective δ-opioid receptor blockade with a δ-antagonist greatly reduced the development of morphine tolerance and dependence [5,6]. In connection with this observation, it is of interest to note that 1 δ-receptors are upregulated upon chronic morphine treatment [66]; 2 administration of an antisense oligodeoxynucleotide to the δ-receptor prevented the development of morphine tolerance and dependence [67]; and 3 morphine retains its analgesic activity in δ-opioid receptor knockout mice without producing tolerance and dependence upon chronic administration [68]. These various observations indicate that δ-receptors play a major role in the development of morphine tolerance and dependence, and provide a rationale for the development of mixed μ-agonist/δ-antagonists, compounds that would act as an agonist at the μ-receptor to produce the analgesic effect and as an antagonist at the δ-receptor to prevent the development of tolerance and physical dependence. Furthermore, it has been shown that a δ-antagonist reversed μ-agonist-induced respiratory depression [7,8] and enhanced colonic propulsion [9]. These results suggest that a mixed μ-agonist/δ-antagonist may also cause less respiratory depression and less inhibition of gastrointestinal transit than a μ-agonist such as morphine.

The first known compound with a mixed μ-agonist/δ-antagonist profile was the tetrapeptide amide H-Tyr-Tic-Phe-Phe-NH_2 (TIPP-NH_2) [27] (Table 3). TIPP-NH_2 was a moderately potent μ-agonist in the GPI assay and a potent δ antagonist in the MVD assay. It showed some δ- versus μ-receptor

TABLE 3 In Vitro Opioid Activity Profiles of Mixed μ-Agonist/δ-Antagonists

Compound	GPI assay IC_{50} [nM]	MVD assay K_e [nM][a]	Receptor binding assays[b] K_i^{μ} [nM]	K_i^{δ} [nM]	Ref.
H-Tyr-Tic-Phe-Phe-NH_2 (TIPP-NH_2)	1700	18.0	78.8	3.0	27
H-Dmt-Tic-Phe-Phe-NH_2 (DIPP-NH_2)	18.2	0.209	1.19	0.118	69,70
H-Dmt-TicΨ[CH_2NH]Phe-Phe-NH_2 (DIPP-NH_2[Ψ])	7.71	0.537	0.943	0.447	69,70
H-Tyr-Tic-NH-$(CH_2)_3$-Ph	PA (e = 0.38)[c]	41.9	160	3.01	70
H-Dmt-Tic-NH-$(CH_2)_3$-Ph	102	1.69	0.386	0.126	70
H-Dmt-Tic-NH-$(CH_2)_2$-Ph	48.0	2.30 (IC_{40})[c]	1.59	0.0577	72
H-Dmt-Tic-NH-(CH_2)-3-In	405	176	78.8	78.9	72
H-Dmt-Tic-NH-$(CH_2)_2$-Ch	268	2.88	4.96	0.676	72
H-Tyr-c[-D-Orn-2-Nal-D-Pro-Gly-]	268	202	5.89	17.2	74
H-Dmt-c[-D-Orn-2-Nal-D-Pro-Gly-]	7.88	2.13	0.460	0.457	75

[a] Determined against DPDPE.
[b] Displacement of [^{3}H]DAMGO (μ-selective) and [^{3}H]DSLET (δ-selective) from rat brain membrane-binding sites.
[c] Partial agonist.

selectivity ($K_i^{\mu}/K_i^{\delta} = 26$) in the binding assays and no binding affinity for κ-receptors at concentrations up to 10 μM. Replacement of Tyr^1 in TIPP-NH_2 with Dmt led to a compound, H-Dmt-Tic-Phe-Phe-NH_2 (DIPP-NH_2), showing a 93-fold μ-agonist potency increase in the GPI assay and an 86-fold enhancement in δ-antagonist activity in the MVD assay [69,70]. The receptor-binding assay data indicate that DIPP-NH_2 is still somewhat δ-selective ($K_i^{\mu}/K_i^{\delta} = 10.1$). Reduction of the peptide bond between Tic^2 and Phe^3 of DIPP-NH_2 resulted in a compound, H-Dmt-TicΨ[CH_2NH]Phe-Phe-NH_2 (DIPP-NH_2[Ψ]), which displayed further increased μ-agonist potency in the GPI assay (IC_{50} = 7.71 nM) and retained very high δ-antagonist activity ($K_e = 0.537$ nM) in the MVD assay [69,70]. DIPP-NH_2[Ψ] displayed subnanomolar binding affinities for both μ- and δ-receptors and thus turned out to be a "balanced" μ-agonist/δ-antagonist ($K_i^{\mu}/K_i^{\delta} = 2.11$). In the rat tail flick test, DIPP-NH_2[Ψ] given ICV produced a potent analgesic effect, being about three times more potent than morphine [70]. Unlike morphine, this compound produced no physical dependence upon chronic administration at high doses. It produced less acute analgesic tolerance than morphine but still a certain level of chronic tolerance. Thus, DIPP-NH_2[Ψ] fulfills to a large extent the expectations based on the mixed μ-agonist/δ-antagonist concept with regard to analgesic activity and the development of tolerance and dependence.

In an effort to develop mixed μ-agonist/δ-antagonists of lower molecular weight that might cross the BBB, dipeptides of the general formula H-Tyr (or Dmt)-Tic-NH-R (R = arylalkyl or alkyl) were developed. The compound H-Tyr-Tic-NH-$(CH_2)_3$-Ph was found to be a moderately potent partial μ-agonist in the GPI assay and a δ-antagonist in the MVD assay [69] (Table 3). Replacement of Tyr^1 in the latter dipeptide derivative with Dmt resulted in a compound, H-Dmt-Tic-NH-$(CH_2)_3$-Ph, that was able to fully inhibit the electrically evoked contractions of the GPI with an IC_{50} of 102 nM and showed 25-fold enhanced δ-antagonist activity ($K_e = 1.69$ nM). This compound displayed very high μ-receptor binding affinity ($K_i^\mu = 0.368$ nM) and extraordinary binding affinity for δ-receptors ($K_i^\delta = 87$ pM). The observed discrepancy between the relatively low μ-agonist potency of this compound in the GPI assay and its high μ-receptor-binding affinity indicates that it is also a partial μ-agonist/δ-antagonist. In this case, the partial μ-agonist effect is not directly apparent in the GPI assay because of the large μ-receptor reserve in the ileum [71]. The analogue H-Dmt-Tic-NH-$(CH_2)_2$-Ph turned out to be a compound with a mixed μ-agonist/partial δ-agonist profile [72]. The dipeptide derivative H-Dmt-Tic-NH-$(CH_2)_2$-3-In (3-In = 3-indole) was found to be a potent *full* μ-agonist/δ-antagonist, whereas the analogue H-Dmt-Tic-NH-$(CH_2)_2$-Ch (Ch = cyclohexyl) was a somewhat less potent μ-agonist but more potent δ-antagonist [72] (Table 3). An H-Dmt-Tiq derivative (Tiq = tetrahydroisoquinoline) containing a benzimidazole moïety at the 3 position of Tiq has also been reported to have mixed μ-agonist/δ-antagonist properties [73].

The cyclic β-casomorphin analogue H-Tyr-c[-D-Orn-2-Nal-D-Pro-Gly-] showed a mixed μ-agonist/δ-antagonist profile as well [74]. Replacement of Tyr^1 of the latter peptide with Dmt produced a compound, H-Dmt-c[-D-Orn-2-Nal-D-Pro-Gly-], that bound to both μ- and δ-receptors with subnanomolar affinities (~0.5 nM), was a potent μ-agonist, and showed 100-fold increased δ-antagonist potency as compared to the parent peptide [75] (Table 3). Ananthan and colleagues [76,77] described a nonpeptidic opioid μ-agonist/δ-antagonist (SoRI 9401) which showed moderate μ agonist potency in the GPI assay and high antagonist potency in the MVD assay. This compound given ICV was a partial agonist in the mouse tail flick test and produced full agonist effects in the mouse writhing assay after ICV or IP administration. In the latter assay, it did not produce analgesic tolerance after repeated IP administration.

ACKNOWLEDGMENT

The author's work described in this review was supported by operating grants from the CIHR (MT-5655) and the NIH (DA-04443).

REFERENCES

1. Arakawa K, Akami T, Okamoto M, Nakajima H, Mitsuo M, Nakai I, Oka T, Nagase H, Matsumoto S. Transplant Proc 1992; 24:696–697.
2. Arakawa K, Akami T, Okamoto M, Oka T, Nagase H, Matsumoto S. Transplantation 1992; 53:951–952.
3. Reid LD, Glick SD, Menkens KA, French ED, Bilsky EJ, Porreca F. NeuroReport 1995; 6:1281–1284.
4. Krishnan-Sarin S, Jing S-L, Kurtz DL, Zweifel M, Portoghese PS, Li TK, Froehlich JC. Psychopharmacology 1995; 120:177–185.
5. Abdelhamid EE, Sultana M, Portoghese PW, Takemori AE. J Pharmacol Exp Ther 1991; 258:299–303.
6. Fundytus ME, Schiller PW, Shapiro M, Weltrowska G, Coderre T. Eur J Pharmacol 1995; 286:105–108.
7. Freye E, Latasch L, Portoghese PS. Eur J Anaesthesiol 1992; 9:457–462.
8. Su YF, McNutt RW, Chang KJ. J Pharmacol Exp Ther 1998; 287:815–823.
9. Foxx-Orenstein AE, Jin JG, Grider JR. Am J Physiol 1998; 275:G979–G983.
10. Hahn EF, Fishman J, Shiwaku Y, Foldes F. Res Commun Chem Pathol Pharmacol 1977; 18:1–9.
11. Belton P, Cotton R, Giles MB, Gormley JJ, Miller L, Shaw JS, Timms D, Wilkinson A. Life Sci 1983; 33:443–446.
12. Shaw JS, Miller L, Turnbull MJ, Gormley JJ, Morley JS. Life Sci 1982; 31:1259–1262.
13. Corbett AD, Gillan MGC, Kosterlitz HW, McKnight AT, Paterson SJ, Robson LE. Br J Pharmacol 1984; 83:271–279.
14. Cotton R, Giles MG, Miller L, Shaw JS, Timms D. Eur J Pharmacol 1984; 97: 331–332.
15. Thornber CW, Shaw JS, Miller L, Hayward CF, Morley JS, Timms D, Wilkinson A. In: Holaday JW, Law P-Y, Herz A, eds. Progress in Opioid Research. National Institute on Drug Abuse Monograph. Vol. 75. Washington: U.S. Government Printing Office, 1986:177–180.
16. Costa T, Herz A. Proc Natl Acad Sci USA 1989; 86:7321–7325.
17. Thornber CW, Shaw JS, Miller L, Hayward CF. In: Holaday JW, Law P-Y, Herz A, eds. Progress in Opioid Research. National Institute on Drug Abuse Monograph. Vol 75. Washington: U.S. Government Printing Office, 1986:181–184.
18. Lovett JA, Portoghese PS. J Med Chem 1987; 30:1144–1149.
19. Lovett JA, Portoghese PS. J Med Chem 1987; 30:1668–1674.
20. Rónai AZ, Botyánski J, Hepp J, Medzihradszky K. Life Sci 1992; 50:1371–1378.
21. Rónai AZ, Magyar A, Orosz G, Borsodi A, Benyhe S, Tóth G, Makó E, Kátay E, Babka E, Medzihradsky K. Arch Int Pharmacodyn 1995; 330:361–369.
22. Lu Y, Weltrowska G, Lemieux C, Chung NN, Schiller PW. Bioorg Med Chem Lett 2001; 11:323–325.
23. Lu Y, Schiller PW. Synthesis 2001:1639–1644.
24. Schiller PW, Lu Y, Weltrowska G, Berezowska I, Wilkes BC, Nguyen TM-D,

Chung NN, Lemieux C. In: Lebl M, Houghten RA, eds. Peptides: The Wave of the Future. Proc 2nd Int Peptide Symp/17th Am Peptide Symp. San Diego: American Peptide Society, 2001:676–678.
25. Hruby VJ, Baratosz-Bechowski H, Davis P, Slaninova J, Zalewska T, Stropova D, Porreca F, Yamamura HI. J Med Chem 1997; 40:3957–3962.
26. Vavrek RJ, Cui R-L, Stewart JM. Life Sci 1982; 31:2249–2252.
27. Schiller PW, Nguyen TM-D, Weltrowska G, Wilkes BC, Marsden BJ, Lemieux C, Chung NN. Proc Natl Acad Sci USA 1992; 89:11871–11875.
28. Marsden BJ, Nguyen TM-D, Schiller PW. Int J Peptide Protein Res 1993; 41:313–316.
29. Schiller PW, Weltrowska G, Nguyen TM-D, Wilkes BC, Chung NN, Lemieux C. J Med Chem 1993; 36:3182–3187.
30. Schiller PW, Nguyen TM-D, Weltrowska G, Wilkes BC, Marsden BJ, Schmidt R, Lemieux C, Chung NN. In: Hodges RS, Smith JA, eds. Peptides: Chemistry, Structure and Biology. Proc 13th American Peptide Symposium. Leiden, Netherlands: ESCOM Science Publishers, 1994:483–486.
31. Hansen DW, Stapelfeld A, Savage MA, Reichman M, Hammond DL, Haaseth RC, Mosberg HI. J Med Chem 1992; 35:684–687.
32. Berezowska I, Lemieux C, Chung NN, Schiller PW. In: Bajusz S, Hudecz F, eds. Peptides 1998, Proc 25th Eur Peptide Symposium. Budapest, Hungary: Académia Kiadó, 1999:718–719.
33. Lee PHK, Nguyen TM-D, Chung NN, Schiller PW, Chang K-J. Eur J Pharmacol 1995; 280:211–214.
34. Schiller PW, Weltrowska G, Schmidt R, Berezowska I, Nguyen TM-D, Lemieux C, Chung NN, Carpenter KA, Wilkes BC. J Receptor Signal Transduc Res 1999; 19:573–588.
35. Tourwé D, Mannekens E, Diem TNT, Verheyden P, Jaspers H, Tóth G, Péter A, Kertész I, Török G, Chung NN, Schiller PW. J Med Chem 1998; 41:5167–5176.
36. Schiller PW, Weltrowska G, Nguyen TM-D, Lemieux C, Chung NN, Zelent B, Wilkes BC, Carpenter KA. In: Kaumaya TP, Hodges RS, eds. Peptides: Chemistry, Structure and Biology. Proc 14th American Peptide Symposium. Kingswinford, England: Mayflower Scientific, 1996:609–611.
37. Schiller PW, Weltrowska G, Berezowska I, Nguyen TM-D, Wilkes BC, Lemieux C, Chung NN. Biopolymers (Peptide Sci) 1999; 51:411–425.
38. Schiller PW, Weltrowska G, Berezowska I, Nguyen TM-D, Chung NN, Lemieux C, Carpenter KA, Wilkes BC. In: Xu X-J, Ye YH, Tam JP, eds. Peptides: Biology and Chemistry. Proc 1996 Chinese Peptide Symposium. Leiden, Netherlands: ESCOM Science Publishers, 1998:138–141.
39. Wilkes BC, Schiller PW. Biopolymers (Peptide Sci) 1995; 37:391–400.
40. Temussi PA, Salvadori S, Amadeo P, Bianchi C, Guerrini R, Tomatis R, Lazarus LH, Picone D, Tancredi T. Biochem Biophys Res Commun 1994; 198: 933–939.
41. Schiller PW, Nguyen TM-D, Berezowska I, Weltrowska G, Schmidt R, Marsden BJ, Wilkes BC, Lemieux C, Chung NN. In: Yanaihara N, ed. Peptide

Chemistry 1992. Proc 2nd Japan Symposium on Peptide Chemistry. Leiden, Netherlands: ESCOM Science Publishers, 1993:337–340.
42. Schiller PW, Schmidt R, Wilkes BC, Weltrowska G, Nguyen TM-D, Chung NN, Lemieux C. In: Lu G-S, Tam JP, Du Y-C, eds. Peptides: Biology and Chemistry. Proc 1994 Chinese Peptide Symposium. Leiden, Netherlands: ESCOM Science Publishers, 1995:140–143.
43. Schmidt R, Chung NN, Lemieux C, Schiller PW. Regul Peptides 1994; 54:259–260.
44. Schiller PW, Weltrowska G, Nguyen TM-D, Wilkes BC, Chung NN, Lemieux C. J Med Chem 1992; 35:3956–3961.
45. Guerrini R, Capasso A, Marastoni M, Bryant SD, Cooper PS, Lazarus LH, Temussi PA, Salvadori S. Bioorg Med Chem 1998; 6:57–62.
46. Salvadori S, Attila M, Balboni G, Bianchi C, Bryant SD, Crescenzi O, Guerrini R, Picone D, Tancredi T, Temussi P, Lazarus LH. Mol Med 1995; 1:678–689.
47. Pagé D, McClory A, Mischki T, Schmidt R, Butterworth J, St-Onge S, Labarre M, Payza K, Brown W. Bioorg Med Chem Lett 2000; 10:167–170.
48. Santagada V, Caliendo G, Severino B, Perissutti E, Ceccarelli F, Giusti L, Mazzoni MR, Salvadori S, Temussi PA. J Peptide Sci 2001; 7:374–385.
49. Salvadori S, Balboni G, Guerrini R, Tomatis R, Bianchi C, Bryant SD, Cooper PS, Lazarus LH. J Med Chem 1997; 40:3100–3108.
50. Nevin ST, Tóth G, Nguyen TM-D, Schiller PW, Borsodi A. Life Sci (Pharmacol Lett) 1993; 53:PL57–PL62.
51. Nevin ST, Tóth G, Weltrowska G, Schiller PW, Borsodi A. Life Sci (Pharmacol Lett) 1995; 56:PL225–PL230.
52. Szatmári I, Tóth G, Kertész I, Schiller PW, Borsodi A. Peptides 1999; 20:1079–1083.
53. Collier TL, Schiller PW, Waterhouse RN. Nucl Med Biol 2001; 28:375–381.
54. Visconti LM, Standifer KM, Schiller PW, Pasternak GW. Neurosci Lett 1994; 181:47–49.
55. Wilkes BC, Schiller PW. Biopolymers 1994; 34:1213–1219.
56. Temussi PA, Salvadori S, Amadeo P, Bianchi C, Guerrini R, Tomatis R, Lazarus LH, Picone D, Tancredi T. Biochem Biophys Res Commun 1994; 198:933–939.
57. Flippen-Anderson JL, George C, Deschamps JR, Reddy PA, Lewin AH, Brine GA. Lett Peptide Sci 1994; 1:107–115.
58. Wilkes BC, Nguyen TM-D, Weltrowska G, Carpenter KA, Lemieux C, Chung NN, Schiller PW. J Peptide Res 1998; 51:386–394.
59. Piñeyro G, Azzi M, DeLéan A, Schiller PW, Bouvier M. Mol Pharmacol 2001; 60:816–827.
60. Hosohata K, Burkey TH, Alfaro-Lopez J, Hruby VJ, Roeske VR, Yamamura H. Eur J Pharmacol 1999; 380:R9–R10.
61. Lia SB, Lin J, Shenderovich MD, Han YL, Hasohata K, Davis P, Qiu W, Porreca F, Yamamura HI, Hruby VJ. Bioorg Med Chem Lett 1997; 7:3049–3052.
62. Labarre M, Butterworth J, St-Onge S, Payza K, Schmidhammer H, Salvadori

S, Balboni G, Guerrini R, Bryant SC, Lazarus LH. Eur J Pharmacol 2000; 406:R1–R3.
63. Mullaney I, Carr IC, Milligan G. Biochem J 1996; 315:227–234.
64. Szekeres PG, Traynor JR. J Pharmacol Exp Ther 1997; 283:1276–1284.
65. Martin NA, Ruckle MB, Vanhoof SL, Prather PL. J Pharmacol Exp Ther 2002; 301:661–671.
66. Holaday JW, Hitzeman RJ, Curell J, Tortella FC, Belenky GL. Life Sci 1982; 31: 2359–2362.
67. Kest B, Lee CY, McLemore GL, Inturrisi CE. Brain Res Bull 1996; 39:185–188.
68. Zhu Y, King MA, Schuller AG, Nitsche JF, Reidl M, Elde RP, Unterwald E, Pasternak GW, Pintar JE. Neuron 1999; 24:243–252.
69. Schiller PW, Weltrowska G, Schmidt R, Nguyen TM-D, Berezowska I, Lemieux C, Chung NN, Carpenter KA, Wilkes BC. Analgesia 1995; 1:703–706.
70. Schiller PW, Fundytus ME, Merovitz L, Weltrowska G, Nguyen TM-D, Lemieux C, Chung NN, Coderre TJ. J Med Chem 1999; 42:3520–3526.
71. Chavkin C, Goldstein A. Proc Natl Acad Sci USA 1984; 81:7253–7257.
72. Schiller PW, Weltrowska G, Nguyen TM-D, Wilkes BC, Lemieux C, Chung NN. In: Fields GB, Tam JP, Barany G, eds. Peptides for the New Millenium. Proc 16th American Peptide Symposium. Dordrecht, Netherlands: Kluwer Academic Publishers, 2000:229–230.
73. Balboni G, Salvadori S, Guerrini R, Bianchi C, Santagada V, Calliendo G, Bryant SD, Lazarus LH. Peptides 2000; 21:1663–1671.
74. Schmidt R, Vogel D, Mrestani-Klaus C, Brandt W, Neubert K, Chung NN, Lemieux C, Schiller PW. J Med Chem 1994; 37:1136–1144.
75. Schmidt R, Chung NN, Lemieux C, Schiller PW. In: Kaumaya TP, Hodges RS, eds. Peptides: Chemistry, Structure and Biology. Proc 14th American Peptide Symposium. Kingswinford, England: Mayflower Scientific, 1996:645–646.
76. Ananthan S, Kezar HS III, Carter RL, Saini SK, Rice KC, Wells JL, Davis P, Xu H, Dersch CM, Bilsky EJ, Porreca F, Rothman RB. J Med Chem 1999; 42:3527–3538.
77. Wells JL, Bartlett JL, Ananthan S, Bilsky EJ. J Pharmacol Exp Ther 2001; 297:597–605.

13

Inverse Agonism at the Delta Opioid Receptors

Eva V. Varga, Keiko Hosohata, Yoshiaki Hosohata, Jennifer Tsang, Thomas Burkey, Josue Alfaro-Lopez, Xuejun Tang, Victor J. Hruby, William R. Roeske, and Henry I. Yamamura

University of Arizona, Tucson, Arizona, U.S.A.

1 INTRODUCTION

According to the traditional concept of agonist activity, agonist occupation of the receptor is a fundamental condition of receptor activation [1]. The ternary complex model furthermore postulates that in order to produce a cellular response, the agonist-receptor complex should interact with a third, guanine nucleotide-sensitive binding partner (guanine nucleotide binding, or G protein) [2]. According to this classical concept, two classes of physiologically active ligands can be anticipated: agonists, and competitive antagonists. Synthetic agonists are thought to interact with the receptors the same way as endogenous hormones, and produce identical physiological response (the "pharmacophore" concept, reviewed in [3]. Competitive antagonists share the receptor binding site with agonists, but are not able to promote the formation of the ternary complex. Therefore, competitive antagonists produce physiological response by competing with the endogenous agonist for the common receptor binding site.

This simple model has been able to explain the majority of experimental data for a number of ligands, in different tissue preparations. However, by the mid-1980s data suggesting the existence of an interaction between receptors and G proteins in the absence of agonists (precoupling) started to accumulate [4]. In 1989, Costa and Herz [5] demonstrated that delta opioid receptors in NG108-15 neuroblastoma-glioma cells were also able to activate G proteins to a certain extent, even in the absence of agonists. In addition, by studying opioid regulated GTPase activity in NG108-15 cell membranes, the authors were able to distinguish two types of competitive antagonists: the first type (MR2266) neither stimulated nor inhibited GTPase activity, while the second type (ICI 174,864) inhibited basal cellular GTPase activity, and thus exhibited an intrinsic activity opposite to the agonists [6]. This second type of ligand, the inverse agonist, was previously observed only for extracellular ligand gated ion channel GABA receptors [7].

However, although Costa and Herz [5] suggested that "the existence of antagonists with negative intrinsic activity may be a general feature of several classes of neurotransmitter or hormone receptors, and calls for a reevaluation of biological effects produced by competitive antagonists," the concept of inverse agonism did not reach general acceptance for the G protein–coupled receptors until subsequent advances in biochemistry and molecular biology led to more sensitive cellular models.

Currently, constitutive receptor activity and inverse agonism have been demonstrated for >40% of the known G protein–coupled receptors [8]. Detailed current reviews are available concerning inverse agonism and constitutive activity in different G protein–coupled receptor systems [8–10]. The pathophysiological importance of inverse agonism has also been recently reviewed [8,9]. The present review, on the other hand, will focus on constitutive activity and inverse agonism as they are manifested in the delta opioid receptor system.

2 DEVELOPMENT OF DELTA OPIOID ANTAGONISTS

Since δ-selective opioid antagonists are thought to exhibit favorable pharmacological properties, in the past decades considerable effort was undertaken to design potent, selective, and stable delta opioid antagonists. The pseudopeptide ICI 174,864 (N,N-diallyl-Tyr-Aib-Aib-Phe-Leu-OH) was designed by Cotton et al. [11], by introducing a synthetic amino acid (Aib = α-aminoisobutyric acid) into the prototype delta opioid antagonist, N, N-diallyl-Leu-enkephalin [12]. Later, structure–activity studies and systematic conformational restriction of the amino acid residues in the enkephalin or deltorphin peptides yielded a number of metabolically stable delta-selective pseudopeptide antagonists, such as TIPP [13], TIPPΨ [14], and Tyr-Tic [15]. Additional steric constrains by introduction of different methylated Tyr analogues into the Tyr-Tic pharmacophore produced the highly potent selective delta antagonists, DMT-L-Tic [16] and TMT-L-Tic [17]. Similarly, investigation of the structure–activity relationships of the morphine derivatives led to the development of nonselective (naloxone, naltrexone) and delta-selective (naltrindole (NTI), benzilidene-naltrexone (BNTX), naltriben (NTB), HS-378) alkaloid opioid antagonists [18,19]. Several potent delta opioid antagonists have also been developed (RTI 5989 derivatives, [20] by modifications of the structure of (±)-3,4-dimethyl-4-(3-hydroxyphenyl)-piperidine (LY272922).

The structures of selected peptide and nonpeptide delta-selective opioid antagonists are shown in Figure 1. However, although the compounds shown in Figure 1 have been introduced as neutral delta opioid antagonists, data from more sensitive functional assay systems have subsequently proven that these compounds are also capable of behaving as inverse agonists at the delta opioid receptor under appropriate conditions.

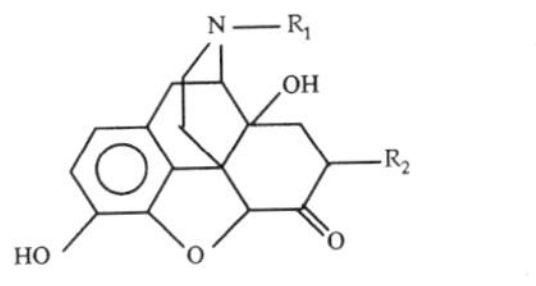

	R_1	R_2
Naloxone	allyl	H
Naltrexone	cyclopropylmethyl	H
BNTX	cyclopropylmethyl	benzilidene

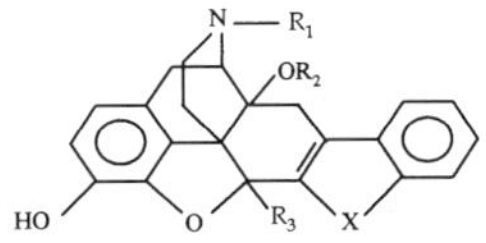

	R_1	R_2	R_3	X
NTI	cyclopropylmethyl	H	H	NH
HS 378	cyclopropylmethyl	C_2H_5	CH_3	NH
NTB	cyclopropylmethyl	H	H	O

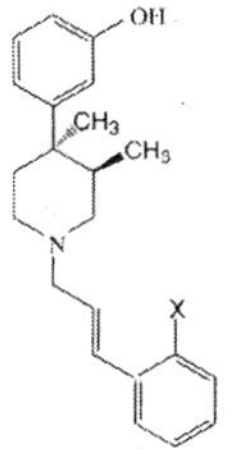

	X
RTI-5989-1	H
RTI-5989-25	CH_3
RTI-5989-23	Cl

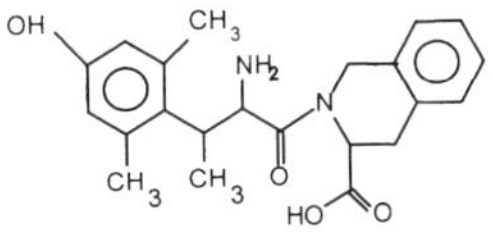

(2S,3R)-TMT-L-Tic
TMT=β-methyl-2'6'-dimethyltyrosine
Tic= 1,2,3,4-tetrahydroisoquinoline

CH_2= CH—CH_2 \
N-Tyr-Aib-Aib-Phe-Leu-OH
CH_2= CH—CH_2 /

ICI 174,864
Aib=α-amino-isobutyric acid

H-Tyr-Tic-Phe-Phe-OH
TIPP
Tic= 1,2,3,4-tetrahydroisoquinoline

H-Tyr-TicΨ-CH_2NH-Phe-Phe-OH
TIPPψ
Tic= 1,2,3,4-tetrahydroisoquinoline

FIGURE 1 Chemical structures of selected delta opioid antagonists/inverse agonists.

TABLE 1 Peptide Delta Opioid Antagonists/Inverse Agonists

Ligand	Cell line	Receptor	Assay	Effect	Ref.
ICI 174, 864	NG108-15	Native	GTPase	–	5
			$[^{35}S]GTP\gamma S$	–	37
	Rat-1	Recombinant	GTPase,	–	34
			$[^{35}S]GTP\gamma S$	–	38
			Cholera toxin	+	
			$[^{35}S]GTP\gamma S$	–	
			cAMP	+	
			cAMP	+	43
	HEK 293	Recombinant	$[^{35}S]GTP\gamma S$,	–	32
			Receptor B_{max}	Up-regulation	
			Receptor B_{max}	Up-regulation	33
			Receptor B_{max}	+/–	31
			$[^{35}S]GTP\gamma S$	–	39
	C-6 glioma	Recombinant	$[^{35}S]GTP\gamma S$	–	28
	CHO	Recombinant	$[^{35}S]GTP\gamma S$	–	40
	Rat brain	Native	GTPase	–	35
TIPP	GH3	Recombinant	Regulation of binding by Na^+ and GDP	Increase	27
			$[^{35}S]GTP\gamma S$, cell membrane	None	
			$[^{35}S]GTP\gamma S$, whole cell	+	
			cAMP	–	
TIPPψ	Rat-1	Recombinant	GTPase,	None	34
			$[^{35}S]GTP\gamma S$	None	
			Cholera toxin	None	
	HEK 293	Recombinant	Receptor B_{max}	Up-regulation	33
DMT-L-Tic-OH	HEK 293	Recombinant	$[^{35}S]GTP\gamma S$	–	39
$N,N(CH_3)_2$-DMT-L-Tic-NH2	HEK 293	Recombinant	$[^{35}S]GTP\gamma S$	None	39
(2S,3R)-TMT-L-Tic-OH	CHO	Recombinant	$[^{35}S]GTP\gamma S$	–	40
	Mouse brain	Native	$[^{35}S]GTP\gamma S$	None	41

3 CELLULAR EFFECTS OF INVERSE DELTA OPIOID AGONISTS

In the 1990s the cDNAs encoding the rodent and the human delta opioid receptors were cloned [21]. Subsequent construction of recombinant cell lines expressing high densities of the cloned receptors as well as site-directed mutagenesis of different amino acids led to the production of mammalian cell lines where physiological receptor activity could be detected in the absence of agonists. In these cell lines some (but not all) of the ligands previously classified as competitive delta opioid antagonists were shown to be able to switch off constitutive delta receptor activity—i.e., behave as inverse agonists. Inverse agonism can be detected in radioligand-binding studies in assays measuring receptor-mediated G protein activation or receptor-mediated regulation of intracellular second-messenger concentrations. Tables 1 and 2 summarize the experimental data indicating inverse

TABLE 2 Nonpeptide Delta Opioid Antagonists/Inverse Agonists

Ligand	Cell line	Receptor	Assay	Effect	Ref.
NTB	C-6 glioma	Recombinant	$[^{35}S]GTP\gamma S$	–	28
	NG108-15	Native	$[^{35}S]GTP\gamma S$ (w/o Na^+)	–	37
			$[^{35}S]GTP\gamma S$ (with Na^+)	None	
	HEK 293	Recombinant	Receptor B_{max}	Up-regulation	33
Clocinnamox	C-6 gloma	Recombinant	$[^{35}S]GTP\gamma S$	–	28
HS-378	HEK 293	Recombinant	$[^{35}S]GTP\gamma S$	–	39
RTI 5989-1 RTI 5989-23 RTI 5989-25	HEK-293	Recombinant	$[^{35}S]GTP\gamma S$, Receptor B_{max}	– +	32
NTI	HEK-293	Recombinant	Receptor B_{max}	Up-regulation	33
	COS-1	Recombinant	$[^{35}S]GTP\gamma S$	–	67
	CHO	Recombinant	$[^{35}S]GTP\gamma S$	None	(Fig. 4)
Naloxone	NG108-15	Native	Receptor B_{max}	Up-regulation	75
	HEK 293	Recombinant	$[^{35}S]GTP\gamma S$, Receptor B_{max}	– Up-regulation	32

agonist properties of peptide and alkaloid delta opioid antagonists in different cellular assay systems.

3.1 Measurement of Inverse Agonist-Mediated Conformational Changes in the Receptor Protein

Biophysical methods, such as fluorescence spectroscopy or plasmon-waveguide resonance spectroscopy, are being used to detect conformational changes associated with activation or inactivation of G protein–coupled receptors. Thus, inverse agonist-mediated changes in the conformation of the purified β_2AR receptor have been recently demonstrated by fluorescence spectroscopy [22]. Similar results were recently obtained by plasmon-waveguide resonance spectroscopy of the purified human delta opioid receptor in the presence and absence of the inverse agonist (2S,3R)TMT-L-Tic [23].

3.2 Radioligand-Binding Studies

Inverse agonism can be detected in radioligand-binding assays by measuring the inhibition of the binding of a radiolabeled antagonist by inverse agonist in the presence and absence of sodium [24] and/or guanine nucleotide [25], or after pretreatment of the cells with pertussis toxin [5]. Inverse agonists have higher affinity for the inactive R receptor conformation; therefore, the presence of agents that shift the equilibrium in the direction of the inactive, uncoupled receptor would shift inverse agonists competition curves to the left. Sodium is an allosteric regulator that stabilizes the inactive receptor conformation for many G protein–coupled receptors [26]. Guanine nucleotides and pertussis toxin pretreatment, on the other hand, increase the affinity of inverse agonists by uncoupling of the receptors from the G protein. Accordingly, increased affinity was detected for the delta-selective peptide ICI 174,864, both in the presence of sodium ions [24] and after pertussis toxin pretreatment [5]. Similar regulation of the affinity of TIPP [27], naltrindole, and clocinnamox [28] by sodium and GDP has also been demonstrated recently. The magnitude of the affinity shift produced by these agents, however, is relatively small, and thus the application of these simple assays is rather limited.

Saturation binding assays after long-term treatment with the ligand can also be used to discriminate between neutral antagonists and inverse agonists. It has been demonstrated that spontaneous formation of receptor active states augments receptor phosphorylation, internalization, and downregulation of many G protein–coupled receptors [29,30]. An inverse agonist, therefore, by stabilizing the inactive receptor conformation, would reduce receptor phosphorylation, internalization, and degradation. Therefore, chronic treatment of the cells with inverse agonists frequently leads to upregulation of cell membrane receptor concentration. Upregulation of membrane-bound delta opioid receptors by chronic treatment with ICI 174,864 [31–33], TIPPΨ (33), NTB [33], and several members of the RTI 5989 series of compounds [32], as well as the nonpeptide antagonists, naltrindole, and naloxone [32], has been demonstrated. On the other hand, Petaja-Repo et al. [33] have recently demonstrated that membrane-bound delta opioid receptor concentrations can be increased by any membrane-permeable ligand, irrespective of their pharmacological efficacy. They suggested that membrane-permeable agonists, antagonists, and inverse agonists are all able to act as chemical chaperones that facilitate the maturation and transport of the newly synthesized receptor protein. Interestingly, it has also been demonstrated that the effects of chronic inverse agonist treatment has differential effects on the maximal number of radiolabeled agonist or inverse agonist binding sites [31].

4 MEASUREMENT OF RECEPTOR-MEDIATED G PROTEIN ACTIVATION

Interaction of the heterotrimeric G protein with the activated R^* conformation of the receptor leads to GDP/GTP exchange in the G protein α-subunit, G protein activation, and signal transduction. Subsequently the intrinsic GTPase activity of the α-subunits eventually lead to the hydrolysis of the bound GTP. Measurement of receptor-mediated regulation of GTPase or guanine nucleotide exchange rate, therefore, is frequently used to show constitutive receptor activity. Inhibition of basal G protein activity by a ligand on the other hand, would indicate inverse agonism. Regulation of the low K_m GTPase activity by the prototype delta-selective inverse agonist ICI 174,864 has been shown in recombinant cell lines [34] as well as in tissues containing endogenous delta opioid receptors (NG108-15 cells [5,6] and rat brain membranes [35]. A novel method was recently identified to enhance the sensitivity of this assay. It was found [36] that addition of certain recombinant regulator of G protein signaling (RGS) proteins to cell membrane preparations significantly augments constitutive $5HT_{1A}$ serotonin receptor activity and agonist/inverse agonist–mediated regulation of the high-affinity GTPase rate. The application of this assay for the delta opioid receptors may help to identify additional inverse agonists in the future.

The most frequently used method to investigate constitutive receptor activity and inverse agonism, however, is the measurement of the binding of the radiolabeled nonhydrolyzable GTP analog [^{35}S]GTPγS, to cell membrane preparations or permeabilized intact cells. Since pertussis toxin (Ptx) abolishes the stimulation of G_i/G_0 proteins by agonist-free (as well as by agonist-occupied) receptors, Ptx treatment can be used to estimate the maximal possible effect of inverse agonists in [^{35}S]GTPγS-binding assays and to calculate the efficacy of inverse agonists in a given cell line or tissue. Thus, as shown in Figure 2, pertussis toxin pretreatment of recombinant Chinese

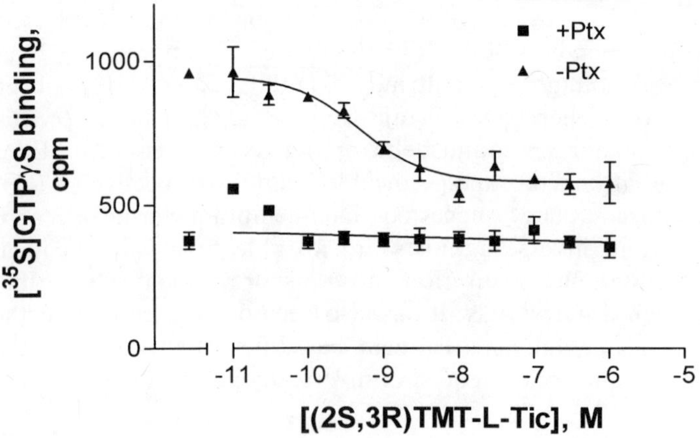

FIGURE 2 Effect of pertussis toxin treatment on the regulation of basal [^{35}S]GTPγS binding by (2S,3R)TMT-L-Tic in hDOR/CHO cell membranes. Chinese hamster ovary cells stably expressing the human delta opioid receptors (hDOR/CHO) were treated in the presence (◆) or absence (■) of 50 ng/mL pertussis toxin for 18 h. The cells were washed and cell membranes prepared as previously described [40]. Membranes were incubated with appropriate concentrations of (2S,3R)TMT-L-Tic in the presence of 0.1 nM [^{35}S]GTPγS (1,250 Ci/mmol) in 1.0 mL of assay buffer (25 mM Tris, 150 mM NaCl, 50 μM GDP, 2.5 mM $MgCl_2$, 1 mM EDTA, 30 μM bestatin, 10 μM captopril, pH = 7.4). After 90 min incubation at 30°C, the reaction was terminated by rapid filtration. The filters were washed with 25 mM Tris/120 mM NaCl, pH 7.4, and bound radioactivity was measured by liquid scintillation spectrophotometry.

hamster ovary (CHO) cells stably expressing the human delta opioid receptors reduced basal membrane [^{35}S]GTPγS binding by ~50%, and completely abolished inhibition of basal [^{35}S]GTPγS binding by the inverse agonist ((2S,3R)TMT-L-Tic).

Reduction of basal [^{35}S]GTPγS binding has been demonstrated for a number of delta-selective peptide, nonpeptide ligands in transfected cell lines expressing moderate to high delta opioid receptor densities (see Table 1). Some of these ligands (such as ICI 174,864) were also able to regulate [^{35}S]GTPγS binding in both recombinant cells [32,34,38,39] and cell membranes, containing endogenous delta opioid receptors in physiological concentrations(NG108-15 cells, [37]), while for others inverse agonism can be detected only in recombinant cell lines expressing a high density of delta opioid receptors. Thus, we have demonstrated earlier that the highly delta-selective constrained opioid peptide (2S,3R)TMT-L-Tic inhibits basal

[^{35}S]GTPγS binding in CHO cells expressing 1800±150 fmol human delta opioid receptors per 10^6 cells [40]. However, subsequently we found that in mouse brain membrane preparations (2S,3R)TMT-L-Tic behaves as a neutral antagonist since it did not change basal [^{35}S]GTPγS binding [41].

Low sodium and GDP concentrations favor constitutive activity and amplify inverse agonism. The sensitivity of the [^{35}S]GTPγS-binding assay, therefore, can be increased by replacing NaCl by KCl and by reducing the concentration of GDP in the assay buffer. Accordingly, BNTX and NTB behave as neutral antagonists in NG108-15 cell membranes at high sodium concentrations, but are able to inhibit basal [^{35}S]GTPγS binding in the absence of sodium chloride [37]. Interestingly, it was also recently shown [27] that although the delta-selective opioid pseudopeptide TIPP does not regulate [^{35}S]GTPγS binding in recombinant GH_3 cell membranes (delta opioid receptor concentration: 2.3 pmol/mg protein); in other assays it behaves either as an inverse agonist (radioligand binding) or as an agonist ([^{35}S]GTPγS binding in permeabilized cells, and inhibition of forskolin stimulated cAMP formation).

One possible explanation for the interesting properties of TIPP is that this compound belongs to yet another novel class of ligands, the so-called protean agonists [42]. According to the prediction of Kenakin [42], the intrinsic activity of agonists that produce a receptor active state that is less efficacious than the constitutively active receptor conformation can be dramatically different under different assay conditions. These ligands were predicted to behave as agonist in cellular systems that do not exhibit constitutive activity. However, if the system is constitutively active, the ligand would reduce signaling activity by promoting a receptor conformation that is less efficacious than the constitutively active receptor. Therefore, in quiescent systems—for example, in the presence of sodium and guanine nucleotides (intact cells)—the ligand would be an agonist, while in constitutively active systems (e.g., in binding assays, in the absence of sodium and guanine nucleotide) it would be an inverse agonist. Finally, receptor mediated guanine nucleotide exchange can be monitored by covalent labeling of the G proteins. Thus, constitutive cholera toxin catalyzed [^{32}P]ADP-ribosylation of Gi and Gs proteins was shown to be attenuated by the inverse agonist ICI 174,864 in transfected Rat-1 fibroblasts [34].

5 MEASUREMENT OF THE ACTIVITY OF RECEPTOR-MEDIATED SIGNAL TRANSDUCTION CASCADES

Other, less frequently applied methods distinguish inverse agonist activity by measuring the concentration of downstream signaling molecules, such as basal- or forskolin-stimulated cAMP formation, in intact cells or in cell membranes. Thus, it has been shown previously that the delta-selective inverse agonist ICI 174,864 augments forskolin-stimulated cAMP formation in both recombinant (HEK 293 or Rat-1) cells stably transfected with the delta opioid receptors [38,43]. We have also demonstrated recently (see Fig. 3) that the novel delta-selective inverse agonist (2S,3R)TMT-L-Tic augments forskolin-stimulated cAMP formation in recombinant CHO cells expressing the human delta opioid receptor.

Measurement of the regulation of cAMP-dependent reporter gene expression, inhibition of constitutive mobilization of intracellular Ca^{2+}, and enhancement of neurotransmitter release have also been used to detect inverse agonism in other Gi/o coupled receptor systems [see 8 for review];

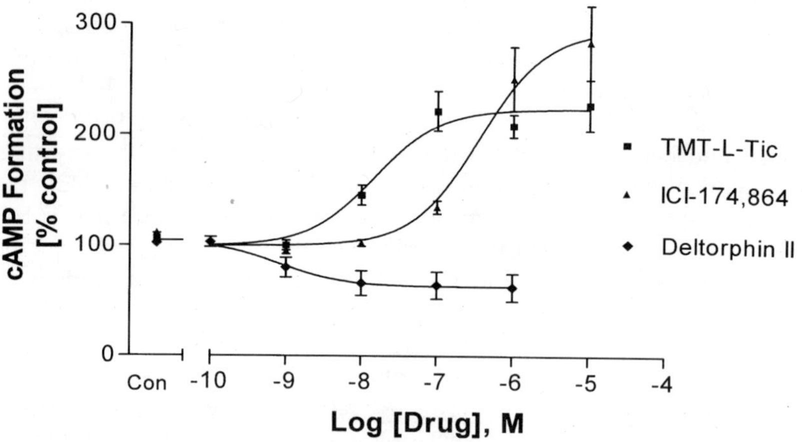

FIGURE 3 Inverse delta opioid agonists augment forskolin-stimulated cAMP formation in hDOR/CHO cells. hDOR/CHO cells were plated in 24-well cell culture dishes 48 h before the assay. The growth medium was replaced with serum-free medium (IMDM) containing 5 mM 3-isobutyl-1-methylxanthine, 100 μM 7-deacetyl-7-(O-N-methylpiperazino)-γ-butyryl-forskolin, and appropriate concentrations of (2S,3R)TMT-L-Tic (■), ICII 74,864 (▲) or deltorphin II (♦). After 20 min incubation at 37°C, the cells were lysed and centrifuged, and the supernatant was incubated with 4 nM [^{3}H]cAMP and 30 μg/mL protein kinase A for 2 h. Activated charcoal was added to adsorb free cAMP. Bound radioctivity was measured by liquid scintillation spectrophotometry. The delta-selective agonist, deltophin II inhibited forskolin stimulated cAMP formation by $38 \pm 5.6\%$ with an EC_{50} value of 0.8 ± 0.4 nM (n = 5). Conversely, the delta-selective inverse agonists (2S,3R)TMT-L-Tic and ICI 174,864 augmented forskolin stimulated cAMP formation by $224 \pm 9\%$ and $294 \pm 19\%$ with EC_{50} values of 15 ± 2 (n = 5) and 357 ± 21 (n = 3) nM, respectively.

however, we are not aware of application of these methods for the delta opioid receptors. It has been demonstrated, however, using the Cytosensor microphysiometer, that agonists and ICI 174,864 change the basal rate of cellular proton exclusion in opposite directions [38].

A further, potentially very important cellular effect of inverse agonists has recently been demonstrated in the cannabinoid receptor system. Cannabinoid receptor inverse agonists were shown to inhibit Gi/o protein–dependent, mitogen-activated protein kinase activation by both tyrosine kinase– and G protein–coupled receptors in CHO cells expressing CB_1 or CB_2 cannabinoid receptors [44,45]. It has been suggested that inverse agonist promote the formation of a novel receptor conformation, $R^{(-)}$. The $R^{(-)}$ receptor conformation is thought to have a high affinity for the G proteins, but to be unable to promote guanine nucleotide exchange. Sequestration of the cellular G protein population by the cannabinoid receptor therefore would antagonize the signaling of other cellular receptor populations. The ability of delta opioid receptor inverse agonist to sequester cellular G proteins and inhibit signal transduction by other receptors, however, has not been investigated.

6 IN VIVO EFFECTS AND PHYSIOLOGICAL RELEVANCE OF DELTA-SELECTIVE INVERSE AGONISTS

Although it has been proven that a number of delta-selective inverse opioid ligands behave as inverse agonists in constitutively active recombinant cell lines, the physiological and therapeutic relevance of the inverse delta opioid agonists is still not completely understood.

The therapeutic utility of the inverse agonists depends on the extent of constitutive delta opioid receptor activity in native tissues and on the role of constitutive delta opioid receptor activity in the pathophysiology of diseases. For example, it was suggested that selective delta opioid antagonists could be therapeutically useful as immunosuppressants in organ transplants [46] and in chronic inflammatory diseases [47]. Interestingly, it was recently shown that although the inverse agonist HS 378 and the neutral antagonists naltrindole interact with the delta opioid receptors with similar affinities, the potency of the inverse agonist HS 378 to regulate T-cell proliferation is 13-fold higher than that of naltrindole [47]. The increased potency of an inverse agonist in this assay indicates that inverse agonists may have a therapeutic advantage above neutral antagonists for the treatment of inflammatory diseases.

Selective delta opioid antagonists have also been shown to modulate the development of tolerance and dependence to morphine [48,49], cocaine

[50], and ethyl alcohol [51]. The possible therapeutic advantage of inverse agonists versus neutral antagonists in these conditions, however, has not been investigated.

Theoretically, inverse agonists may be useful agents for the treatment of disease symptoms caused by constitutive receptor activity. Covalent modifications of the receptor protein, pathological increase in receptor or G protein densities, or inherited or somatic point mutations may lead to increased constitutive receptor activity. For example, certain inherited point mutations in rhodopsin have been shown to cause retinitis pigmentosa. The involvement of inherited, constitutively activating receptor point mutations has also been demonstrated in certain forms of hyperthyroidism, in male precocious puberty, and in Jansen-type metaphyseal chondrodysplasia [8]. Constitutive G protein activity may also be an important factor in autoimmune diseases [9].

In some conditions, however, inverse agonist activity can also prove to be a disadvantage. For example, inverse agonist activity of the CB_1 cannabinoid receptor ligand SR 141,716 has been demonstrated in recombinant in vitro assay systems [52], and was subsequently confirmed in a number of in vivo assays. Thus, SR141716A was shown to enhance memory [53] and immune function [54], suppress appetite [55], and reduce ethanol-seeking behavior in laboratory animals [56], all of which can be considered desirable pharmacological properties. On the other hand, intrathecal administration of SR141716A evokes thermal hyperalgesia in mice (57)—clearly an unwanted property for a therapeutic agent.

Conversely, although we have shown that (2S,3R)TMT-L-Tic is an inverse agonist in both [^{35}S]GTPγS binding [40] and the forskolin stimulated cAMP formation (Fig. 3) assays in recombinant CHO cells expressing the human delta opioid receptor, the ligand did not modulate withdrawal latencies in either the warm water tail flick or radiant heat paw withdrawal assays in mice [41], indicating that (2S,3R)TMT-L-Tic-OH does not exhibit hyperalgesic properties in vivo in mice. The difference between the in vivo properties of (2S,3R)TMT-L-Tic and SR174716A is likely due to an approximately 10-fold difference between cannabinoid and opioid receptor densities in rodent brain [58].

Another pathological condition where inverse agonists may not be beneficial is the treatment of drug addiction and drug overdose by opioid antagonists. Evidence is accumulating that chronic opioid agonist treatment may increase constitutive delta [58] and mu [59–62] opioid receptor activity. The mechanism of constitutive activation of these receptors by chronic opioid agonist treatment is unclear, but the involvement of protein kinases has been indicated [61,62]. Furthermore, chronic DPDPE treatment of GH3 cells stably expressing the delta opioid receptors has been shown to convert some (but not all) neutral antagonists (such as naloxone, naltriben) into inverse agonists [58]. Since it was shown that neutral antagonists produce significantly fewer withdrawal symptoms than inverse agonists, sensitive cellular assays are necessary to screen for inverse agonist properties in opioid antagonists intended for use in the treatment of drug addiction.

7 PHARMACOLOGICAL THEORIES OF G PROTEIN–COUPLED RECEPTOR ACTIVATION AND INVERSE AGONISM

According to the traditional concept [1], drugs are thought to bind to their cellular receptor reversibly in a process characterized by the affinity constant (K), according to the law of mass action:

$$H + R \overset{K}{\leftrightarrows} HR \Rightarrow\Rightarrow \text{cellular response} \qquad (1)$$

Later, to account for the binding properties of the β-adrenergic ligands in frog erythrocyte membrane preparations, this classical model of receptor action evolved into the "ternary complex" model [2]. The ternary complex model postulates that in order to produce a cellular response, the agonist-receptor complex should interact with a third, guanine nucleotide–sensitive binding partner (guanine nucleotide binding, or G protein).

$$\begin{array}{l} H + R \overset{\alpha K}{\leftrightarrows} HR + X \\ \qquad\qquad \downarrow\uparrow L \\ \qquad\qquad HRX \Rightarrow\Rightarrow \text{cellular response} \end{array} \qquad (2)$$

The discovery of partial agonists led to further refinement of the ternary complex model by introducing the allosteric constant, α. For full agonists $\alpha = 1$, while partial agonists have α values <1.

Later, in order to account for the effects of a point mutation on the activity of the β_2-adrenergic receptor, Samama et al. [29] have proposed an extended version of the ternary complex model. In this model the receptor molecule exists in an equilibrium between the inactive R and the active R* conformations. In the absence of ligand, the ability of the receptor to spontaneously convert from the inactive to the active conformation is determined by the isomerization constant, J. The active R* conformation is the molecular species that enters into productive interaction with the G protein, described by the equilibrium constant M. The values of both J and M are dependent only on the receptor–G protein system, and are independent of the presence or absence of ligand. The ability of different ligands to perturb this equilibrium is gauged by the ligand-specific equilibrium constant β, the microscopic equivalent of efficacy. Ligands with β values >1 are agonists (partial or full agonists, depending on the actual β value) and shift the equilibrium into the direction of the active R* conformation and simultaneously into the direction of the signaling HR*G ternary complex. Neutral agonists ($\beta = 1$) have no effect on the equilibrium, while inverse agonists ($\beta < 1$), having higher affinity to the inactive R conformation, will shift the equilibrium into the opposite direction.

$$\begin{array}{ccc} R & \overset{K}{\longleftrightarrow} & LR \\ \Big\updownarrow J & & \Big\updownarrow \beta J \\ R^* & \overset{\beta K}{\longleftrightarrow} & LR^* \\ \Big\updownarrow M & & \Big\updownarrow \alpha M \\ R^*G & \overset{\alpha\beta K}{\longleftrightarrow} & LR^*G \end{array} \qquad (3)$$

In this model measurable constitutive activity (elevated levels of [R*G]) can be achieved by increasing the concentration of R and/or G. Another way in which constitutive activity can be produced is through alteration of the J or M allosteric constants—for example, by the removal of sodium ions or by introducing different point mutations into the receptor protein.

In a further, thermodynamically more complete version of the ternary complex model (cubic ternary complex [63]), the interaction of the inactive receptor (R) with the G protein has also been taken into consideration. For many receptors the interaction between the inactive R and LR receptor conformations with G proteins is negligible, and the extended ternary complex model is adequate to describe receptor behavior. On the other hand, if a large fraction of R and/or LR complex interacts with the G protein, as these complexes are not able to promote guanine nucleotide exchange, the receptor would act as a dominant negative "trap." Sequestration of the cellular G protein population in their inactive conformation by these receptors therefore will antagonize cellular signaling by other cellular receptor populations. The experimental data indicate that cannabinoid receptors may behave this way both in the presence and absence of inverse agonists [44,45].

Formally, the extended ternary complex model is a two-state model: different agonists apparently produce cellular response by causing (quantitatively different) enrichment of the (qualitatively identical) R* active receptor conformation. However, upon more careful examination it becomes clear that the allosteric constants α and/or β can theoretically be specific for each ligand [42]. Under these circumstances the ternary complex model principally can accommodate an infinite number of receptor active states. Therefore, in this model only the unliganded R* and the inactive R species are regarded as a homogeneous population.

Recent data from our laboratory and from other investigators [66], however, indicate a more complex picture. We have reported earlier that mutation of a tryptophan residue (W^{284}) in the sixth transmembrane domain of the human delta opioid receptor (hDOR) to leucine differentially effects the affinities of various delta-selective opioid agonists [64] as well, as their potencies to stimulate [^{35}S]GTPγS binding [65] in membranes of CHO cells stably expressing the mutant W^{284}L/hDOR. Importantly, the effects of the mutation on the affinities and potencies were independent of its effects on the intrinsic activities of the agonists. Recent computer modeling indicates a crucial trigger role of the third extracellular loop in receptor activation [67]. Furthermore, site-directed mutagenesis experiments by other investigators indicates an important role for residues homologous to W^{284}, or the third extracellular loop as a whole, in the activation of the delta-opioid [66,68] and other G protein–coupled (such as the nociceptin [69], muscarinic [70], adrenergic [71] receptors in receptor–G protein coupling.

Recently, we have investigated the effects of the W^{284}L mutation on the intrinsic activities of different inverse agonists to inhibit basal [^{35}S] GTPγS binding in W^{284}L/CHO cells. We found that the mutation had differential effects on the intrinsic activities of two delta-selective inverse agonists, ICI 174,864 and (2S,2R)TMT-L-Tic. Both ICI 174,864 and (2S, 3R)-TMT-L-Tic were inverse agonists in CHO cells expressing wild-type hDOR. However, while ICI 174,864 retained its ability to inhibit basal [^{35}S]GTPγS binding in CHO cell membranes expressing the W^{284}L mutant hDOR, (2S,3R)-TMT-L-Tic was converted into a neutral antagonist under the same experimental conditions. (Fig. 4). The difference was not due to differences in receptor densities, since [^{3}H]NTI saturation binding assays indicated similar B_{max} values for cells expressing the wild-type (1460 $\pm$ 200) and mutant (2050 $\pm$ 300 fmol/mg protein) human delta opioid receptors, respectively. NTI and naltrexone, on the other hand, remained neutral antagonists in both cell lines.

If the inactive R receptor population is homogeneous, as was hypothesized in the original versions of the ternary complex model, it is very hard to explain how the W^{284}L point mutation can convert TMT-L-Tic into a neutral antagonist while having minimal effect of the intrinsic activity of ICI 174,864. The data can be better interpreted using the probabilistic model of Onaran and Costa [72]. In this model, the empty receptors can exist in a discrete set of conformational states. Some of these receptor conformations are able to activate the G proteins, while others are not. Binding of a ligand (agonist, inverse agonist, neutral antagonist, or G protein) to the receptor

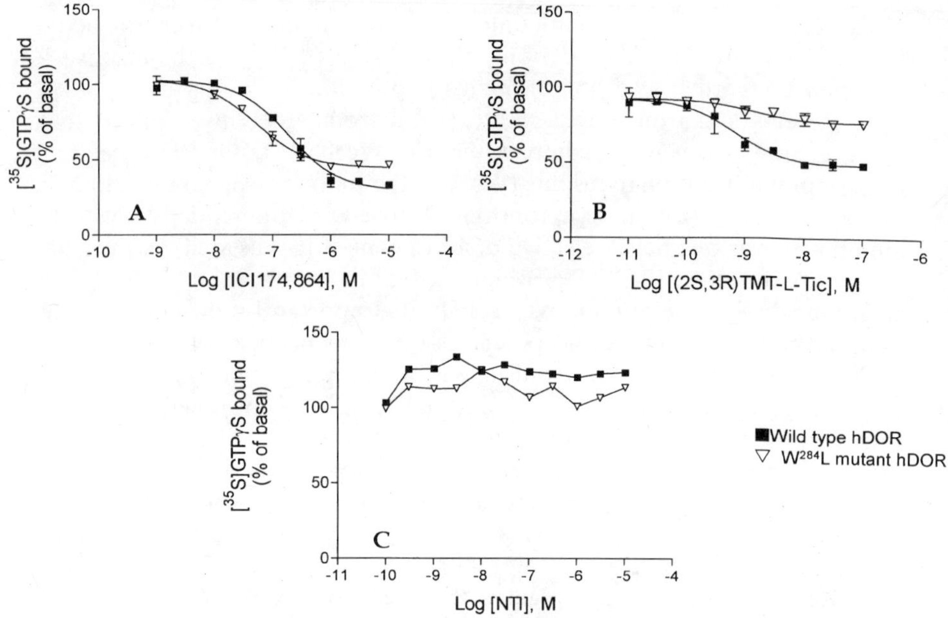

FIGURE 4 Regulation of basal [^{35}S]GTPγS binding by delta-selective antagonists/inverse agonists in CHO cell membranes expressing the wild-type or $W^{284}L$ mutant human delta opioid receptors. Recombinant wild-type (hDOR/CHO (■)) or mutant ($W^{284}L$/CHO, (∇)) cells were incubated with 0.1 nM [^{35}S]GTPγS (1,250 Ci/mmol) in the absence and presence of appropriate concentrations of (A) ICI 174,864, (B) (2S,3R)TMT-L-Tic, and (C) NTI in 1 mL assay buffer (25 mM Tris, 150 mM NaCl, 50 μM GDP, 2.5 mM $MgCl_2$, 1 mM EDTA, 30 μM bestatin, 10 μM captopril, pH = 7.4) for 90 min, at 30°C. The reaction was terminated by rapid filtration, and bound radioactivity was measured by liquid scintillation spectrophotometry. In hDOR/CHO cell membranes the maximal inhibition of basal [^{35}S]GTPγS binding (E_{max}) by ICI 174,864, (2S,3R)TMT-L-Tic and was: 70.5 ± 2.6, 55.5 ± 1.3, respectively. [^{35}S]GTPγS binding in the presence of NTI was 129 ± 11.4 of basal in hDOR/CHO cells. On the other hand, the respective E_{max} values in the mutant W284L/CHO cell membranes were: 56 ± 1.3, 25 ± 1.1, and 110 ± 11.3 for the three ligands, respectively. The EC_{50} values were 18.7 ± 0.8 nM and 53 ± 0.6 nM for ICI 174,864; and 0.6 ± 0.1 nM and 1.3 ± 0.2 nM for (2S,3R)TMT-L-Tic, in hDOR/CHO and W284L/CHO cell membranes, respectively.

changes this resting conformational distribution. Furthermore, different ligands may have different effects on the conformational distribution. To explain our data, we hypothesize that different inverse agonists stabilize different subpopulations of inactive receptor conformations. Thus, inverse agonists (such as (2S,3R)TMT-L-Tic) that reduce the probability of R/R* transitions that have a reduced probability in the mutant receptor will no longer have an effect on the R/R* equilibrium and will behave as neutral antagonists in the mutant $W^{284}L$/CHO cell membranes. Conversely, inverse agonists that preferentially stabilize other constraining bonds in the wild-type receptor (such as ICI 174,864) will retain their inverse agonist function at the mutant receptor.

8 SUMMARY

Inverse agonistic properties of several delta-selective and nonselective delta-opioid antagonists have been demonstrated in both native tissues and in recombinant cell lines heterologously expressing the delta-opioid receptors. Constitutive activity for the delta-opioid receptors has been observed at both endogenous receptors in NG108-15 cells and in recombinant cell lines expressing moderate to high delta-opioid receptor densities. The in vivo importance and role of constitutive delta-opioid receptor activity in different tissues, however, are not understood, because there is insufficient information about cellular delta-opioid receptor concentrations in physiologically important cell populations. Constitutive delta-opioid receptor activity may, however, regulate basal neuronal activity in brain, and may have cardioprotective or immunomodulatory functions in other tissues. Several disease-causing genetic mutants that augment constitutive receptor activity have been identified for other G protein–coupled receptors [8,9]. However, a large number of clinically used drugs were found to behave as inverse agonists at different G protein–coupled receptors [73–75]. It has also been demonstrated that inverse agonist activity parallels clinical efficacy for antipsychotic [73,74] and β_1-adrenergic [75] drugs. Therefore the design, synthesis, and characterization of novel, therapeutically effective inverse agonists constitute an emerging new frontier of drug discovery.

REFERENCES

1. Langley JN. J Physiol 1878; 1:339–369.
2. DeLean A, Stadel JM, Lefkowitz RJ. J Biol Chem 1980; 255:7108–7117.
3. Coop A, Jacobson A. Bioorg Med Chem Lett 1999; 9:357–362.
4. Lee TW, Sole MJ, Wells JW. Biochemistry 1986; 25:7009–7020.
5. Costa T, Herz A. Proc Natl Acad Sci USA 1989; 86:7321–7325.

6. Costa T, Lang J, Gless C, Herz A. Mol Pharmacol 1990; 37:383–394.
7. Ehlert FJ, Roeske WR, Braestrup C, Yamamura SH, Yamamura HI. Eur J Pharmacol 1981; 70:593–595.
8. Seifert R, Wenzel-Seifert K. Naunyn-Schmiedebergs Arch Pharmacol 2002; 366:381–416.
9. De Ligt RAF, Kourounakis AO, Ijzerman AP. Br J Pharmacol 2000; 130:1–12.
10. Strange PG. Trends Pharmacol Sci 2002; 23:89–95.
11. Cotton R, Shaw JS, Miller L, Giles MG, Timms D. Eur J Pharmacol 1984; 97:331–339.
12. Shaw JS, Gormley JJ, Turnbull JJ, Miller L, Morley JS. Life Sci 1982; 31:1259–1268.
13. Schiller PW, Nguyen TMD, Weltrowska G, Wilkes BC, Marsden BJ, Lemieux C, Chung NN. Proc Natl Acad Sci USA 1992; 89:11871–11875.
14. Schiller PW, Weltrowska G, Nguyen TMD, Wilkes BC, Chung NN, Lemieux CJ. J Med Chem 1993; 36:3182–3187.
15. Temussi PA, Salvadori S, Amodeo P, Bianchi C, Guerrini R, Tomatis R, Lazarus LH, Picone D, Tancredi T. Biochem Biophys Res Commun 1994; 198:933–939.
16. Salvadori S, Attila M, Balboni G, Bianchi C, Bryant SD, Crescenzi O, Guerrini R, Picone D, Tancredi T, Temussi PA. Mol Med 1995; 1:678–689.
17. Liao S, Lin J, Shenderovich MD, Han Y, Hosohata K, Davis P, Qiu W, Porreca F, Yamamura HI, Hruby VJ. Bioorg Med Chem Lett 1997; 7:3049–3052.
18. Portoghese PS. Farmaco 1993; 48:243–251.
19. Schmidhammer H, Krassnig R, Greiner E, Schutz J, White A, Berzetei-Gurske IP. Helv Chim Acta 1998; 81:1064–1069.
20. Thomas JB, Mascarella SW, Rothman RB, Partilla JS, Xu H, McCullough KB, Dersch CM, Cantrell BE, Zimmerman DM, Carroll FI. J Med Chem 1998; 41: 1980–1990.
21. Quock RM, Burkey TH, Varga E, Hosohata Y, Hosohata K, Cowell SM, Slate CA, Ehlert FJ, Roeske WR, Yamamura HI. Pharmacol Rev 1999; 51: 503–532.
22. Kobilka BK, Gether U. Methods Enzymol 2002; 343:170–182.
23. Salamon Z, Cowell S, Hruby VJ, Tollin G. J Peptide Res 2002; 60:322–328.
24. Appelmans N, Carrol JA, Rance MJ, Simon EJ, Traynor JR. Neuropeptides 1986; 7:139–143.
25. Liu JG, Prather PL. J Pharmacol Exp Ther 2002; 302:1070–1079.
26. Gierschik P, Sidiropoulos D, Steisslinger M, Jacobs KH. Eur J Pharmacol 1989; 172:481–492.
27. Martin NA, Ruckle MB, Vanhoof SL, Prather PL. J Pharmacol Exp Ther 2002; 301:661–671.
28. Neilan CL, Akil H, Woods JH, Traynor JR. Br J Pharmacol 1999; 128:556–562.
29. Samama P, Pei G, Costa T, Cotecchia S, Lefkowitz RJ. Mol Pharmacol 1993; 45:390–394.
30. Leurs R, Smit MJ, Alewijnse AE, Timmerman H. TIBS 1998; 23:418–422.

31. Pineyro G, Azzi M, De Lean A, Schiller P, Bouvier M. Mol Pharmacol 2001; 60:816–827.
32. Zaki PA, Keith DE, Thomas JB, Carroll FI, Evans CJ. J Pharmacol Exp Ther 2001; 298:1015–1020.
33. Petaja-Repo UE, Hogue M, Bhalla S, Laperriere A, Morello JP, Bouvier M. EMBO J 2002; 21:1628–1637.
34. Mullaney I, Carr IC, Milligan G. Biochem J 1996; 315:227–234.
35. Georgoussi Z, Zioudrou C. Biochem Pharmacol 1993; 45:2405–2410.
36. Welsby PJ, Kellett E, Wilkinson G, Milligan G. Mol Pharmacol 2002; 61: 1211–1221.
37. Szekeres PG, Traynor JR. J Pharmacol Exp Ther 1997; 283:1276–1284.
38. Merkouris M, Mullaney I, Georgoussi Z, Milligan G. J Neurochem 1997; 69:2115–2122.
39. Labarre M, Butterworth J, St-Onge S, Pyza K, Schmidhammer H, Salvadori S, Balboni G, Guerrini R, Bryant SD, Lazarus LH. Eur J Pharmacol 2000; 406: R1–R3.
40. Hosohata K, Burkey TH, Alfaro-Lopez J, Hruby VJ, Roeske WR, Yamamura HI. Eur J Pharmacol 1999; 380:R9–R10.
41. Hosohata K, Varga EV, Alfaro-Lopez J, Tang X, Vanderah TW, Porreca F, Hruby VJ, Roeske WR, Yamamura HI, J Pharmacol Exp Ther 2003; 304:683–688.
42. Kenakin T. FASEB J 2001; 15:598–611.
43. Chiu TT, Yung LY, Wong YH. Mol Pharmacol 1996; 50:1651–1657.
44. Bouaboula M, Perrachon S, Milligan L, Canat X, Rinaldi-Carmona M, Portier M, Barth F, Calandra B, Pecceu F, Lupker J, Maffrand JP, Le Fur G, Casellas P. J Biol Chem 1997; 272:22330–22339.
45. Bouaboula M, Desnoyer N, Carayon P, Combes T, Casella P. Mol Pharmacol 1999; 55:473–480.
46. House RV, Thomas PT, Kozak JT, Bhargava HN. Neurosci Lett 1993; 198:119–122.
47. Spetea M, Erlandsson Harris H, Berzetei-Gurske IP, Klareskog L, Schmidhammer H. Life Sci 2001; 69:1775–1782.
48. Abdelhamid EE, Portoghese PS, Sultana M, Takemori AE. J Pharmacol Exp Ther 1991; 258:299–303.
49. Fundytus ME, Weltrowska G, Shapiro M, Schiller PW, Coderre TJ. Eur J Pharmacol 1995; 286:105–108.
50. Menkens K, Reid LD, Portoghese PS, Wild KD, Bilsky EJ, Porreca F. Eur J Pharmacol 1992; 219:345–346.
51. Froehlich JC, McCullough DE, Zink RW, Badia-Elder NE, Portoghese PS. J Pharmacol Exp Ther 1998; 287:284–292.
52. Landsman RS, Burkey TH, Consroe P, Roeske WR, Yamamura HI. Eur J Pharmacol 1997; 334:R1–R2.
53. Terranova JP, Storme JJ, Lafon N, Perio A, Rinaldi-Carmona M, Le Fur G, Suobrie P. Psychopharmacology 1996; 126:165–172.
54. Gross A, Terraza A, Marchant J, Bouaboula M, Ouahrani-Bettache S, Liautard JP, Casellas P, Dornand J. J Leukoc Biol 2000; 67:334–335.

55. Freedland CS, Poston JS, Porrino LJ. Pharmacol Biochem Behav 2000; 67:265–270.
56. Freedland CS, Sharpe AL, Samson HH, Porrino LJ. Alcohol Clin Exp Res 2001; 25:277–282.
57. Richardson JD, Aanonsen L, Hargreaves LKM. Eur J Pharmacol 1997; 319: R3–R4.
58. Sim LJ, Selley DE, Xiao R, Childers SR. Eur J Pharmacol 1996; 1996 307(97):105–112.
59. Liu JG, Prather PL. Mol Pharmacol 2001; 60:53–62.
60. Wang Z, Bilsky EJ, Wang D, Porreca F, Sadee W. Eur J Pharmacol 1999; 371:1–9.
61. Cruz SL, Villareal JE, Volkow ND. Life Sci 1996; 58:PL381–P389.
62. Bilsky EJ, Bernstein RN, Wang Z, Sadee W, Porreca F. J Pharmacol Exp Ther 1996; 277:484–490.
63. Onaran OH, Costa T, Rodbard D. Mol Pharmacol 1992; 43:245–256.
64. Varga EV, Li X, Stropova D, Zalewska T, Landsman RS, Knapp RJ, Malatynska E, Kawai K, Mizusura A, Nagase H, Calderon SN, Rice K, Hruby VJ, Roeske WR, Yamamura HI. Mol Pharmacol 1996; 50:1619–1624.
65. Hosohata Y, Varga EV, Stropova D, Li X, Knapp RJ, Hruby VJ, Rice KC, Nagase H, Roeske WR, Yamamura HI. Life Sci 2001; 68:2233–2239.
66. Meng F, Wei Q, Hoverstein MT, Taylor LP, Akil H. J Biol Chem 2000; 275: 21939–21945.
67. Chaturvedi K, Jiang X, Christoffers KH, Chinen N, Bandari P, Raveglia LF, Ronzoni S, Dondio G, Howells RD. Brain Res Mol Brain Res 2000; 80:166–176.
68. Mouledous L, Topham CM, Moisand C, Mollereau C, Meunier JC. Mol Pharmacol 2000; 57:495–502.
69. Spalding TA, Burstein ES, Brauner-Osborne H, Hill-Eubanks D, Brann MR. J Pharmacol Exp Ther 1995; 275:1274–1279.
70. Zhao MM, Gaivi RJ, Perez DM. Mol Pharmacol 1998; 53:524–529.
71. Onaran HO, Costa T. Ann N.Y. Acad Sci 1997; 812:98–115.
72. Weiner DM, Burstein ES, Nash N, Croston GE, Currier EA, Vanover KE, Harvey SC, Donohue E, Hansen HC, Andersson CM, Spalding TA, Gibson DFC, Krebs-Thomson K, Powell SB, Geyer MA, Hacksell U, Brann MR. J Pharmacol Exp Ther 2001; 299:268–276.
73. Strange PG. Pharmacol Rev 2001; 53:119–133.
74. Engelhardt S, Grimmer Y, Fan GH, Lohse MJ. Mol Pharmacol 2001; 60:712–717.
75. Barg J, Levy R, Simantov R. Neurosci Lett 1984; 50:133–137.

14

Mixed Opioid Receptor Agonists as a New Class of Agents for the Treatment of Moderate to Severe Pain

Peter J. Gengo and Kwen-Jen Chang
Ardent Pharmaceuticals, Inc., Durham, North Carolina, U.S.A.

1 INTRODUCTION

Pain management is a major therapeutic challenge for which opioid analgesics remain the mainstay in the treatment of moderate to severe pain [1–4]. These agents, which have powerful analgesic action, have been used for well over 200 years in spite of their narrow therapeutic index and their participation in deleterious drug-drug interactions. Adverse effects commonly associated with the use of narcotic analgesics include respiratory depression, nausea and vomiting, constipation, bradycardia, hypotension, hallucinations, euphoria, tolerance, dependence, and addiction potential [2,5,6]. The most life-threatening of these adverse effects is respiratory depression, which accounts for a majority of the resulting deaths, linked to the use of narcotic analgesics [2,6].

Nucleotide sequences for three distinct pertussis toxin–sensitive heterotrimeric GTP binding protein–coupled opioid receptors have been reported with ~65% homology existing between their amino acids [8–13]. These distinct opioid receptors, termed mu, delta, and kappa, have been shown to elicit a variety of pharmacologic actions; the most notable is analgesia [5,7].

Opioids, like endogenous endorphins, enkephalins, and dynorphin peptides, produce their effects by acting at one or more of these opioid receptors [7]. They alter synaptic transmission by modulating the presynaptic release of neurotransmitters such as acetycholine, norepinephrine, serotonin, dopamine, and substance P [14]. Changes in receptor-operated potassium currents, adenylyl cyclase activity, and intracellular free ionized calcium concentrations have all been reported to contribute to these changes in synaptic transmission [15,16]. To date, hyperpolarization of membrane potential and the corresponding effects on voltage-sensitive calcium channels [17] together represent the most plausible mechanisms for explaining the actions of opioid receptor active ligands in blunting the cellular excitability of neurons. Whatever the specific molecular mechanism, by modulating the release of these neurotransmitters, opioids inhibit responses to painful stimuli.

Opioid receptors are widely distributed throughout the central and peripheral nervous system and play a fundamental neuromodulatory role in the perception of pain [7,13]. Today, most clinically available opioid drugs act by targeting the mu subclass of the opioid receptor for mediating the relief of moderate to severe pain, reflecting their similarity to morphine [2–4]. These clinically available narcotic analgesics that utilize mu opioid receptors for their pharmacologic action, of which morphine and fentanyl remain the most widely used, also produce respiratory depression at therapeutic doses. The rostrodorsal surface of the pons, the nucleus tractus solitarius and the nucleus ambiguous are specific sites linked to respiratory depression stemming from the use of opioids [3,4]. Studies have compared the ratio of analgesia to respiratory depression in a series of morphine-like opioids. These studies have demonstrated that when equivalent analgesic doses are compared, the degree of respiratory depression is similar in magnitude to that measured with morphine [18].

Despite numerous efforts to develop analgesics with improved analgesic to respiratory depressant safety profiles, morphine and fentanyl remain the choice of clinicians for treating the analgesic needs of their patients. Recently, the analgesic activity of nonpeptide delta opioid receptor agonists has been described [20], and there is experimental evidence to support the potential clinical advantages of a compound with combined mu and delta opioid receptor agonist activities [21]. In studies with laboratory mice, additive analgesic actions were measured when fentanyl was administered in combination with the delta opioid receptor agonist, (+)-BW373U86, (+)-4-((α-R)-α-(2S,5R)-4-allyl-2,5-dimethyl-1-piperazinyl)-3-hydroxybenzyl)-N,N-diethylbenzamide. Not only was there additive analgesic activity in these studies, but coadministration was also shown to diminish fentanyl-induced muscle rigidity (Straub tail) and attenuate BW373U86-mediated seizures.

Other investigators have demonstrated that prolonged coadministration of morphine and BW373U86 attenuated the development and expression of morphine-induced dependence and tolerance in rats [22]. Perhaps most importantly, it has been reported that alfentanil-mediated antinociception was not altered by coadministration with BW373U86, and in these studies the alfentanil-mediated respiratory depression was significantly attenuated [23]. Taken together, these data may suggest that a compound with mixed mu and delta opioid receptor agonist activity may have utility in achieving a similar, or even a greater degree of analgesia and fewer adverse effects than fentanyl or morphine. To this end, a series of piperazinyl methylbenzamide analogues were synthesized in an attempt to define compounds that 1) possess both mu and delta opioid receptor affinity and intrinsic activity; 2) have strong analgesic activity; 3) have a long duration of action; and 4) do not cause excessive respiratory depression. One such compound, DPI-3290 [chemical name, (+)-3-((α-R)-α-((2S,5R)-4-allyl-2,5-dimethyl-1-piperazinyl)-3-hydroxybenzyl)-N-(3-fluorophenyl)-N-methylbenzamide], appears to fulfill these criteria. In this chapter we will review briefly the studies that support the rationale for the use of mixed opioid receptor agonists and the data from DPI-3290 that has now begun to test this approach.

2 PHARMACOLOGICAL OUTCOME OF CROSSTALK BETWEEN DELTA AND MU OPIOID RECEPTORS

The evidence linking the complex interaction between delta and mu opioid receptors now dates back almost a quarter of a century yet the precise molecular mechanism(s) that underlie(s) these effects has not been fully elucidated. The first investigators to report the interaction between opioid receptors were Vaught and Takemori [24], who demonstrated the enhanced antinociceptive action of morphine following the intracerebroventricular injection of [Leu^5]enkephalin. Since these landmark studies, several laboratories throughout the world have reported the interaction between opioid receptors. Furthermore, the introduction of nonpeptide delta opioid receptor agonists such as BW373U86, SNC-80 and TAN-67 has also aided the investigation of these interactions. In particular, studies by Dykstra [25] demonstrated BW373U86 potentiation of morphine and L-methadone antinociception in an electric shock titration assay in monkeys. Shortly after these studies were published, O'Neill [21] reported the enhanced antinociceptive action of fentanyl when coadministered with BW373U86. These investigators also reported that the seizurelike behavior associated with BW373U86 was attenuated when coadministered with fentanyl. In all instances, the effects of

BW373U86 were dose dependent and blocked by selective delta opioid receptor antagonists.

3 PHARMACOLOGY OF A MIXED OPIOID RECEPTOR AGONIST: DPI-3290

Following our initial reports that outline the synthesis and pharmacology of the selective, nonpeptide, delta opioid receptor agonist BW373U86, a series of piperazinyl methylbenzamide analogues with activity at multiple opioid receptors were synthesized and studied [26,27; Chap. 8]. Table 1 summarizes the binding affinity at delta, mu, and kappa opioid receptors of a limited number of these molecules. From this series of compounds, DPI-3290, DPI-130 [chemical name 3-((αR)-α-((2S,5R)-4-benzyl-2,5-dimethyl-1-piperazinyl)-3-hydroxybenzyl)-N-(3-fluorophenyl)-N-methylbenzamide] and DPI-125 [chemical name (-)-3-((S)-((2S,5R)-4-allyl-2,5-dimethyl-1-piperazinyl(3-thienyl)methyl)phenol] are of particular interest since they have high-affinity binding properties at delta and mu opioid receptors and a lower affinity at the kappa opioid receptor. These compounds with mixed opioid receptor binding affinity show an improved analgesic to respiratory depressant safety profile in rats when compared to morphine and fentanyl (Table 2). The safety ratio varies from 14.4, 18.2, to 26.9 for DPI-125, DPI-3290, and DPI-130, respectively, suggesting they may be safer analgesics than current narcotic analgesics such as morphine, fentanyl, their derivatives, and analogues.

To date, the compound that has been studied in both preclinical and clinical studies in greatest detail is DPI-3290. The structure of DPI-3290 is illustrated in Table 1. In saturation equilibrium binding studies performed at 25°C using membranes from rat brain or guinea pig cerebellum, the K_i values measured for DPI-3290 at delta, mu, and kappa opioid receptors were 0.18 ± 0.02 nM, 0.46 ± 0.05 nM, and 0.62 ± 0.09 nM, respectively. The three chiral centers in this molecule create 8 theoretical diastereoisomers. The RSR configuration (DPI-3290) yields a compound with the highest binding affinity for mu, delta, and kappa opioid receptors when compared with the remaining seven possible diastereoisomers. The rank order for binding affinity at mu opioid receptors was RSR > RSS > RRS > RRR > SSS > SSR > SRS > SRR [26]. The binding affinity of these eight diastereoisomers at delta opioid receptors was distinct from that measured at mu opioid receptors, yielding the following rank order of binding activity RSR > RRR > RRS > RSS > SRS > SRR > SSS > SSR [26].

In studies measuring opioid receptor–mediated inhibition of tension development in vas deferens isolated from mice, DPI-3290 is an extremely potent mixed opioid receptor agonist as summarized in Table 3. The intrinsic

TABLE 1 K_i Values for Mixed Opioid Receptor Agonists at Delta ([^{3}H]DPDPE), Mu ([^{3}H]DAMGO), or Kappa ([^{3}H]U69593) Opioid Receptors[a]

	[^{3}H]DPDPE (nM)	[^{3}H]DAMGO (nM)	[^{3}H]U69593 (nM)
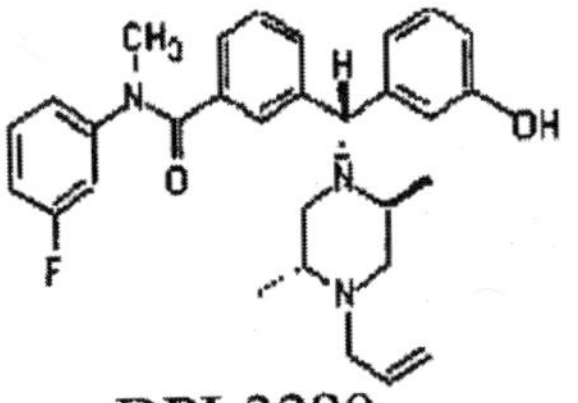 DPI-3290	0.18 ± 0.02	0.46 + 0.05	0.62 + 0.09
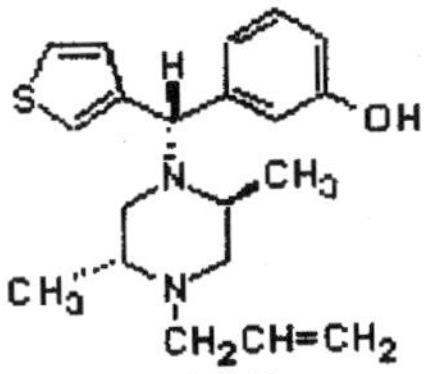 DPI-125	0.97 ± 0.30	0.36 ± 0.10	1.36 ± 0.39
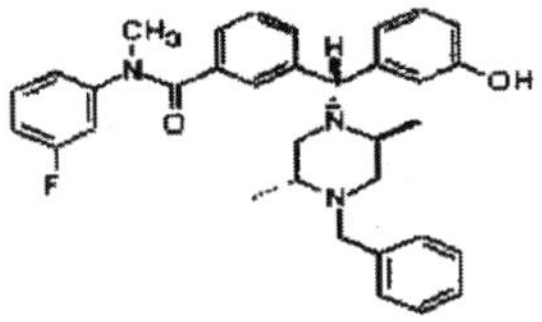 DPI-130	0.40 ± 0.24	1.58 ± 0.63	21.8 ± 4.6
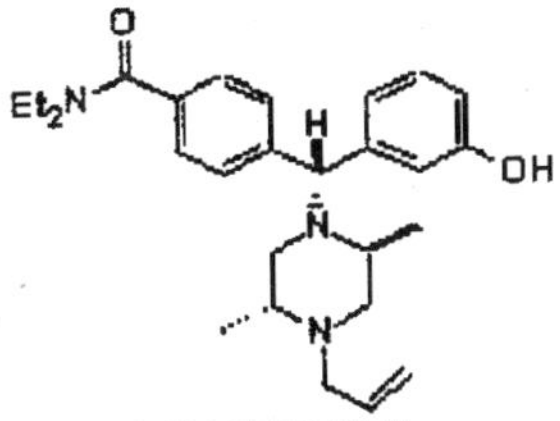 BW373U86	1.8 ± 0.40	15 ± 2.8	34 ± 3.0

[a] The results summarized represent the mean ± S,E,M from 3–4 separate experiments. DPI-3290 [chemical name (+)-3-((α-R)-α-((2S,5R)-4-allyl-2,5-dimethyl-1-piperazinyl)-3-hydroxybenzyl)-N-(3-fluorophenyl)-N-methylbenzamide], DPI-130 [chemical name 3-((αR)-α-((2S, 5R)-4-benzyl-2,5-dimethyl-1-piperazinyl)-3-hydroxybenzyl)-N-(3-fluorophenyl)-N-methylbenzamide], and DPI-125 [chemical name (-)-3-((S)-((2S,5R)-4-allyl-2,5-dimethyl-1-piperazinyl)(3-thienyl)methyl)phenol].

TABLE 2 ED_{50} Values for Opioid-Mediated Antinociception and Hypercapnia in Conscious Laboratory Rats[a]

Compound	Tail pinch (mg/kg)	PCO_2 (mg/kg)	Safety ratio PCO_2:tail pinch
DPI-3290	0.05±0.007	0.91±0.22	18.2[b]
DPI-130	0.08±0.007	2.15±0.80	26.9[b]
DPI-125	0.05±0.009	0.72±0.21	14.4[b]
Morphine	2.01±0.0005	4.23±0.72	2.1[c]
Fentanyl	0.0034±0.0002	0.0127±0.0035	3.7

[a] The results summarized represent the mean ± S.E.M. from 6–8 independent animals.
[b] $P < .05$ vs. morphine or fentanyl.
[c] $P > .05$ vs. fentanyl.

activity of DPI-3290 at delta opioid receptors was assessed in vas deferens in the presence of the highly selective mu opioid receptor antagonist CTOP (cyclic[D-Phe-Cys-Tyr-D-Trp-Orn-Thr-Pen-Thr-NH2]) (1 μM) and the kappa opioid receptor antagonist nor-BNI (nor-binaltorphimine) (15 nM). DPI-3290 produced delta opioid receptor–mediated, concentration-dependent inhibition of electrically stimulated smooth muscle contractions with a corresponding IC_{50} value of 1.0 ± 0.3 nM.

Further studies in mouse vas deferens indicated that DPI-3290 is also active at mu opioid receptors. The intrinsic activity of DPI-3290 at mu opioid receptors was determined in the presence of the highly selective delta opioid receptor antagonist TIPP (H-Tyr-Tic-Phe-Phe-OH) (3 μM) and the selective kappa opioid receptor antagonist nor-BNI (15 nM). Under these conditions, DPI-3290 again caused concentration-dependent inhibition of muscle contraction with a corresponding IC_{50} value of 6.2±2.0 nM. Although far less potent at kappa opioid receptors in comparison to its intrinsic activity at mu

TABLE 3 IC_{50} Values for Delta, Mu, or Kappa Opioid-Mediated Inhibitory Effects on the Contractility of Electrically Stimulated Mouse Vas Deferens[a]

Compound	Delta (nM)	Mu (nM)	Kappa (nM)
DPI-3290	1.0 ± 0.3	6.2 ± 2.0	25 ± 3.3
Morphine	19,700 ± 170	1,090 ± 30	37,750 ± 140
Fentanyl	492 ± 15	14 ± 3	868 ± 19
BW373U86	0.49 ± 0.03	134 ± 9	201 ± 7
U69593	5,200 ± 1,200	10,000 ± 2,000	24 ± 2.2

[a] The results summarized represent the mean ± S.E.M. from 6–8 independent tissues.

or delta opioid receptors, DPI-3290 elicited kappa opioid receptor-mediated inhibition of muscle contraction in mouse vas deferens. The IC_{50} was shifted 30-fold to the right of delta opioid receptor activity and sevenfold rightward from the mu opioid receptor–mediated inhibition curves. When compared to fentanyl, morphine, or U69593, the activity of DPI-3290 was 600–24,000 times more potent at delta opioid receptor–mediated changes and 2–1600 times more potent at mu opioid receptors in eliciting changes in tension development in the mouse vas deferens (Table 3).

The activity of DPI-3290 was further studied in guinea pig ileum, a tissue previously shown to have a high density of mu opioid receptors that are linked to strong effects on muscle contractility. The guinea pig ileum has also been reported to have an intermediate kappa opioid receptor density, with little to no delta opioid receptor density or influence on muscle contraction. In the guinea pig ileum, increasing concentrations of DPI-3290 again produced significant and concentration-related decreases in electrically stimulated muscle contractions. The mu opioid receptor–mediated IC_{50} value in this tissue was 3.4 ± 1 [26].

DPI-3290 has nanomolar binding affinity and strong agonist intrinsic activity at mu opioid receptors, characteristics it shares with narcotic analgesics. To compare and differentiate the actions of DPI-3290 with morphine and fentanyl, studies were conducted to simultaneously measure the antinociceptive properties and respiratory depressant activity of these compounds in laboratory rats. When DPI-3290 was administered intravenously to conscious rats the most striking effect was a dose-related increase in antinociception. The ED_{50} value for DPI-3290-mediated antinociception was 0.05 ± 0.007 mg/kg. In addition, this compound also produced an increase in blood pCO_2 concentration, but changes in pCO_2 concentration resulted at markedly higher doses in relationship to those that elicited antinociception, especially when compared to morphine or fentanyl. For example, in conscious rats, the ED_{50} value for hypercapnia mediated by DPI-3290 was 0.91 ± 0.22 mg/kg, a dose that was 18.2-fold higher than its ED_{50} value for antinociception. In comparison, the ED_{50} values for morphine and fentanyl-mediated antinociception in conscious rats were 2.01 ± 0.0005 mg/kg and 0.0034 ± 0.0002 mg/kg, respectively; doses that were only 2.1-fold lower than the ED_{50} value for hypercapnia with morphine and 3.7-fold lower for the ED_{50} value for hypercapnia with fentanyl (Table 2).

Because selective delta opioid peptides, DPDPE (cyclic [D-Pen^2, D-Pen^5]enkephalin) and deltorphin-II, and nonpeptide (BW373U86) agonists have been shown to inhibit the hypercapnia induced by the continuous infusion of the selective mu opioid analgesic alfentanil [23], it has been proposed that delta opioid receptor agonists mitigate the hypercapnia resulting from narcotic analgesics. Measuring the actions of DPI-3290 on alfentanil-mediated hypercapnia and antinociception in rats tested the possibility that a mixed opioid receptor agonist may have these same effects. A 6 μg/kg/min intravenous infusion of alfentanil in conscious rats maintained full antinociception and increased pCO_2 concentrations from 35 ± 4 mm Hg to 55 ± 8 mm Hg in rats. These changes in radiant tail flick latency (expressed as a maximal percent effect, MPE) and blood pCO_2 concentration induced by alfentanil reached steady state in 20–25 min and were maintained throughout the 60-min duration of the study (Fig. 1). Bolus intravenous doses of DPI-3290 that ranged from 0.2 mg/kg to 1.0 mg/kg resulted in no change in radiant tail flick latency when coadministered during the alfentanil infusion. Bolus intravenous doses of DPI-3290 (ranging from 0.2 mg/kg to 1.0 mg/kg) reversed the alfentanil-induced elevation in pCO_2 concentration by ~50% at all doses studied. Thirty minutes after the alfentanil infusion had been terminated, antinociceptive response (MPE) and arterial blood gases (pCO_2) both returned to baseline values.

The downregulation of opioid receptors and loss of analgesic activity following chronic administration of morphine, a phenomenon commonly termed tolerance, desensitization, or tachyphylaxis, often limit the clinical efficacy of narcotic analgesics. Tolerance to the antinociceptive actions of DPI-3290 was assessed in laboratory rats. Following 5 days of DPI-3290 treatment given twice daily at a dose of 0.5 mg/kg, tolerance did develop to tail pinch latency. As illustrated in Figure 2, over 5 days of DPI-3290 treatment, a time-dependent decline in antinociceptive action was measurable at day 2 and maximal by day 5. This pattern of tolerance is very similar to that measured in rats administered morphine twice daily for 5 days at a dose of 5 mg/kg. The striking difference between the tolerance that develops after chronic exposure to DPI-3290 or morphine is the "asymmetric" nature of the cross-tolerance that develops between these two agents. As would be expected with any opioid receptor analgesic, following 5 days of morphine treatment, significant tolerance to the antinociceptive actions of DPI-3290 was reported. Quite unexpected was the asymmetric or lack of tolerance to morphine's antinociceptive actions following 5 days of treatment with DPI-3290.

Studies examined the physical dependence associated with chronic treatment of rats with DPI-3290 and compared this with the physical dependence measured in rats treated for 3 days with morphine. Physical dependence was facilitated by injection with naloxone. Illustrated in Figure 3 is the dose-response relationship for DPI-3290 and morphine induced abstinence (physical dependence). At all doses tested, the magnitude of DPI-3290-mediated abstinence was markedly smaller than the abstinence associated with morphine. Most interestingly, at the higher doses of DPI-3290, abstinence appears to plateau whereas the abstinence associated with morphine continues to increase with dose.

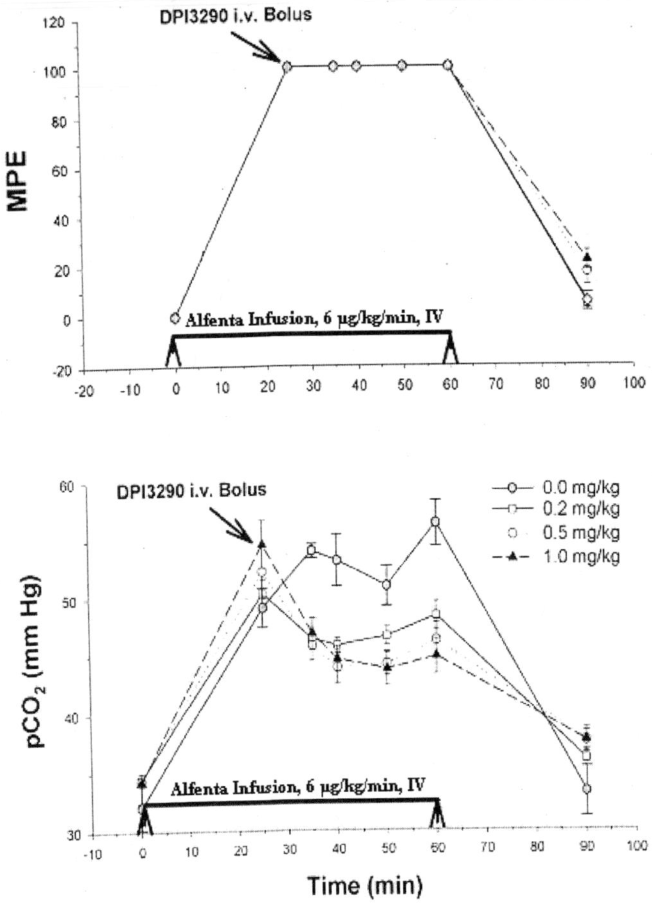

FIGURE 1 Time course for effects of the mixed opioid agonist DPI-3290 on antinociception and blood pCO_2 levels in alfentanil-infused rats. Alfentanil was intravenously infused at 6 μg/kg/min. At the time points outlined in the figure, radiant tail flick testing and arterial blood samples were collected and analyzed by standard methods. Antinociception was expressed as maximal percent effect (MPE).

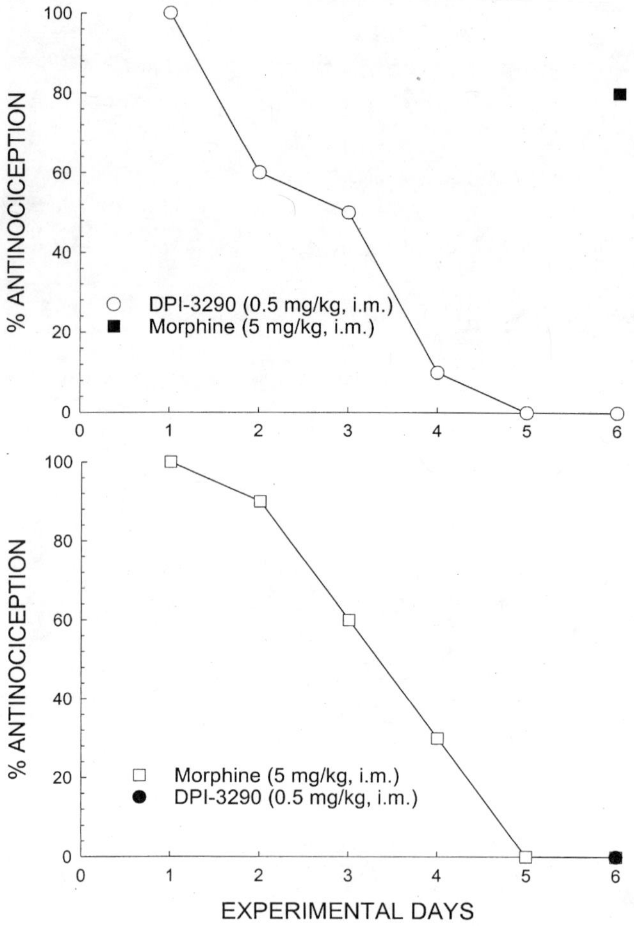

FIGURE 2 One-way cross-tolerance between DPI-3290 and morphine. Top panel illustrates data from rats receiving DPI-3290 (0.5 mg/kg, IM) subchronically (5 days) and subsequently challenged with DPI-3290 or morphine at day 6 (N=6). Data illustrated in the bottom panel was obtained from rats that received morphine (5 mg/kg, IM) subchronically (5 days) and were challenged with morphine or DPI-3290 at day 6. Antinociceptive responses were assessed by tail pinch test with an artery clamp. Pain responsiveness was assessed at both 20 and 30 min after drug administration.

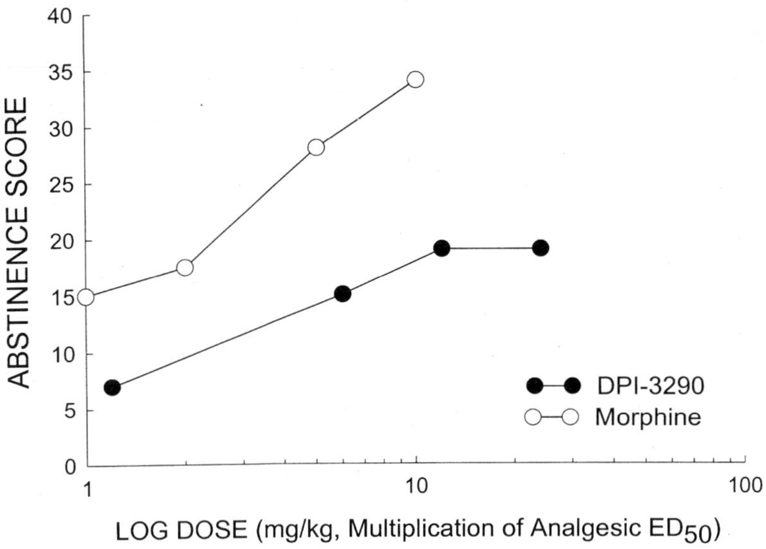

FIGURE 3 Naloxone precipitated withdrawal abstinence scores in rats subchronically treated with DPI-3290 or morphine (N = 6). For direct dose comparison the figure illustrates abstinence scores in rats receiving DPI-3290 or morphine subchronically at doses that are equated by multiplication of the analgesic ED_{50}s for each compound. Physical dependence in rats was induced as described elsewhere [22,28]. Morphine sulfate or DPI-3290, dissolved in a 5% dextrose solution was administrated IM twice daily (8 AM and 6 PM) for 3 days with increasing doses on each day (2nd day = 2 × dose of 1st day; 3rd day = 3 × dose of 1st day). Initial morphine doses were 2, 5, 10, and 20 mg/kg; initial DPI-3290 doses were 0.2, 1, 2, and 5 mg/kg, IM. At 8 AM on the 4th day, the animals were given a single administration of the same dose as injected on the 3rd day and then were challenged with naloxone (10 mg/kg, IP) 3 h later. Signs of abstinence [29,30] were monitored 30 min before and after the naloxone injection. Intensity of abstinence was assessed by a point-scoring technique modified by weighting the signs [22]. Two classes of signs were distinguished: counted and checked signs. Checked signs included irritability, diarrhea, salivation, licking penis, ptosis, and weight loss. Counted signs included jumping, wet dog shake, forelimb tremor, digging, teeth chattering, and writhing. The scores of these signs increased with the frequency of the incidence. Scores for each of the signs were summed to give a grand total that represented the intensity of the abstinence syndromes precipitated by naloxone.

4 ACTIONS IN MAN

Consistent with the preclinical pharmacological profile of DPI-3290, recent early clinical studies with this compound have indicated strong analgesic activity in man. These actions are dose dependent and rapid in onset, with less apparent adverse changes in saturated O_2 or emesis when compared with equivalent analgesic doses of morphine or fentanyl. Patients appeared to tolerate doses of DPI-3290 well in these early studies with no dose-limiting adverse events.

5 CONCLUSION

Defining the differences and mechanisms of action of opioid receptor active agents and their crosstalk between opioid receptors is germane to understanding their pharmacology and anticipating their clinical utility. This is of particular interest for compounds interacting not only at one opioid receptor, but even more so, for those that interact with multiple opioid receptors. Today, the literature is replete with reports that describe the distinct pharmacology among delta, mu, and kappa opioid receptors [19]. More recent is the interest in understanding the outcome of simultaneously activating multiple opioid receptors. It is difficult to argue that activation of multiple opioid receptors results in differing pharmacology. What is still not fully elucidated is if these effects are consistent across all target organ systems. It also remains uncertain what particular second-messenger system(s) crosstalk between activation of these receptors. Recent reports that define the striking differences in the pharmacology of opioid receptor homo- and heterodimers further add to the complexity of understanding the actions of opioid receptors [31,32].

To date, one of the most direct means for examining this crosstalk between opioid receptors is characterizing the actions of nonpeptide mixed opioid receptor agonists such as DPI-3290. The in vitro and in vivo pharmacological and radioligand binding data are consistent with the strong antinociceptive actions of DPI-3290. The mechanism(s) of action responsible for this activity is (are) centered at opioid receptors since opioid receptor antagonists block this effect [26,27]. The pharmacological profile of this agent is consistent with studies that coadminister combinations of compounds that act at distinct opioid receptors and further support the results of these studies.

Since opioids are the principal agents used for the treatment of moderate to severe pain, there are several interesting implications to the data that defines the pharmacological consequences of activating multiple opioid receptors. The establishment that mixed opioid receptor agonist activity can be contained within a single chemical entity with varying degrees of antinociception and respiratory depression opens the possibility for further advances in the treatment of severe pain. The characteristic that defines the paramount difference between DPI-3290 and narcotic analgesics like morphine or fentanyl is the marked difference in its antinociceptive and respiratory depressant activities. Because the most life-threatening adverse effect associated with the use of narcotic analgesics for moderate to severe pain is respiratory depression, a drug with an appropriate separation between analgesia and hypercapnic activities could relieve severe pain with a broader therapeutic index. In this regard, the mixed opioid receptor agonist activity of DPI-3290 and its antinociceptive and hypercapnic pharmacologies is evidence of the likelihood for achieving such a goal. Needless to say, this potential breakthrough in the management of pain could be the first in a series of pharmacophores with mixed opioid receptor agonist activity that produce strong, safe, and effective analgesia.

REFERENCES

1. Mason P. Curr Opin Neurobiol 1999; 9:436–441.
2. Inturrisi CE. Opiate analgesic therapy in cancer pain. In: Forley KM, ed. Advances in Pain Research and Therapy. Vol. 16. New York: Raven Press, 1990:133–154.
3. Clotz MA, Nahata MC. Clin Pharm 1991; 10:581–593.
4. Holder KA, Dougherty TB, Chiang JS. Postoperative pain management. Cancer Bull 1995; 47:43–51.
5. Chang K-J. Opioid receptors: multiplicity and sequelae of ligand-receptor interactions. In: Conn M, ed. The Receptors. New York: Academic Press, 1984:1–81.
6. Resine T, Pasternak G. Opioid analgesics and antagonists. In: Molinof PB, Ruddon RW, eds. The Pharmacological Basis of Therapeutics. Vol. 1. New York: Macmillan, 1993:521–555.
7. Gutstein HB, Akil H. Opioid analgesics. In: Hardman JG, Limbird LE, Gilman AG, eds. Goodman & Gilman's The Pharmacological Basis of Therapeutics. 10th ed. New York: McGraw-Hill, 2001;569–619.
8. Evans CJ, Keith DE, Morrison H, Magendzo K, Edwards RH. Science, 1992; 258, 1952–1955.
9. Kieffer BL, Befort K, Gaveriaux-Ruff C, Hirth CG. Proc Natl Acad Sci USA 1992; 89:12048–12052.
10. Childers SR. Life Sci 1991; 48:1991–2003.
11. Chen Y, Mestek A, Liu J, Hurley JA, Yu L. Mol Pharmacol 1993; 44:8–12.
12. Li S, Zhu J, Chen C, Chen Y-W, Deriel JK, Ashby B, Liu-Chen L-Y. Biochem J 1993; 295:629–633.
13. Quock RA, Burkey TH, Varga E, Hosohata Y, Hosohata K, Cowell SM, Slate CA, Ehlert FJ, Roeske WR, Yamamura YI. Pharmacol Rev 1999; 51:503–532.
14. Hagelberg N, Kajander JK, Nagren K, Hinkka S, Hietala J, Scheinin H. Synapse 2002; 45(1):25–30.
15. Murthy DS, Makhlouf GM. Mol Pharmacol 1996; 50:870–877.

16. Fan S, Crain SM. Brain Res 1995; 696:97–105.
17. Tang T, Kiang JG, Cox BM. J Pharmacol Exp Ther 1994; 270:40–46.
18. Hurle MA, Mediavilla A, Florez J. Neuropharmacology 1985; 24:597–606.
19. Morin-Surun M-P, Boudinot E, Gacel G, Champagnat J, Roques BP, Denavit-Saube M. Eur J Pharmacol 1984; 98:235–240.
20. Chang K-J, Rigdon GC, Howard JL, McNutt RW. J Pharmacol Exp Ther 1993; 267:852–857.
21. O'Neill SJ, Collins MA, Pettit HO, McNutt RW, Chang K-J. J Pharmacol Exp Ther 1997; 282:271–277.
22. Lee PHK, McNutt RW, Chang K-J. J Pharmacol Exp Ther 1993; 267:883–887.
23. Su YF, McNutt RW, Chang K-J. J Pharmacol Exp Ther 1998; 287:815–823.
24. Vaught JL, Takemori AE. J Pharmacol Exp Ther 1979; 208:86–90.
25. Dykstra LA, Schoenbaum GM, Yarbrough J, McNutt RW, Chang K-J. J Pharmacol Exp Ther 1993; 267:875–882.
26. Gengo PJ, Pettit HO, O'Neill SJ, Wei K, McNutt RW, Chang K-J. J Pharmacol Exp Ther. In press.
27. Gengo PJ, Pettit HO, O'Neill SJ, Su YF, McNutt RW, Chang K-J. J Pharmacol Exp Ther. In press.
28. Cerletti C, Keinath SH, Reidenberg MM, Adler JW. Pharmacol Biochem Behav 1976; 4:323–327.
29. Wei E, Loh HH, Way EL. J Pharmacol Exp Ther 1973; 184:398–403.
30. Blasig J, Herz A, Reihold K, Zieglansberger S. Psychopharmacology (Berl) 1973; 33:19–38.
31. Jordan BA, Devi LA. Nature 1999; 399:697–700.
32. Gomes I, Jordan BA, Gupta A, Trapaidze N, Nagy V, Devi LA. J Neurosci 2000; 20:1–5.

15

Biphalin: A Multireceptor Opioid Ligand

Andrzej W. Lipkowski
Medical Research Centre, Polish Academy of Sciences, Warsaw, Poland

Daniel B. Carr and Iwona Bonney
Tufts–New England Medical Center, Boston, Massachusetts, U.S.A.

Aleksandra Misicka
Warsaw University, Warsaw, Poland

1 INTRODUCTION

This book focuses on biological and pharmacological effects of ligands interacting with delta opioid receptors. Selective or even specific ligands for one type or subtype of the receptor are tools to separate and characterize the roles of particular receptors in the functions of living organism. Nevertheless, most if not all functions of endogenous opioids reflect their concerted action upon processes mediated by a multiplicity of opioid receptors. Therefore, it is not unexpected that endogenous opioid peptides express a broad spectrum of affinity to neuropeptide G protein–coupled receptors. The prospective development of new opioid drugs has taken into account such multitarget, coordinated action. This chapter surveys the biological effects resulting from biphalin, a compound with significant binding to the delta opioid receptor as well as to other types of opioid receptors.

2 CHEMISTRY

Biphalin, first synthesized by Lipkowski et al. [1], is a tetrapeptide dimer with the chemical structure shown in Figure 1. The compound is a "head-to-head" hybrid of two opioid active tetrapeptide fragments of enkephalin analogues connected with hydrazide bridge. Standard crystallographic analysis of the structure of biphalin revealed a tetrapeptide sequence (Fig. 2) with two structurally equal peptide "arms" [2] that suggested the term "palindromic sequence." Tetrapeptide fragments are flexible and may adapt themselves to complementary structures of receptor-binding sites. In contrast, the hydrazide bridge connecting the two tetrapeptide arms is very rigid. The two carbonyls of phenylalanine residues and two nitrogens of hydrazide may exist in two forms. The first structural form is a planar location of all elements of the bridge [3]. Minimization of energy during crystallization and charged ionic force may place the hydrazide bridge into another form, in which its two planar halves are located in approximately 60° of torsion across the N-N bond [4].

In its compact crystal structure, the two opioid tetrapeptide pharmacophores of biphalin are not conformationally equivalent. One tetrapeptide, which has a steric similarity with the delta-selective peptide DADLE, folds into a random coil. The contralateral tetrapeptide, sterically similar to the mu-selective peptide D-TIPP-NH_2, exhibits a fairly normal type III′ β bend [4]. These conformational features suggest that under physiological conditions, biphalin may easily bind to these respective opioid receptors. This duality of binding affinity is probably the reason that biphalin is able to interact with all opioid receptor types.

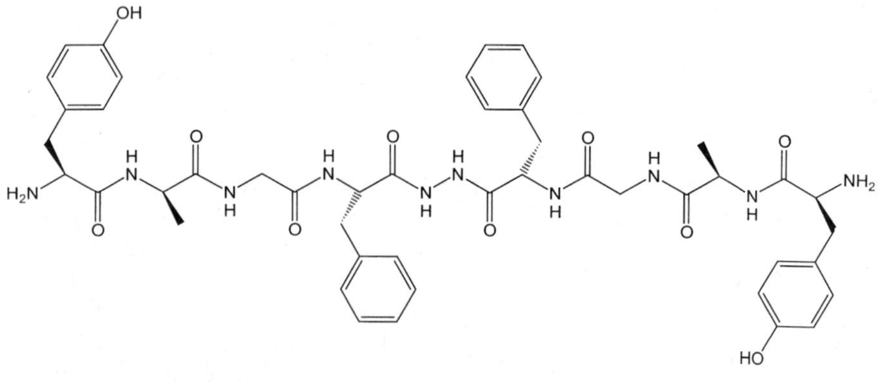

FIGURE 1 Chemical structure of biphalin.

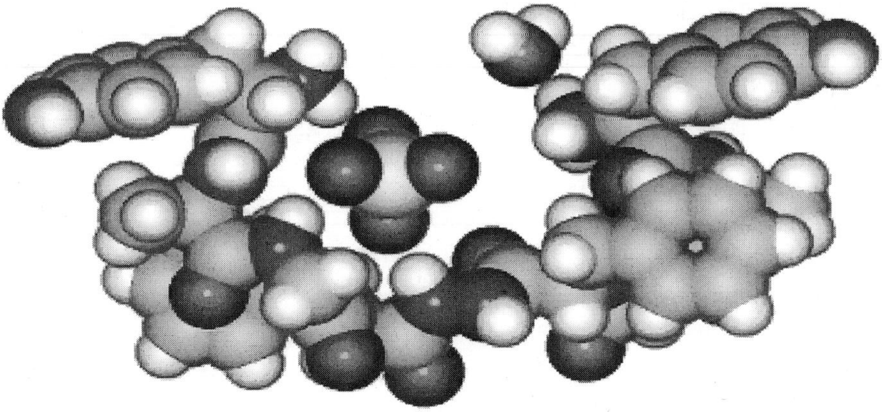

FIGURE 2 Space-filling model of the dimeric opioid peptide biphalin as determined by a single crystal x-ray study [4]. The peptide cation co-crystallizes with a sulfate anion and several molecules of water. The illustration shows how the peptide interacts with the sulfate anion and one of the water molecules.

Biphalin, in standard receptor-binding assays, shows twice the affinity to mu than delta receptors (Table 1). Structure-activity relationships of a series of symmetric biphalin analogues show that replacing phenylalanine residues with *p*-chlorophenylalanine increases the affinity for delta receptors and decreases it for mu receptors, resulting in more potent and more

TABLE 1 Biphalin Analogues with Modifications at the Phenylalanine Residues

	IC_{50} (nM)	
Compound	delta	mu
[Tyr-D-Ala-Gly-Phe-NH-]$_2$ Biphalin	2.6	1.4
[Tyr-D-Ala-Gly-Phe(*p*Cl)-NH-]$_2$	0.54	2.44
[Tyr-D-Ala-Gly-Phe(*p*F)-NH-]$_2$	0.31	0.64
[Tyr-D-Ala-Gly-Phe(*p*I)-NH-]$_2$	5.20	24.5
[Tyr-D-Ala-Gly-Phe(*p*NO_2)-NH-]$_2$	0.63	0.94
[Tyr-D-Ala-Gly-Phe(*p*NH_2)–NH-]$_2$	120	10 μM (0%)
[Tyr-D-Ala-Gly-(2*S*,3*R*)Phe(β-Me)-NH-]$_2$	110	1.3
[Tyr-D-Ala-Gly-(2*S*,3*S*)Phe(β-Me)-NH-]$_2$	11	3.0
[Tyr-D-Ala-Gly-1′-Nal-NH-]$_2$	6.4	0.79
[Tyr-D-Ala-Gly-2′-Nal-NH-]$_2$	7.4	1.7
[Tyr-D-Ala-Gly-Phe(F_5)-NH-]$_2$	7.8	0.91

Source: Refs. 5, 6.

TABLE 2 Biological Activity of Truncated Biphalin and Analogues

	Binding IC50 (nM)		Bioassay EC_{50} (nM)	
Compound	delta	mu	MVD[a]	GPI[a]
[Tyr-D-Ala-Gly-Phe-NH-]$_2$ Biphalin	2.6	1.4	27	8.8
Tyr-D-Ala-Gly-Phe-NH-NH_2	230	4.7	290	90
Tyr-D-Ala-Gly-Phe-NH-NH < -Phe	15	0.74	27	2.7
Tyr-D-Ala-Gly-Phe-NH-NH < -D-Phe	30	0.88	32	9.8
Tyr-D-Ala-Gly-Phe-NH-NH < -Nle	71	5.9	95	5.2
Tyr-D-Ala-Gly-Phe-NH-NH < -D-Nle	21	1.3	20	24
Tyr-D-Ala-Gly-Phe-NH-NH < -Tyr	16	1.6	45	15
Tyr-D-Ala-Gly-Phe-NH-NH < -Trp	29	2.0	15	7.1
Tyr-D-Ala-Gly-Phe-NH-NH < -DNS				

[a] MVD, mouse vas deferens; GPI, guinea pig ileum.
Source: Refs. 7, 8.

delta-selective analogues [5]. Substitution of phenylalanine with *p*-nitro- or *p*-fluorophenylalanine increases affinity to both delta and mu receptor types. Other substitutions at the phenylalanine residue with aromatic amino acids, such as β-methyl-phenylalanines or naphthylalanines, increase selectivity for mu receptor types, mainly by decreasing affinity for delta receptors (Table 2) [6].

3 PHARMACOLOGY IN VITRO

When the efficacy of biphalin-stimulated G protein activation was examined (Table 3) in delta opioid receptor–transfected CHO cells, an efficacy ratio of 0.42 was determined as compared with deltorphin-II and DPDPE, the latter a reference delta-selective agonist. Such low efficacy values suggest that biphalin does not efficiently stimulate the G protein through the delta receptor [9]. Relative affinities of biphalin and morphine for mu, delta, and kappa binding sites in guinea pig brain membranes are shown in Table 4.

Shen and Crain [13] tested the effects of biphalin on naive and chronic morphine-treated dorsal root ganglion (DRG) neurons in cell culture. At low (pM-nM) concentrations, most mu, delta, or kappa opioid peptides as well as morphine and other opioid alkaloids elicit dose-dependent excitatory prolongation of the calcium-dependent component of the action potential duration (APD) of many mouse sensory DRG neurons. In contrast, application of the same opioids at higher (μM) concentrations results in inhibitory shortening of the APD [14]. Biphalin at a low concentration elicits only dose-

TABLE 3 Relative Efficacies of Biphalin and Delta-Selective Agonists at the Wild-Type Cloned Human Opioid Receptor

	Agonist		
	DPDPE	Deltorphin-II	Biphalin
K_i^a (nM)	85.5 ± 7.2	42.7 ± 9.7	46.5 ± 1.5
EC_{50}(nM)[a]	19.1 ± 7.2	9.3 ± 4.2	34.0 ± 13.1
E_{max}[a]	82 ± 2	96 ± 2	98 ± 10
Relative efficacy	2.74	2.80	1.18
Efficacy ratio	0.98	1.00	0.42

Ki values were determined from [^{3}H]naltrindole competitive inhibition experiments using the equation of Cheng and Prusoff [10]. EC_{50} values were determined from [^{35}S]GTPγS stimulation experiments as previously reported by Quock et al. [11]. Relative efficacy (e_{rel}) was determined using the equation: $e_{rel} = 0.5 \times E_{max}/E_{max}\text{-sys} \times (1 + K_D/EC_{50})$ with $E_{max}/E_{max\text{-sys}} = 1$.
[a] Values are mean ± SE.
Source: Ref. 9.

dependent inhibitory (APD-shortening) effects on DRG neurons. Furthermore, at pM concentrations biphalin acts as a selective antagonist of opioid excitatory (APD-prolonging) functions. This antagonist action upon excitatory opioid receptors and agonist action upon inhibitory opioid receptors of DRG neurons in culture was limited to mu and delta, but not kappa, subtypes. These authors suggested that this property of biphalin may account for its unexpectedly high antinociceptive potency in vivo. Chronic treatment of DRG neurons with high (mM) concentrations of biphalin resulted neither in supersensitivity to the excitatory effects of naloxone nor tolerance to opioid inhibitory effects, in contrast to the excitatory opioid supersensitivity and

TABLE 4 Relative Affinities of Biphalin and Morphine for Mu, Delta, and Kappa Binding Sites in Guinea Pig Cell Membranes[a]

	Agonist	
K_i(nM)*	Morphine	Biphalin
Delta	510	4.6
Mu	38	12
Kappa	1900	270

[a] Binding was performed with 100 mM NaCl.
Source: Ref. 12.

tolerance that develop in chronic morphine or DADLE-treated neurons. These results may explain the low dependence liability of biphalin in vivo.

3.1 In Vivo

The primary study of biphalin revealed its antinociceptive, naloxone-reversible effectiveness after intraperitoneal application in mice (Fig. 3) [1] that was similar to morphine. Also similar to morphine, an antinociceptive effect was found after intravenous application in rats. However, biphalin applied intrathecally in rats showed a much stronger effect than morphine [15]. Biphalin expresses affinity to three major opioid receptor types. The question of which opioid receptor type is responsible for biphalin's biological effects has been answered systematically in studies by Porreca and colleagues [16]. After intraventricular application in mice, biphalin has been shown to be one of the most potent opioids ever identified in eliciting antinociception; the potency of this compound in the tail flick test was almost seven times greater than that of intracerebroventricular (ICV) etorphine; 12-fold greater than that of ICV DAMGO; at least 2 orders of magnitude greater than that of ICV carfentanil, sufentanil, fentanyl, or PL017; and 3 orders of magnitude greater than the antinociceptive potency of ICV morphine or alfentanil [16]. Interestingly, biphalin ICV antinociception was significantly antagonized by

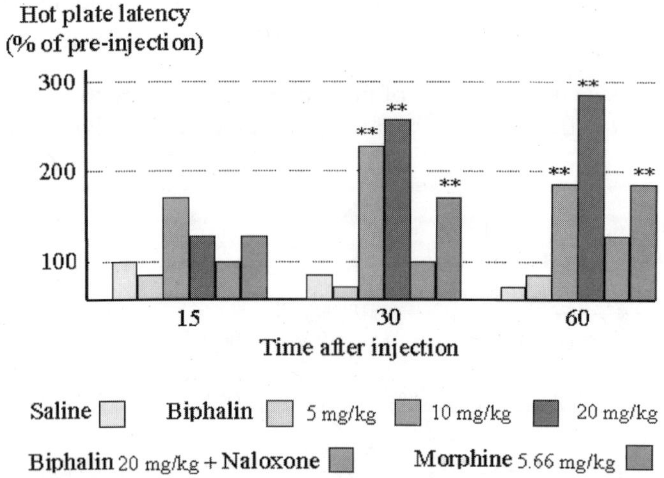

FIGURE 3 Pain threshold in mice after intraperitoneal administration of biphalin (5, 10, 20 mg/kg body weight) and morphine hydrochloride (5.66 mg/kg). **Significantly different from preinjection pain threshold, $P < .01$.

receptor-selective doses of β-funaltrexamine (mu antagonist), naloxonazine (mu_1 antagonist), ICI 174,864 (delta antagonist) and [D-Ala^2, Cys^4]deltorphin ($delta_2$ antagonist), but not by [D-Ala^2, Leu^5, Cys^6]enkephalin ($delta_1$ antagonist) or nor-binaltorphimine (kappa antagonist). These findings suggest that the surprisingly high and unusual profile of antinociceptive activity of ICV biphalin results from simultaneous activation of mu and $delta_2$ receptors in a functional, and possibly physical, complex.

3.2 Blood-Brain Barrier Permeability

The penetration of endogenous substances through the blood-brain barrier (BBB) is one of the most important elements of communication between the periphery and the brain. Hormones and neuromediators, particularly peptides, play crucial roles in homeostasis in both the peripheral and the central nervous systems [17]. The endothelial cells within brain capillaries, which constitute the BBB, are both a physical barrier and a complex biochemical interface containing many biological functions for the specific uptake and exclusion of endogenous neuropeptides. Within the endothelial and/or epithelial cytosolic compartments are peptidases, which degrade peptides [18]. In the case of opioid peptides an important consideration affecting the CNS distribution of peptides is the presence of a brain-to-blood influx system for N-tyrosinated peptides [19]. An in situ brain perfusion study showed that biphalin, [^{125}I-Tyr]biphalin, and its chlorohalogenated analogue, [^{125}I-Tyr]-Phe(pCl)$^{4,4'}$]biphalin, enter significantly into the CNS, to reach both the brain and cerebrospinal fluid [20,21].

Biphalin enters the CNS via two mechanisms. The first mechanism is saturable, and follows Michaelis-Menten kinetics with a K_m of 2.6 μM and V_{max} of 14.6 pmol $min^{-1}g^{-1}$. The second mechanism is nonsaturable diffusion across membranes with K_d of 0.568 mL $min^{-1}g^{-1}$. The possibility of nonsaturable diffusion has also been deduced from observations on the permeability and partitioning through a model membrane composed of neutral phospholipids and cholesterol, representing the major lipid components of vascular membranes [22]. These in vitro data are consistent with analgesic effects of biphalin that are similar to morphine after systemic (IV) administration [15]. Interestingly, biphalin uptake into the CNS was not significantly altered by concurrent administration of opioid peptides such as DPDPE or Leu-enkephalin, but was significantly reduced by ligands of the large neutral amino acid (LNAA) carrier [23].

Peripheral inflammation, commonly associated with acute and chronic painful conditions such as arthritis or infection, greatly influences the blood-brain barrier permeability to peptides [24,25]. Therefore, it might be predicted that the low BBB permeability of biphalin in healthy normals will increase

TABLE 5 Changes in Biphalin Intravenous Antinociception in Lewis Rats During Development of EAE

Day after immunization	% MPE Time after injection			
	5 min	15 min	30 min	60 min
1	12	18	19	6
7	47	62	48	4
13	68	83	35	21
27	26	38	28	9

during pathological pain associated with inflammation. Indeed, in an experimental model of thermal injury [26] or allergic encephalomyelitis [27], antinociceptive potency of intravenously injected biphalin increased in proportion to the magnitude of systemic inflammation. Behavioral symptoms of EAE are maximal 12 days after immunization. The analgesic potency of intravenously administered biphalin correlates well with the progression of EAE (Table 5, Fig. 4). One infers that systemically administered biphalin, and opioid peptides in general, may be much more effective in gaining access to the

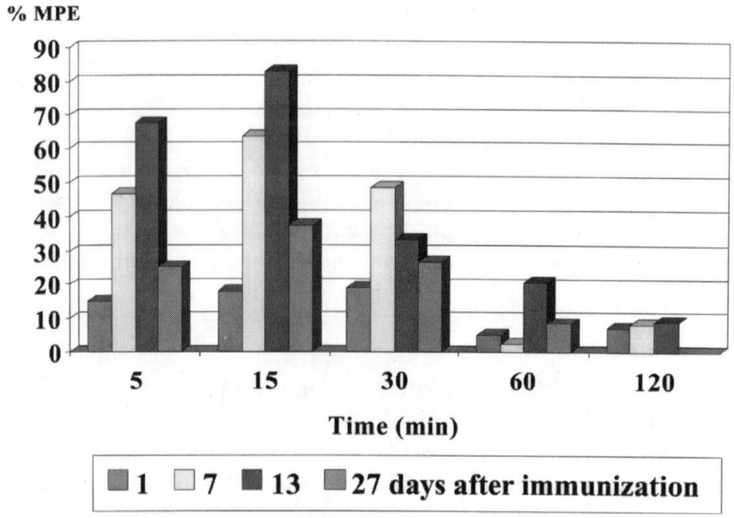

FIGURE 4 Antinociceptive effects of biphalin administered intravenously to rats with encephalomyelitis (EAE).

central nervous system during disease states—when they may be clinically indicated to secure analgesia.

3.3 Intrathecal Analgesia

During the past 20 years, advances in preclinical pain research have led to the recognition of the spinal cord as a key target for inhibition of nociception and preemption of "pain memory" [28]. This preclinical knowledge has strongly influenced progress in techniques of clinical pain treatment. New site-directed techniques of drug application directly to the spinal cord, such as intrathecal infusion pumps with implantable reservoirs, are increasingly used in the treatment of chronic pain [29]. Surprisingly, this knowledge has not evoked new additions in our arsenal of analgesics. This gap is a result of a number of things. The first reason is an obvious delay in the introduction of any new drug into the market, often >10 years. The second reason is more complicated, and relates to early disappointments in the first clinical trials of new analgesics based on endogenous opioid substances. Systemic analgesic properties such as high permeability across membranes and/or enzymatic stability, which are desirable for agents given orally or parenterally, are disadvantages in site-directed techniques. Such analgesics may easily "escape" from the target site and, unchanged or via active metabolites, may produce side effects in distant organs. Opioids with low permeability across membranes, which are rapidly cleared from systemic circulation and the cerebrospinal fluid, would be more suitable for site-directed techniques. Therefore, these modern techniques of anesthesia have created a need for a new type of analgesics in which "natural" peptide properties such as low membrane permeability, low receptor selectivity, and susceptibility to enzymatic degradation become advantages [30].

The preliminary studies of biphalin applied intrathecally in rodents yielded inconsistent results. Intrathecal injection of biphalin through previously implanted catheters in rats showed analgesic potency approximately 100-fold greater than morphine [15]. In contrast, direct injection of biphalin using single needle injections in the mouse showed much less effective antinociception [16]. Recent reevaluation of these data indicated that biphalin and other opioid peptides with low permeability across biological barriers are much more sensitive than alkaloids such as morphine to small errors in the site of drug injection. Subsequent analyses in which catheter location was scrutinized showed that intrathecal biphalin is 500 times more potent than morphine when both are given by this route (Fig. 5) [31].

Biphalin has a broad therapeutic window. Intrathecal morphine has a 50% MPE for tail flick analgesia at a dose of 2.5 μg per rat. A 10-fold higher dose produces rigidity and respiratory depression, causing death. In contrast, biphalin has a 75% MPE of 0.005 μg per rat. Increasing the biphalin dose

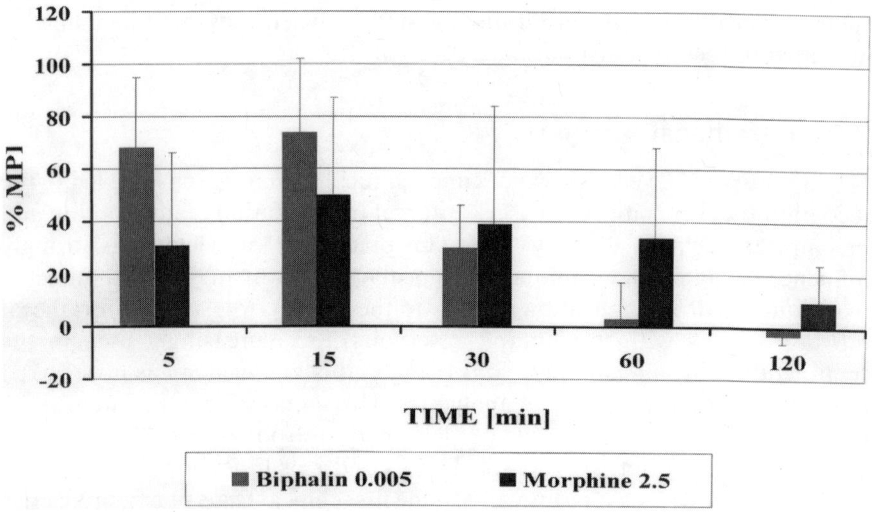

FIGURE 5 Antinociceptive effects of biphalin (0.005 μg) and morphine (2.5 μg) administered intrathecally to rats measured by tail flick test. There is a significantly greater difference (*$P < .05$; t-test) in the antinociceptive effect of morphine as compared to biphalin 120 min after injection.

400 times increased antinociception but without any evident behavioral abnormalities. Increasing the biphalin dose 4000-fold (20 μg/animal) resulted in rigidity but not respiratory depression. Animals given the highest doses completely recovered from rigidity within several hours.

3.4 Interaction of Biphalin with Tachykinin and NMDA Receptors

Insight into spinal cord pathophysiology and pharmacology has spurred novel drug discovery and rekindled interest in spinal delivery of established drugs [32]. Aggressive treatment of serious conditions in current medical practice typically relies on multiple forms of therapy delivered simultaneously. Opioid-NMDA interactions [33], exemplified by coadministration of clinically available NMDA antagonists dextromethorphan or ketamine, are under intense study in many clinical trials currently under way [34]. Ketamine itself, after intrathecal application, has very low antinociceptive activity. Nevertheless, intrathecal administration of the combination of ketamine and biphalin greatly enhances antinociception in thermal tail flick testing in rats (Fig. 6). These results indicate that NMDA antagonists could be highly effective potentiators of biphalin analgesia and that much lower doses of biphalin are needed when coadministered with ketamine. Therefore, a slower rate of tolerance development for the biphalin/ketamine combination might also be expected. Yet to our surprise, rats given an equianalgesic dose of biphalin developed tolerance slower than these exposed to the biphalin/ketamine combination [31].

The tachykinin substance P plays a major role as a nociceptive transmitter in pain processing [35]. Therefore, the search for new analgesics based upon substance P antagonists is an obvious direction. Unfortunately, clinical trials of tachykinin antagonists have produced disappointing results [36]. However, substance P antagonists do produce transient antinociception. Coadministration of a substance P antagonist ([D-Pro^2,D-Trp^{7,9}]SP) with biphalin synergistically increased opioid antinociception [37].

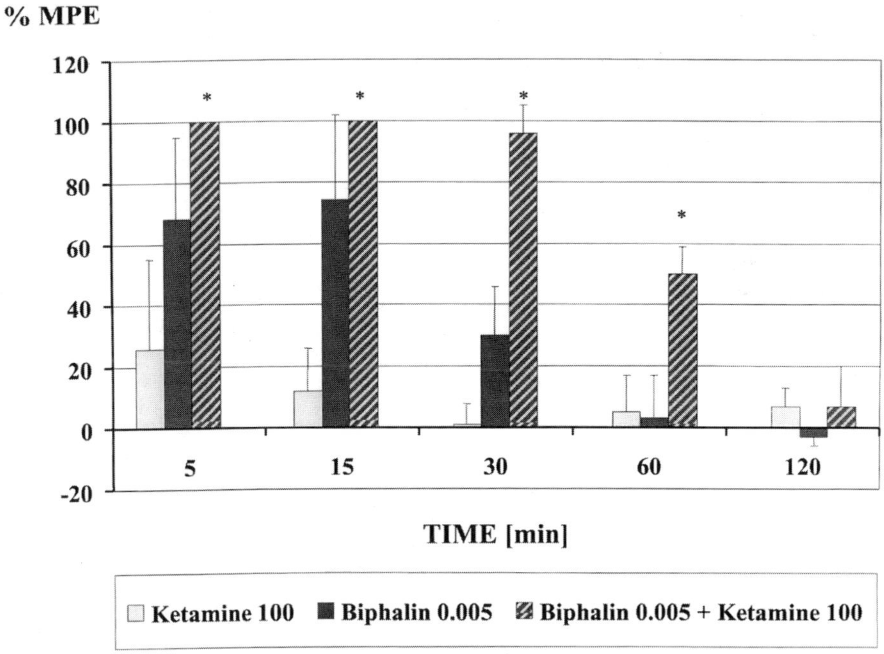

FIGURE 6 Coadministration of ketamine (100 μg) with biphalin (0.005 μg) significantly (*P < .05; t-test) potentiates and prolongs the antinociceptive effects of biphalin as compared to biphalin or ketamine injected alone.

3.5 Adverse Effects

From the therapeutic point of view, the side effects and adverse effects that limit doses of analgesics are critical barriers for prospective application of such compounds as drugs [38,39]. Certain side effects may completely disqualify compounds otherwise effective for analgesia. Other side effects may be desirable in some applications and undesirable in others.

Inhibition of gastrointestinal transit (constipation) is one of the general and typically negative opioid side effects. Supraspinal opioid-mediated inhibition of gastrointestinal propulsion is associated with opioid mu, but not delta or kappa, receptors [40]. Morphine applied intracerebroventricularly is approximately equipotent for both antinociceptive and gastrointestinal endpoints. In contrast, doses of ICV biphalin needed to inhibit gastrointestinal transit were eightfold higher than those to produce antinociception [16]. This discrepancy between analgesic effect of biphalin and a key side effect additionally supports the hypothesis that antinociception is mediated through activation of several specific types of opioid receptors.

Pain treated by an opioid is frequently the result of a pathological condition. Activity of the immune system is a major element which should be considered in therapy of patients with tumors or infections. For almost 20 years it has been known that the all three major types (mu, delta, kappa) of opiate receptors are located on peripheral blood lymphocytes [41,42]. Opioid receptors on white blood cells are involved in stimulation or suppression of immune processes in mammals receiving opioids [43]. Opioid modulatory effects upon the immune system are complicated and only partially understood. The opioid receptor agonist-antagonist classification does not correlate well with immunosuppressive/immunostimulating properties. In the in vivo lymphocyte migration test in rats, intravenous morphine decreased lymphocyte extravasation, but intrathecal morphine administration decreased lymphocyte migration to mesenteric lymph nodes. Intravenous biphalin increased lymphocyte extravasation, but decreased lymphocyte homing to lymph nodes and their release into lymph fluid. Intrathecal administration of biphalin has similar effects on both lymphocyte migration and distribution [44]. In vitro biphalin stimulates T-cell proliferation, natural killer cell cytotoxicity, and interleukin-2 production, and inhibits tumor necrosis factor [45].

Opioids are known to have complex effects on body temperature in animals [46]. Depending on the receptor selectivity of an opioid, but also on several other factors such as dose, route of administration, the gender of the animal, or handling procedures, it may produce either a rise or a fall in body temperature. Biphalin, after peripheral administration in mice, produced a dose-dependent (in the range 0.1–20 mg/kg) hypothermic effect that was naloxone reversible [47].

4 CONCLUSIONS

In the closing decades of the last millennium, several groups synthesized a number of peptide, alkaloid and mixed peptide-alkaloid dimeric opioid ligands [48]. The possibility of simultaneous bridging of two receptors was the main motivation to develop such molecules. In most cases, pharmacological screening of these novel molecules did not reveal spectacular properties, such as great receptor specificity or highly prolonged action. Therefore these novel compounds were not developed for clinical use. Recent evidence of homo- and heterodimerization and multimerization of opioid receptors may produce a renaissance of interest in multivalent opioid ligands. From the beginning, biphalin and its analogues displayed unique properties. However, biphalin is a peptide with a broad spectrum of receptor selectivity. Therefore, for many years the compound was not attractive to pharmacologists interested in developing receptor-selective opioid ligands. Current knowledge of pain indicates that modulation of nociceptive signals is a very complex process in which all opioid receptors participate. This insight provides a rationale for the search for new opioid analgesics that are deliberate multi-target ligands. Unwanted side effects can be minimized through application of modern techniques to deliver drugs to their site of action. The pharmacological properties of biphalin may situate it at the forefront of the next generation of novel analgesic drugs.

REFERENCES

1. Lipkowski AW, Konecka AM, Sroczynska I. Peptides 1982; 3:697–700.
2. Hettiarachchi K, Ridge S, Thomas DW, Olson L, Obi CR, Singh D. J Peptide Res 2001; 57:151–161.
3. Urbanczyk-Lipkowska Z, Krajewski JW, Gluzinski P, Lipkowski AW, Argay G. J Mol Struct 1986; 140:151–157.
4. Flippen-Anderson JL, Deschamps JR, George C, Hruby VJ, Misicka A, Lipkowski AW. J Peptide Res 2002; 59:123–133.
5. Misicka A, Lipkowski AW, Horvath R, Davis P, Porreca F, Yamamura HI, Hruby
VJ. Life Sci 1997; 60:1263–1269.
6. Li G, Haq W, Xiang L, Lou BS, Hughes R, De Leon IA, Davis P, Gillespie TJ, Romanowski M, Zhu X, Misicka A, Lipkowski AW, Porreca F, Davis TP, Yamamura HI, O'Brien DF, Hruby VJ. Bioorg Med Chem Lett 1998; 8:555–560.
7. Lipkowski AW, Misicka A, Davis P, Stropova D, Janders J, Lachwa M, Porreca F, Yamamura HI, Hruby VJ. Bioorg Med Chem Lett 1999; 9:2763–2766.
8. Lipkowski AW, Misicka A, Kosson D, Kosson P, Lachwa-From M, Brodzik-Bienkowska A, Hruby VJ. Life Sci 2002; 70:893–897.

9. Quock RM, Burkey TH, Varga E, Hosohata Y, Hosohata K, Cowell SM, Slate CA, Ehlert FJ, Roeske WR, Yamamura HI. Pharmacol Rev 1999; 51: 503–532.
10. Cheng YC, Prusoff WH. Biochem Pharmacol 1973; 22:3099–3108.
11. Quock RM, Hosohata Y, Knapp RJ, Burkey TH, Hosohata K, Zhang X, Rice KC, Nagase H, Hruby VJ, Porreca F, Roeske WR, Yamamura HI. Eur J Pharmacol 1997; 326:101–104.
12. Lipkowski AW, Konecka AM, Sroczynska I, Przewlocki R, Stala L, Tam SW. Life Sci 1987; 40:2283–2288.
13. Shen KF, Crain SM. Brain Res 1995; 701:158–166.
14. Crain SM, Shen KF. Trends Pharmacol Sci 1990; 11:77–81.
15. Silbert BS, Lipkowski AW, Cepeda MS, Szyfelbein SK, Osgood PF, Carr DB. Agents Actions 1991; 33:382–387.
16. Horan PJ, Mattia A, Bilsky EJ, Weber S, Davis TP, Yamamura HI, Malatynska E, Appleyard SM, Slaninova J, Misicka A. J Pharmacol Exp Ther 1993; 265: 1446–1454.
17. Lipkowski AW, Carr DB. In: Gutte B, eds. Peptides: Synthesis, Structure and Applications. New York: Academic Press, 1995:287–320.
18. Thompson SE, Audus KL. Peptides 1994; 15:109–116.
19. Banks WA, Kastin AJ. Brain Res Bull 1985; 15:287–292.
20. Abbruscato TJ, Williams SA, Misicka A, Lipkowski AW, Hruby VJ, Davis TP. J Pharmacol Exp Ther 1996; 276:1049–1057.
21. Abbruscato TJ, Thomas SA, Hruby VJ, Davis TP. J Neurochem 1997; 69: 1236–1245.
22. Romanowski M, Zhu X, Ramaswami V, Misicka A, Lipkowski AW, Hruby VJ, O'Brien DF. Biochim Biophys Acta 1997; 1329:245–258.
23. Egleton RD, Abbruscato TJ, Thomas SA, Davis TP. J Pharm Sci 1998; 87: 1433–1439.
24. Abbott NJ. Cell Mol Neurobiol 2000; 20:131–147.
25. Huber JD, Witt KA, Hom S, Egleton RD, Mark KS, Davis TP. Am J Physiol Heart Circ Physiol 2001; 280:HI241–HI248.
26. Silbert BS, Lipkowski AW, Cepeda MS, Szyfelbein SK, Osbood PF, Carr DB. Anesth Analg 1991; 73:427–433.
27. Kosson P, Kosson D, Maszczynska Bonney I, Kwiatkowska-Patzer B, Misicka A, Carr DB, Lipkowski AW. Opioid peptide analogues as a potential analgesics in pathophysiological conditions. The increase of analgesic activity of intravenous biphalin in rats with experimental allergic encephalomyelitis (EAE) [Abstract 20]. 33rd International Narcotic Research Conf. Asilomar, July 9–14.
28. Carr DB. JAMA 1998; 279:1114–1115.
29. Carr DB, Cousins MJ. In: Cousins MJ, Bridenbaugh PO, eds. Neural Blockade in Clinical Anesthesia and Management of Pain. Philadelphia: Lippincott-Raven, 1998:915–983.
30. Lipkowski AW, Misicka A, Hruby VJ, Carr DB. Pol J Chem 1994; 68:907–912.
31. Kosson D, Kosson P, Bonney I, Misicka A, Carr DB, Lipkowski AW. Intra-

thecal antinociceptive interaction of biphalin and ketamine. 10th World Congress on Pain, San Diego, Aug. 17–22. Abstract 1479-P27.
32. Walker SM, Goudas LC, Cousins MJ, Carr DB. Anesth Analg 2002; 95:674–715.
33. Mao J. Brain Res Rev 1999; 30:289–304.
34. Choe H, Choi YS, Kim YH, Ko SH, Choi HG, Han YJ, Song HS. Anesth Analg 1997; 84:560–563.
35. Maszczynska I, Lipkowski AW, Carr DB, Kream RM. Analgesia 1998; 3:259–268.
36. Mills SG. Annu Rep Med Chem 1997; 32:51–60.
37. Misterek K, Maszczynska I, Dorociak A, Gumulka SW, Carr DB, Szyfelbein SK, Lipkowski AW. Life Sci 1994; 54:939–944.
38. Edwards JE, McQuay HJ, Moore RA, Collins SL. J Pain Symptom Manage 1999; 18:427–437.
39. McNicol E, Horowicz-Mehler N, Fisk RA, Bennett K, Gialeli-Goudas M, Chew PW, Lau J, Carr DB. J Pain 2003; 4:231–256.
40. Porreca F, Mosberg HI, Hurst R, Hruby VJ, Burks TF. J Pharmacol Exp Ther 1984; 230:341–348.
41. Mehrishi JN, Mills IH. Clin Immunol Immunopathol 1983; 27:240–249.
42. Wybran J, Applelbloom T, Famaey JP, Govaerts A. J Immunol 1979; 123: 1068–1070.
43. Sibinga NE, Goldstein A. Annu Rev Immunol 1988; 6:219–249.
44. Maksymowicz M, Kosson D, Lipkowski AW, Olszewski WL. Transplant Proc 2000; 32:1395–1396.
45. Mehrotra S, Prajapati RK, Haq W, Singh VK. Immunopharmacol Immunotoxicol 2002; 24:83–96.
46. Adler MW, Geller EB, Rosow CE, Cochin J. Annu Rev Pharmacol Toxicol 1988; 28:429–439.
47. Konecka AM, Sroczynska I, Lipkowski AW. Peptides 1987; 8:431–435.
48. Ronsisvalle G, Pappalardo MS, Pasquinucci L, Vittorio F, Salvadori S, Spampinato S, Cavicchini E, Ferri S. Eur J Med Chem 1990; 25:29–33.

16

Binding and Activity of Opioid Ligands at the Cloned Human Delta, Mu, and Kappa Receptors

Kemal Payza
AstraZeneca R&D Montreal, Ville St-Laurent, Quebec, Canada

1 INTRODUCTION

The cloning of human delta [1], mu [2], and kappa [3] opioid receptors has facilitated drug discovery efforts at these therapeutic targets. Delta receptors have been highlighted as potential targets for treatment of chronic pain without producing the adverse effects typical of mu or kappa opioids [4]. The first nonpeptide delta agonist, BW373U86 [5], was a significant advance despite its poor selectivity [6]. Further developments in this field were realized after a related compound, SNC-80, was found to have high delta selectivity [7]. Since then, a significant amount of SAR information has been generated around various series of nonpeptide delta agonists (Chap. 8, this volume) [8–10].

To develop highly potent, selective, and efficacious delta agonists of potential therapeutic value, the Molecular Pharmacology Department at AstraZeneca R&D Montreal has established and implemented binding and functional assays for all three human opioid receptors [11]. These assays were used to provide in vitro pharmacological data to support the design, synthesis, and analysis of newer generations of delta selective compounds. In the literature, many ligands have been characterized at opioid receptors in a wide variety of assays from brain membranes and isolated tissues [12] to cloned receptors [13], but interpretation of the results from different laboratories is often difficult. In binding assays, for example, the choice of radioligand, receptor source, or buffer composition can affect binding potencies. We have used a uniform set of competitive radioligand displacement assays to characterize the binding activity of a range of reference compounds at human opioid receptors. In functional assays, the potency and efficacy of agonists can be affected by the endpoint measured, reflecting the biological amplification in the system (e.g., biochemical response in a cell or tissue vs. agonist-induced GTP binding in membranes). Even in membrane-based functional assays, such factors as receptor expression level and concentration of sodium or GDP can affect the relative potency and efficacy of opioid agonists [14]. We have used GTPγ[^{35}S] binding assays to characterize agonist activity at each opioid receptor, including two assays for delta agonism: one using membranes with high receptor expression, and one with a lower, more physiological level. Thus, this chapter is a summary of results obtained with our assay systems, and is intended to be more in the style of a research report rather than an extensive review of the literature. The results in this chapter were presented previously in GPCRDB, a database of in vitro pharmacological data on GPCRs (for the internet link to the GPCRDB, see [15]).

2 ABBREVIATIONS

AR-M100613: [I]-Dmt-c[-D-Orn-2-Nal-D-Pro-D-Ala-]: ARM390: N,N-Diethyl-4(phenyl-piperidin-4-ylidene-methyl)-benzamide; BNTX: (E)-7-benzylidene-naltrexone; BUBU: Tyr-D-Ser(OtBu)-Gly-Phe-Leu-Thr(OtBu); (+) BW373U86: (+)-4-[(αR)-α-((2S,5R)-4-allyl-2,5-dimethyl-1-piperazinyl)-3-hydroxybenzyl]-N,N-diethylbenzamide; Cha: cyclohexyl-L-Ala; CTOP: D-Phe-Cys-Tyr-D-Trp-Orn-Thr-Pen-Thr-NH_2; DADLE: [D-Ala^2,D-Leu^5]-enkephalin; DAMGO: Tyr-D-Ala-Gly-N-Me-Phe-glycinol; deltorphin II: Tyr-D-Ala-Phe-Glu-Val-Val-Gly-amide; Dmt: (2,6-dimethyl)-Tyr; DIPP-NH_2[Ψ]: Dmt-TicPsi[CH_2NH]Phe-Phe-NH_2; Dmt-TicPsi [CH_2NH]Phe-Phe-NH_2; DSLET: [D-Ser^2]-Leu-enkephalin-Thr; Dynantin: [(2S)-Mdp^1]Dyn A(1-11)-NH_2; FK 33-824: Tyr-D-Ala-Gly-methylPhe-Met(O)-ol; GNTI: 5′-guanidinonaltrindole; HK 08144: (+)-4-[(αR)-α-((2S, 5R)-4-allyl-2,5-dimethyl-1-piperazinyl)-3-iodobenzyl]-N,N-diethylbenzamide; HS 378: (5β-methyl, 14-O-ethyl)-naltrindole; ICI-174,864: [N,N-diallyl-Tyr^1, Aib^{2,3}, Leu^5]enkephalin; n.t.: not tested; SNC 80: (+)-4-[(αR)-α-((2S, 5R)-4-allyl-2,5-dimethyl-1-piperazinyl)-3-methoxybenzyl]-N,N-diethylbenzamide; super-DALDA: Dmt-D-Arg-Phe-Lys-NH_2; Tic: tetrahydroisoquinoline-3-carboxylic acid; TICP[Ψ]: Tyr-TicPsi[CH_2NH]Cha-Phe-OH; TIPP:

Tyr-Tic-Phe-Phe-OH; TIPP[Ψ]: Tyr-TicPsi[CH_2NH]Phe-Phe-OH; (-)U-50, 488: *trans*-3,4-dichloro-N-methyl-N-[2-(1-pyrrolidinyl)-cyclohexyl]-benzeneacetamide; U-69,593: (5α, 7α, 8α)-(-)-N-methyl-N-(7-(1-pyrrolidinyl-1-oxaspiro(4.5)dec-8yl)benzeneacetamide.

3 ASSAYS FOR HUMAN DELTA, MU AND KAPPA RECEPTORS

3.1 Radioligands

The peptides deltorphin II, FK 33-824, and D-Pro^{10}-dynorphin A [1–11] were iodinated with Na[^{125}I] and chloramine T, then purified by reverse-phase high-performance liquid chromatography using a C_{18} analytical column with an acetonitrile/TFA solvent system. The ^{125}I-labeled peptides were purified to apparent homogeneity (2200 Ci/mmole) based on their elution 3–4 min after their noniodinated precursors, and based on their co-elution with their respective nonradioactive monoiodopeptides. Radiolabeled guanosine 5′-(γ-thio) triphosphate (GTPγ[^{35}S], 1250 Ci/mmol) was obtained commercially.

3.2 Biological Test Systems

3.2.1 Cell Lines

Human HEK-293S cells expressing cloned human opioid receptors and geneticin resistance were obtained through stable transfection with hDelta (GenBank #U07882; isoform 27C) [1], hMu (GenBank #L25119, isoform 40N) [2], and hKappa (GenBank #NM_000912) [3] cDNA subcloned in pcDNA3.0 (Invitrogen). The clones, coded 293S-hD11, 293S-hM46, and 293S-hK7, expressed the hDelta (B_{max} = 9.7 pmol/mg), hMu (0.29 pmol/mg), and hKappa (3.4 pmol/mg) receptors, respectively. We also established an additional cell line (293S-hD02) that expressed delta receptors at a lower level (0.63 pmol/mg; see Sec. 3.2.2). All the cell lines were grown in suspension at 37°C and 5% CO_2 in spinner flasks or a bioreactor containing calcium-free DMEM (BIO-Whittaker, Walkersville, MD, U.S.A., or Wisent, St-Bruno, Quebec, Canada), 5% bovine calf serum (Hyclone, Logan, UT, U.S.A., or Wisent, St-Bruno, Quebec, Canada), 0.1% Pluronic F-68, and 600 μg/mL or 300 μg/mL geneticin (Gibco BRL, Burlington, Ontario, Canada; BIO-Whittaker, Walkersville, MD, U.S.A., or Wisent, St-Bruno, Quebec, Canada).

3.2.2 Use of Cell Membranes in Assays of Agonist Activity

The 293S-hD11, 293S-hM46, and 293S-hK7 cells were used to produce membranes for receptor binding and GTPγ[^{35}S] binding assays. Since the delta receptors in 293S-hD11 cells were highly expressed (9.7 pmol/mg) compared to the level in rat brain (0.076 pmol/mg), we also used membranes from the cell line having lower expression (293S-hD02, 0.63 pmol/mg), to assess delta agonist activity of compounds. Membranes from 293S-hD11 and 293S-hD02 were used to create "high-response" and "low-response" GTPγ[^{35}S] binding assays, respectively, as described in Section 3.3.2.

3.2.3 Membrane Preparation

Cells were harvested at 1 million to 1.2 million cells/mL, pelleted, and resuspended in ice-cold lysis buffer (50 mM Tris, pH 7.0, 2.5 mM EDTA, with phenylmethylsulfonyl fluoride added just prior to use to 0.5 mM from a 0.5 M stock in 100% DMSO). After lysis on ice for 15 min, the cells were homogenized with a polytron for 30 sec. The suspension was spun in a centrifuge at 1000g for 10 min at 4°C. The supernatant was saved on ice, and the pellets were resuspended and spun as before. The supernatants from both spins were combined and spun at 46,000g for 30 min or 22,400g for 45 min. The pellets were resuspended in cold Tris buffer (50 mM Tris/Cl, pH 7.0) and spun again. The final pellets were resuspended in membrane buffer (50 mM Tris, 0.32 M sucrose, pH 7.0). Aliquots (0.5 or 1 mL) in polypropylene tubes were frozen in dry ice/ethanol and stored at −80°C until use. The protein concentration was determined using bovine serum albumin (BSA) as standard in a modified Lowry assay with sodium dodecyl sulfate. Before use in receptor binding or functional assays, membranes were thawed at 37°C, cooled to 4°C, passed three times through a 25-gauge blunt-end needle, and diluted into the appropriate buffers as described below.

3.3 Experimental Procedures

3.3.1 Receptor-Binding Assays

Competitive binding experiments were used to measure the binding of a single concentration of radiolabeled ligand in the presence of various concentrations of unlabeled test ligands. The concentration of unlabeled ligand was varied across 4–6 orders of magnitude, and the IC_{50} and K_i values of competing ligands were determined. For each opioid receptor subtype, we used [^{125}I]- rather than [^{3}H]radioligands, in order to obtain the highest specific binding per microgram cell membrane protein. The delta receptor binding assay utilized [^{125}I]-labeled D-Ala^{2}-deltorphin II, which was based on the report of [^{125}I]deltorphin I [16]. We had also tested other radioligands for delta binding: [^{125}I]AR-M100613 [17], [^{125}I]TIPP[Ψ] [18], [^{125}I]HK 08144 [19,20], [^{3}H]DPDPE, [^{3}H]diprenorphine, and [^{3}H]naltrindole. We selected [^{125}I]deltorphin II because of its high signal quality (95% SB/TB) and because it possessed the desired functional activity, i.e., agonism. The mu- and kappa-binding assays used [^{125}I]FK 33-824 [21], and [^{125}I]D-Pro^{10}-dynorphin A [1–11,22], respectively.

In 96-well plates, membranes (2 μg delta, 20 μg mu, or 1 μg kappa) were combined with test compounds and ~0.07 nM of the appropriate radioligand [^{125}I][D-Ala2]-deltorphin (K_d = 0.93 nM at delta), [^{125}I]FK 33-824 (Kd = 1.1 nM at mu), or [^{125}I]D-Pro^{10}-dynorphin A (1-11) (K_d = 0.16 nM at kappa) in a total of 300 μL of 50 mM Tris, 3 mM $MgCl_2$, 1 mg/mL BSA, pH 7.4. Compounds were tested in duplicates across a range spanning 4–6 orders of magnitude. The total and nonspecific binding were determined in the absence and presence of 10 μM naloxone, respectively. The plates were mixed and incubated at room temperature for 60–75 min, after which time the contents were rapidly vacuum filtered through Packard GF/B UniFilter 96 plates presoaked for at least 1 h in 0.1% polyethyleneimine. The filters were subjected to three rapid washes, 1 mL each, of ice-cold wash buffer (50 mM Tris, 3 mM $MgCl_2$, pH 7.0 measured at 22°C). The UniFilter 96 plates were dried in an oven at 55°C for 1 h, then counted in the 2.9-100 keV window in a TopCount (Packard) after adding 65 μL MicroScint-20 (Packard) scintillation fluid per well. The IC_{50} values for ligands displacing specifically bound radioligand were obtained from two-parameter logistic fits. Since none of the test compounds showed partial displacement, the maximum and the minimum binding were set to 100% and 0% specific binding, respectively. The K_i values were determined by correcting the IC_{50} values according to the Cheng-Prusoff equation: $K_i = IC_{50}/(1 + L/K_d)$, where L and K_d are the values of radioligand concentration and its equilibrium dissociation constant, respectively.

3.3.2 GTPγ[^{35}S] Binding Assays

Agonists at $G_{i/o}$-coupled receptors induce GDP/GTP exchange, which is measured by binding of a radioactive GTP analogue to a cell membrane preparation [23,24]. This method has advantages of speed and throughput compared to measurements of adenylate cyclase activity in membranes, or of cAMP accumulation in whole cells. We used this assay to measure the potency (EC_{50}) and maximal effect (E_{max}) of test compounds on GTPγ[^{35}S] binding to membranes expressing cloned human opioid receptors.

The GTPγ[^{35}S] assays for mu, kappa, and delta ("high response," with 293S-hD11 membranes; see Sec. 3.2.2) were performed in 50 mM Hepes, 20 mM NaOH, 100 mM NaCl, 1 mM EDTA, 5 mM $MgCl_2$, 1 mM DTT, 15 μM GDP, and 0.1% BSA. The "low-response" GTPγ[^{35}S] assay for delta (with 293S-hD02) was performed in 50 mM Hepes, 20 mM NaOH, 200 mM NaCl, 1 mM EDTA, 5 mM $MgCl_2$, 1 mM DTT, 3 μM GDP, and 0.5% BSA, pH 7.4. In all cases, the radioligand was used at ~0.15 nM and the total volume was 300 μL per well in a 96-well plate. To define 100% E_{max}, we used SNC 80

(3 μM), DAMGO (30 μM), and U-69,593 (3 μM) as reference agonists at delta, mu, and kappa receptors, respectively. The assay, containing test compounds (10 serial dilutions, threefold apart) was incubated 1 h at room temperature. Filter plates (UniFilter GF/B) were presoaked in deionized water 1 h prior to filtration. Filtration was performed with a Packard cell harvester using 3 × 1 mL ice-cold wash buffer (50 mM Tris, 5 mM $MgCl_2$, 50 mM NaCl, pH 7.0 measured at 22°C). Filter plates were then dried at 55°C for 1.5 h before adding 65 μL Microscint 20 (Packard Biosciences) fluid scintillation. Filter plates were counted in a Packard TopCount, using a ^{35}S-counting window. The effect on GTPγ[^{35}S] binding observed for each concentration of compound was expressed as a percentage of maximal effect elicited by SNC 80 DAMGO and U-69,593 for delta, mu, and kappa receptors, respectively. Dose-response curves were fitted using a three-parameter logistic fit to solve for the EC_{50}, E_{max}, and Hill slope; the E_{min} (GTPγ[^{35}S] binding in absence compound) was set to zero percent.

4 ACTIVITY OF LIGANDS AT HUMAN DELTA, MU, AND KAPPA RECEPTORS

The results of competitive binding studies were used to group compounds according to their selectivity for the three opioid receptor subtypes (Tables 1–4).

4.1 Delta-Selective Ligands

At the cloned human delta receptor, (3′-iodo Tyr^1)-deltorphin II had a potent K_i value of 1.2 nM (Table 1), which was similar to the K_d we determined for its radioactive analog (0.93 nM). Of the delta-selective ligands, TIPP was the most selective antagonist, with selectivity of 38,000- and 21,000-fold for delta over mu and kappa receptors, respectively. TIPP was also among the top three delta antagonists with highest binding affinity, along with naltriben and naltrindole. The selectivity of naltrindole was only ~25-fold for delta over mu. An improved delta/mu selectivity was observed in the N1′(methyl)-derivative of naltrindole, which has been used as a PET ligand for imaging delta receptors in vivo [25,26]. Of the naltrindole analogues, HS 378 (compound 2 in Ref. [27]) had the highest delta selectivity over mu and kappa. The naltrexone derivative BNTX did not show any delta selectivity over mu (see Sec. 4.4). Of all the compounds showing delta agonism in GTPγ[^{35}S] assays, ARM390 (compound 6a in Ref. [9]) showed the highest delta receptor binding selectivity: 4700-fold and 6500-fold over mu and kappa, respectively. The most potent delta agonist of all compounds tested was (+)BW373U86 (Table 1). This may in part be due to the ability of this compound to retain high affinity binding in the presence of high sodium and guanine nucleotides [6,28], which generally shift agonist potencies to low affinity. It is un-

TABLE 1 Activity of Delta-Selective Ligands in Opioid Receptor Binding Assays and Delta GTPγ[^{35}S] Assays

	Binding affinity (K_i, nM)			δ-Selectivity		δ-Agonism (high response)		δ-Agonism (low response)	
Compound	δ	μ	κ	vs. μ	vs. κ	EC_{50}	%E_{max}	EC_{50}	%E_{max}
TIPP	0.13	5,000	2,800	38,000	21,000	> 100,000	0	> 90,000	0
TIPP[Ψ]	0.55	4,000	> 7,000	7,300	> 12,000	0.67	−11	> 30,000	0
ARM390[a]	0.85	4,000	5,500	4,700	6,500	8.1	107	160	93
TICP[Ψ]	0.62	1,900	4,700	3,000	7,500	> 30,000	0	> 30,000	0
Dmt-Tic-OH[b]	1.7	890	2,800	540	1,700	6.4	−19	> 30,000	0
(3′-iodo Tyr[1])-Deltorphin	1.2	550	> 700	470	> 600	n.t.	n.t.	144	44
$N(Me)_2$-Dmt-Tic-NH_2	2.4	890	1400	370	590	2.7	−36	> 3,000	0
SNC 80	1.2	330	2,900	270	2,400	3.0	100	45	100
HK 08144	1.0	200	890	200	880	1.4	102	92	110
ICI-174,864	34	6,300	> 7,000	190	> 200	9.3	−40	> 30,000	0
HS 378[b]	0.30	49	32	170	110	0.54	−13	> 30,000	0
Deltorphin II	1.2	210	> 3,500	170	> 2,800	2.4	74	16	52
(-) TAN 67	0.22	30	380	140	1,700	2.1	86	4.9	36
Deltorphin I	0.84	79	> 3,500	94	> 4,000	9.3	88	26	90
BUBUC	0.35	27	1,900	77	5,400	5.6	84	16	87
DPDPE	0.95	68	> 2,800	72	> 2,900	14	88	92	59
Naltriben	0.11	6.7	20	63	190	> 1,000	0	> 30,000	0
p-CI-DPDPE	0.25	13	> 7,000	55	> 28,000	0.44	76	28	83
Dmt-Tic-NH_2[b]	2.8	130	3,500	46	1,200	> 30,0pt00	0	> 30,000	0
N_1′(methyl)-naltrindole	0.21	9.3	7.5	43	35	n.t.	n.t.	> 30,000	0
BUBU	0.38	14	5,000	36	13,000	3.6	83	39	86
DSLET	0.15	4.5	4,900	29	32,000	0.49	82	27	90
Naltrindole	0.18	4.6	11	25	61	> 30,000	0	> 30,000	0
Tyr-Tic-NH-CH_2-$CH(Ph)_2$	0.27	4.5	1,800	17	6,700	0.52	101	5.3	84
Buprenorphine	0.16	2.1	0.78	14	5.0	4.9	32	> 3,000	0
AR-M100613	0.16	1.5	55	9.1	350	0.82	40	> 30,000	0
(+)BW373U86	0.21	1.9	13	8.9	63	0.060	91	0.90	120
DADLE	0.23	1.9	1,200	8.1	5,200	5.0	90	31	78

[a] Adapted from Ref. 9.
[b] From Ref. 32.

known whether the structural features responsible for this unique property of BW373U86 will aid the development of exceptionally potent agonists in other chemical series. The delta peptides Leu-Enkephalin and Met-Enkephalin showed moderate delta agonist potency but low (five- to sevenfold) selectivity for delta over mu (Table 1); they acted as full agonists at the human mu receptor with EC_{50} values of 300 and 200 nM, respectively. Agonists generally exhibited higher potency and E_{max} in the high-response GTPγ[^{35}S] assay condition (Table 1), due to the higher receptor B_{max} and lower sodium, compared to the low-response condition (see Sec. 3.2.2 and 3.3.2). For instance, the EC_{50} of SNC 80 was shifted 15-fold to the right in the low-response assay. A shift of 10-fold was observed for the potent TIPP-derived delta agonist Tyr-Tic-NH-CH_2-$CH(Ph)_2$ [29]. The partial agonist TAN-67 [30] retained high agonist potency but had 50% reduced E_{max} in the low-response conditions. Buprenorphine was reported as an antagonist or low-efficacy partial agonist at delta receptors in vitro and in vivo [31]; this characterization fits with our observed E_{max} values of 0% and 32% in the low- and high-response assays, respectively (Table 1). Since the basal activity of the delta receptor is lower when the expression level is low, Dmt-Tic-OH, $N(Me)_2$-Dmt-Tic-NH_2, ICI-174,864, and HS 378 [32] acted as neutral antagonists with low delta receptor expression, whereas higher levels of inverse agonism were observed in high expression (Table 1). These observations, in which a receptor ligand can show different levels of agonism depending on the receptor expression, are consistent with theoretical models of receptor agonism [33–35]. To select the appropriate level of receptor expression and assay conditions, the activity of delta agonists in vitro needs to be benchmarked with their activity in vivo for the particular physiological or behavioral endpoint of interest.

4.2 Mu-Selective Ligands

At the cloned human mu receptor, (3′-iodo Tyr^1)-FK 33-824 had a potent K_i value of 0.53 nM (Table 2), similar to the affinity of its radioactive analogue (1.1 nM K_d) and of uniodinated FK 33-824 [36]. Of the mu-selective ligands, CTOP was the most selective antagonist, with 3500- and 3400-fold selectivity for mu over delta and kappa, respectively. A nonpeptide mu antagonist, cyprodime [37], displayed mu binding selectivity of 130-fold over delta, and 3.4-fold over kappa. The most potent mu agonists in the GTPγ[^{35}S] binding assay were etonitazine [38] and super-DALDA [39]. Endomorphin 2 [40] was the most selective agonist for mu over delta, with a binding selectivity of 10,000-fold. Both endomorphins 1 and 2 showed partial mu agonism, consistent with previous reports [41–43]. Compared to literature values, the EC_{50} of these and other compounds tended to be less potent in our mu GTPγ[^{35}S] assay, likely due to our low receptor expression (0.29 pmol/mg) and relatively high GDP concentration (15 μM). Under these conditions, we were able to distinguish partial agonism among many traditional mu opioid compounds (Table 2).

4.3 Kappa-Selective Ligands

The peptide (3′-iodo Tyr^1)-D-Pro^{10}-dynorphin A [1–11] had a potent K_i value of 0.21 nM (Table 3), which was similar to the K_d of its radioactive analog (0.16 nM). Of the kappa-selective ligands, U-50,488 and U-69,593 were

TABLE 2 Activity of Mu-Selective Ligands in Opioid Receptor Binding Assays and Mu GTPγ[35S] assay

	Binding affinity (K_i, nM)			μ-Selectivity		μ-Agonism	
Compound	δ	μ	κ	vs. δ	vs. κ	EC_{50}	$\%E_{max}$
Endomorphin-2	9,100	0.90	2,700	10,000	3,000	330	67
CTOP	5,300	1.5	5,200	3,500	3,400	> 10,000	0
Morphiceptin (D-Pro4)	>9,000	5.3	2,800	>1,600	530	1,000	65
Endomorphin-1	1,100	0.74	3,100	1,500	4,200	340	74
Super-DALDA	200	0.16	4.1	1,200	26	9.2	100
R(-) Methadone	180	0.20	1,500	900	7,500	140	72
Etonitazine	190	0.26	530	740	2,000	3.5	120
Fentanyl	160	0.28	260	570	930	200	64
DAMGO	290	0.51	750	570	1,500	180	100
Morphiceptin	>9,000	22	>1,000	>400	> 45	5,700	31
Oxymorphone	55	0.15	70	370	470	44	53
S(+) Methadone	1,400	4.2	1,200	340	290	1400	48
Normorphine	100	0.36	35	280	98	510	78
(3-iodo Tyr1)-FK 33-824	140	0.53	503	270	960	220	95
Morphine	140	0.53	120	260	220	330	56
FK 33-824	21	0.74	280	28	378	64	90
Hydrocodone bitartrate	480	2.8	760	170	280	1,000	31
Cyprodime	950	7.1	25	134	3.5	>100,000	0
Codeine	>9,000	99	10,500	>90	110	26,000	44
Nalbuphine	82	0.98	9.3	83	9.5	>30,000	0
Meperidine	3,900	63	3,700	62	58	9,400	28
Diphenoxylate HCl	200	3.6	120	55	33	520	59
Levorphanol	4.8	0.12	2.8	39	23	30	45
Morphine-3-β-D-glucuronide	>9,000	290	>7,000	>30	>20	>30,000	0
Dermorphin	43	1.6	1,500	27	940	87	96

the most selective, and both showed full agonism in GTPγ[35S] binding. The dynorphins and their analogues all showed kappa selectivity to varying extents, and they generally showed partial agonism under our assay conditions. The kappa antagonists dynantin [44] and GNTI [45] were ~30-fold selective for kappa over mu, and the latter was over 150-fold selective over delta. The widely used kappa antagonist nor-binaltorphimine [46] had

TABLE 3 Activity of Kappa-Selective Ligands in Opioid Receptor Binding Assays an[d] Kappa GTPγ[^{35}S] Assay

Compound	Binding affinity (K_i, nM)			κ-Selectivity		κ-Agonism	
	δ	μ	κ	vs. δ	vs. μ	EC_{50}	$\%E_{ma[x]}$
(-) U-50,488	2300	130	0.36	6500	360	6.8	108
U-69,593	4500	270	2.0	2300	140	21	100
Dynorphin B	3.6	10	0.14	26	72	0.80	58
Dynorphin A	3.1	0.89	0.017	180	53	0.44	54
Dynantin	116	101	3.0	39	34	n.t.	n.t.
GNTI	20	3.7	0.12	167	31	n.t.	n.t.
(3′-iodo Tyr[1])-D-Pro[10]-Dynorphin A (1–11)	100	8.3	0.21	480	40	18	88
Nor-Binaltorphimine	14	11	0.42	34	25	> 30,000	0
Dynorphin A (1–10)	380	88	3.5	110	25	11	26
Dynorphin A (D-Arg[8] 1–13)	2.5	0.21	0.011	230	20	0.34	65
Dynorphin A amide	1.9	0.53	0.030	63	18	0.22	61
Dynorphin A (1–9)	1.6	0.55	0.037	43	15	1.0	67
D-Pro[10]-Dynorphin A (1–11)	1.9	0.21	0.016	120	13	1.4	64
Dynorphin A (1–13)	2.2	0.29	0.023	93	13	0.24	66
Dynorphin A (1–7)	1.2	1.4	0.15	8.0	9.2	3.2	69
Dynorphin A (1–8)	1.3	0.82	0.12	11	6.9	1.4	66

34- and 25-fold selectivity for kappa over delta and mu, respectively (Table 3).

4.4 Nonselective Ligands

Some ligands showed preferential binding to two of the opioid receptor subtypes, with clearly lower binding to a third; there were also examples of truly nonselective ligands that bound with roughly equal affinity to all three receptor subtypes (Table 4). The dimeric opioid peptide agonist biphalin [47] (see also Chap. 15, this volume), showed preferential binding to mu and delta receptors compared to kappa. Similarly, both Leu-enkephalin-Arg-Phe and Met-enkephalin-Arg-Phe bound with higher affinity to mu and delta than to kappa receptors. The antagonist BNTX has been described as a selective

TABLE 4 Activity of Nonselective Ligands in Opioid Receptor Binding and GTPγ[35S] Assays

Compound	Binding affinity (K_i, nM)			δ-Agonism (high response)		δ-Agonism (low response)		μ-Agonism		κ-Agonism	
	δ	μ	κ	EC_{50}	$\%E_{max}$	EC_{50}	$\%E_{max}$	EC_{50}	$\%E_{max}$	EC_{50}	$\%E_{max}$
β-Endorphin	1.9	4.8	10	37	71	260	60	260	77	n.t.	n.t.
Biphalin	0.50	0.33	27	8.0	85	89	61	46	104	990	69
β-Neoendorphin	0.63	2	0.22	n.t.	n.t.	n.t.	n.t.	n.t.	n.t.	11	68
BNTX	2.5	2.4	24	> 3,000	0	> 30,000	0	> 90,000	0	> 10000	0
Bremazocine	0.94	0.36	0.2	2.6	80	> 30,000	0	> 90,000	0	0.4	92
DIPP-NH_2[Ψ]	0.28	1.5	5.5	> 30,000	0	> 30,000	0	13	19	n.t.	n.t.
Diprenorphine	0.22	0.30	0.26	0.53	59	> 30,000	0	> 10,000	0	0.31	54
DPI 3290	0.36	0.91	1.8	1.0	100	16	118	4.3	126	3.4	98
Leu-Enkephalin-Arg-Phe	1.2	2.1	36	3.9	81	143	59	74	26	75	75
Leu-Enkephalin-Arg-Phe-amide	0.46	0.20	0.37	3.2	85	101	66	180	97	9.5	95
Met-Enkephalin-Arg-Phe	0.69	2.9	23	14	91	77	63	43	94	130	83
Met-Enkephalin-Arg-Phe-amide	0.52	0.37	0.48	13	91	82	62	25	89	4.5	83
Morphine-6-β-D-glucuronide	17	4.0	1600	1,063	97	1605	32	300	49	n.t.	n.t.
Nalorphine	51	0.36	2	152	33	> 30,000	0	> 90,000	0	18	64
Naloxonazine	5.6	0.60	3.9	n.t.	n.t.	> 30,000	0	> 30,000	0	> 10000	0
Naloxone	21	0.66	2.7	> 30,000	0	> 30,000	0	> 30,000	0	8.1	13
Naltrexone	9.2	0.17	0.9	n.t.	n.t.	n.t.	n.t.	> 30,000	0	n.t.	n.t.
Peptide E	6.6	1.4	0.68	14	80	157	37	2.0	90	2.0	76
Tonazocine mesylate	1.5	0.098	0.32	9.4	40	> 30,000	0	> 90,000	0	> 10000	0

$delta_1$ receptor ligand [48] with 27-fold selectivity for mouse delta receptors compared to rat mu receptors [13]. At human opioid receptors, it was reported to have delta K_i values in the range of 1–3 nM [1,49], but it showed no selectivity for delta over mu [49]. Our findings are consistent with the latter report: BNTX bound with equal affinity to human mu and delta receptors, and it had weak binding to kappa (Table 4). The dual mu/delta peptide DIPP-NH_2[Ψ] showed partial mu agonism and delta antagonism (Table 4); it was reported to act as a full mu agonist in guinea pig ileum and to elicit potent antinociception in rats [50]. Nalorphine showed preference for binding to mu and kappa over delta; it had little or no activity at mu and delta, but had significant partial agonism at kappa. Naloxone had little or no agonism at any opioid receptor subtype. Diprenorphine had equal affinity for all three opioid receptors; it showed mu antagonism, partial kappa agonism, and either partial or no agonism at delta depending on the assay conditions (Table 4). The C-terminal extended enkephalins, Leu-enkephalin-Arg-Phe-amide and Met-enkephalin-Arg-Phe-amide, were promiscuous agonists at all three opioid receptors; these peptides also bind to (nonopioid) NPFF receptors in rat spinal cord membranes because of their C-terminal Arg-Phe-NH_2 [51]. Bremazocine was also very nonselective in binding to all opioid receptors. It showed potent kappa agonism and mu antagonism, and at delta it showed partial agonism or no activity in high- or low-response conditions, respectively. We saw a similar pattern of results at delta with tonazocine [52], a compound that was completely devoid of agonist activity at mu and kappa (Table 4). Finally, the nonselective opioid agonist DPI 3290 (Chap. 14, this volume) showed nonselective high-affinity binding, and exceptionally potent and full agonist activity, at all three opioid receptor subtypes (Table 4).

5 CONCLUSION

In this chapter, I have reported the affinity, selectivity, and agonist activity of 90 ligands at cloned human mu, delta, and kappa opioid receptors in vitro, as determined in my laboratory. Our methods for the binding and functional assays were given in detail. As mentioned previously, we preferred ^{125}I-labeled radioligands to the more traditional [^{3}H]radioligands because of the higher signal quality and lower amount of membrane protein required per assay point. The results from the GTPγ[^{35}S] assays illustrated that differences in delta agonist activity can be observed for the same ligand under different assay conditions and receptor levels. The binding and functional assays described herein can be used, in 96-well format, to develop potent opioid ligands with agonist or antagonist activity at one or more receptor subtype. Based on the results from these binding and functional assays, the following conclusions can be drawn. The most selective agonist and antagonist at the human delta receptor were ARM390 and TIPP, respectively; ICI-174864 was a delta-selective inverse agonist. The most potent delta agonist was (+)BW373U86; however, it had poor selectivity over mu. Of the mu-selective compounds, the most selective agonist and antagonist were endormorphin-2 and CTOP, respectively. Of the kappa ligands tested, the most selective agonist was U-50488, and the only selective antagonist was nor-binaltorphimine. The non-selective agonist DPI 3290 showed potent, full agonism at all three opioid receptor subtypes.

ACKNOWLEDGMENTS

The following scientists from AstraZeneca R&D Montreal were involved in generating the pharmacological data presented in this chapter: Martin Coupal, Stephane St-Onge, Maryse Labarre, Claude Godbout, Dominic Salois, Mylène Gosselin, Lejla Hodzic, Joanne Butterworth, and Lynda Adam. The following scientists were involved in producing biological materials or chemical compounds used in the experiments: Zhong-Yong Wei, Ralf Schmidt, Mélanie Duchesne, Manon Valiquette, and Huy K. Vu. We gratefully acknowledge the gifts of TAN-67 from G. Massimo Dondio, HK 08144 from Kenner C. Rice, DMT-Tic compounds from Lawrence H. Lazarus, and various compounds from Peter W. Schiller (super-DALDA, TIPP and DIPP analogues; GNTI, and dynantin).

REFERENCES

1. Knapp RJ, Malatynska E, Fang L, Li X, Babin E, Nguyen M, Santoro G, Vargal EV, Hruby VJ, Roeske WR. Life Sci 1994; 54:L463–L469.
2. Wang JB, Johnson PS, Persico AM, Hawkins AL, Griffin CA, Uhl GR. FEBS Lett 1994; 338:217–222.
3. Mansson E, Bare L, Yang D. Biochem Biophys Res Commun 1994; 202:1431–1437.
4. Rapaka RS, Porreca F. Pharm Res 1991; 8:1–8.
5. Chang KJ, Rigdon GC, Howard JL, McNutt RW. J Pharmacol Exp Ther 1993; 267:852–857.
6. Wild KD, Fang L, McNutt RW, Chang KJ, Toth G, Borsodi A, Yamamura HI, Porreca F. Eur J Pharmacol 1993; 246:289–292.
7. Calderon SN, Rothman RB, Porreca F, Flippen-Anderson JL, McNutt RW, Xu H, Smith LE, Bilsky EJ, Davis P, Rice KC. J Med Chem 1994; 37:2125–2128.
8. Knapp RJ, Santoro G, De Leon IA, Lee KB, Edsall SA, Waite S, Malatynska E, Varga E, Calderon SN, Rice KC, Rothman RB, Porreca F, Roeske WR, Yamamura HI. J Pharmacol Exp Ther 1996; 277:1284–1291.
9. Wei ZY, Brown W, Takasaki B, Plobeck N, Delorme D, Zhou F, Yang H, Jones

P, Gawell L, Gagnon H, Schmidt R, Yue SY, Walpole C, Payza K, St-Onge S, Labarre M, Godbout C, Jakob A, Butterworth J, Kamassah A, Morin PE, Projean D, Ducharme J, Roberts E. J Med Chem 2000; 43:3895–3905.
10. Plobeck N, Delorme D, Wei ZY, Yang H, Zhou F, Schwarz P, Gawell L, Gagnon H, Pelcman B, Schmidt R, Yue SY, Walpole C, Brown W, Zhou E, Labarre M, Payza K, St-Onge S, Kamassah A, Morin PE, Projean D, Ducharme J, Roberts E. J Med Chem 2000; 43:3878–3894.
11. Payza K, St-Onge S, LaBarre M, Roberts E, Fraser G, Valiquette M, Vu HK, Calderon SN, Rice KC, Walker P, Wahlestedt C. IASP Abstracts 8th World Congress on Pain, Vancouver, 1996:46.
12. Corbett AD, Peaterson SJ, Kosterlitz HW. Selectivity of ligands for opioid receptors. In: Herz A, ed. Opioids I: Handbook of Experimental Pharmacology. Vol. 104. Berlin: Springer-Verlag, 1993:645–679.
13. Raynor K, Kong H, Chen Y, Yasuda K, Yu L, Bell GI, Reisine T. Mol Pharmacol 1994; 45:330–334.
14. Selley DE, Cao CC, Liu Q, Childers SR. Br J Pharmacol 2000; 130:987–996.
15. Horn F, Weare J, Beukers MW, Hörsch S, Bairoch A, Chen W, Edvardsen Ø, Campagne F, Vriend G. Nucleic Acids Res 1998; 26:277–281.
16. Dupin S, Tafani JA, Mazarguil H, Zajac JM. Peptides 1991; 12:825–830.
17. Fraser GL, Labarre M, Godbout C, Butterworth J, Clarke PB, Payza K, Schmidt R. Peptides 1999; 20:1327–1335.
18. Collier TL, Schiller PW, Waterhouse RN. Nucl Med Biol 2001; 28:375–381.
19. Pou C, Lee KS, Kayakiri H, Rice K, Payza K. Soc Neurosci Abstr 1997; 27:396.
20. Calderon SN, Rice KC, Rothman RB, Porreca F, Flippen-Anderson JL, Kayakiri H, Xu H, Becketts K, Smith LE, Bilsky EJ, Davis P, Horvath R. J Med Chem 1997; 40:695–704.
21. Moyse E, Pasquini F, Quirion R, Beaudet A. Peptides 1986; 7:351–355.
22. Gairin JE, Jomary C, Pradayrol L, Cros J, Meunier JC. Biochem Biophys Res Commun 1986; 134:1142–1150.
23. Traynor JR, Nahorski SR. Mol Pharmacol 1995; 47:848–854.
24. Lorenzen A, Fuss M, Vogt H, Schwabe U. Mol Pharmacol 1993; 44:115–123.
25. Smith JS, Zubieta JK, Price JC, Flesher JE, Madar I, Lever JR, Kinter CM, Dannals RF, Frost JJ. J Cereb Blood Flow Metab 1999; 19:956–966.
26. Madar I, Lever JR, Kinter CM, Scheffel U, Ravert HT, Musachio JL, Mathews WB, Dannals RF, Frost JJ. Synapse 1996; 24:19–28.
27. Spetea M, Harris HE, Berzetei-Gurske IP, Klareskog L, Schmidhammer H. Life Sci 2001; 69:1775–1782.
28. Childers SR, Fleming LM, Selley DE, McNutt RW, Chang KJ. Mol Pharmacol 1993; 44:827–834.
29. Schiller PW, Weltrowska G, Berezowska I, Nguyen TMD, Wilkes BC, Lemieux C, Chung NN. Biopolymers 1999; 51:411–425.
30. Knapp RJ, Landsman R, Waite S, Malatynska E, Varga E, Haq W, Hruby VJ, Roeske WR, Nagase H, Yamamura HI. Eur J Pharmacol 1995; 291:129–134.
31. Negus SS, Bidlack JM, Mello NK, Furness MS, Rice KC, Brandt MR. Behav Pharmacol 2002; 13:557–570.

32. Labarre M, Butterworth J, St-Onge S, Payza K, Schmidhammer H, Salvadori S, Balboni G, Guerrini R, Bryant SD, Lazarus LH. Eur J Pharmacol 2000; 406: R1–R3.
33. Kenakin T. Pharmacol Rev 1996; 48:413–463.
34. Leff P, Scaramellini C, Law C, McKechnie K. Trends Pharmacol Sci 1997; 18:355–362.
35. Kenakin T. Trends Pharmacol Sci 1999; 20:400–405.
36. Stacher G, Bauer P, Steinringer H, Schreiber E, Schmierer G. Pain 1979; 7:159–172.
37. Marki A, Monory K, Otvos F, Toth G, Krassnig R, Schmidhammer H, Traynor JR, Roques BP, Maldonado R, Borsodi A. Eur J Pharmacol 1999; 383:209–214.
38. Yu Y, Zhang L, Yin X, Sun H, Uhl GR, Wang JB. J Biol Chem 1997; 272: 28869–28874.
39. Neilan CL, Nguyen TM, Schiller PW, Pasternak GW. Eur J Pharmacol 2001; 419:15–23.
40. Zadina JE, Hackler L, Ge LJ, Kastin AJ. Nature 1997; 386:499–502.
41. Narita M, Mizoguchi H, Oji GS, Tseng EL, Suganuma C, Nagase H, Tseng LF. Eur J Pharmacol 1998; 351:383–387.
42. Alt A, Mansour A, Akil H, Medzihradsky F, Traynor JR, Woods JH. J Pharmacol Exp Ther 1998; 286:282–288.
43. Hosohata K, Burkey TH, Alfaro-Lopez J, Varga E, Hruby VJ, Roeske WR, Yamamura HI. Eur J Pharmacol 1998; 346:111–114.
44. Lu Y, Nguyen TM, Weltrowska G, Berezowska I, Lemieux C, Chung NN, Schiller PW. J Med Chem 2001; 44:3048–3053.
45. Jones RM, Portoghese PS. Eur J Pharmacol 2000; 396:49–52.
46. Portoghese PS, Lipkowski AW, Takemori AE. Life Sci 1987; 40:1287–1292.
47. Horan PJ, Mattia A, Bilsky EJ, Weber S, Davis TP, Yamamura HI, Malatynska E, Appleyard SM, Slaninova J, Misicka A. J Pharmacol Exp Ther 1993; 265: 1446–1454.
48. Portoghese PS, Sultana M, Nagase H, Takemori AE. Eur J Pharmacol 1992; 218:195–196.
49. Parkhill AL, Bidlack JM. Eur J Pharmacol 2002; 451:257–264.
50. Schiller PW, Fundytus ME, Merovitz L, Weltrowska G, Nguyen TM, Lemieux C, Chung NN, Coderre TJ. J of Med Chem 1999; 42:3520–3526.
51. Payza K, Akar CA, Yang HY. J Pharmacol Exp Ther 1993; 267:88–94.
52. Hudzik TJ, Howell A, Payza K, Cross AJ. Eur J Pharmacol 2000; 396:101–107.

17

Inhibitors of Enkephalin-Inactivating Enzymes and Delta Opioid Responses

Bernard P. Roques and Florence Noble
UFR des Sciences Pharmaceutique et Biologiques, Paris, France

1 INTRODUCTION

The endogenous opioid system includes four distinct neuronal pathways that are widely distributed throughout the central nervous system (CNS). The endogenous opioid peptides are mainly derived from four precursors—proenkephalin, proopiomelanocortin, prodynorphin, and pronociceptin/orphanin FQ—and exert their physiological actions by interacting with various classes of opioid receptor types—mu, delta, kappa, and Noc/ORL_1—present on both pre- and postsynaptic membranes of opioid and opioid-target neurons. The most important opioid peptides seems to be the pentapeptides enkephalins [1], which interact with both mu and delta receptors, their affinities being significantly better for the latter. This has triggered intensive studies to elucidate the role of enkephalins in the brain and to develop putative novel effective treatments mainly in the field of analgesia and CNS disorders [2]. Moreover, the presence of peripheral enkephalin-controlled peripheral systems such as intestinal motility and fluid secretion, or heart rhythm, constitute interesting clinical targets. The most simple and efficient means to investigate the respective physiological roles of mu and delta receptors is to analyze the responses generated by an enhancement in the levels of synaptic enkephalins and their prevention by selective antagonists of both kinds of binding sites. The specific delta responses observed by this approach based on the use of enkephalin-inactivating enzymes are summarized in this paper.

2 ENDOGENOUS ENKEPHALINS

[Met]- and [Leu]enkephalin, Tyr-Gly-Gly-Phe-Met(Leu) have high affinities for delta receptors, 10-fold weaker for mu receptors and a negligible affinity for kappa receptors. The DOR-1 gene is the only delta receptor gene cloned to date [3,4]. However, a subdivision into $delta_1$ and $delta_2$ subtypes was proposed primarily on the basis of in vivo pharmacological studies [5,6]. In vitro experiments have shown that enkephalins are the endogenous effectors

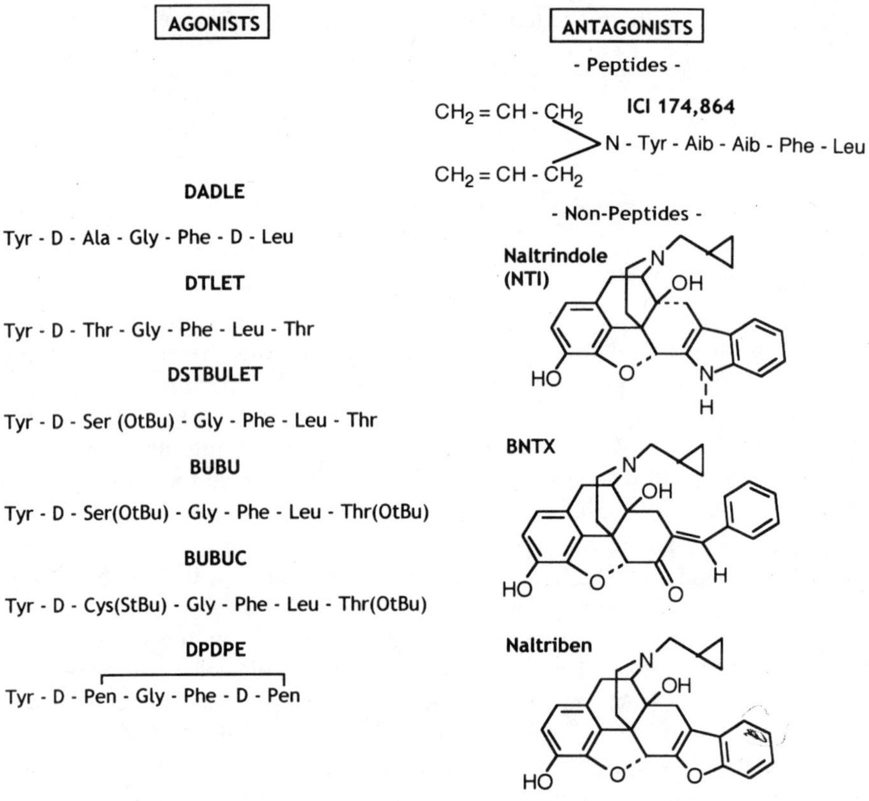

FIGURE 1 Delta opioid receptor agonists and antagonists.

of $delta_1$ and $delta_2$ opioid receptor subtypes. Thus, inhibition of the adenylyl cyclase activity induced by these peptides in mouse caudate-putamen was reduced by the $delta_2$-selective antagonist (naltriben) or the $delta_1$-selective antagonist (BNTX) alone (Fig. 1), but completely blocked by the association of both antagonists or by the mixed antagonist naltrindole [7]. In addition numerous selective agonists active by ICV or systemic routes (Fig. 1) have been developed to analyze the role of the delta receptors.

3 PEPTIDASE-DEPENDENT INACTIVATION OF ENKEPHALINS: DESIGN OF DUAL INHIBITORS OF NEP/APN

Early studies on the enkephalins showed that they have a short half-life in both in vivo and in vitro preparation. A weak and transient analgesia is obtained only for high doses (i.e., 100 μg per mouse) of intracerebroventricularly (ICV) administered Met-enkephalin or Leu-enkephalin [8]. In contrast to catecholamines and amino acid transmitters, which are essentially cleared from the extracellular space by reuptake mechanisms, neuropeptides appear to be rapidly inactivated by peptidases cleaving the biologically active peptides into inactive fragments. This process was clearly demonstrated for the enkephalins, which are degraded by the concomitant action of two peptidases: neutral endopeptidase (NEP), and aminopeptidase N (APN). The former enzyme cleaves the Gly^3-Phe^4 bond of the pentapeptides, and APN the Tyr^1-Gly^2 bond [review in 9].

This well-admitted interruption by the two peptidases of the messages conveyed by enkephalins is in good agreement with the demonstration of a colocalization of NEP and APN in brain areas where opioid peptides and receptors are present [10–12]. This mechanism of enkephalins inactivation was used to develop a new physiological approach to enhance the actions of the peptides by inhibiting their enzymatic catabolism. As discussed above, the concomitant use of mu and delta antagonists allowed the responses given by the enkephalins protected from their degradation to be attributed to one or both opioid receptors.

Several classes of potent and selective inhibitors of NEP and APN have been rationally designed. The specificity of NEP is essentially ensured by the S_1' subsite, which interacts preferentially with aromatic or large hydrophobic moieties, whereas the S_2' subsite has a poor specificity [13]. These observations were used to design thiorphan [14] and retrothiorphan [15], which were the first described potent synthetic NEP inhibitors. Protection of the thiol and carboxyl groups of thiorphan led to acetorphan, a compound able to cross the blood-brain barrier (BBB) after systemic administration. Various natural APN inhibitors have been isolated, some of them exhibiting relatively good affinity ($\sim 10^{-8}$ M) for the enzyme [16,17]. However, these compounds present a poor selectivity versus other aminopeptidases. An improved selectivity for APN was reached with simple β-aminothiols endowed with K_is in the 10^{-8} M range such as PC18 [18]. However, the first highly potent and selective APN inhibitor which belongs to the class of α-aminophosphonic compounds, RB 129 (2(S)-benzyl-3[hydroxyl(l′(R)-aminoethyl)phosphinyl]-propanoyl-L-3-iodotyrosine), was recently designed [19]. It has nanomolar inhibitory potency toward APN and is highly selective versus other metallopeptidases of the same family.

Owing to the complementary roles of NEP and APN in enkephalin inactivation, selective inhibitor of only one of the two peptidases gives weak antinoceptive effects even after ICV administration. This led us to propose the concept of mixed inhibitors, that is, compounds able to simultaneously block NEP and APN activities [reviews in 9,20]. This was possible owing to the fact that these two membrane-bound enzymes belong to the superfamily of zinc metallopeptidases.

The early dual inhibitors were built by using the hydroxamate group as zinc-chelating moiety, hypothesizing that the strength of its coordination to the metal should counterbalance a less than perfect fit of the inhibitor side chains to the active sites of the two enzymes [9,13]. This was obtained with compounds such as kelatorphan or RB 38A designed by incorporating side chains able to recognize the S_1'-S_2' subsites of NEP and APN in short pseudodipeptide structures (Fig. 2). These compounds were shown to completely block enkephalin metabolism in vivo without changes in the secretion of peptides [21]. Using this new concept, a large number of bidentate inhibitors with nanomolar affinities both for NEP and for APN have been synthesized. However, their high water solubility prevents them from crossing the BBB. New lipophilic, systemically active prodrugs such as RB 101 (Fig. 2) were therefore developed [22]. In these compounds, highly potent and hydrophobic thiol-containing APN and NEP inhibitors were linked by a disulfide bond. One of the main advantages of these mixed inhibitors is the relative stability of the disulfide bond in plasma, contrasting with its rapid biologically dependent clivage in brain, generating APN and NEP inhibitors endowed with potencies in the nanomolar range toward their respective target enzymes [22].

Very recently, the first true dual inhibitors belonging to the family of aminophosphinic compounds such as RB 3005 (Fig. 2) were designed [19, 23,24]. These compounds recognize with nanomolar affinities the two peptidases and were shown under prodrugs (RB 3007) to possess a longer duration of action than RB 101 [24,25].

These dual inhibitors, keltorphan, RB 101, or RB 3007, completely inhibit NEP and APN in vivo [26] and increase the extracellular concentra-

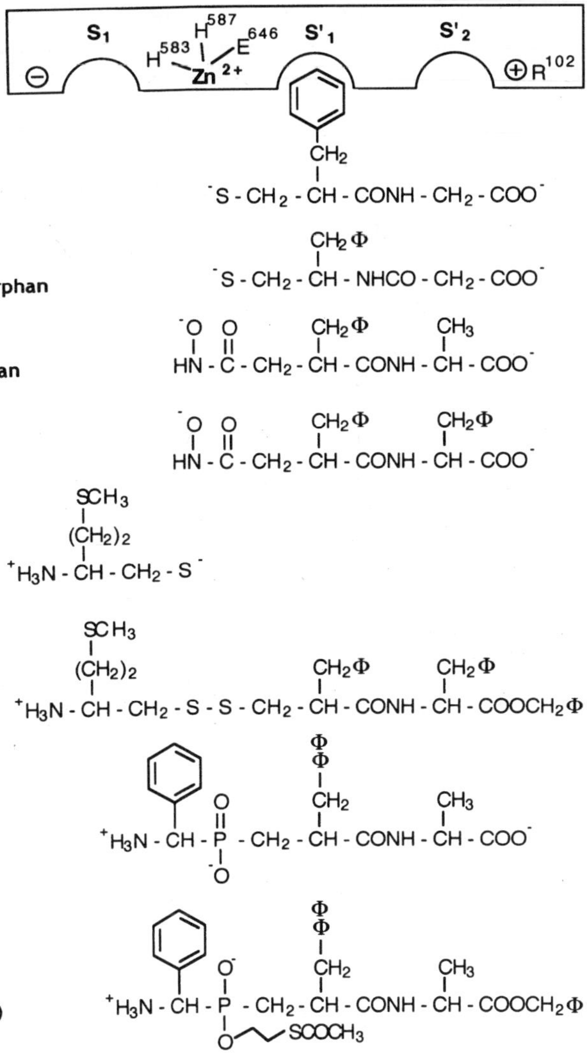

Compound	K_I (nM) NEP	K_I (nM) APN
Thiorphan	4	140
Retrothiorphan	2.3	>10,000
Kelatorphan	1.8	380
RB 38A	0.9	120
PC18	>10,000	8
RB 101 (prodrug)	>10,000	>10,000
RB 3005	2.0	4.8
RB 3007 (prodrug)	>10,000	>10,000

FIGURE 2 Mixed inhibitors of NEP/APN.

tions of Met-enkephalin in brains of freely moving rats, in structures involved in the control of pain (periaqueductal gray matter and spinal cord) [21,25], or in motivated and reward processes (nucleus accumbens) [27]. Consistent with the physiological role of these peptidases to control the pain stimuli, invalidation of the gene encoding NEP led to an increased threshold in regard to a thermal nociceptive stimulus [28].

4 PHYSIOLOGICAL ANTINOCICEPTIVE EFFECTS OF DELTA VERSUS MU RECEPTORS USING DUAL NEP/APN INHIBITORS

The enkephalins are found in high concentration in the spinal cord, especially in the substantia gelatinosa, a region also enriched in mu and delta opioid receptors and in NEP and APN [10,12,29–31].

The antinociceptive properties of kelatorphan, locally infused onto the spinal cord, were inhibited by the selective delta opioid antagonist ICI 174,864 [32] and are additive with those of the mu-selective agonist DAMGO, but not with those of the selective delta-agonist DSTBULET [33], which confirms that endogenous enkephalins, protected from their enzymatic degradation, and delta-selective agonists act on a common binding site to produce spinal antinociception. Kelatorphan and DSTBULET have also been tested on the response of convergent dorsal horn neurons to a more prolonged chemical noxious stimulus elicited by an SC injection of 5% formalin. Both compounds completely abolished C fiber–evoked responses, and these effect were selectively antagonized by ICI 174,864 [34]. These results could be of interest in the case of morphine tolerance, which is probably one of the main interest of delta agonists in analgesia. However, the antinociceptive responses obtained with such compounds remain largely inferior to those given by mu agonists. This has been clearly demonstrated by using mice with deletion of the mu or delta opioid receptor [2,35,36].

Complete inhibition of enkephalin metabolism, by coadministration of selective APN and NEP inhibitors or by dual inhibitors, induced antinociceptive responses in all the various assays commonly used to select analgesics. Thus, it has been shown that endogenous enkephalins completely protected from metabolizing enzymes after central or peripheral administration are able to elicit pain suppressive effects not only in tests in which naloxone produces pronociceptive effects, but more generally in morphine-sensitive assays. In all of the tests used, the pain-alleviating effects of dual inhibitors were suppressed by the opioid antagonist naloxone, but not by the delta-selective antagonist naltrindole, except in the tail flick (Fig. 3) and the motor response

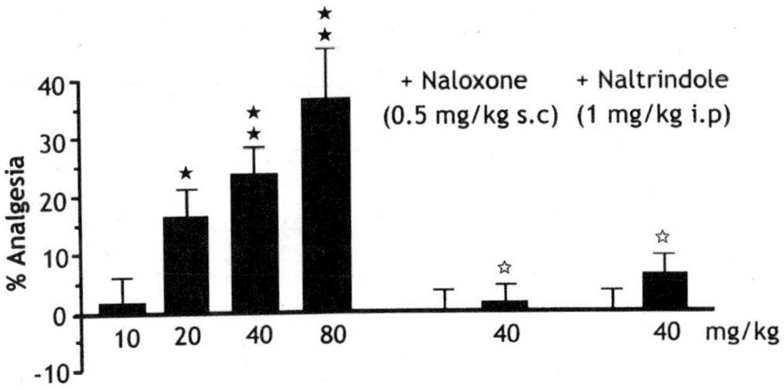

FIGURE 3 Antinociceptive response observed after IV administration of RB 101 and antagonism by naloxone and naltrindole in the rat tail flick test. The results are expressed as percent analgesia ± SEM (n = 8) for each group. ★$P < .05$; ★★$P < .01$ as compared to control; ☆$P < .05$ as compared to the same dose without antagonist (Newman-Keuls test).

to tail electric stimulation. All these results support a preferential involvement of mu receptors in supraspinal analgesia at least regarding thermal nociceptive stimuli, and delta receptors, at least partly, in spinal analgesia [review in 37]. In this case a possible modulation of the analgesic effects through mu/delta receptor interaction is suggested, as discussed below.

The cloning and sequencing of the opioid receptors demonstrated that these binding sites correspond to independent and structurally different entities, with several mu opioid receptor splice variants (MorlA-1F), which have predicted amino acid sequences identical to mu receptor through the transmembrane regions, differing only at the intracellular carboxyl tail [38]. However, numerous studies support the occurrence of interactions between mu (type or isotypes) and delta opioid receptors in a variety of pharmacological tests [review in 39]. Thus, several authors have shown that the antinociceptive effects of mu agonists could be potentiated by delta opioid agonists [40–44]. Moreover, observations in mu receptor knockout mice show that delta opioid receptor–mediated analgesia and respiratory control were reduced and abolished, respectively [35,45].

The noxiously evoked expression of c-Fos in the substantia gelatinosa is reduced by analgesics [46]. In the spinal cord, a possible functional interaction between mu and delta receptors in the analgesic responses of RB 101 has been recently suggested using an immunohistochemical quantification of c-Fos. Thus, both selective mu and delta antagonists were able to block the reducing effects of systemic RB 101 in carrageenan-induced c-Fos protein expression [47]. This effects observed in inflammation-related pain did not occur for all nociceptive stimuli. For instance, NTI has been shown to be inefficient on RB 101 antinociceptive effects on hot plate or writhing tests [26].

5 ROLE OF DELTA OPIOID RECEPTORS IN THE BEHAVIORAL EFFECTS OF DUAL INHIBITORS—INTERACTION WITH THE DOPAMINERGIC SYSTEM

It has been suggested that the endogenous enkephalins might be involved in the etiology of depression [48]. Accordingly, the behavioral responses triggered by forced swimming, conditioned suppression of motility, and learned helplessness tests were attenuated by treatment with kelatorphan or RB 101 or with antidepressant drugs, which suggests a potential role of endogenous enkephalins in depressive syndromes [review in 2,49]. These tests have demonstrated that the inhibitors could modulate the functioning of the mesocorticolimbic and nigrostriatal dopaminergic systems, which are implicated in mood control and connected with enkephalin pathways. In line with this finding, an increase in the levels of endogenous enkephalins by RB 101 induced antidepressant-like effects, which were suppressed by both the delta opioid antagonist naltrindole and the dopamine D_1 antagonist SCH-23390 [50].

Several pharmacological studies have been performed to clarify the role of opioids in dopaminergic systems. The nucleus accumbens (N Acc) plays a key role in reward and coping processes. This structure contains all the components of the enkephalinergic systems (enkephalins, mu and delta receptors, NEP and APN) and contains the terminals of dopamine neurons issuing from the ventral tegmental area (VTA). Local administration of delta agonists in N Acc produces an increase in locomotor activity reversed by naloxone, but not by D_2 antagonists [51]. A similar effect was observed with kelatorphan, supporting a phasic control of locomotor activity by endogenous enkephalins through stimulation of delta receptors as illustrated by the lack of responses in presence of ICI-174,864 [52]. Moreover, deafferentation of the mesolimbic pathway by 6-OHDA, or neuroleptics acting as antagonists of D_2 receptors such as sulpiride, facilitates the responses induced by enkephalins, which was maximum 3 weeks after the administration of sulpiride (Fig. 4). At this time, a large increase in the levels of preproenkephalin and enkephalins without changes in NEP were reported [53 and Refs. cited herein]. A similar increase of $\sim 40\%$ of preproenkephalin expression was observed in mice with invalidation of the dopamine D_2 receptor [54]. Taken together, these results suggest that alterations in the opioidergic systems, very likely through its interrelationships with the dopaminergic pathways, could

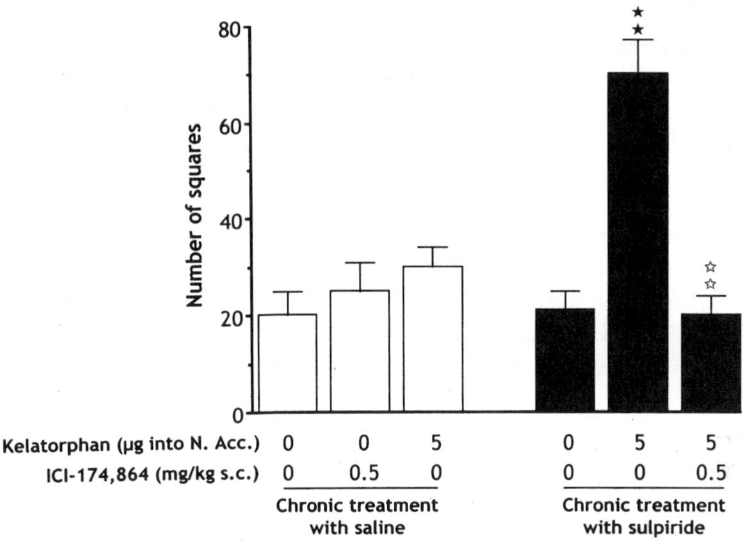

FIGURE 4 Effects of chronic treatment with sulpiride (100 mg/kg, IP, once daily for 3 weeks) on the behavioral activation induced by administration of kelatorphan, and prevention by the delta opioid antagonist ICI 174,864. Number of squares crossed in 6 min in the open field. ★★$P < .01$ vs. control rats; ☆☆$P < .01$ vs. kelatorphan-treated rats.

be taking place in the neuronal system, critically involved in the control of mood [48,55].

In line with this hypothesis, the mechanism of dopamine control in the mesolimbic pathway by delta receptor recruited by exogenous or endogenous opioids was investigated. An increase in dopamine metabolism in the rat N Acc and a potentiation of the behavioral effect of dopamine injected into this region were observed following injection of enkephalin-degrading enzyme inhibitor in the VTA [56], suggesting a phasic control of the dopamine mesolimbic pathway by endogenous enkephalins, through activation of delta opioid receptor, as these effects were suppressed by the delta selective antagonist ICI 174,864. This was confirmed using in vivo microdialysis with simultaneous measurement of motor activity. Under this condition, micro-injection of the dual inhibitor kelatorphan into the VTA produced a dose-related increase in both motor activity and extracellular dopamine levels in the nucleus accumbens, and these effects were blocked by ICI 174,864 [57]. However, as expected from the limited increase of dopamine produced by the protected enkephalins as compared to exogenous morphine, the responses were lower than those induced by the alkaloid.

The specific involvement of delta opioid receptors to induce hyperactivity following protection of enkephalin catabolism has been confirmed after systemic RB 101 administration [50]. As expected, in anxiogenic situations evoked by forced swim and conditioned suppression of motility tests, mice with deletion of the gene encoding the delta opioid receptor displayed behaviors that contrast with those previously observed following administration of delta agonists or RB 101. Interestingly, in the same behavioral tests, mu opioid receptor knockout mice showed responses that were roughly opposite to those found with delta opioid receptor knockout, i.e., antidepressant-like effects which were blocked by naltrindole [55]. These results suggest that, in wild-type animals tonic or phasic stimulation of the mu opioid receptor by the endogenous enkephalins reduce the proposed positive control of mood ensured by delta opioid receptor activation [48]. Therefore, stimulation of delta receptors as this occurs with dual inhibitors could have interesting clinical applications in depressive syndromes.

6 GASTROINTESTINAL EFFECTS OF ENKEPHALIN DEGRADING ENZYME INHIBITORS: ROLE OF DELTA OPIOID RECEPTOR IN THE ANTISECRETORY EFFECT OF TIORFAN

The well-known antidiarrheal effect of opioids has been therapeutically exploited for many years. Diarrhea involves both an increase in the motility of the gastrointestinal tract and a decrease in the absorption of fluid and thus a loss of electrolytes (particularly sodium) and water.

Pharmacological studies with selective agonists have shown that opioid control of intestinal electrolyte transport is predominantly mediated by delta opioid receptors [58], while the gastrointestinal propulsion is under the control of mu receptors [59,60]. The antidiarrheal effects of NEP inhibitors, such as acetorphan, the prodrug of thiorphan, have been compared to those of an opiate agonist, loperamide, in a model of castor oil–induced diarrhea in rats. When administered peripherally, they produced a delayed onset of diarrhea with no reduction in the gastrointestinal transit [61,62], as is commonly observed with loperamide [63].

This difference of action may be due to the fact that endogenous enkephalins are tonically released from the submucosal plexus neurons which contains mainly delta opioid receptors, but sparsely from the myenteric plexus neurons where mu receptors are present [64–66].

Accounting for all these results, the selective NEP inhibitor prodrug acetorphan, (TIORFAN) was introduced on the market as an antisecretory agent. It could be noticed that no significant drawback was reported even after chronic treatment with this compound. This seems to indicate that, as hypothesized, the limited stimulation of opioid receptors by protected endogenous enkephalins strongly reduces the side effects issuing from ubiquitous recruitment of receptors by exogenous opiates such as morphine [9].

7 REDUCTION OR SUPPRESSION OF OPIATE-INDUCED SIDE EFFECTS BY ENDOGENOUS STIMULATION OF DELTA RECEPTORS

Mice with a deletion of the gene encoding the mu receptor have shown that both the physical and psychic dependencies to morphine are essentially due to the hyperstimulation of this receptor. This is consistent with previous studies assessed with delta-selective agonists or with dual inhibitors such as RB38A, which were unable to trigger significant dependence syndromes (Fig. 5) as well as tolerance [67,68]. In the case of dual inhibitors, this indicates that the levels of enkephalins produced by protection of NEP and APN are never enough to overstimulate the mu receptors localized in structures such as the locus coeruleus or the limbic pathways, these two structures being involved in physical and psychic dependence to morphine, respectively. The same phenomenon occurs at the levels of the nucleus tractus solitarius, where mu and delta agonists are involved in the control of respiratory volume and rhythm, respectively [69,70]. Thus, systemic injection or local administration in the fourth ventricle of dual inhibitors in doses range that induce potent analgesia do not depress ventilation in awake, anesthetized, or arthritic animals, indicating that there is a low tonic endogenous opioid release in this brain region [71].

On the other hand, a number of studies have shown that morphine and opioid peptides could have cardioprotective effects toward ischemic processes and may be able to reduce the size of infarct [72,73]. These effects seem to involve the activation of delta opioid receptors, the localization of which remains unknown. In addition, enkephalin-degrading enzyme inhibitors, such as acetorphan and, particularly, RB 101, have also been demonstrated to decrease the susceptibility to the arrhythmogenic action of epinephrine. Thus RB 101 completely prevented the ventricular tachycardia, fibrillation, and repetitive ventricular extrasystoles induced by epinephrine (Maslov et al., unpublished data). These effects were reversed by the selective delta antagonist ICI 174,864.

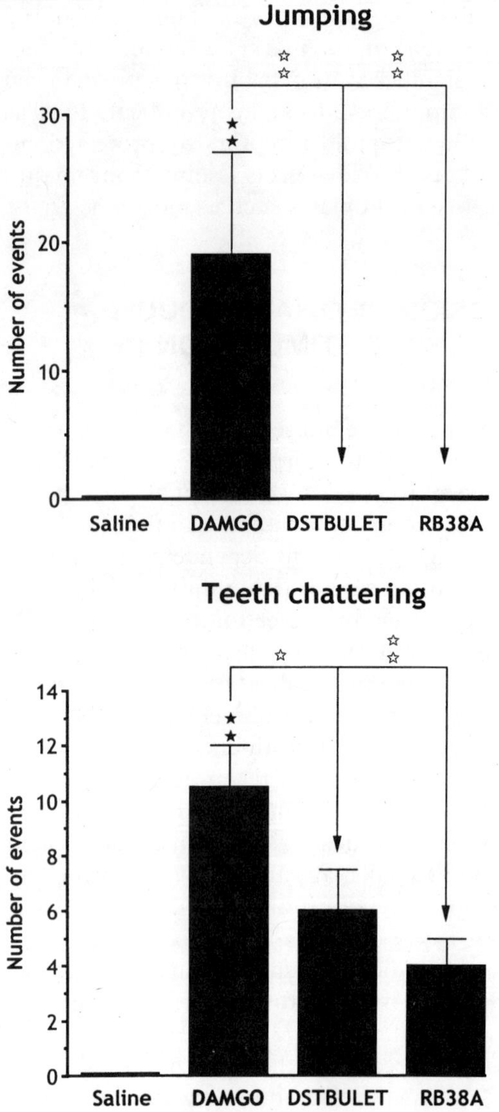

FIGURE 5 Naloxone-precipitated withdrawal signs (jumping and teeth chattering) in rats chronically treated with saline, DAMGO (0.18 μg/μL/h), DSTBULET (66.5 μg/μL/h) or RB 38A (40 μg/μL/h) for 6 days. ★★$P < .01$ vs. control rats; ☆$P < .05$, ☆☆$P < .01$ vs. DAMGO-treated rats.

The NEP and APN levels are moderate on heart [74,75] while the concentration of both peptidases is higher in vascular endothelium or vagus nerve terminals [76–78]. However, the mechanisms and site of action (central or peripheral) involved in the cardioprotective effects of the endogenous opioid peptides remain unknown. Nevertheless, owing to their lack of narcotic effects, inhibition of endogenous enkephalin catabolism and subsequent stimulation of delta receptor could have interesting clinical applications in the cardiovascular domain.

8 INTERACTION BETWEEN ENDOGENOUS ENKEPHALINS AND CHOLECYSTOKININ SYSTEMS: RESPECTIVE ROLES OF MU AND DELTA OPIOID RECEPTORS

A large body of evidence has now been accumulated supporting physiological interactions between cholecystokinin (CCK) and enkephalins [review in 79]. There is an overlapping distribution of the neuropeptides CCK_8 and endogenous opioid peptides and their respective receptors in numerous brain regions and spinal cord. In the case of CCK receptors, this is the CCK_2 receptor, which is the most abundant in the CNS. Regulatory loops between enkephalin and CCK systems have been proposed (Fig. 6) [80].

Biochemical studies have shown in vitro and in vivo differential inhibitory/stimulatory modulation of spinal and supraspinal CCK release by mu and delta opioid agonists. Thus, delta opioid agonists enhance the release of CCK, whereas stimulation of mu opioid receptors reduces its release [81,82]. Moreover, it has been shown that activation of CCK_1 receptors potentiates the analgesic responses induced by mu opioid agonists or by endogenous enkephalins, protected from their catabolism by the dual inhibitor RB 101, while activation of CCK_2 receptors reduces them [80].

Intravenous injection of RB 101 or the delta-selective agonist BUBUC produced a dose-dependent inhibition of the in vivo binding of [^{3}H]pBC 264 in brain, very likely by increasing the extracellular levels of endogenous CCK [82], through a preferential activation of delta opioid receptors. This was confirmed by using the selective delta opioid antagonist naltrindole, which eliminates the inhibition process.

As enkephalins are able to activate both $delta_1$ and $delta_2$ receptors, their respective roles in CCK release have also been investigated. As previously mentioned, the analgesic responses obtained in the hot plate test and resulting from the protection of endogenous enkephalins are predominantly associated with mu receptor stimulation [26,83]. Nevertheless, it has been shown that the antinociceptive effects observed in this test in mice following

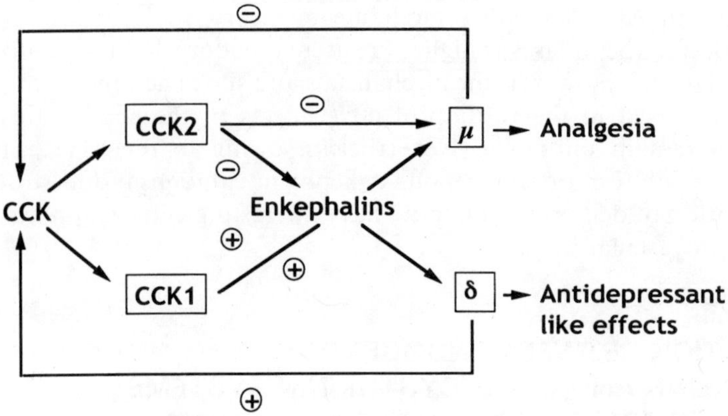

FIGURE 6 Hypothetical model of the interactions between CCK, via CCK_1 and CCK_2 receptors, and the opioid system via delta opioid and mu opioid receptors. CCK agonists, endogenous and/or exogenous, stimulate the CCK_2 and/or the CCK_1 receptors, which can modulate the opioidergic systems either directly (via binding of opioid agonists or via C fiber–evoked activity) or indirectly (via the release of endogenous enkephalins). In addition, activation of mu opioid receptors could negatively modulate the release of endogenous CCK, whereas delta opioid receptors may enhance it.

IV administration of RB 101 were selectively antagonized by prior administration of the $delta_2$ opioid antagonist naltriben, while they were selectively potentiated by the $delta_1$ opioid antagonist BNTX. This latter effect could be related to an enhancement in endogenous CCK release. Thus, the facilitating effect of mu-mediated antinociception by blockade of $delta_1$ opioid receptor (BNTX) was completely reversed by prior administration of the selective CCK_1 antagonist L-364,718. This suggests that, in the presence of BNTX, endogenous enkephalins, through $delta_2$ opioid receptor activation, increase endogenous CCK release, resulting in potentiation of mu opioid analgesia [7].

In contrast, blockade of $delta_2$ opioid receptor by naltriben reduced RB 101–induced antinociceptive response. Thus, it could be argued that activation of $delta_1$ opioid receptors by endogenous enkephalins reduced endogenous CCK release, resulting in a decrease in CCK_1 receptor activation with subsequent reduction of the facilitatory effect of CCK system on opioid system, as CCK_1 activation seems to be essential for the potentiation of mu opioid analgesia [42,80].

CCK_1 and CCK_2 receptors play opposing roles on opioid mediated behavioral and antinociceptive responses. However, when CCK_1 and CCK_2

antagonists were administered alone, they were found to be inactive in antinociceptive studies, in contrast to the behavioral effects induced by the CCK_2 antagonist L-365,260, which is able to induce antidepressant-like effects, suggesting a physiological tonic control of endogenous CCK on CCK_2 receptors on behavior, but not on antinociception. Interestingly, these effects were blocked by the delta opioid antagonist naltrindole, like those induced by the dual inhibitor of enkephalin-degrading enzyme inhibitor RB 101 [50,84].

These results support the proposed existence of a reciprocal modulation between endogenous CCK and enkephalin systems, with in particular an increased release of CCK peptide following activation of delta opioid receptors, as previously suggested [82,85,86].

9 CONCLUSION

The physiological roles of delta opioid receptors have been investigated with dual inhibitors of NEP and APN that increase the circulating levels of the endogenous ligands, enkephalins. Met- and Leu-enkephalins have a better affinity for the delta receptor than for the mu receptor, and Leu-enkephalin is often considered as the true endogenous agonist of the delta site. The highly potent and selective delta antagonists and the use of mice with specific invalidation of mu or delta receptors have provided with dual NEP/APN inhibitors, important tools to clarify the role of each binding site.

Regarding analgesia, the delta receptors may be recruited by enkephalins to slightly reduce the nociception generated by thermal or inflammatory stimuli. The strenght of responses induced by endogenously or exogenously stimulated delta receptors appears potentiated by activation of mu receptors. This could occur through mu-delta complexes hypothesized to trigger stronger intracellular messages resulting in a largest reduction of neural transmission of pain. Therefore, one cannot eliminate a putative clinical interest of delta agonists for the treatment of pain in particular in morphine-tolerant patients.

Moreover, as clearly shown with dual NEP/APN inhibitors, the main interest of delta receptors seems to be their critical role in the control of mood mainly at the mesoaccumbens level. Thus, local injection of selective delta agonists (DSTBULET, BUBU, or DTLET) or kelatorphan in the VTA induced similar naltrindole-reversible hyperactivity in a familiar (actimeter), unfamiliar (four-hole box), and a fear-inducing (open-field) environment. In contrast, in the same conditions, DAMGO produced the reverse effects assumed to result from an enhanced emotionality evidenced by the decrease in number of visits and time spent in open arms of the elevated plus maze [56]. This seems to indicate that different neuronal pathways are recruited by selective stimulation of mu or delta receptors in the rat VTA. Accordingly, enkephalin immunoreactive fibers establish contacts with dopamine and nondopamine neurons [87], mu receptor being localized on nondopaminergic neurons, whereas delta receptors could be present on axon terminals in the VTA [88].

Enkephalins could be implicated in dopamine mesoaccumbens-dependent appetitive motivation and behavioral positive reinforcement by separated mechanisms involving both mu and delta receptors. Activation of the former in heroin abusers produces sedation and overstimulation, lack of vigilance, and possibly fear. These effects cannot be reached with dual inhibitors, and one can speculate that this is related to an increased level of endogenous enkephalins leading to a preferential interaction with delta receptors for both binding and anatomical reasons.

Defects in the enkephalinergic system may be involved in the pathogenesis of various mental diseases, making delta agonists and enkephalin degrading enzyme inhibitors of interest as putative new antidepressants [48]. Such a possible critical role of neuropeptides in behavioral adaptations is reinforced by the recent clinical demonstration of the very promising antidepressant effects resulting from the inhibition of substance P binding to NK_1 receptors [89].

REFERENCES

1. Hughes J, Smith TW, Kosterlitz HW, Fotherghill LA, Morgan BA, Morris HR. Nature 1975; 258:577–579.
2. Roques BP. Trends Pharmacol Sci 2000; 21:475–483.
3. Kieffer BL, Befort K, Gaveriaux-Ruff C, Hirth CG. Proc Natl Acad Sci USA 1992; 89:12048–12053.
4. Evans CJ, Keith DE, Morrison H, Magendzo K, Edwards RH. Science 1992; 258:1952–1955.
5. Jiang Q, Takemori AE, Sultana M, Portoghese PS, Bowen WD, Mosberg HI, Porreca F. J Pharmacol Exp Ther 1991; 257:1069–1075.
6. Sofuoglu M, Portoghese PS, Takemori AE. J Pharmacol Exp Ther 1991; 257: 676–680.
7. Noble F, Fournié-Zaluski MC, Roques BP. Neuroscience 1996; 75:917–926.
8. Belluzi JD, Grant N, Garsky V, Sarantakis D, Wise CD, Stein D. Nature (Lond) 1976; 260:625–626.
9. Roques BP, Noble F, Daugé V, Fournié-Zaluski MC, Beaumont A. Pharmacol Rev 1993; 45:87–146.
10. Waksman G, Hamel E, Fournié-Zaluski MC, Roques BP. Proc Natl Acad Sci USA 1986; 83:1523–1527.
11. Mansour A, Watson SJ. In: Herz A, Akil H, Simon AJ, eds. Handbook of Experimental Pharmacology Opioids I. Berlin: Springer-Verlag, 1993:79–105.

12. Noble F, Banisadr G, Jardinaud F, Popovici T, Lai-Kuen R, Chen H, Bischoff L, Melik Parsadaniantz S, Fournié-Zaluski MC, Roques BP. Neuroscience 2001; 105:479–488.
13. Fournié-Zaluski MC, Perdrisot R, Gacel G, Swerts JP, Roques BP, Schwartz JC. Biochem Biophys Res Commun 1979; 91:130–135.
14. Roques BP, Fournié-Zaluski MC, Soroca E, Lecomte JM, Malfroy B, Llorens C, Schwartz JC. Nature 1980; 288:286–288.
15. Roques BP, Lucas-Soroca E, Chaillet P, Costentin J, Fournié-Zaluski MC. Proc Natl Acad Sci USA 1983; 80:3178–3182.
16. Umezawa H, Aoyagi T, Tanaka T, Suda H, Okuyama A, Naganawa H, Hamada M, Takeuchi T. J Antibiot (Tokyo) 1985; 38:1629–1630.
17. Rich DH, Moon BJ, Harbeson S. J Med Chem 1984; 27:417–422.
18. Fournie-Zaluski MC, Coric P, Turcaud S, Bruetschy L, Lucas E, Noble F, Roques BP. J Med Chem 1992; 35:1259–1266.
19. Chen H, Roques BP, Fournié-Zaluski MC. Bioorg Med Chem Lett 1999; 9:1511–1516.
20. Fournié-Zaluski MC, Chaillet P, Bouboutou R, Coulaud A, Chérot P, Waksman G, Costentin J, Roques BP. Eur J Pharmacol 1984; 102:525–528.
21. Bourgoin S, Le Bars D, Artaud F, Clot AM, Bouboutou R, Fournié-Zaluski MC, Roques BP, Hamon M, Cesselin F. J Pharmacol Exp Ther 1986; 238:360–366.
22. Fournié-Zaluski MC, Coric P, Turcaud S, Lucas E, Noble F, Maldonado R, Roques BP. J Med Chem 1992; 35:2474–2481.
23. Chen H, Noble F, Coric P, Fournié-Zaluski MC, Roques BP. Proc Natl Acad Sci USA 1998; 95:12028–12033.
24. Chen H, Noble F, Roques BP, Fournie-Zaluski MC. J Med Chem 2001; 44:3523–3530.
25. Le Guen S, Mas Nieto M, Canestrelli C, Chen H, Fournié-Zaluski MC, Cupo A, Maldonado R, Roques BP, Noble F. Pain. In press.
26. Noble F, Soleilhac JM, Soroca-Lucas E, Turcaud S, Fournié-Zaluski MC, Roques BP. J Pharmacol Exp Ther 1992; 261:181–190.
27. Daugé V, Mauborgne A, Cesselin F, Fournié-Zaluski MC, Roques BP. J Neurochem 1996; 67:1301–1308.
28. Saria A, Hauser KF, Traurig HH, Turbek CS, Hersh L, Gerard C. Neurosci Lett 1997; 234:27–30.
29. Zajac JM, Charnay Y, Soleilhac JM, Sales N, Roques BP. FEBS Lett 1987; 216:118–122.
30. Sales N, Charnay Y, Zajac JM, Dubois PM, Roques BP. J Chem Neuroanat 1989; 2:179–188.
31. Besse D, Lombard MC, Zajac JM, Roques BP. Brain Res 1990; 521:15–22.
32. Dickenson AH, Sullivan A, Feeney C, Fournié-Zaluski MC, Roques BP. Neurosci Lett 1986; 72:179–182.
33. Dickenson AH, Sullivan AF, Roques BP. Eur J Pharmacol 1988; 148:437–439.
34. Sullivan AF, Dickenson AH, Roques BP. Br J Pharmacol 1989; 98:1039–1049.

35. Matthes HWD, Smadja C, Valverde O, Vonesch JL, Foutz AS, Boudinot E, Denavit-Saubié M, Severini C, Negri L, Roques BP, Maldonado R, Kieffer BL. J Neurosci 1998; 18:7285–7295.
36. Kieffer BL. Trends Pharmacol Sci 1999; 20:19–26.
37. Roques BP, Noble F, Fournie-Zaluski MC. In: Stein C, ed. Opioids in Pain Control: Basic and Clinical Aspects. New York: Cambridge University Press, 1999:21–45.
38. Pan YX, Xu J, Bolan E, Abbadie C, Chang A, Zuckerman A, Rossi G, Pasternak GW. Mol Pharmacol 1999; 56:396–403.
39. Traynor JR, Elliott J. Trends Pharmacol Sci 1993; 14:84–86.
40. Lee NM, Leybin L, Chang JK, Loh HH. Eur J Pharmacol 1980; 68:181–185.
41. Schmidt BL, Tambeli CH, Levine JD, Gear RW. Eur J Neurosci 2002; 15:861–868.
42. Noble F, Smadja C, Roques BP. J Pharmacol Exp Ther 1994; 271:1127–1134.
43. Porreca F, Jiang Q, Tallarida RJ. Eur J Pharmacol 1990; 179:463–468.
44. Porreca F, Takemori AE, Sultana M, Portoghese PS, Bowen WD. J Pharmacol Exp Ther 1992; 263:147–152.
45. Morin-Surun MP, Boudinot E, Dubois C, Matthes HW, Kieffer BL, Denavit-Saubie M, Champagnat J, Foutz AS. Eur J Neurosci 2001; 13:1703–1710.
46. Hunt SP, Pini A, Evan G. Nature 1987; 328:632–634.
47. Le Guen S, Catheline G, Fournié-Zaluski MC, Roques BP, Besson JM, Buritova J. Brain Res 2003; 967:106–112.
48. Roques BP, Daugé V, Gacel G, Fournié-Zaluski MC. Biol Psychiatry 1985; 7: 287–289.
49. Noble F, Roques BP. In: Maldonado R, ed. Molecular Biology of Drug Addiction. Totowa, NJ: Humana Press, 2002:61–75.
50. Baamonde A, Daugé V, Ruiz-Gayo M, Fulga IG, Turcaud S, Fournié-Zaluski MC, Roques BP. Eur J Pharmacol 1992; 216:157–166.
51. Pert A, Sivit C. Nature 1977; 265:645–647.
52. Dauge V, Rossignol P, Roques BP. Psychopharmacology (Berl) 1988; 96:343–352.
53. Maldonado R, Dauge V, Feger J, Roques BP. Neuropharmacology 1990; 29: 215–223.
54. Baik JH, Picetti R, Saiardi A, Thiriet G, Dierich A, Depaulis A, Le Meur M, Borrelli E. Nature 1995; 377:424–428.
55. Filliol D, Ghozland S, Chluba J, Martin M, Matthes HW, Simonin F, Befort K, Gaveriaux-Ruff C, Dierich A, LeMeur M, Valverde O, Maldonado R, Kieffer BL. Nat Genet 2000; 25:195–200.
56. Calenco-Choukroun G, Daugé V, Gacel G, Féger J, Roques BP. Psychopharmacology 1991; 103:493–502.
57. Dauge V, Kalivas PW, Duffy T, Roques BP. Brain Res 1992; 599:209–214.
58. Sheldon RJ, Riviere PJ, Malarchik ME, Moseberg HI, Burks TF, Porreca F. J Pharmacol Exp Ther 1990; 253:144–151.

59. Sbacchi M, La Regina A, Petrillo P, Tavani A. NIDA Res Monogr 1986; 75:520–523.
60. Heyman JS, Williams CL, Burks TF, Mosberg HI, Porreca F. J Pharmacol Exp Ther 1988; 245:238–243.
61. Roge J, Baumer P, Berard H, Schwartz JC, Lecomte JM. Scand J Gastroenterol 1993; 28:352–354.
62. Salazar-Lindo E, Santisteban-Ponce J, Chea-Woo E, Gutierrez M. N Engl J Med 2000; 343:463–467.
63. Kachel G, Ruppin H, Hagel J, Barina W, Meinhardt M, Domschke W. Gastroenterology 1986; 90:85–93.
64. Binder HJ, Laurenson JP, Dobbins JW. Am J Physiol 1984; 247:G432–G436.
65. Surprenant A, North RA. Neuroscience 1985; 16:425–430.
66. Mihara S, North RA. Br J Pharmacol 1986; 88:315–322.
67. Maldonado R, Féger J, Fournié-Zaluski MC, Roques BP. Brain Res 1990; 520:247–254.
68. Meucci E, Delay-Goyet P, Roques BP, Zajac JM. Eur J Pharmacol 1989; 171:167–178.
69. Sales N, Riche D, Roques BP, Denavit-Saubie M. Brain Res 1985; 344:382–386.
70. Shook JE, Watkins WD, Camporesi EM. Am Rev Respir Dis 1990; 142:895–909.
71. Boudinot E, Morin-Surun MP, Foutz AS, Fournié-Zaluski MC, Roques BP, Denavit-Saubié M. Pain 2001; 90:7–13.
72. Schultz JJ, Hsu AK, Gross GJ. J Mol Cell Cardiol 1997; 29:2187–2195.
73. Schultz JE, Gross GJ. Pharmacol Ther 2001; 89:123–137.
74. Dutriez I, Sales N, Fournie-Zaluski MC, Roques BP. Experientia 1992; 48:290–300.
75. Jardinaud F, Banisadr G, Chen H, Noble F, Melik Parsadaniantz S, Rostene W, Roques BP, Popovici T. Eur J Biochem. Submitted.
76. Soleilhac JM, Lucas E, Beaumont A, Turcaud S, Michel JB, Ficheux D, Fournie-Zaluski MC, Roques BP. Mol Pharmacol 1992; 41:609–614.
77. Llorens-Cortes C, Huang H, Vicart P, Gasc JM, Paulin D, Corvol P. J Biol Chem 1992; 267:14012–14018.
78. Ladic L, Buchan A. Neurosci Lett 1997; 222:41–44.
79. Roques BP, Noble F. Neurochem Res 1996; 21:1397–4010.
80. Noble F, Derrien M, Roques BP. Br J Pharmacol 1993; 109:1064–1070.
81. Benoliel JJ, Bourgoin S, Mauborgne A, Legrand JC, Hamon M, Cesselin F. Neurosci Lett 1991; 124:204–207.
82. Ruiz-Gayo M, Durieux C, Fournié-Zaluski MC, Roques BP. J Neurochem 1992; 59:1805–1811.
83. Chaillet P, Coulaud A, Zajac JM, Fournié-Zaluski MC, Costentin J, Roques BP. Eur J Pharmacol 1984; 101:83–90.
84. Derrien M, Durieux M, Roques BP. Br J Pharmacol 1994; 111:956–960.
85. Benoliel JJ, Collin E, Mauborgne A, Bourgoin S, Legrand JC, Hamon M, Cesselin F. Brain Res 1994; 653:81–91.

86. Becker C, Hamon M, Cesselin F, Benoliel JJ. Synapse 1999; 34:47–54.
87. Sesack SR, Pickel VM. J Neurosci 1992; 12:1335–1350.
88. Dilts RP, Kalivas PW. Brain Res 1989; 488:311–327.
89. Kramer MS, Cutler N, Feighner J, Shrivastava R, Carman J, Sramek JJ, Reines SA, Liu G, Snavely D, Wyatt-Knowles E, Hale JJ, Mills SG, MacCoss M, Swain CJ, Harrison T, Hill RG, Hefti F, Scolnick EM, Cascieri MA, Chicchi GG, Sadowski S, Williams AR, Hewson L, Smith D, Rupniak NM. Science 1998; 281:1640–1645.

18

The Delta Opioid Receptor Subtypes and Pain Modulation

Michael H. Ossipov, Josephine Lai, Todd W. Vanderah, and Frank Porreca
University of Arizona, Tucson, Arizona, U.S.A.

1 INTRODUCTION

The existence of endogenous receptors for opioids in 1973 [1–3] ignited the vigorous search for endogenous substances that act at these receptor sites. Furthermore, the existence of these receptors generated great interest in the exploration of endogenous pain mechanisms and modulation by inhibitory systems, as well as into biological mechanisms of opiate dependence and withdrawal. Recognition of the strong likelihood of multiple subtypes of opioid receptors came from the work of Martin [4]. It was found that vastly different behavioral syndromes were elicited in spinalized dogs by morphine, ketocyclazocine, and SKF-10,047 leading to the nomenclature of mu, kappa, and sigma opiate receptors, respectively [4,5]. The observation of differential binding affinities for [Met^5]enkephalin and β endorphin against [^{3}H][Leu^5] enkephalin and [^{3}H]naloxone in guinea pig brain and differential agonistic activity profiles of [Leu^5]enkephalin and [Met^5]enkephalin in the isolated mouse vas deferens and guinea pig ileum strongly supported the existence of the δ-opioid receptor [6–8]. Over time, the existence of additional opioid receptor subtypes has been hypothesized to include the epsilon receptor [9] and possibly others. The actual characterization of these additional receptors has not been achieved, and with the advent of molecular cloning of the opioid receptors, only three subtypes are generally accepted; namely, the mu, kappa, and delta (δ) subtypes [10]. The existence of subtypes of these opioid receptors, including the δ-opioid receptor have also been suggested, based on a broad array of approaches including pharmacological and receptor binding studies. Although different subtypes have not been cloned, it is possible for example that posttranscriptional events may occur to produce these subtypes [10]. Here, the pharmacology and physiology of the δ-opioid receptors is discussed.

The enkephalins, in combination with select antagonists, were useful for characterization studies performed in isolated tissue and binding assays. The enkephalin analogue [D-Ala^2,D-Leu^5]enkephalin (DADLE) was among the first "selective" δ-opioid agonists, although it also possessed some agonistic activity at the μ-opioid receptor [11]. The peptide Tyr-D-Ser-Gly-Phe-Leu-Thr (DSLET) was also identified as a highly selective δ-opioid agonist based on its agonistic activities in the stimulated guinea pig ileum and mouse vas deferens assays (GPI/MVD) [12]. These peptides proved to be fairly labile in vivo, however, and were not suitable, at least for most types of pharmacological studies. The conformationaly constrained penicillamine derivatives [D-Pen^2,D-Pen^5]enkephalin (DPDPE) and [D-Pen^2,L-Pen^5]enkephalin (DPLPE) were among the first analogues for the δ receptor which displayed high selectivity and stability; these constrained agonists have been widely employed in attempts to understand the pharmacology and physiology of opioid δ receptors [13–15]. Naturally occurring peptides with selectivity for the δ-opioid receptor have also been identified on the basis of binding studies and competition assays in the GPI/MVD assays. The highly selective δ-opioid agonistic peptides [D-Ala^2,Asp^4]deltorphin (deltorphin I) and [D-Ala^2,Glu^4]deltorphin (deltorphin II) were isolated from the skin of the amphibian *Phyllomedusa bicolor*, which also produces the potent μ-opioid agonist dermorphin [16,17]. These compounds have shown δ-opioid-selective antinociceptive activity in several models of nociception in mice and rats (*vide infra*).

2 NONPEPTIDIC δ-OPIOID AGONISTS

Several nonpeptidic δ-opioid agonists have been synthesized, although none are currently available clinically. These include BW373U86 which is ~10 and 20 times more selective for δ-opioid receptor over the μ and kappa opioid receptor, respectively, in receptor-binding assays [18]. BW 373U86 also demonstrated high potency (ED_{50} of 0.2 ± 0.06 nM) in the MVD assay, and its activity was blocked by naltrindole, a selective δ-opioid antagonist [18]. This compound demonstrated a 100-fold greater potency than DSLET in inhibiting adenylyl cyclase activity in rat striatal membranes, and its activity was blocked by naltrindole but not by the selective μ-opioid antagonist D-Phe-Cys-Tyr-D-Trp-Orn-Thr-Pen-Thr [19]. Moreover, BW 373U86 was antinociceptive in mice when given intrathecally, but not intracerebroventricularly (ICV) or systemically, although it was active systemically and intracisternally in a model of visceral pain in the rat [20,21]. Ultimately, this compound was discovered to produce brief convulsions in mice [22]. This effect was suggested to be mediated through δ-opioid receptors, since naloxone and naltrindole produced rightward shifts in the convulsant dose-effect curves [22].

The chiral methylether derivative of BW373U86, SNC 80, was found to be a highly selective agonist at the δ-opioid receptor [23–25]. Receptor-binding studies performed with rat brain tissue revealed that SNC 80 displayed nanomolar affinity for the δ-opioid and micromolar affinity for the μopioid receptor, indicating a high degree of selectivity [24]. SNC 80 was antinociceptive after central and systemic administration in mice and rats [26,27]. Structure-activity studies based on the SNC 80 molecule revealed a family of compounds with nanomolar affinity for the δ-opioid receptor and negligible μ opioid or kappa opioid affinity [28]. Interestingly, SNC80, like BW373U86, also produced convulsions in mice again linking the δ-opioid receptor to possible toxicity. However, it should be noted that the convulsive actions of closely related structural derivatives of SNC80 which did not bind to the opioid δ receptor, also produced convulsive actions (Bilsky and Porreca, unpublished observations), suggesting the possibility of a structural, rather than a mechanistic, link to convulsive activity.

3 SUBTYPES OF δ-OPIOID RECEPTORS

As is the case with the μ-opioid and the kappa opioid receptors, only a single gene for the δ-opioid receptor has been identified and cloned in the mouse, rat, or human [29–32]. In contrast, however, a significant number of reports, chiefly antinociceptive studies in mice, have provided functional evidence for the existence of subtypes of opioid δ receptors, termed δ_1 and δ_2 [33,34]. The existence of these subtypes has been postulated based on a number of studies indicating differential antagonism and two-way lack of cross-tolerance between DPDPE and [D-Ala^2,Glu^4]deltorphin. For example, it was found that the irreversible δ-opioid antagonist [D-Ala^2,Leu^5,Cys^6] enkephalin (DALCE) [35] blocked the antinociceptive effect of DPDPE given ICV but did not attenuate the antinociceptive action of [D-Ala^2,D-Glu^4]deltorphin [34].

In contrast, the nonequilibrium $\delta_2$antagonist naltrindole 5′-isothiocyanate (5′-NTII) [36] blocked the antinociceptive action of ICV [D-Ala^2,

D-Glu^4]deltorphin but not that of DPDPE [34]. Furthermore, the antinociceptive effect of DSLET was blocked by 5′-NTII but not by DALCE [34]. Although, unlike DPDPE or [D-Ala^2,D-Glu^4]deltorphin, DSLET was also partly antagonized by the μ-opioid antagonist β-funaltrexamine (β-FNA), the combination of β-FNA and 5′-NTII provided a profound blockade of the activity of DSLET [34]. These studies led to the postulation of the existence of the δ_1- (DALCE-sensitive) and δ_2- (5′-NTII-sensitive) opioid sites [34]. In another study, the benzofuran derivative of naltrindole (NTI), naltriben (NTB), antagonized the antinociceptive action of ICV or ITH DSLET but did not block the action of ICV or ITH DPDPE [37]. In contrast, NTI blocked the effect of both DSLET or DPDPE given either ICV or ITH [38]. Similarly, the novel δ_1- antagonist BNTX given ITH or systemically blocked the antinociceptive effect of ITH DPDPE but not that of [D-Ala^2,D-Glu^4]deltorphin or morphine, whereas NTB blocked the antinociceptive action of [D-Ala^2,D-Glu^4]deltorphin but not those of DPDPE or morphine, again demonstrating differential antagonism [39]. Similar differential antagonism of DPDPE and [D-Ala^2,D-Glu^4]deltorphin by NTB was observed after ITH injection in the rat [40]. Moreover, the observation that naltrindole was more potent against ITH DPDPE than against DADLE in rats provided further support for the concept of δ-opioid receptor subtypes [41].

In a study performed after ITH injections in mice, it was also found that antinociception induced by δ-opioid agonists given spinally was blocked by 5′-NTII but not DALCE, also given spinally [42]. Based on the premise that a sulfhydryl group on or near the delta opioid–binding site formed a disulfide bridge with DALCE to produce its noncompetitive antagonism, Cys^4 and Ser^4 derivatives of [D-Ala^2,D-Glu^4]deltorphin were synthesized [43]. It was expected that the action of [D-Ala^2,D-Glu^4]deltorphin would be blocked by the Cys^4, but not the Ser^4, derivative. Unexpectedly, both [D-Ala^2,D-Cys^4]deltorphin and [D-Ala^2,D-Ser^4]deltorphin blocked the antinociceptive action of [D-Ala^2,D-Glu^4]deltorphin, but not that of DPDPE, given ICV [43]. It was therefore interpreted that the δ_2, unlike the δ_1, opioid receptor is unlikely to have a thiol group at or near the binding site, indicating possible structural differences between these subtypes of the δ-opioid receptor [43].

Another important consideration promoting the view that subtypes of the δ-opioid receptor exist is the remarkable lack of cross-tolerance between agonists at the δ_1 and δ_2 opioid receptors. Furthermore, no cross-tolerance exists between δ -and μ-opioid receptors [33,38,44,45]. In one of the earliest studies, acute tolerance to spinal DPDPE or to DSLET was elicited in β-FNA-pretreated (i.e., μ receptors blocked) mice [45]. Pretreatment with DPDPE attenuated the antinociceptive effect of subsequent spinal DPDPE but not that of DSLET, whereas pretreatment with DSLET attenuated the effect of DSLET and not DPDPE [45]. A similar lack of cross-tolerance between DSLET and DPDPE was observed after ICV injections in mice [38]. In a more detailed study, mice received repeated ICV injections of DPDPE, [D-Ala^2,D-Glu^4]deltorphin, or DAMGO and were challenged with each of the agonists on day 3 [33]. As expected, pretreatment with the μ-opioid agonist DAMGO did not attenuate the antinociceptive effects of either DPDPE or [D-Ala^2,D-Glu^4]deltorphin [33]. However, pretreatment with DPDPE attenuated that antinociceptive action of DPDPE but not of [D-Ala^2,D-Glu^4]-deltorphin and, conversely, pretreatment with [D-Ala^2,D-Glu^4]deltorphin produced tolerance to [D-Ala^2,D-Glu^4]deltorphin but not to DPDPE [33]. Competition binding studies performed with rat brain membranes and employing [^{3}H]DPDPE or [^{3}H][D-Ala^2,Asp^4]deltorphin revealed the existence of two subpopulations of the δ-opioid receptor [46]. Binding studies performed in mouse vas deferens and brain where [^{3}H]NTI was displaced by DPDPE or [D-Ala^2,Glu^4]deltorphin also revealed δ-opioid receptor heterogeneity [47]. Saturation binding studies in rat brain membranes demonstrated that pretreatment with DALCE reduced the number of δ_1-binding sites without affecting the composition of the δ_2-opioid binding sites [48]. These findings have been replicated and continuously supported in a variety of functional assays by many laboratories throughout the world [for review, see 49].

On the other hand, the underlying basis for the distinction between different classes of δ-selective agonists remains obscure as to date; no molecular evidence has yet appeared to support the existence of opioid δ receptor subtypes. Although there is no evidence of different genes for the δ-opioid receptor subtypes, it is possible that a single gene could produce variants of a receptor. Splice variants of the MOR and ORL-1 receptors have been identified [50,51]. Variants of the μ-opioid receptor, termed MOR1 and MOR1B, have been transfected into CHO cells and demonstrate differential agonistic activities [52,53]. A variant of the δ-opioid receptor, with a deletion of the third cytoplasmic loop, has found to occur in the human neuroblastoma cell line SH-SY5Y, although its relevance to δ-opioid receptor subtypes is unknown [54]. Alternatively, posttranslational events, including glycosylation, palmitoylation, or proteolysis, may produce different tertiary structures, which would also alter binding affinities and/or receptor transduction [for review see 10].

4 CLONING AND LOCALIZATION OF THE δ-OPIOID RECEPTORS

The successful cloning of the δ-opioid receptor from cells derived from mice was announced almost simultaneously by Evans and colleagues [30] and by Kieffer and colleagues [29]. Evans and colleagues [30] isolated cDNA libraries from NG108-15 cells, which are derived from mouse neuroblastoma × rat glioma cell line known to express high concentrations (>10^4) of δ-opioid receptors per cell. The cDNAs were transfected into COS cells, and the δ-opioid receptors were expressed. Competition studies employing [^{125}I]DADLE as the radioligand competing with a number of ligands revealed a binding profile consistent with that of the δ-opioid receptor [30]. Furthermore, transfected COS cells demonstrated a naloxone-sensitive inhibition of forskolin-stimulated cAMP accumulation, indicating functional transduction by this cloned receptor [30]. A similar approach was employed by Kieffer and colleagues [29]. The cDNA libraries were obtained from NG108-15 cells and expressed in COS cells. Competition studies against [^{3}H]DTLET revealed binding profiles indicative of δ-opioid receptors. Importantly, however, in these studies, the expressed cDNA was compared to mouse and rat genomic DNA, revealing that the expressed protein was in fact a mouse δ-opioid receptor [29].

These discoveries were followed by the isolation of cDNA for two opioid receptors derived from rat brain, identified as ROR-A and ROR-B [55]. Competition studies performed with DSLET, DPDPE, and DAMGO in transfected CHO cells determined that the ROR-A corresponds to the δ-opioid receptor and ROR-B corresponds to the μ-opioid receptor [55]. Finally, human cDNA for the δ-opioid was expressed in COS-7 cells [56]. Again, competition studies confirmed the presence of the δ-opioid receptor. Moreover, it was determined that the cloned receptor represents the δ_2-opioid receptor, since NTB showed an approximately eightfold greater affinity than BNTX [56]. The molecular biology of the cloned opioid receptors has since been reviewed [57]. It has been subsequently determined that the cloned δ-opioid receptor may correspond to the δ_2-isoform [57].

5 SUPRASPINAL DISTRIBUTION OF THE DELTA OPIOID RECEPTOR

Once the genomic sequences for the δ-opioid receptors have been identified, it became possible to engineer mRNA probes corresponding to the genetic structure of the genes in order to allow for in situ hybridization techniques to identify the distribution of neurons that actually express the receptors [58]. Moreover, RT-PCR could also be employed to determine the general distribution of the δ-opioid receptors throughout the nervous system. For example, RT-PCR and Northern blot analysis performed in human brain sections showed a high concentration of δ-opioid receptors in the cortical and hippocampal sections and the caudate nucleus and the nucleus accumbens and little or no signal in the medullary and mesencephalic sections [59]. These findings were consistent with the proposed existence of an enkephalinergic system between the striatum and the globus pallidus [59]. More detailed analysis were performed with in situ hybridization techniques in the rat [60]. Message for the rat δ-opioid receptor was found in the cortex, hippocampus, and hypothalamic nuclei as well as the pontine nuclei and the inferior olive [60]. Surprisingly, no message was detected in the caudate nucleus, which did not correspond to findings obtained with autoradiography [61].

In an early study employing autoradiography with [^{3}H]DTLET, δ-opioid receptors were found to be chiefly in the olfactory bulb, caudate-putamen, and the cortex, whereas the μ opioid receptors were distributed in the caudate-putamen, nucleus accumbens, cingulate cortex, and various thalamic nuclei [62]. Similarly, differential distributions were found when [^{3}H]DAMGO and [^{3}H]DPDPE were employed as the radioligands, with the δ-opioid receptors having a more limited distribution than the δ-opioid receptors [61]. Similarly, autoradiography performed in rat brain sections with [^{3}H]DPDPE also showed highest densities of the δ-opioid receptor in the amygdala, caudate-putamen, nucleus accumbens, and olfactory bulb, and low binding density in the cerebellum, medulla oblongata, and the dorsal horn of the spinal cord [63]. Recent receptor autoradiography and in situ hybridization studies have clearly demonstrated the presence of δ-opioid receptors or of δ-opioid receptor mRNA consistent with the neuroanatomy of ascending and descending pain pathways [58,64]. Studies employing immunoreactivity also confirmed a distribution of δ-opioid receptors in structures essential to pain processing. Immunostaining for the δ-opioid receptor was found in the trigeminal nuclei, raphe and parabrachial nuclei, and the periaqueductal gray (PAG) [65]. In addition, immunostaining for the δ-opoid receptor was also found at pre- and postsynaptic sites within the hippocampus and the dentate gyrus [66,67]. In general, there was consistent agreement between δ-opioid receptor distributions identified though autoradiography or in situ hybridization [68].

Neuroanatomical sites important to pain processing and modulation that were found to express the δ-opioid receptor included the PAG, raphe nuclei, the nucleus gigantocellularis, and the lateral reticular nucleus [58,64]. In contrast, however, an autoradiographic survey of the rat brain with [^{125}I][D-Ala^2,Asp^4]deltorphin showed a sparse concentration of δ-opioid receptors within the PAG [69]. Although not common, differences in δ-opioid receptor distribution determined by these techniques were found on occasion. It was suggested that these differences may either be due to transport of a receptor from the cell body, where it is synthesized, or may reflect the higher sensitivity of the in situ techniques [58]. Surprisingly, although there was a fair degree of correlation between δ-opioid receptor distributions found through immunohistochemistry and either quantitative autoradiography or in situ hybridization, there were also instances of discrepancies [70]. For example, the distribution of δ-opioid-binding sites in the mouse brain appeared dense in the cerebral cortex, caudate-putamen, and hippocampus, whereas immunohistochemical labeling was sparse in these regions [71].

More recently, studies employing both light and electron microscopy revealed the presence of intensely labeled cell discrete populations within hypothalamus, PAG, dorsal raphe, and n. gigantocellularis as well as moderate staining throughout the medullary raphe nuclei and the parabrachial nucleus [70]. It was also found that the δ-opioid receptor occurs in two immunological forms, corresponding to a membrane and intracellular form of the receptor [70]. Studies employing [^{14}C]deoxyglucose have also demonstrated increased glucose utilization after ICV DPDPE administration in motor regions, limbic forebrain, and the ventroposterolateral thalamic nucleus, which is involved in modulation of sensory and nociceptive information to the cortex [68].

6 SPINAL DISTRIBUTION OF THE DELTA OPIOID RECEPTOR

A number of studies employing quantitative autoradiography, in situ hybridization and immunohistochemical methods have demonstrated the clear presence of the δ-opioid receptor concentrated in the superficial laminae of the dorsal horn of the spinal cord [58,64,65,72,73]. Autoradiography employing [^{3}H]DAMGO and [^{3}H]DTLET showed that the δ-opioid receptor was found in laminae I and II of the dorsal horn of the spinal cord, and accounted for only 23% of the total opioid receptor population, whereas the μ-opioid receptor population accounted for ~70% of the total [74,75]. However, when [^{3}H]sufentanil and [^{3}H]DPDPE were employed as the radioligands, the contribution of the μ-opioid receptors to the total in lamina I–II was estimated to be 90% and that of the δ-opioid receptor was estimated at 7% [76]. The relative proportions were 70% and 27% in lamina V and 65% and 33% about the central canal for the μ-opioid and δ-opioid receptors, respectively [76]. In contrast, autoradiographic studies performed with [^{125}I][D-Ala^{2},Aps^{4}]deltorphin demonstrated the presence of the δ-opioid receptor throughout the dorsal horns of the spinal cord [69]. Dorsal rhizotomy produced dramatic changes in the dorsal horn, reducing the presence of δ-opioid receptors by up to 70%, indicating that the majority of opioid receptors reside presynaptically on primary afferent terminals [74,75,77]. In contrast, Stevens and Seybold [78] found a 23% decrease in δ-opioid-binding sites in the superficial dorsal horn after rhizotomy. In situ hybridization studies demonstrated a high concentration of mRNA for the δ-opioid receptor in the DRG, partiçularly the large diameter neurons, and an almost complete absence of message in the superficial dorsal horns of the spinal cord, a finding consistent with the observations indicating a preponderance of receptors on the primary afferent nerve terminals [58,64]. Similarly, no mRNA for the δ-opioid receptor was found in either the ventral or dorsal horns of the human spinal cord [79]. Immunofluorescence methods showed that δ-opioid receptors were found in a dense plexus of small-profile fibers in laminae I and II of the spinal dorsal horns [80]. Dorsal rhizotomies and neonatal capsaicin treatments both produced significant reductions in immunostaining for the δ-opioid receptor in laminae I and II of the dorsal horn [81]. Based on these studies, it was reported that approximately 50% of the μ-opioid and 67% of the δ-opioid receptors exist presynaptically on primary afferent fibers [70,81]. Studies performed with electron micrographs showed a significant degree of immunostaining for the δ-opioid receptor on unmyelinated axons and axon terminals [70]. It was found that, in the rat and monkey, there was substantial colocalization of immunoreactivity for the δ-opioid receptor with reactivity for substance P or CGRP in neuronal cell bodies the DRG [82]. In addition to its presence on nerve terminals, the δ-opioid receptors have been identified on neuronal soma and dendrites in the spinal cord [70]. Taken together, these studies provide strong evidence that the δ-opioid receptor is anatomically linked to pain modulatory pathways.

7 δ-OPIOID-MEDIATED ANTINOCICEPTION

7.1 Spinal Antinociception

That the δ-opoid receptors exist on the terminals of primary afferents in the spinal cord strongly suggests that agonists acting at these sites would be effective antinociceptive agents [83,84]. Shock titration tests performed in rats and monkeys demonstrated that δ-opioid agonists produce antinociception when given intrathecally (ITH), even in morphine-tolerant animals [85]. Both DPDPE and DPLPE given ITH produced dose-dependent antinociception after ITH injection in the mouse [86]. Furthermore, δ-mediated antinociception induced by DPDPE or DPLPE were blocked by the selective μ-opioid antagonist ICI 174864 [87]. In other studies, ICI 174,864 produced a rightward shift in the antinociceptive dose-response curve for spinal DPDPE but not morphine, whereas the μ-opioid antagonist β FNA blocked morphine but not DPDPE [88]. Inhibition of the tail flick reflex by applying a noxious conditioning stimulus to the hindpaw of the rat was blocked by the spinal injection of β FNA or the δ-opioid antagonist H-Tyr-Tic psi[CH_2NH]Phe-Phe-OH, suggesting that endogenous ligands for the μ- and δ-opioid receptors are released in response to nociceptive inputs [89]. Further evidence for an endogenous compound acting at the spinal δ-opioid receptor site was indicated when the peptidase inhibitor kelatorphan did not further enhance the ability of the δ-opioid agonist Tyr-D-Ser(Otbu)-Gly-Phe-Leu-Thr (DSTBULET) to attenuate dorsal horn unit activity in response to electrical stimulation at C-fiber intensities [90].

In contrast, there was an enhanced response to μ-agonists, indicating that separate endogenous substrates act at each of these receptors [90]. The spinal administration of NTI produced a shift to the right of the dose-response curve for DPDPE, but not DAMGO, in the rat tail flick and hot plate tests [91]. Furthermore, apparent pA_2 values obtained from these studies provided additional evidence that NTI acts as a selective δ-opioid antagonist in vivo [91]. Similar to ITH bolus injecions, spinal infusions of [D-Ala^2,Glu^4] deltorphin also produced NTI and naloxone-sensitive antinociception in the rat [92]. Later studies showed that although ICI 174,864 abolished the antinociceptive effect of spinal DPDPE and [D-Ala^2,Glu^4]deltorphin, DPDPE was blocked by DALCE and not 5′-NTII whereas [D-Ala^2,Glu^4] deltorphin was blocked by 5′-NTII and not DALCE [42]. Moreover, neither morphine nor DAMGO was antagonized by these doses, indicating selective antagonistic activity of DALCE and 5′-NTII at the δ_1- and δ_2- opioid receptors, respectively [42].

In a detailed pharmacological study, it was found that ITH [D-Ala^2, Glu^4]deltorphin was ~10-fold more potent than DPDPE in the rat tail flick test [40]. Naltriben increased the ED_{50} of [D-Ala^2,Glu^4]deltorphin 25-fold without altering that of DPDPE [40]. Surprisingly, ITH [D-Ala^2,Glu^4]deltorphin did not produce antinociception in the hot plate test at doses that did not also produce motor impairment, whereas DPDPE was effective in the latter test [40]. These results were taken as pharmacological evidence of differential actions of spinal δ_1- and δ_2-opioid receptors in the spinal cord. Later studies by the same group also showed the participation of spinal δ_1-opioid receptors in modulation of nociception [93]. The selective δ_1-antagonist, 7-benzylidenenaltrexone (BNTX) blocked the antinociceptive effect of ITH DPDPE in the tail flick and hot plate tests without inhibiting the actions of either [D-Ala^2,Glu^4]deltorphin or DAMGO [93]. Interestingly, blockade of the δ_1-opioid receptor enhanced the antinociceptive effect of [D-Ala^2,Glu^4]deltorphin [93]. A different profile was revealed when antinociception was produced by glutamate-induced stimulation of the nucleus raphe magnus (NRM) or the nucleus reticularis gigantocellularis pars alpha (NGCpα) [94]. The ITH injection of BNTX ($\delta_1$antagonist) did not block antinociception elicited by either site, whereas naltriben blocked the antinociception induced by glutamate in the NRM, but not the NGCpα. It was concluded that antinociception mediated through bulbospinal pathways depends on δ_2-, and not δ_1-, opioid receptor activation [94,95].

Studies employing endpoints other than tail flick and hot plate have shown antinociceptive activity mediated through spinal δ-opioid receptor sites. Low (i.e., δ-selective) concentrations of DPDPE or morphine inhibited K^+-evoked release of substance P from trigeminal slices. Furthermore, this inhibitory effect was blocked by naloxone or ICI 174,864, but not by β-FNA or BNI ($_6$antagonist) or naloxonazine (μ_1 antagonist) [95]. Mechanical and thermal noxious stimuli evoked the release of substance P measured by microdialysis in the spinal cord, and this action was attenuated by spinal infusions of [Met^5]enkephalin [96]. Naltrindole not only blocked the effect of exogenous [Met^5]enkephalin, but, when given alone, it elicited a 75% increase in the basal release of substance P–like immunoreactivity, suggesting the presence of an inhibitory tone [96]. Similar observations were made with the δ-opioid agonists Tyr-D-Thr-Gly-Phe-Leu-Thr and dermenkephalin [97]. Behavioral demonstration of an endogenous δ-opioid receptor–mediated inhibitory tone that modulates nociceptive inputs was shown when the ITH administration of naltrindole or of antisera to [Leu^5]enkephalin, a putative endogenous δ-opioid agonist, produced marked, significant increases in the tonic phase of formalin-induced flinching in rats [98].

In contrast, neither β-FNA nor antiserum to [Met^5]enkephalin enhanced formalin-induced flinching behavior in the same study [98]. In the same study, ITH DPDPE and DAMGO blocked formalin-induced flinching, and these effects were antagonized by naltrindole and β-FNA, respectively [98]. In similar studies, the spinal administration of DPDPE or [D-Ala^2,Glu^4]deltorphin produced an inhibition of formalin-induced flinching that was blocked by selective antagonists at δ_1 and δ_2 receptors, respectively [94]. Surprisingly, DPDPE produced only a moderate inhibition of Fos expression in spinal dorsal horn neurons, and [D-Ala^2,Glu^4]deltorphin did not produce any suppression of Fos expression [94]. In contrast, DAMGO produced dramatic decreases in flinching behavior and completely prevented Fos expression in response to nociceptive stimulus. It was surmised that the predominant presynaptic site of action of the δ-opioid receptors allowed a sufficient number of postsynaptic second-order neurons to express Fos [94].

7.2 Supraspinal Antinociception

There is considerable evidence that δ_1- and δ_2-opioid receptors act at supraspinal sites to modulate nociceptive responses. Direct ICV injection of the putative δ-opioid agonists DPDPE or DPLPE in the mouse produced dose-dependent antinociception that was blocked by ICI 174,864 but not by β-FNA [87,88]. Furthermore, δ-opioid selective doses of DPDPE given ICV blocked nociception, but not gastrointestinal transit, in mice [99].The ICV administration of DPDPE has been shown to produce dose-dependent antinociception against mechanical nociception (Randall-Selitto paw withdrawal test) in rats at δ-opioid-selective doses, and that effect was blocked by ICI 174,864 [100]. It was also shown that [D-Ala^2,Glu^4]deltorphin given ICV was 13-fold more potent than DPDPE and equipotent to morphine [101]. Additionally, [D-Ala^2,Glu^4]deltorphin was blocked by ICI 174,864 but not by β-FNA [101]. Later studies revealed that ICV [D-Ala^2,Glu^4]deltorphin was blocked by 5′-NTII and not DALCE whereas DPDPE was blocked by DALCE and not 5′-NTII, thus indicating that DPDPE and [D-Ala^2,Glu^4]-deltorphin act as agonists at δ_1- and δ_2-opioid receptors, respectively [101]. Dose-effect curves constructed with ICV DPDPE and [D-Ala^2,Glu^4]deltorphin in the presence of 5′-NTII and DALCE confirmed these conclusions but also demonstrated that, at higher doses, DPDPE might also act as a partial agonist at the δ_2-opioid receptor [102].

Stress-induced antinociception also appears to be mediated through δ-opioid receptors. Tail shock–induced antinociception was blocked by ICV naltrindole, thus showing a δ-opioid-mediated component [103]. Enhancement of morphine-mediated antinociception by cold-water swim stress (CWSS) was blocked by ICV ICI 174,864 [104,105]. Moreover, the ICV administration of β-FNA or of the δ_1-antagonist DALCE did not block this effect, whereas 5′-NTII did, indicating that this modulatory action is mediated specifically through the δ_2-opioid receptor [104,105]. In spite of the abundant support for supraspinally mediated antinociception produced by δ-opioid agonists in mice, the antinociceptive action of supraspinally administered δ-opioid agonists has not been universally confirmed in the rat. For example, ICV DPDPE given in doses of up to 300 μg demonstrated antinociceptive activity in the cold water (−3 °C), but not the hot water (50 °C), tail flick tests [106]. Furthermore, at doses that produced δ-selective, naltrindole- reversible increases in locomotion and rearing behavior (i.e., up to 38 nmole), [D-Ala^2,Glu^4]deltorphin given ICV did not produce any observable antinociception in the tail flick test [46,107]. One limitation may be that many studies have studies the effects of ICV δ-opioid agonists, where penetration into deep brain structures was limited. Support for this interpretation is provided by studies that demonstrate limited diffusion of ^{3}H-labeled DPLPE, DSLET, and DAMGO from the site of ICV injection [108].

In contrast, direct microinjection into brain parenchyma provided evidence of δ-opioid-mediated antinociception arising from supraspinal sites. In an early study aimed at determining the relative contribution of opioid subtypes to antinociception arising from different brain regions, morphine, DADLE, and ethylketocyclazocine (EKC) were microinjected directly into the PAG, NGC, and the NRM in the rat [109]. It was found that DADLE was more potent than morphine by 1 order of magnitude in the PAG and NRM, and less potent in the NGC [109]. Furthermore, the potency of DADLE among these sites was consistent [109]. In contrast, EKC was weakly active among these supraspinal sites. These data indicated that the degree of antinociceptive activity of the μ-opioid and δ-opioid receptors varied among brain regions [109].

Similarly, the microinjection of μ-opioid (morphine, sufentanil) and δ-opioid (DSLET, DADLE) into the PAG produced complete blockade of thermal and visceral chemical nociception, while only the δ-opioid agonists produced a complete blockade of thermal nociception when given into the medullary reticular formation MRF [110]. A detailed characterization of the role of the δ_1- and δ_2-opioid receptors of the PAG and the MRF in the modulation of nociception was reported [111]. It was found that neither DPDPE nor [D-Ala^2,Glu^4]deltorphin was effective against thermal nociception when microinjected into the PAG. When given into the MRF, however, [D-Ala^2,Glu^4]deltorphin, but not DPDPE, inhibited nociceptive responses to noxious heat [111]. In addition, ICV [D-Ala^2,Glu^4]deltorphin was fully active, but DPDPE was <50% active, against thermal nociception. Importantly, DPDPE but not [D-Ala^2,Glu^4]deltorphin was antagonized by DALCE and conversely, [D-Ala^2,Glu^4]deltorphin but not DPDPE was blocked by [Cys^4]deltorphin, indicating selective activities at the δ_1- and δ_2-opioid receptors, respectively [111].

In contrast, morphine and the mixed μ/δ-opioid agonist DSLET produced antinociception in all assays and through each route of administration [111]. It was concluded that supraspinal δ-opioid receptors can be activated to elicit antinociception in the rat, and that opioid δ_2 receptors predominate in this effect. Slightly different results were seen when it was observed that microinjections into the rostroventromedial medulla (RVM) of DPDPE or of [D-Ala^2,Glu^4]deltorphin was inactive or weakly active against the 55° C hot plate test in the rat [112]. In contrast, DPDPE and [D-Ala^2,Glu^4]deltorphin were approximately equipotent against the radiant heat tail flick test [112]. Furthermore, DPDPE was selectively antagonized by naltriben, and [D-Ala^2,Glu^4]deltorphin was selectively blocked by BNTX, indicating the participation of δ_1- and δ_2-opioid receptors in this region [112]. Furthermore, since neither antagonist given alone enhanced nociceptive responses, it was concluded that the RVM does not receive a tonic enkephalinergic input from more rostral sites (e.g., PAG) [112]. The finding of δ-opioid receptor–mediated antinociceptive activity arising from the PAG and MRF suggests the participation of these receptors in this major antinociceptive pathway. A particularly interesting observation was that antinociception produced by morphine microinjected into the PAG or MRF was blocked by naltrexone, β-FNA, or naltrindole microinjected into the MRF [113]. It was concluded that μ- and δ_2-opioid receptors mediate the enkephalinergic inputs from the PAG to produce antinociception [113]. The existence of δ-opioid-mediated descending inhibitory pathways was also suggested by the observation that the antinociceptive effect of ICV DPDPE was attenuated by spinal administration of bicuculline or 2-hydroxysaclofen ($GABA_A$ and $GABA_B$ antagonists) [114].

More recent studies provided firm evidence of a δ-opioid-mediated antinociceptive pathway from the MRF. Unilateral lesions of the dorsolateral funiculus (DLF), which represents the major bulbospinal pathway from the MRF, completely blocked the ability of [D-Ala^2,Glu^4]deltorphin to attenuate formalin-induced flinching in rats [115]. Furthermore, DLF lesions also blocked the pain-induced expression of Fos in the dorsal horns of the spinal cord [115]. Moreover, the infusion of [D-Ala^2,Glu^4]deltorphin into the RVM inhibited both the tail flick reflex and the accompanying increase in pronociceptive "ON" cells of the RVM, while reducing the pause in antinociceptive "off" cell activity [116]. These effects were reversed by the selective δ_2-opioid antagonist naltriben. Thus, medullary δ_2-opioid receptors, like μ-opioid receptors, mediate antinociception through enhanced "off" cell activity [116].

8 STUDIES WITH ANTISENSE OLIGODEOXYNUCLEOTIDES

An important consequence of decoding the gene for the δ-opioid receptor is the ability to design antisense strands of oligodeoxynucleotides (ODN) complementary to portions of mRNA responsible for synthesis of the protein [for reviews see 57,117]. Thus, sequestration of the mRNA for coding for the δ-opioid receptor by antisense ODN prevents the expression of the receptor peptide, indicated by a "knock down" or significant loss of the receptor in cells and whole animals [57,117]. The construction of an antisense ODN, 18–20 bases in length, reduced the expression of δ-opioid receptors in cultured NG108-15 cells by 50% without altering expression of μ- or κ-opioid receptors [118,119]. Furthermore, the ITH infusion of the antisense ODN to mice resulted in a reversible down-regulation of δ-opioid receptor binding and a loss of antinociceptive activity of DPDPE but not of μ- or κ-opioid agonists [118,119].

In a contemporary study, twice-daily injections of antisense ODN were given to mice ICV for 3 days, and antinociceptive dose-effect curves for DPDPE and [D-Ala^2,Glu^4]deltorphin were generated [120]. Antisense treatment significantly attenuated the antinociceptive effect of [D-Ala^2,Glu^4] deltorphin but not of DPDPE, strengthening the conclusion that the cloned δ-opioid receptor corresponds to the δ_2 subtype [120]. Similar results were obtained when antisense ODN was given either ITH or ICV [121]. The antisense, but not the mismatch, ODN given ITH blocked the antinociceptive effect of spinally administered [D-Ala^2,Glu^4]deltorphin or DPDPE, but not of DAMGO or of the κ-opioid agonist U69,593 [121]. In contrast, the ICV injection of the antisense ODN selectively blocked the antinociceptive effect of [D-Ala^2,Glu^4]deltorphin [121].

Based on these and similar results, it was interpreted that the δ_2-opioid receptor represents the cloned species of this opioid subtype, but also that it is the predominant δ-opioid receptor in the spinal cord [121]. The loss of antinociceptive activity of ICV [D-Ala^2,Glu^4]deltorphin by ICV treatment with antisense, but not mismatch, ODN was time sensitive in that it persisted while antisense ODN was administered but was reversed within 3 days of terminating the injections [122]. Moreover, saturation studies performed with [^{3}H]naltrindole on brain tissue from antisense ODN-treated mice revealed a significant (30–40%) decrease in δ-opioid-binding sites that returned to baseline values 3 days after termination of the ODN injections [122]. Moreover, in the same study, the administration of an antisense ODN designed to be complementary to a region of the genome conserved among the μ-, κ-, and δ-opioid receptors demonstrated knockdown and attenuation of antinociceptive action of all three opioid receptor subtypes [122]. Taken together, these data confirm the existence of subtypes of the δ-opioid receptor [123,124].

To determine if the same gene may code for the δ_1- and δ_2-opioid receptor subtypes, five different antisense sequences based on three exons were produced [125]. It was found that all five sequences blocked the antinociceptive effects of DPDPE and of [D-Ala^2,Glu^4]deltorphin when given ITH and only [D-Ala^2,Glu^4]deltorphin when given ICV [125]. Only two antisense probes, derived from exon 3, blocked the antinociceptive effect of ICV DPDPE, while those derived from exons 1 and 2 were inactive [125]. These results indicate that "antisense mapping" may be used to discriminate genetic coding for receptor subtypes [125]. A more recent study provided some intriguing results suggesting the possibility of yet another subtype of the δ-opioid receptor. It was found that ICV antisense treatment blocked the antinociceptive effects of SNC-80, pCl-DPDPE, and [D-Ala^2,Glu^4]deltorphin, but did not attenuate that of DAMGO or DPDPE [126]. Moreover, the μ-opioid antagonist CTOP antagonized the action of ICV DAMGO and DPDPE, and partly blocked that of pCl-DPDPE but did not attenuate the effect of SNC-80 or [D-Ala^2,Glu^4]deltorphin [126]. It was concluded that the supraspinal antinociceptive effect of DPDPE is mediated through a δ-opioid receptor different from the native receptor, either by acting directly at the μ-opioid receptor or through a novel δ-opioid receptor that directly or indirectly modulates the action of a μ-opioid receptor [126]. These observations are not inconsistent with the intriguing report of a "novel" δ-opioid antinociceptive system [127]. Gene targeting based on exon 2 of the δ-opioid gene resulted in mice devoid of δ_1- and δ_2-opioid receptors as determined by immunocytochemical and receptor-binding studies [127]. Surprisingly, these knockout mice demonstrated antinociceptive activity to ICV DPDPE and enhanced antinociceptive activity to BW373U69, and this activity was only partly blocked by naltrindole [127].

Additional studies were performed where antisense ODN to the δ-opioid receptor was conjugated to Texas Red dye in order to visualize neuronal uptake of the ODN [124]. The labeled ODN was shown to be taken up into cultured NG108-50 cells after 4 days of incubation, and these neurons expressed a knockdown of the δ-opioid receptor [124]. Additionally, ITH injection of the same ODN resulted in a significant loss of immunostaining for the δ-opioid receptors in the outer laminae of the spinal cord in treated mice [124]. Determination of the presence of δ-opioid receptors was performed through standard immunofluorescence protocols, using FITC as the fluorophore [124]. Quantitative fluorescent image analyses coupled with electron microscopy allowed the direct measurement of ODN uptake and δ-opioid receptor density on individual cells. It was found that neuronal uptake of antisense ODN was time and concentration dependent, and that it was not unvarying within a neuronal population. Most importantly, however, it was found that there was a direct, inverse correlation between intracellular antisense ODN fluorescent intensity and density of immunostaining for the δ-opioid receptor [124]. Importantly, uptake of mismatch ODN was not associated with a decrease in opioid receptor staining. More recently, the ITH injection of conformational constrained, stable antisense ODNs to the δ-opioid receptor produced an attenuation of the antinociceptive effect of ITH [D-Ala^2,Glu^4]deltorphin in rats [128]. It is generally accepted that the δ-opioid receptors act predominantly by inhibiting neurotransmitter release from nerve terminals of primary afferent neurons. Thus, δ-opioid agonists may block nociceptive inputs centrally, in the spinal cord, or peripherally, at the site of primary afferent excitation. Accordingly, the spinal administration of antisense ODN to the δ-opioid receptor blocked the antinociceptive action of ITH and intrapaw [D-Ala^2,Glu^4]deltorphin against the formalin-induced flinch in mice [129]. This treatment did not affect the antinociceptive activity of μ- and κ-opioid agonists, nor did it result in a knockdown of spinal μ- and κ-opioid receptors [129]. These data indicate that δ-opioid receptors may be synthesized in the neuronal cell bodies and transported to both the central and peripheral terminals, and that spinally administered agents may affect the production of peripheral proteins [129].

9 SPINAL/SUPRASPINAL SYNERGY OF δ-OPIOID ANTINOCICEPTION

The fact that δ-opioid receptors act at spinal and supraspinal sites increases the possibility that, like morphine, δ-opioid agonists may exhibit a synergistic interaction between the spinal and supraspinal sites of action. It has been clearly established that the concurrent administration of ICV and ITH morphine results in a multiplicative antinociceptive interaction [130,131]. It is this supraspinal/spinal multiplicative nature of morphine and its clinical analgesic utility at tolerable doses. Early studies with mice indicated only an additive interaction between ICV and ITH DPDPE against thermal endpoints in mice [131]. In contrast, however, the concomitant ICV and ITH administration of DPDPE in rats demonstrated a synergistic interaction against mechanical nociception [132,133]. However, one difficulty with these studies was that either the ITH or ICV dose was held constant while the doses given at the alternate site were altered to produce dose-effect curves [132,133]. This paradigm would cause changes in dose ratio, possibly complicating the data interpretation. More recently, fixed ratios of agonists given spinally and suprsapinally were employed to allow for rigorous isobolographic interpretation of additivity or synergy [134–136]. It was found that over low dose ranges, the interaction between [D-Ala^2,Glu^4]deltorphin given ITH and into the RVM was synergistic, and became additive at higher dose ranges [134]. Because of its relative lack of efficacy, a dose-effect curve for RVM DPDPE was not constructed in these studies, although an additive interaction was surmised based on a modeling isobolograms when a single component is effective [134,137]. More detailed results were obtained when less intense noxious stimuli were applied (e.g., 52EC vs. 55EC hot plate, 52EC hot water vs. radiant heat tail flick; radiant heat paw withdrawal test) [136]. A synergistic interaction was detected between RVM and spinal [D-Ala^2, Glu^4]deltorphin in all tests and between spinal and RVM DPDPE in all tests, with the exception of the hot plate test [136]. The synergistic interaction between RVM and ITH [D-Ala^2,Glu^4]deltorphin was abolished by the ITH administration of the α_2-adrenergic antagonist yohimbine [135]. Furthermore, the concomitant spinal injection of [D-Ala^2,Glu^4]deltorphin with the α_2-adrenergic dexmedetomidine produced a synergistic antinociceptive interaction [135]. It was hypothesized that activation of the supraspinal δ_2-opioid receptors resulted in a spinal release of endogenous norepinephrine, which acts to enhance the inhibition of nociceptive inputs. Interestingly, mice made tolerant to morphine lost μ-opioid spinal/supraspinal synergy, while DADLE, a relatively selective δ-opioid agonist, actually developed a multiplicative interaction, leading to the speculation that δ-opioid agonists may be clinically useful in conditions of opioid tolerance [138].

10 INTERACTIONS BETWEEN δ-OPIOID AND μ-OPIOID AGONISTS

An important aspect regarding the exploration of μ/δ-opioid interactions is that synergism between agonists acting at these sites presents the potential for significant therapeutic benefits. An initial report of such functional modulation came from the observations when the putative δ-opioid agonist [Leu^5] enkephalin given ICV at doses that did not produce antinociception enhanced the antinociceptive potency of morphine in mice [139]. Additional studies also demonstrated an enhanced antinociceptive effect of ICV morphine or DAMGO by ICV DPDPE, which was blocked by ICI 174,864 [87,100,140, 141]. As would be expected in drug interaction studies, the level of synergy or additivity was found to be sensitive to the drug ratio employed and to the test used [106]. For example, although a 1:5 ICV ratio of DPDPE to morphine was synergistic in a cold-water tail flick test, a 2:5 ratio was not [106]. The ICV administration of fixed ratios of morphine with either DPDPE or [D-Ala^2, Glu^4]deltorphin produced a synergistic antinociceptive effect in the mouse [142]. In contrast, the combination of [Met^5]enkephalin with morphine was found to be subadditive in the same study [142]. A particularly impressive demonstration of such synergy was shown when systemic [Leu^5]enkephalin, inactive alone, produced synergy with systemic morphine, as quantified by isobolographic analyses [143]. The synergy, but not the effect of morphine, was blocked by ICV ICI 174,864 [143].

The spinal administration of low doses of DPDPE enhanced the antinociceptive effect of DAMGO while actually decreasing the DAMGO-induced loss of motor coordination [144]. Detailed isobolographic analysis performed over several levels of effect demonstrated a clear spinal synergism between δ-opioid (DPDPE, DADLE) and μ-opioid (morphine, PL-017, DAMGO) [145]. Synergy was reversed in a dose-dependent fashion by naloxone, naltriben, and naltrindole, indicating a requirement for occupancy at both populations of opioid receptors [145]. An interesting observation, indicating a complex interaction among descending inhibitory pathways, was the observation that DAMGO and [D-Ala^2,Glu^4]deltorphin produced antinociceptive synergy when one was given into the PAG and the other into the RVM, but not when both were administered to the same site [146]. It was therefore suggested that the resulting synergy was due to interactions among pathways rather than occurring at the cellular or molecular level [146]. Consistent with this interpretation, it was found that antinociception produced by morphine given into the PAG was blocked by naltrindole microinjected into the RVM, demonstrating a δ_2-opioid medullary link in this descending inhibitory pathway [113].

Alternatively, it was also shown that μ/δ-opioid synergy may occur at the cellular level [147]. The addition of DAMGO and DPDPE to cultures of undifferentiated SH-SY5Y cells demonstrated synergistic activity determined by real-time extracellular acidification rate [147]. Furthermore, the coadministration of DAMGO and DPDPE also attenuated receptor desensitization [148]. Similarly, the addition of ohmefentanyl and DPDPE to cultures of differentiated SH-SY5Y cells with functional μ- and δ-opioid receptors also demonstrated synergism [148]. An important interaction between the μ- and δ-opioid receptors has recently been described that provides a mechanistic interpretation through which such synergy may occur. Prior exposure of cortical neurons in culture to δ-opioid agonists (morphine, fentanyl) resulted in robust enhancement of δ-opioid receptor mediated internalization of fluodeltorphin [73]. Electron microscopy revealed a marked increase in cell surface density of δ-opioid receptors on the cell bodies of these neurons [73]. Moreover, it was found that this increase was not due to de novo synthesis, but to mobilization from intracelluar stores [73]. A similar translocation of the δ-opioid receptors to the neuronal membranes in response to δ-opioid receptor activation was observed with dorsal horn neurons of the spinal cord [73].

Potentiation of morphine-mediated antinociception through the activation of endogenous δ-opioid systems has been reported. It was found that CWSS produced antinociception that was mediated by supraspinal δ_2-opioid receptors [104]. Furthermore, CWSS also potentiated the antinociceptive effect of ICV morphine, producing a significant shift to the left of the morphine dose-response curve [105]. Importantly, however, the potentiation of the antinociceptive effect of ICV morphine was blocked by 5′-NTII and by antiserum to [Leu^5]enkephalin [105]. In contrast, the potentiation of morphine was not blocked by β-FNA or by antiserum to [Met^5]enkephalin [105]. These results suggested that CWSS increases release of the endogenous δ-opioid agonist [Leu^5]enkephalin, which acts at the δ_2-opioid receptor to enhance the antinociceptive effect of exogenous opioids [105].

The development of compounds with agonistic activity at both the μ-and δ-opioid receptors, and thus demonstrating "self-synergy," have been reported [149]. For example, the compound D-Tca-Cys-Tyr-D-Trp-Arg-Thr-Pen-Thr-NH2 ([D-Tca^1]CTAP) was determined through competition binding and isolated tissue assays to have weak agonistic activity at the μ-opioid and δ-opioid receptors [149–151]. However, [D-Tca^1]CTAP demonstrated a remarkably enhanced antinociceptive potency, inconsistent with the in vitro profiles [149]. For example, it showed to be equipotent to DPDPE, although its binding affinity for the δ-opioid receptor was 230 times weaker than that of DPDPE and its binding affinity for the δ-opioid receptor was 144-fold weaker than that of morphine [149]. Additional studies with receptor-selective antagonists revealed that [D-Tca^1]CTAP was an agonist at the μ- and δ_2-opioid receptors [149]. A similar unexpected potency and "self-potentiation" was observed with the novel opioid dipeptide, biphalin (Try-D-Ala-Gly-Phe-NH)-2) [150,151].

11 δ-OPIOID RECEPTORS AND INFLAMMATORY PAIN

It has been consistently and repeatedly reported that the inflammatory process is accompanied by a seemingly paradoxical enhancement of the antinociceptive activity of opioids [152]. For example, carrageenan-induced inflammation elicited a 30-fold increase in the ability of morphine to inhibit evoked activity of C-fibers [153]. The antinociceptive effect of morphine was similarly enhanced in arthritic rats [154]. Further, the effect of ITH DAMGO and U50,488H, but not DTLET, was enhanced in arthritic rats [155]. The ICV administration of DAMGO, morphine, [D-Ala^2,Glu^4]deltorphin, or SNC80 to rats made arthritic with complete Freund's adjuvant (CFA) demonstrated an enhanced antihyperalgesic response of these compounds [156]. Inflammation has been associated with an upregulation of spinal opioid peptides or of mRNA for the peptides [157,158]. Enhanced axonal transport of δ-opioid and μ-opioid receptors to terminals of the sciatic nerve was demonstrated, indicating a mechanism for enhanced antinociception of centrally and systemically administered δ-opioid and μ-opioid agonists [159].

More recently, Hammond and colleagues demonstrated significant potentiation of the antinociceptive action of DAMGO or of [D-Ala^2,Glu^4]-deltorphin administered into the RVM of rats with inflammation induced by CFA [160,161]. Moreover, based on observations that naltriben in the RVM alone enhanced hyperalgesia, and that [Met^5]enkephalin and [Leu^5]enkephalin levels increased in the RVM after inflammation, it was hypothesized that endogenous enkephalins mediate the enhanced antinociceptive effects of opioids during persistent inflammation [160,161]. These studies correlate with the fact that blockade of endogenous δ-opioidergic tone enhanced hyperalgesia after formalin injection [98]. An alternative explanation for enhanced δ-opioid antinociceptive activity after inflammation retrograde axonal transport of receptors to the terminals [159]. Binding studies and autoradiography demonstrated a profound increase in opioid receptors in the periphery in rats with CFA-induced inflammation [159]. Moreover, the time course was consistent with the inflammatory process and with receptor synthesis and axonal transport [159]. Further evidence was provided when ligation of the sciatic nerves after CFA treatment resulted in an accumulation of receptor protein proximal to the ligature, and a decrease of the receptor populations in the periphery [159]. It was also found that the δ-opioid receptors are found in large, dense-core vesicles of primary afferent fibers, and are colocalized with substance P and/or CGRP [82]. Moreover, these receptors embed into the plasma membranes of primary afferent terminals, and regulate exocytotic release of substance P and/or CGRP from these terminals, likely in response to [Leu^5]enkephalin released from adjacent neurons [82].

12 CCK AS A MEDIATOR OF ENHANCED OPIOID ACTIVITY IN INFLAMMATION

Although the mechanisms through which enhanced opioid activity may occur are not known, a complex interaction among μ-opioid, δ-opioid, and cholecystokinin (CCK) receptors has been proposed [153,162]. A basis for such interactions may lie in reports that CCK antagonizes morphine antinociception and that CCK antagonists potentiate morphine antinociception [163–168]. Numerous studies demonstrate that CCK acts in a pro-nociceptive fashion through the CCK_B receptor. Spinal or systemic CCK blocked endogenously opioid-mediated footshock-induced as well as morphine-induced antinociception [169]. CCK antagonists elicited an enhancement of morphine-induced antinociception while producing no antinociceptive activity when given alone [153,164,165,167,169–172]. Moreover, during conditions of chronic inflammation, endogenous levels of CCK decrease, thus removing the CCK-mediated antianalgesic mechanism and consequently enhancing endogenous opioid activity [153,162]. Evidence was presented that a decrease in endogenous CCK levels during inflammation may mediate the enhanced antinociceptive effect of morphine. Inflammation induced by carrageenan injection into a hindpaw produced increased C-fiber-evoked responses of dorsal horn units [162]. The antinociceptive effect of morphine to attenuate C-fiber-evoked responses was enhanced in the inflamed rats. Moreover, the administration of the CCK_B receptor antagonist L365,260 enhanced the effect of morphine in normal, but not in inflamed, rats [162]. Conversely, CCK attenuated the effect of morphine in inflamed, but not normal, rats [162]. It was suggested that, in the normal state, spinal morphine elicits a release of spinal CCK, which acts to modulate its own effect, and therefore exogenous CCK did not further reduce the effect of morphine [162]. Further, there exists a reduction in morphine-induced release of spinal CCK in the inflamed state, leading to an enhanced antinociceptive potency [162]. Notably, the CCK_B antagonist L365,260 significantly enhanced the antinociceptive effect of systemic or i.th. morphine in rats and mice, although it is inactive alone [173,174]. Finally, antinociception mediated by endogenous opioids by blockade of enkephalinases was also enhanced by CCK antagonists [175–178].

A possible mechanism through which CCK may act as an "antianalgesic" agent is by counteracting the opioid-induced inhibition of depolarization-induced Ca^{2+} influx into primary afferent neurons by eliciting a mobilization of Ca^{2+} from intracellular stores [153]. The net effect would be to maintain nociceptive neurotransmitter release from primary afferent fibers [153]. It is suggested that there exists a balance between endogenous opioid and CCK activity such that they modulate each other's activity [153,179]. Opioids promote CCK release, which in turn modulates the antinociceptive activity of the opioid, acting as an endogenous regulatory system [153,179]. Critically, a number of studies have suggested that modulatory actions of CCK_B antagonists against the antinociceptive potency of opioids is mediated through the δ-opioid receptors. For example, the CCK_B antagonist L365,260 given either ITH or SC enhanced the antinociceptive effect of morphine in rats [173]. This enhancement was completely blocked by δ-opioid selective doses of naltrindole given ITH or SC [173]. Importantly, naltrindole alone did not alter nociceptive responses, nor did it alter the antinociceptive effect of morphine alone, indicating a δ-opioid mediated link in this interaction [173]. Similarly, carrageenan-induced inflammation in the rat resulted in an enhanced antinociceptive effect of morphine, and this enhancement of morphine was blocked by naltrindole [180]. Importantly, the same dose of naltrindole was inactive against morphine in noninflamed rats [180]. It was suggested that CCK, acting via CCK_B receptors, causes a tonic increase in the release and/or availability of an endogenous ligand of opioid δ receptors, which is likely to be [Leu^5]enkephalin [173,180]. This interpretation is supported by the fact that the systemic administration of an inactive dose of [Leu^5]enkephalin produced a definitive synergistic interaction with systemic morphine [143]. In a detailed series of studies, the antinociceptive dose-response curve for ICV morphine was shifted sixfold to the left by ICV treatment with antisense, but not mismatch, ODN to CCK_B receptor [176].

The addition of naltrindole or of antisera to [Leu^5]enkephalin returned the dose-effect curve for morphine to its baseline level [176]. Importantly, the doses of NTI or of antiserum to [Leu^5]enkephalin were inactive alone, and did not alter the antinociceptive effect of morphine in mice not receiving antisense ODN. In a second, related series of experiments, the ITH, ICV, or SC administration of L365,260 produced leftward shifts in the antinociceptive dose-response curves of morphine given by the same route [174]. In all cases, the enhancement of morphine was blocked by SC naltrindole, and that of ICV morphine was also blocked by ICV antiserum to [Leu^5]enkephalin but not by antiserum to [Met^5]enkephalin [174]. In addition, the repeated administration of L365,260 to mice resulted in a decreased enhancement of the antinociceptive effect of morphine by L365,260, thus mimicking tolerance. Interestingly, however, this treatment also produced a significant rightward shift in the dose-effect curve for ICV [D-Ala^2,Glu^4]deltorphin [174]. Likewise, "tolerance" to repeated ICV [Leu^5]enkephalin also abolished the enhancement of morphine seen with L365,260. Moreover, the ICV administration of thiorphan, which was inactive alone, along with L365,260 produced an antinociception that was blocked by naltrindole or antiserum to [Leu^5]enkephalin, but not to [Met^5]enkephalin, and was not present in mice made tolerant to [D-Ala^2,Glu^4]deltorphin [174]. Taken together, these data provide strong evidence that activation of the CCK_B receptor by CCK results in a reduction in the release or availability of endogenous δ-opioid agonists, likely to be [Leu^5]enkephalin [174]. Thus, blockade of the CCK_B receptor may result in enhanced availability of endogenous [Leu^5]enkephalin, which is then able to potentiate the antinociceptive effect of morphine through the synergistic action mediated through δ-opioid and μ-opioid receptors [174]. This aspect of activation of CCK receptors is also critical in the reduced antinociceptive effects of morphine in states of peripheral nerve injury, as described below. Several studies employing microdialysis have recently clearly demonstrated that morphine promotes the release of CCK in vivo within the CNS [181–184]. Importantly, the morphine-induced release of CCK is blocked by naltrindole or naltriben, but not by CTOP or nor-BNI, indicating a uniquely δ-opioid mediated effect [182,183]. Furthermore, in vivo release of CCK was also elicited by δ-opioid agonists [181,183]. Finally, morphine-evoked release of CCK was blocked by the δ_2-opioid antagonist naltriben but not by the δ_1-opioid antagonist 7-benzylidenenaltrexone, leading to the suggestion that CCK release is mediated through the δ_2-opioid receptors [181].

13 NEUROPATHIC PAIN

It has been repeatedly demonstrated, in several animal models of nerve injury and clinically, that morphine is either ineffective or markedly attenuated against behavioral signs of neuropathic pain [185–192]. In addition, spinal morphine has been shown to be completely inactive against tactile hypersensitivity after L5/L6 spinal nerve ligation (SNL), even at doses that normally produce antinociception [187,188,193]. However, there are some inconsistencies in the literature; moreover, the effects of peripheral nerve injuries on the antinociceptive activity opioid agonists other than morphine have not been resolved. For example, in one study, systemic morphine was shown to elevate vocalization thresholds in animals with chronic constriction injury (CCI) of the sciatic nerve [194]. Rats with CCI also demonstrated attenuation of mechanical nociception in response to IV DAMGO or to the putative δ-opioid agonist BUBU [195]. Similarly, systemic BUBUC or DTLET also produced equivalent antinociception when the injured or contralateral hindpaw was tested [196]. In contrast, a three- to fivefold shift to the right in the spinal antinociceptive effect of DPDPE against thermal endpoints was reported in rats with CCI [197]. The spinal administration of [D-Ala^2,Glu^4]deltorphin within the dose range active against acute thermal nociceptive stimuli was shown to produce dose-dependent, naltrindole-sensitive blockade of tactile hypertsensitivity in rats with SNL [198]. Moreover, morphine, which was inactive alone, was active when given in the presence of an inactive dose of [D-Ala^2,Glu^4]deltorphin, and this potentiation of morphine was blocked by naltrindole [198]. In contrast, DPDPE given either ITH or ICV, and morphine given ITH, did not attenuate tactile hypersensitivity in rats with SNL [187]. Most recently, the ITH injection of DPDPE and of [D-Ala^2,Glu^4]deltorphin in rats with a crush injury of the sciatic nerve produced dose-dependent reversal of tactile hypersensitivity and of cold allodynia [199]. Furthermore, the effects of DPDPE and of [D-Ala^2,Glu^4]deltorphin were blocked by BNTX and 5NTII, respectively, indicating selective actions at the

δ_1- and δ_2-opioid receptors, respectively [199]. Interestingly, the ITH injection of BNTX alone enhanced cold allodynia in this model, unmasking a δ-opioid-mediated inhibitory tone [199]. Recently, a nonpeptidic δ-opioid agonist, SB 235863, was reported to be active against behavioral signs of neuropathic pain [200].

The possibility that peripheral nerve injuries resulted in a downregulation of spinal opioid receptors, thus resulting in diminished opioid antinociceptive activity, was investigated and produced conflicting conclusions. Quantitative autoradiography showed that rats with CCI presented an initial increase, then decrease, in μ-opioid-binding sites in the spinal dorsal horn [201]. The same injury produced a gradual decrease in δ-opioid receptor binding concentrations in laminae I, II, V, and X of the spinal dorsal horn [201]. In a similar study, a reduction of 24% in numbers of δ-opioid sites was determined 2 weeks after CCI, with the δ-opioid receptor population returning to control levels within 4 weeks [202]. More sensitive immunohistochemical techniques demonstrated only modest reductions in immunoreactivity for the δ-opioid receptor and for substance P in the outer laminae of the dorsal horns of the spinal cord within 7 days of either nerve transection or crush [203]. Moreover, staining intensity for the δ-opioid receptor appeared to be maximal at 4 weeks postinjury and return to baseline levels within 32 weeks of the nerve injuries [203]. Competition studies were performed in brain tissue of rats with diabetes-produced neuropathy induced with streptozotocin. Although the rats demonstrated a definitive loss of antinociceptive potency of morphine against mechanical, thermal, and chemical nociceptive stimuli, the K_D and B_{max} values for the μ-opioid and the δ-opioid receptors remained unchanged, indicating no loss of numbers or of binding affinities of these opioid receptors [204]. These findings are consistent with the report that, at least for the μ-opioid receptors, peripheral nerve injury does not result in receptor downregulation in the spinal cord [205].

It is probable that δ-opioid agonists retain antinociceptive activitity in conditions of neuropathic pain because of the interactions among the opioid and CCK receptors. Importantly, peripheral nerve injury results in an increase, rather than a decrease, in endogenous CCK levels, thus promoting nociception [153,162,172]. For example, sciatic nerve section was associated with autotomy behavior, an upregulation of CCK mRNA in primary sensory neurons of the rat dorsal root ganglia and a loss of sensitivity of ITH morphine [206]. The coadministration of morphine with the CCK_B antagonist CI988 blocked signs of neuropathic pain [206]. In another study, it was found that CI988 relieved tactile hypersensitivity in a naloxone-sensitive fashion, suggesting that neuropathic pain results from a disruption of a normal tonic opioidergic control of nociceptive impulses secondary to an upregulation of an endogenous spinal CCK [207]. In a more recent study, it was found that, although ITH injections of morphine or of the CCK_B antagonist L365,260 alone did not modify tactile hypersensitivity in rats with SNL, the coadministration of ITH morphine and L365,260 produced a significant antiallodynic action [198,208]. Importantly, this enhancement of the antinociceptive effect of morphine was blocked by NTI, again showing that CCK may function to inhibit the availability of an endogenous δ-opioid agonist, perhaps [Leu^5]enkephalin. In support of this hypothesis, it was also demonstrated that an inactive dose of [D-Ala^2, Glu^4]deltorphin also produced an NTI-sensitive enhancement of ITH morphine in rats with SNL [198]. Interestingly, while the CCK antagonist alone did not elicit an antiallodynic action, coadministration of the CCK antagonist with thiorphan elicited a significant blockade of tactile hypersensitivity which was readily antagonized by NTI or by antisera to [Leu^5]enkephalin [208]. Thus, CCK_B receptor blockade may enhance endogenous enkephalin actions resulting in enhancement of morphine efficacy. Consistent with this hypothesis, it was shown that although exogenously administered CCK blocked the antinociceptive effect of μ-opioid (PL-017) or κ-opioid (66A-078), it was absolutely inactive against DPDPE [209]. Thus, the lack of effect of CCK against δ-opioid agonists, along with the fact that activation of δ-opioid receptors result in reduced binding of CCK agonists to their receptors [210], indicates that δ-opoid agonists are ideally suited as therapeutic targets for the treatment of neuropathic pain states. These findings support the concept that blockade of enhanced levels or activity of CCK, which appears to be associated with the neuropathic state, can result in a significant benefit to the potency and efficacy of substances that act at opioid receptors, especially those that act at opioid δ receptors, and supports the hypothesis that a molecule with these characteristics would be useful in treatment of neuropathic pain states.

14 CONCLUSIONS

The existence of the δ-opioid receptor as an antinociceptive mediator has been well established. Although only a single gene for the δ-opioid receptor has been cloned, considerable pharmacologic evidence establishes the presence of δ_1- and δ_2-opioid receptor subtypes, which may represent splice variants of a single gene. The δ-opioid receptors are distributed throughout the CNS in a manner consistent with anatomically defined pain-processing pathways. Moreover, the δ-opioid receptors are present at pre- and postsynaptic sites, and may synergize with μ-opioid receptors to enhance their antinociceptive actions. The possibility that the δ-opioid receptors may promote the release of endogenous CCK, which may mediate enhanced pain after nerve injury or antinociceptive tolerance to morphine, indicates that an increased understanding of the physiologic functions of the δ-opioid receptors may promote the development of novel therapeutic avenues for the treatment of these conditions.

REFERENCES

1. Pert CB, Snyder SH. Science 1973; 179:1011–1014.
2. Pert CB, Snyder SH. Proc Natl Acad Sci USA 1973; 70:2243–2247.
3. Simon EJ. Am J Med Sci 1973; 266:160–168.
4. Martin WR, Eades CG, Thompson JA, Huppler RE, Gilbert PE. J Pharmacol Exp Ther 1976; 197:517–532.
5. Martin WR. Pharmacol Rev 1983; 35:283–323.
6. Waterfield AA, Lord JA, Hughes JKosterlitz HW. Eur J Pharmacol 1978; 47.
7. Waterfield AA, Leslie FM, Lord JA, Ling N, Kosterlitz HW. Eur J Pharmacol 1979; 58:11–18.
8. Lord JA, Waterfield AA, Hughes J, Kosterlitz HW. Nature 1977; 267:495–499.
9. Schulz R, Wuster M, Herz A. J Pharmacol Exp Ther 1981; 216:604–606.
10. Zaki PA, Bilsky EJ, Vanderah TW, Lai J, Evans CJ, Porreca F. Annu Rev Pharmacol Toxicol 1996; 36:379–401.
11. Pfeiffer A, Herz A. Eur J Pharmacol 1982; 77:359–361.
12. Ward SJ, Portoghese PS, Takemori AE. Eur J Pharmacol 1982; 85:163–170.
13. Mosberg HI, Hurst R, Hruby VJ, Galligan JJ, Burks TF, Gee K, Yamamura HI. Life Sci 1983; 32:2565–2569.
14. Mosberg HI, Hurst R, Hruby VJ, Gee K, Akiyama K, Yamamura HI, Galligan JJ, Burks TF. Life Sci 1983; 33(suppl 1):447–450.
15. Akiyama K, Gee KW, Mosberg HI, Hruby VJ, Yamamura HI. Proc Natl Acad Sci USA 1985; 82:2543–2547.
16. Erspamer V, Melchiorri P, Falconieri-Erspamer G, Negri L, Corsi R, Severini C, Barra D, Simmaco M, Kreil G. Proc Natl Acad Sci USA 1989; 86:5188–5192.
17. Kreil G, Barra D, Simmaco M, Erspamer V, Erspamer GF, Negri L, Severini C, Corsi R, Melchiorri P. Eur J Pharmacol 1989; 162:123–128.
18. Chang KJ, Rigdon GC, Howard JL, McNutt RW. J Pharmacol Exp Ther 1993; 267:852–857.
19. Childers SR, Fleming LM, Selley DE, McNutt RW, Chang KJ. Mol Pharmacol 1993; 44:827–834.
20. Craft RM, Henley SR, Haaseth RC, Hruby VJ, Porreca F. J Pharmacol Exp Ther 1995; 275:1535–1542.
21. Wild KD, McCormick J, Bilsky EJ, Vanderah T, McNutt RW, Chang KJ, Porreca F. J Pharmacol Exp Ther 1993; 267:858–865.
22. Comer SD, Hoenicke EM, Sable AI, McNutt RW, Chang KJ, De Costa BR, Mosberg HI, Woods JH. J Pharmacol Exp Ther 1993; 267:888–895.
23. Calderon SN, Rothman RB, Porreca F, Flippen-Anderson JL, McNutt RW, Xu H, Smith LE, Bilsky EJ, Davis P, Rice KC. J Med Chem 1994; 37:2125–2128.

24. Calderon SN, Rice KC, Rothman RB, Porreca F, Flippen-Anderson JL, Kayakiri H, Xu H, Becketts K, Smith LE, Bilsky EJ, Davis P, Horvath R. J Med Chem 1997; 40:695–704.
25. Schetz JA, Calderon SN, Bertha CM, Rice K, Porreca F. J Pharmacol Exp Ther 1996; 279:1069–1076.
26. Cao CQ, Hong Y, Dray A, Perkins M. Eur J Pharmacol 2001; 418:79–87.
27. Bilsky EJ, Calderon SN, Wang T, Bernstein RN, Davis P, Hruby VJ, McNutt RW, Rothman RB, Rice KC, Porreca F. J Pharmacol Exp Ther 1995; 273: 359–366.
28. Zhang X, Rice KC, Calderon SN, Kayakiri H, Smith L, Coop A, Jacobson AE, Rothman RB, Davis P, Dersch CM, Porreca F. J Med Chem 1999; 42: 5455–5463.
29. Kieffer BL, Befort K, Gaveriaux-Ruff C, Hirth CG. Proc Natl Acad Sci USA 1992; 89:12048–12052.
30. Evans CJ, Keith DE Jr, Morrison H, Magendzo K, Edwards RH. Science 1992; 258:1952–1955.
31. Keith DE Jr, Anton B, Evans CJ. Proc West Pharmacol Soc 1993; 36:299–306.
32. Malatynska E, Wang Y, Knapp RJ, Santoro G, Li X, Waite S, Roeske WR, Yamamura HI. Neuroreport 1995; 6:613–616.
33. Mattia A, Vanderah T, Mosberg HI, Porreca F. J Pharmacol Exp Ther 1991; 258:583–587.
34. Jiang Q, Takemori AE, Sultana M, Portoghese PS, Bowen WD, Mosberg HI, Porreca F. J Pharmacol Exp Ther 1991; 257:1069–1075.
35. Bowen WD, Hellewell SB, Kelemen M, Huey R, Stewart D. J Biol Chem 1987; 262:13434–13439.
36. Portoghese PS, Sultana M, Takemori AE. J Med Chem 1990; 33:1547–1548.
37. Sofia RD, Nalepa SD, Harakal JJ, Vassar HB. J Pharmacol Exp Ther 1973; 186:646–655.
38. Sofuoglu M, Portoghese PS, Takemori AE. J Pharmacol Exp Ther 1991; 257: 676–680.
39. Sofuoglu M, Portoghese PS, Takemori AE. Life Sci 1993; 52:769–775.
40. Stewart PE, Hammond DL. J Pharmacol Exp Ther 1993; 266:820–828.
41. Tiseo PJ, Yaksh TL. European Journal of Pharmacology 1993; 236:89–96.
42. Mattia A, Farmer SC, Takemori AE, Sultana M, Portoghese PS, Mosberg HI, Bowen WD, Porreca F. J Pharmacol Exp Ther 1992; 260:518–525.
43. Horan PJ, Wild KD, Misicka A, Lipkowski A, Haaseth RC, Hruby VJ, Weber SJ, Davis TP, Yamamura HI, Porreca F. J Pharmacol Exp Ther 1993; 265:896–902.
44. Kalso EA, Sullivan AF, McQuay HJ, Dickenson AH, Roques BP. J Pharmacol Exp Ther 1993; 265:551–558.
45. Sofuoglu M, Portoghese PS, Takemori AE. Life Sci 1991; 49:PL153–PL156.
46. Negri L, Potenza RL, Corsi R, Melchiorri P. Eur J Pharmacol 1991; 196:335–336.
47. Fang L, Knapp RJ, Horvath R, Matsunaga TO, Haaseth RC, Hruby VJ, Porreca F, Yamamura HI. J Pharmacol Exp Ther 1994; 268:836–846.

48. Rothman RB, Bykov V, Jacobson AE, Rice KC, Long JE, Bowen WD. Peptides 1992; 13:691–694.
49. Porreca F, Bilsky EJ, Lai J. In: Tseng LF, ed. The Pharmacology of Opioid Peptides. Singapore: Harwood Academic Publishers, 1995:219–248.
50. Bare LA, Mansson E, Yang D. FEBS Lett 1994; 354:213–216.
51. Xie GX, Meuser T, Pietruck C, Sharma M, Palmer PP. Life Sci 1999; 64:2029–2037.
52. Zimprich A, Simon T, Hollt V. Neuroreport 1995; 7:54–56.
53. Zimprich A, Simon T, Hollt V. FEBS Lett 1995; 359:142–146.
54. Mayer P, Tischmeyer H, Jayasinghe M, Bonnekoh B, Gollnick H, Teschemacher H, Hollt V. FEBS Lett 2000; 480:156–160.
55. Fukuda K, Kato S, Mori K, Nishi M, Takeshima H. FEBS Lett 1993; 327:311–314.
56. Kim DH, Fields HL, Barbaro NM. Brain Res 1990; 516:37–40.
57. Knapp RJ, Malatynska E, Collins N, Fang L, Wang JY, Hruby VJ, Roeske WR, Yamamura HI. FASEB J 1995; 9:516–525.
58. Mansour A, Fox CA, Akil H, Watson SJ. Trends Neurosci 1995; 18:22–29.
59. Simonin F, Befort K, Gaveriaux-Ruff C, Matthes H, Nappey V, Lannes B, Micheletti G, Kieffer B. Mol Pharmacol 1994; 46:1015–1021.
60. Bzdega T, Chin H, Kim H, Jung HH, Kozak CA, Klee WA. Proc Natl Acad Sci USA 1993; 90:9305–9309.
61. Mansour A, Khachaturian H, Lewis ME, Akil H, Watson SJ. J Neurosci 1987; 7:2445–2464.
62. Quirion R, Zajac JM, Morgat JL, Roques BP. Life Sci 1983; 33(supp 1):227–230.
63. Gulya K, Gehlert DR, Wamsley JK, Mosberg H, Hruby VJ, Yamamura HI. J Pharmacol Exp Ther 1986; 238:720–726.
64. Mansour A, Fox CA, Burke S, Meng F, Thompson RC, Akil H, Watson SJ. J Comp Neurol 1994; 350:412–438.
65. Arvidsson U, Dado RJ, Riedl M, Lee JH, Law PY, Loh HH, Elde R, Wessendorf MW. J Neurosci 1995; 15:1215–1235.
66. Commons KG, Milner TA. J Comp Neurol 1997; 381:373–387.
67. Commons KG, Milner TA. Brain Res 1996; 738:181–195.
68. Ableitner A. Eur J Pharmacol 1994; 271:213–222.
69. Gouarderes C, Tellez S, Tafani JA, Zajac JM. Synapse 1993; 13:231–240.
70. Cahill CM, McClellan KA, Morinville A, Hoffert C, Hubatsch D, O'Donnell D, Beaudet A. J Comp Neurol 2001; 440:65–84.
71. Bausch SB, Patterson TA, Appleyard SM, Chavkin C. J Chem Neuroanat 1995; 8:175–189.
72. Besse D, Lombard MC, Besson JM. Brain Res 1991; 548:287–291.
73. Cahill CM, Morinville A, Lee MC, Vincent JP, Collier B, Beaudet A. J Neurosci 2001; 21:7598–7607.
74. Besse D, Lombard MC, Zajac JM, Roques BP, Besson JM. Brain Res 1990; 521:15–22.

75. Besse D, Lombard MC, Zajac JM, Roques BP, Besson JM. Prog Clin Biol Res 1990; 328:183–186.
76. Stevens CW, Lacey CB, Miller KE, Elde RP, Seybold VS. Brain Res 1991; 550:77–85.
77. Zajac JM, Lombard MC, Peschanski M, Besson JM, Roques BP. Brain Res 1989; 477:400–403.
78. Stevens CW, Seybold VS. Brain Res 1995; 687:53–62.
79. Peckys D, Landwehrmeyer GB. Neuroscience 1999; 88:1093–1135.
80. Honda CN, Arvidsson U. Neuroreport 1995; 6:1025–1028.
81. Abbadie C, Lombard MC, Besson JM, Trafton JA, Basbaum AI. Brain Res 2002; 930:150–162.
82. Zhang X, Bao L, Arvidsson U, Elde R, Hokfelt T. Neuroscience 1998; 82: 1225–1242.
83. Yaksh TL. Eur J Anaesthesiol 1984; 1:171–199.
84. Yaksh TL. Acta Anaesthesiol Scand Suppl 1987; 85:25–37.
85. Yaksh TL. J Pharmacol Exp Ther 1983; 226:303–316.
86. Porreca F, Mosberg HI, Hurst R, Hruby VJ, Burks TF. J Pharmacol Exp Ther 1984; 230:341–348.
87. Porreca F, Heyman JS, Mosberg HI, Omnaas JR, Vaught JL. J Pharmacol Exp Ther 1987; 241:393–400.
88. Heyman JS, Mulvaney SA, Mosberg HI, Porreca F. Brain Res 1987; 420:100–108.
89. Pitcher GM, Yashpal K, Coderre TJ, Henry JL. J Pharmacol Exp Ther 1995; 273:1428–1433.
90. Dickenson AH, Sullivan AF, Roques BP. Eur J Pharmacol 1988; 148:437–439.
91. Drower EJ, Stapelfeld A, Rafferty MF, de Costa BR, Rice KC, Hammond DL. J Pharmacol Exp Ther 1991; 259:725–731.
92. Improta G, Broccardo M. Peptides 1992; 13:1123–1126.
93. Hammond DL, Stewart PE, Littell L. J Pharmacol Exp Ther 1995; 274:1317–1324.
94. Hammond DL, Wang H, Nakashima N, Basbaum AI. J Pharmacol Exp Ther 1998; 284:378–387.
95. Suarez-Roca H, Maixner W. Eur J Pharmacol 1992; 229:1–7.
96. Zachariou V, Goldstein BD. Brain Res 1996; 736:305–314.
97. Collin E, Mauborgne A, Bourgoin S, Chantrel D, Hamon M, Cesselin F. Neuroscience 1991; 44:725–731.
98. Ossipov MH, Kovelowski CJ, Wheeler-Aceto H, Cowan A, Hunter JC, Lai J, Malan TP Jr, Porreca F. J Pharmacol Exp Ther 1996; 277:784–788.
99. Galligan JJ, Mosberg HI, Hurst R, Hruby VJ, Burks TF. J Pharmacol Exp Ther 1984; 229:641–648.
100. Miaskowski C, Taiwo YO, Levine JD. Eur J Pharmacol 1991; 205:247–252.
101. Jiang Q, Mosberg HI, Porreca F. Life Sci 1990; 47:PL43–PL47.
102. Vanderah T, Takemori AE, Sultana M, Portoghese PS, Mosberg HI, Hruby VJ, Haaseth RC, Matsunaga TO, Porreca F. Eur J Pharmacol 1994; 252:133–137.

103. Watkins LR, Wiertelak EP, Maier SF. Brain Res 1992; 582:10–21.
104. Vanderah TW, Wild KD, Takemori AE, Sultana M, Portoghese PS, Bowen WD, Mosberg HI, Porreca F. J Pharmacol Exp Ther 1992; 262:190–197.
105. Vanderah TW, Wild KD, Takemori AE, Sultana M, Portoghese PS, Bowen WD, Hruby VJ, Mosberg HI, Porreca F. J Pharmacol Exp Ther 1993; 267: 449–455.
106. Adams JU, Tallarida RJ, Geller EB, Adler MW. J Pharmacol Exp Ther 1993; 266:1261–1267.
107. Negri L, Noviello V, Angelucci F. Eur J Pharmacol 1991; 209:163–168.
108. Dauge V, Petit F, Rossignol P, Roques BP. Eur J Pharmacol 1987; 141:171–178.
109. Satoh M, Kubota A, Iwama T, Wada T, Yasui M, Fujibayashi K, Takagi H. Life Sci 1983; 33(suppl 1):689–692.
110. Jensen TS, Yaksh TL. Brain Res 1986; 372:301–312.
111. Ossipov MH, Kovelowski CJ, Nichols ML, Hruby VJ, Porreca F. Pain 1995; 62:287–293.
112. Thorat SN, Hammond DL. J Pharmacol Exp Ther 1997; 283:1185–1192.
113. Kiefel JM, Rossi GC, Bodnar RJ. Brain Res 1993; 624:151–161.
114. Holmes BB, Fujimoto JM. Pharmacol Biochem Behav 1994; 49:675–682.
115. Kovelowski CJ, Ossipov MH, Hruby VJ, Porreca F. Pain 1999; 83:115–122.
116. Harasawa I, Fields HL, Meng ID. Pain 2000; 85:255–262.
117. Quock RM, Burkey TH, Varga E, Hosohata Y, Hosohata K, Cowell SM, Slate CA, Ehlert FJ, Roeske WR, Yamamura HI. Pharmacol Rev 1999; 51: 503–532.
118. Standifer KM, Chien CC, Wahlestedt C, Brown GP, Pasternak GW. Neuron 1994; 12:805–810.
119. Standifer KM, Jenab S, Su W, Chien CC, Pan YX, Inturrisi CE, Pasternak GW. J Neurochem 1995; 65:1981–1987.
120. Lai J, Bilsky EJ, Rothman RB, Porreca F. Neuroreport 1994; 5:1049–1052.
121. Bilsky EJ, Bernstein RN, Pasternak GW, Hruby VJ, Patel D, Porreca F, Lai J. Life Sci 1994; 55:PL37–PL43.
122. Bilsky EJ, Bernstein RN, Hruby VJ, Rothman RB, Lai J, Porreca F. J Pharmacol Exp Ther 1996; 277:491–501.
123. Lai J, Bilsky EJ, Porreca F. J Recept Signal Transduct Res 1995; 15:643–650.
124. Lai J, Crook TJ, Payne A, Lynch RM, Porreca F. J Pharmacol Exp Ther 1997; 281:589–596.
125. Rossi GC, Su W, Leventhal L, Su H, Pasternak GW. Brain Res 1997; 753:176–179.
126. Fraser GL, Pradhan AA, Clarke PB, Wahlestedt C. J Pharmacol Exp Ther 2000; 295:1135–1141.
127. Zhu Y, King MA, Schuller AG, Nitsche JF, Reidl M, Elde RP, Unterwald E, Pasternak GW, Pintar JE. Neuron 1999; 24:243–252.
128. Wahlestedt C, Salmi P, Good L, Kela J, Johnsson T, Hokfelt T, Broberger C, Porreca F, Lai J, Ren K, Ossipov M, Koshkin A, Jakobsen N, Skouv J,

Oerum H, Jacobsen MH, Wengel J. Proc Natl Acad Sci USA 2000; 97:5633–5638.
129. Bilsky EJ, Wang T, Lai J, Porreca F. Neurosci Lett 1996; 220:155–158.
130. Yeung JC, Rudy TA. J Pharmacol Exp Ther 1980; 215:633–642.
131. Roerig SC, Fujimoto JM. J Pharmacol Exp Ther 1989; 249:762–768.
132. Miaskowski C, Taiwo YO, Levine JD. Brain Res 1993; 608:87–94.
133. Miaskowski C, Levine JD. Brain Res 1992; 595:32–38.
134. Hurley RW, Grabow TS, Tallarida RJ, Hammond DL. J Pharmacol Exp Ther 1999; 289:993–999.
135. Grabow TS, Hurley RW, Banfor PN, Hammond DL. Pain 1999; 83:47–55.
136. Kovelowski CJ, Bian D, Hruby VJ, Lai J, Ossipov MH, Porreca F. Brain Res 1999; 843:12–17.
137. Tallarida RJ, Porreca F, Cowan A. Life Sci 1989; 45:947–961.
138. Roerig SC, Hoffman RG, Takemori AE, Wilcox GL, Fujimoto JM. J Pharmacol Exp Ther 1991; 257:1091–1099.
139. Vaught JL, Takemori AE. J Pharmacol Exp Ther 1979; 211:280–283.
140. Heyman JS, Jiang Q, Rothman RB, Mosberg HI, Porreca F. Eur J Pharmacol 1989; 169:43–52.
141. Jiang Q, Mosberg HI, Porreca F. Prog Clin Biol Res 1990; 328:449–452.
142. Horan P, Tallarida RJ, Haaseth RC, Matsunaga TO, Hruby VJ, Porreca F. Life Sci 1992; 50:1535–1541.
143. Porreca F, Jiang Q, Tallarida RJ. Eur J Pharmacol 1990; 179:463–468.
144. Miaskowski C, Sutters KA, Taiwo YO, Levine JD. Pain 1992; 49:137–144.
145. Malmberg AB, Yaksh TL. J Pharmacol Exp Ther 1992; 263:264–275.
146. Rossi GC, Pasternak GW, Bodnar RJ. Brain Res 1994; 665:85–93.
147. Chen ZW, Yang K, Wang Y, Han JS. Neuroreport 2001; 12:845–849.
148. Chen ZW, Yang K, Wang Y, Han JS. Neurosci Lett 2001; 298:199–202.
149. Horan PJ, Wild KD, Kazmierski WM, Ferguson R, Hruby VJ, Weber SJ, Davis TP, Fang L, Knapp RJ, Yamamura HI. Eur J Pharmacol 1993; 233:53–62.
150. Shen KF, Crain SM. Brain Res 1995; 701:158–166.
151. Horan PJ, Mattia A, Bilsky EJ, Weber S, Davis TP, Yamamura HI, Malatynska E, Appleyard SM, Slaninova J, Misicka A. J Pharmacol Exp Ther 1993; 265:1446–1454.
152. Stanfa L, Dickenson A. Inflamm Res 1995; 44:231–241.
153. Stanfa L, Dickenson A, Xu XJ, Wiesenfeld-Hallin Z. Trends Pharmacol Sci 1994; 15:65–66.
154. Kayser V, Guilbaud G. Brain Res 1983; 267:131–138.
155. Neil A, Kayser V, Gacel G, Besson JM, Guilbaud G. Eur J Pharmacol 1986; 130:203–208.
156. Fraser GL, Gaudreau GA, Clarke PB, Menard DP, Perkins MN. Br J Pharmacol 2000; 129:1668–1672.
157. Dubner R, Ruda MA. Trends Neurosci 1992; 15:96–103.
158. Noguchi K, Dubner R, Ruda MA. Neuroscience 1992; 46:561–570.
159. Hassan AH, Ableitner A, Stein C, Herz A. Neuroscience 1993; 55:185–195.

160. Hurley RW, Hammond DL. J Neurosci 2000; 20:1249–1259.
161. Hurley RW, Hammond DL. J Neurosci 2001; 21:2536–2545.
162. Stanfa LC, Dickenson AH. Br J Pharmacol 1993; 108:967–973.
163. Wiesenfeld-Hallin Z, de Arąuja Lucas G, Alster P, Xu XJ, Hokfelt T. Brain Res 1999; 848:78–89.
164. Watkins LR, Kinscheck IB, Mayer DJ. Brain Res 1985; 327:169–180.
165. Watkins LR, Kinscheck IB, Kaufman EF, Miller J, Frenk H, Mayer DJ. Brain Res 1985; 327:181–190.
166. Xu XJ, Hoffmann O, Wiesenfeld-Hallin Z. Neuropeptides 1996; 30:203–206.
167. Dourish CT, O'Neill MF, Coughlan J, Kitchener SJ, Hawley D, Iversen SD. Eur J Pharmacol 1990; 176:35–44.
168. O'Neill MF, Dourish CT, Iversen SD. Neuropharmacology 1989; 28:243–247.
169. Faris PL, Komisaruk BR, Watkins LR, Mayer DJ. Science 1983; 219:310–312.
170. Hughes J, Hunter JC, Woodruff GN. Neuropeptides 1991; 19(suppl):85–89.
171. Suh HH, Tseng LF. Eur J Pharmacol 1990; 179:329–338.
172. Stanfa LC, Sullivan AF, Dickenson AH. Pain 1992; 50:345–354.
173. Ossipov MH, Kovelowski CJ, Vanderah T, Porreca F. Neurosci Lett 1994; 181:9–12.
174. Vanderah TW, Bernstein RN, Yamamura HI, Hruby VJ, Porreca F. J Pharmacol Exp Ther 1996; 278:212–219.
175. Noble F, Derrien M, Roques BP. Br J Pharmacol 1993; 109:1064–1070.
176. Vanderah TW, Lai J, Yamamura HI, Porreca F. Neuroreport 1994; 5:2601–2605.
177. Maldonado R, Derrien M, Noble F, Roques BP. Neuroreport 1993; 4:947–950.
178. Valverde O, Maldonado R, Fournie-Zaluski MC, Roques BP. J Pharmacol Exp Ther 1994; 270:77–88.
179. Wiesenfeld-Hallin Z, Xu XJ. Regul Pept 1996; 65:23–28.
180. Ossipov MH, Kovelowski CJ, Porreca F. Neurosci Lett 1995; 184:173–176.
181. Becker C, Hamon M, Cesselin F, Benoliel JJ. Synapse 1999; 34:47–54.
182. Becker C, Pohl M, Thiebot MH, Collin E, Hamon M, Cesselin F, Benoliel JJ. Neuropharmacology 2000; 39:161–171.
183. Gustafsson H, Afrah AW, Stiller CO. J Neurochem 2001; 78:55–63.
184. Gustafsson H, Afrah A, Brodin E, Stiller CO. J Neurochem 1999; 73:1145–1154.
185. Arner S, Meyerson BA. Pain 1988; 33:11–23.
186. Payne R. Clin J Pain 1986; 2:59–73.
187. Lee YW, Chaplan SR, Yaksh TL. Neurosci Lett 1995; 199:111–114.
188. Bian D, Nichols ML, Ossipov MH, Lai J, Porreca F. Neuroreport 1995; 6:1981–1984.
189. Wegert S, Ossipov MH, Nichols ML, Bian D, Vanderah TW, Malan TP Jr, Porreca F. Pain 1997; 71:57–64.
190. Mao J, Price DD, Mayer DJ, Lu J, Hayes RL. Brain Res 1992; 576:254–262.
191. Mao J, Price DD, Mayer DJ. Pain 1995; 61:353–364.
192. Yamamoto T, Yaksh TL. Neurosci Lett 1992; 135:67–70.

193. Bian D, Ossipov MH, Ibrahim M, Raffa RB, Tallarida RJ, Malan TP Jr, Lai J, Porreca F. Brain Res 1999; 831:55–63.
194. Attal N, Bouhassira D. Acta Neurol Scand Suppl 1999; 173:12–24 (discussion 48–52).
195. Desmeules JA, Kayser V, Guilbaud G. Pain 1993; 53:277–285.
196. Catheline G, Kayser V, Idanpaan-Heikkila JJ, Guilbaud G. Eur J Pharmacol 1996; 318:273–281.
197. Yamamoto T, Yaksh TL. Anesthesiology 1991; 75:817–826.
198. Nichols ML, Bian D, Ossipov MH, Lai J, Porreca F. J Pharmacol Exp Ther 1995; 275:1339–1345.
199. Mika J, Przewlocki R, Przewlocka B. Eur J Pharmacol 2001; 415:31–37.
200. Dondio G. Farmaco 2000; 55:178–180.
201. Stevens CW, Kajander KC, Bennett GJ, Seybold VS. Pain 1991; 46:315–326.
202. Besse D, Lombard MC, Perrot S, Besson JM. Neuroscience 1992; 50:921–933.
203. Robertson B, Schulte G, Elde R, Grant G. Eur J Pain 1999; 3:115–129.
204. Courteix C, Bourget P, Caussade F, Bardin M, Coudore F, Fialip J, Eschalier A. J Pharmacol Exp Ther 1998; 285:63–70.
205. Porreca F, Tang QB, Bian D, Riedl M, Elde R, Lai J. Brain Res 1998; 795:197–203.
206. Xu XJ, Puke MJ, Verge VM, Wiesenfeld-Hallin Z, Hughes J, Hokfelt T. Neurosci Lett 1993; 152:129–132.
207. Xu XJ, Hao JX, Seiger A, Hughes J, Hokfelt T, Wiesenfeld-Hallin Z. Pain 1994; 56:271–277.
208. Nichols ML, Bian D, Ossipov MH, Malan TP Jr, Porreca F. Neurosci Lett 1996; 215:161–164.
209. Wang XJ, Wang XH, Han JS. Brain Res 1990; 523:5–10.
210. Ruiz-Gayo M, Durieux C, Fournie-Zaluski MC, Roques BP. J Neurochem 1992; 59:1805–1811.

19

Delta Opioid Receptor–Mediated Antinociception/Analgesia

Minoru Narita and Tsutomu Suzuki
Hoshi University School of Pharmacy and Pharmaceutical Sciences, Tokyo, Japan

1 IDENTIFICATION AND CLONING OF DELTA OPIOID RECEPTORS

Opioid analgesic drugs and endogenous opioid peptides exert a wide spectrum of physiological and behavioral effects on pain perception, mood, motor control, and autonomic function. The myriad opioid-induced effects are mediated via a family of the specific membrane-bound receptor located on nervous tissue, namely the opioid receptor. Early studies of the bindings of various ligands have suggested the existence of a multitude of distinct types of opioid receptors that can interact with opioid drugs or endogenous peptides. More evidence for the existence of opioid receptor types has recently been confirmed by molecular cloning. The three major opioid receptor types—mu, delta, and kappa—have been cloned and sequenced [1–12].

The relatively large number of opioid peptides isolated in recent years is a reflection of the complexity of the endogenous opioid system. The major opioid peptides are cleavage products of three distinct proteins, which are the primary products of three genes. These precursor proteins are proenkephalin

(PENK), prodynorphin (PDYN), and pro-opiomelanocortin (POMC) [13–15]. PENK was originally discovered in bovine adrenal cortex and pig brain, where enkephalin biosynthesis was elucidated [13]. It contains one copy of Leu-enkephalin, four copies of Met-enkephalin, and one copy each of a Met-enkephalin C-terminal-extended heptapeptide and octapeptides, all flanked by basic dipeptides, where processing generally takes place. The primary identification of delta opioid receptors was a direct consequence of the discovery of these enkephalins [16]. These peptides were first examined for their potency at opioid receptors present in the guinea pig ileum and mouse vas deferens bioassays. The result of these studies shows that enkephalins are acting at a different receptor in the mouse vas deferens than the guinea pig ileum and this receptor for enkephalins was defined as the elimination of the delta opioid receptor [17].

In late 1992, a delta opioid receptor was the first such gene to be cloned by two independent groups, Evans et al. [1] and Kieffer et al. [2]. Delta opioid receptors present on NG 108-15 cells are monomeric membrane proteins having seven transmembrane domains. The two receptors described in these studies show some differences in nucleotide sequence; one encodes 371 amino acid [2] while the second has 372 [1]. The two receptors cloned from NG 108-15 cDNA were expressed in COS cell and show high binding affinity for delta opioid receptor ligands but low affinity for mu or kappa opioid receptor selective ligands. Evans et al. [1] showed that the expressed delta opioid receptor mediates agonist-induced inhibition of adenylyl cyclase, while Kieffer et al. [2] showed that the receptor was from the mouse component of the hybrid NG 108-15 cell genome. Subsequent studies have in the mouse delta opioid receptor [8]. The rat delta opioid receptor has 97% amino acid sequence identity relative to the mouse delta opioid receptor. Both receptors have ~60% amino acid identity to the rat mu opioid receptor. Following the cloning of the delta opioid receptor, mu and kappa opioid receptors have been cloned in the past several years [3–12]. A cDNA clone encoding a structurally related receptor with amino acid similarity to these three receptor genes as high as 65% has been reported by several groups [18–20]. Although all three major types of opioid receptors are able to mediate antinociception/analgesia, their individual binding profiles and other pharmacological activities clearly distinguish one from another.

2 ANATOMICAL DISTRIBUTION OF DELTA OPIOID RECEPTORS

The cloning of the opioid receptors has enabled the direct visualization of the cells that synthesize the receptors by *in situ* hybridization of the receptor mRNA in relation to the final distribution of the binding site as determined using receptor autoradiography. A high correlation between delta opioid receptor mRNA expression and binding is observed in such regions as the anterior olfactory nucleus, neocortex, caudate-putamen, nucleus accumbens, olfactory tubercle, diagonal band of Broca, globus pallidus and ventral pallidum, septal nuclei, amygdala, and the pontine nuclei [21,22], suggesting local receptor synthesis. Regions of high delta opioid receptor mRNA expression and comparatively low receptor binding include the internal granular layer of the olfactory bulb and ventromedial nucleus of the hypothalamus [21,22]. The apparent discrepancy between delta opioid receptor mRNA expression and delta opioid receptor binding in several brainstem nuclei and in the cerebellum is, in part, due to the increased sensitivity of *in situ* hybridization methods and high levels of nonspecific binding observed with delta-selective ligands [21,22]. Differences in distributions indicative of receptor transport are also observed in the substantia gelatinosa of the spinal cord, the external plexiform layer of the olfactory bulb, and the superficial layer of the superior colliculus [21,22].

In the superficial layers of the rat spinal cord, the delta opioid receptor is present principally within axon terminals in laminae I and II [23]. Within laminae I and II, enkephalin-like immunoreactivity is intensely localized to axonal varicosities presynaptic to dendrites or cell bodies [24]. The cells expressing the delta opioid receptor mRNA are distributed in the laminae of the dorsal and ventral horns. The delta opioid receptor mRNA is expressed in laminae IV, V, and VII–X, with a few cells in lamina III of the thoracic cord [22]. The expression of delta opioid receptor mRNA is noted in motor neurons of the ventral horn in the spinal cord [22]. In dorsal root ganglia (DRG), delta opioid receptor mRNA is expressed with comparatively fewer levels [22]. The cells expressing delta opioid receptor mRNA in DRG are predominantly large-diameter neurons [22].

3 SELECTIVE DELTA OPIOID RECEPTOR AGONISTS AND THEIR ANTINOCICEPTIVE ACTIONS

Currently, most narcotic analgesics used clinically act through mu opioid receptors, but their use is limited by side effects, such as vomiting, respiratory depression, constipation, and dependence [25]. In contrast, delta opioid receptor agonists can produce antinociception with mild incidence of several somatic signs of withdrawal, indicating that delta opioid receptors remain potentially important therapeutic targets for the development of novel analgesic agents [26].

The ability of delta opioid receptors to modulate pain perception in mice is illustrated by the antinociceptive activity of a number of delta-selective peptides as well as nonpeptide alkaloids [27,28]. Delta opioid receptor–

selective compounds have been shown to produce antinociception that can be blocked by delta-selective antagonists [29–31] but not by mu-selective antagonists [29].

Successful development of highly selective agonists has been crucial in determining the physical and chemical binding requirements specifically attributed to the delta opioid receptor. [D-Pen^2,D-Pen^5]Enkephalin (DPDPE) was the first highly selective agonist ligand for delta opioid receptors [32] and its value as a tool for the investigation of delta opioid receptor pharmacology has been immense. DPDPE is about 3000-fold more potent at delta relative to either mu or kappa opioid receptors when measured in the mouse vas deferens and guinea pig ileum bioassays [32,33]. DPDPE can provide a profound antinociception when administered intracerebroventricularly (ICV) or intrathecally (ITH) [34,35].

Deltorphins are endogenous linear heptapeptides, isolated from skin extracts of frogs belonging to the genus *Phyllomedusa*, that have a higher affinity and selectivity for delta opioid–binding sites than any other natural compound known [36]. Two deltorphins with the sequence Tyr-Ala-Phe-Asp (or Glu)-Val-Val-Gly-NH_2 have been identified and well characterized [36].

BW373U86 ((±)-4-[(αR)-α-((2S,5R)-4-allyl-2,5-dimethyl-1-piperazinyl)-3-hydroxybenzyl]-N,N-diethylbenzamide) was described as the first systemically active, nonpeptidic agonist selective for delta opioid receptors [28]. BW373U86 produces a significant antinociception in mice, but it is less effective than peptidic delta opioid receptor agonists [28]. BW373U86 also produces antinociceptive effects in a model of bradykinin-induced hyperalgesia in treating pain associated with inflammation [37]. However, in primates as in rodents, BW373U86 cannot produce antinociception when high-intensity noxious thermal and electrical stimuli are used, although it produces convulsant effects [38,39].

SNC80 ((+)-4-[(αR)-α-((2S,5R)-4-allyl-2,5-dimethyl-1-piperazinyl)-3-methoxy-benzyl]-N,N-diethylbenzamide), the O-methylated derivative of (+)-BW373U86, has an affinity of 1 nM for delta opioid receptors. Moreover, SNC80 has more than 500-fold selectivity for delta versus mu opioid receptors, which is much greater than the receptor selectivity of the parent compound BW373U86 and similar to that of the most selective peptidic delta opioid receptor agonists [40–42]. SNC80 produces weak but replicable antinociceptive effects.

The effects of enantiomorphs of TAN-67 (2-methyl-4*a*α-(3-hydroxyphenyl)-1,2,3,4,4a,5,12,12aα-octahydro-quinolino[2,3,3-g]isoquinoline), (-)TAN-67 and (+)TAN-67 on antinociceptive response has been recently investigated. (-)TAN-67 produces a significant antinociceptive response when given ITH. [43]. The antinociceptive response induced by (-)TAN-67 can be blocked by pretreatment with delta but not by mu or kappa opioid receptor antagonist [43]. Pretreatment with (-)TAN-67 given ITH attenuates the antinociception induced by subsequent ITH administration of (-)TAN-67 and by DPDPE [43]. However, the antinociceptive effect induced by mu and kappa opioid receptor agonists cannot be affected by (-)TAN-67 pretreatment [43]. Conversely, ITH pretreatment with DPDPE attenuates the antinociception induced by subsequent ITH administration of (-)TAN-67 and by DPDPE. These findings indicate a high delta specificity of (-)TAN-67. Unlike (-)TAN-67, (+)TAN-67 even at smaller doses given ITH produces hyperalgesia [43]. At higher doses given ITH, (+)TAN-67 produces scratching and biting pain-like responses.

The antagonists designed to block the delta opioid receptor have been employed as pharmacological probes to investigate the mechanism of delta opioid receptor binding and explore receptor-specific function. The enkephalin analogue N,N-diallyl-Tyr-Aib-Aib-Phe-Leu-OH (ICI174864) was the first pure delta opioid receptor antagonist but possessed only moderate delta opioid receptor potency [44]. The nonpeptide delta opioid receptor antagonist naltrindole (NTI) has been shown to be highly potent [45]. Naltriben (NTB) [46] and 7-benzylidenenaltrexone (BNTX) [47] display elevated delta opioid receptor selectivity relative to NTI. The tetrapeptide Tyr-Tic-Phe-Phe (TIPP) has been recently introduced as a highly selective delta opioid receptor antagonist [48,49]. This unusual peptide consists entirely of aromatic amino acids which include a conformationally constrained form of Phe, tetrahydro-3-isoquinoline (Tic), in the 2 position. Relative to NTI, TIPP has about 10-fold lower affinity for delta opioid receptors but is about 10-fold more selective for delta opioid receptor over mu opioid receptors. The antinociceptive actions induced by delta opioid receptor agonists including DPDPE, deltorphin, BW373U86, and SNC80 are almost blocked by the treatment with those delta opioid receptor–selective antagonists [32–42]. These data have been taken to indicate specific delta opioid receptor-mediated antinociception distinct from classical mu opioid receptor agonist–induced antinociception.

Pharmacological studies have suggested the expression of at least two delta opioid receptor subtypes. The delta$_1$ receptor subtype is preferentially activated by DPDPE and antagonized by [D-Ala^2,Leu^5,Cys^6]enkephalin (DALCE) and BNTX, while the delta$_2$ receptor is preferentially activated by the [D-Ala^2,D-Glu^4]deltorphin [43,50–56] and blocked by naltorindole-5′-isothilcyanate (5′NTII) and NTB. This classification is further supported by delta$_1$- and delta$_2$-mediated antinociceptive action [43,50–56] and adenylyl cyclase [57]. Both subtypes are implicated in the modulation of nociception in the spinal cord of the rat [52,58]. It has been more recently reported that the disruption of exon 2 of the cloned delta opioid receptor (DOR-1) abolishes DOR-1 immunoreactivity, (DOR-LI), and both delta$_1$ and delta$_2$ bindings [59]. This genetic analysis of DOR-1 knockout mice can support the idea that

DOR-1 encodes both $delta_1$ and $delta_2$ receptor subtypes. Therefore, the different properties of the $delta_1$ and $delta_2$ receptor subtypes could result from differential posttranslational modification of the DOR-1-encoded receptor protein, alterations in the molecular environment of the receptor protein. Alternatively, the understanding of the existed splice variants for DOR-1 must be important to accept the concept of the $delta_1$ and $delta_2$ receptor subtypes, as suggested by DOR-1 antisense mapping studies [60].

4 SPINAL DELTA OPIOID RECEPTOR–MEDIATED ANTINOCICEPTION

There is considerable evidence that delta opioid receptor agonists act in the spinal cord to produce antinociception. This evidence includes pharmacological investigations of the antinociceptive effects of ITH- administered delta opioid receptor agonists [51,52,61,62], electrophysiological characterization of the effects of these agonists on the response properties of dorsal horn neurons [63–66], and neurochemical determinations of their effects on the release of neurotransmitters from the spinal cord [67–70].

Molecular cloning of the receptors and spectacular advances in recombinant DNA methods now make it possible to address the issue by a genetic approach. The activity of known genes can be modified in vivo using gene-targeting technology. An antisense oligodeoxynucleotide is a short piece of synthetic DNA with a nucleotide sequence that is the reverse of and complementary to a part of mRNA. It therefore hybridizes to mRNA and inhibits the synthesis of the encoded protein. It has been demonstrated that ITH treatment with antisense oligodelxynucleotides targeted at DOR-1 can selectively attenuate the spinal antinociception produced by delta, but not mu or kappa, opioid receptor agonists [71,72]. These findings provide further evidence at the molecular level that selective delta opioid receptor agonists such as DPDPE and deltorphins act predominantly through spinal delta opioid receptors to produce spinal antinociception.

More recently, mice lacking opioid peptides or receptors have been generated by homologous recombination. Multiple components of the murine opioid system, including genes encoding the cloned mu opioid receptor (MOR-1) [73–77], DOR-1 [59], the cloned kappa opioid receptor (KOR-1) [78], the pre-PENK precursor [79], and the beta-endorphin domain of POMC precursor [80], have been disrupted by gene targeting. Analysis of these mutant strains of mice thus far suggests that each component has a distinct role in a wide range of physiological functions, including antinociception/analgesia, stress responses, hematopoiesis, and reproduction.

Knockout mice with MOR-1 display profound gene dose-dependent reductions in the morphine-induced antinociception [73–75]. The availability of transgenic MOR-1 knockout mice allows us to determine the extent to which MOR-1 gene products are necessary for the expression of pharmacological actions by delta opioid receptor agonists. The mice with homozygous mutation of MOR-1 exhibit spinal antinociceptive response to selective delta opioid receptor–selective agonists, indicating a lack of involvement of mu opioid receptors in spinal antinociception induced by delta opioid receptor agonists.

Genetic disruption of DOR-1 does not influence the expression of other opioid peptide genes [59]. No significant differences in the levels of mu opioid receptor binding are observed among the wild-type, heterozygous, or homozygous DOR-1 knockout mice [59]. These homozygous DOR-1 knockout mice display a marked suppression of spinal antinociception induced by ITH treatment with either DPDPE or deltorphin [59]. These findings confirm previous pharmacological studies at the molecular level, indicating a distinct delta opioid receptor for antinociception in the spinal cord.

Ultrastructural immunolabeling shows prominent presynaptic vesicular localization of delta opioid receptors within both [Met^5]enkephalin- and nonenkephalin-containing axon terminals in the superficial layers of the rat cervical spinal cord [23]. As well as presynaptic localization of delta opioid receptors in the dorsal horn of the spinal cord, DOR-LI is detected occasionally at postsynaptic sites opposing terminals without detectable [Met^5]enkephalin-like immunoreactivity (ME-LI) and to nonsynaptic sites along the plasmalemma of dendrites [23]. These findings suggest the possibility that [Met^5]enkephalin is derived from a more distal terminal.

5 INVOLVEMENT OF G PROTEINS AND ION CHANNELS IN SPINAL DELTA OPIOID RECEPTOR–MEDIATED ANTINOCICEPTION

Molecular analysis of the opioid receptors indicates that they conform to the structural motif of the G protein receptor family. All three types of opioid receptors contain cysteine residues believed to be involved in disulfide bonds, and a cysteine fatty acid attachment site in locations common with other G protein–coupled receptors. All three opioid receptors contain consensus asparagine-linked glycosylation sites in the extracellular N-terminal domain as well as many consensus protein kinase sites in the first and third intracellular loops, and in the C-terminal domain.

Receptors that interact with G proteins produce an increase in GTP hydrolysis, ultimately via an increase in the GTPase activity of Gα, but initially by stimulating the binding of GTP to Gα [81]. Thus, a study of the ability of various receptor agonists to stimulate GTPase activity is useful to determine the nature of the interaction between the G protein and receptor.

The high-affinity GTPase activity in the mouse spinal cord is increased in a concentration-dependent manner by [D-Ala^2]deltorphin II [82]. This increase of GTPase activity induced by [D-Ala^2]deltorphin II is completely blocked by coincubation with a selective delta opioid receptor antagonist NTB [82].

The activation of G proteins by the opioid receptor agonist can be also measured by assessing agonist-induced stimulation of membrane binding of the nonhydrolyzable analogue of GTP, guanosine-5′-*O*-(3-[^{35}S]thio)triphosphate ([^{35}S]GTPγS) [55,83–87]. [^{35}S]GTPγS addition results in accumulation of a stable Gα-[^{35}S]GTPγS complex in spinal cord membranes. Using this procedure, both DPDPE and [D-Ala^2]deltorphin II produce a robust stimulation of [^{35}S]GTPγS binding in membranes of the mouse spinal cord [88]. These effects are reversed by delta opioid receptor antagonists. The levels of [^{35}S]GTPγS binding stimulated by DPDPE and [D-Ala^2]deltorphin II in membranes of the spinal cord obtained from both heterozygous and homozygous MOR-1 knockout mice are similar to those found in wild-type mice [88]. Homozygous DOR-1 knockout mice display markedly reduced spinal antinociception by delta opioid receptor agonists [59]. These data strongly support the idea that the spinal delta opioid receptor is functionally coupled to G protein, and the activation of this G protein–associated receptor by agonists can produce spinal antinociception.

Identification of the G protein α subunits coupled to a specific receptor subtype is a complex process and often requires a number of approaches. Some receptors interact with a multitube of different α subunits, making identification even more difficult. The most straightforward approach is to isolate the receptor/G proteins complex and identify the α subunit component by immunoblotting. However, this is not be easy. Distinct G protein α subunits are thought to be inactivated by pretreatment with toxins, antisera or antisense oligodeoxynucleotides, and the subsequent loss of function assessed. This approach is particularly useful when studying native receptor-mediated functions in vivo and in vitro. Recently, antisense approach is used to explore the G protein α subunits responsible for transducing delta opioid receptor-mediated antinociception [89]. Mice receiving antisense oligodeoxynucleotides to $G_{i\alpha 1}$, $G_{i\alpha 2}$, $G_{i\alpha 3}$, $G_{o\alpha}$, $G_{s\alpha}$, $G_{q\alpha}$, or $G_{x/z\alpha}$ subunits show an impaired antinociceptive response to spinal delta opioid receptor agonists [90]. These findings support the idea that spinal delta opioid receptor can interact $G_{i\alpha 1}$, $G_{i\alpha 2}$, $G_{i\alpha 3}$, $G_{o\alpha}$, $G_{s\alpha}$, $G_{q\alpha}$, or $G_{x/z\alpha}$ subunits to produce spinal antinociception. As well as the G protein α subunits, we found in the preliminary study that βγ subunit is also implicated in the delta opioid receptor–mediated spinal antinociception (unpublished observation).

The interaction of delta opioid receptor–selective ligands with its respective receptor has been reported to hyperpolarize neurons. Hyperpolarization of neurons in the delta opioid–sensitive pathway may play a major role in the antinociception produced by delta opioids [91–96]. The K^+ channels represent the largest and most diverse group of any ion channel family that has been identified in cells. Delta opioid receptor agonists have been found to hyperpolarize the membrane potential [91,97–102] and reduce action potential duration in many different neurons [103,104]. The hyperpolarization is accompanied by an increase in resting membrane conductance, varies with changes in extracellular K^+ concentration and can be blocked by Cs^+ and/or Ba^+, two cations known to block K^+ channels [91,99]. DPDPE reduces the duration of action potentials and decreases voltage-dependent outward K^+ currents in both cultured DRG [105] and F11 (a neuroblastoma × DRG hybrid) neurons [106]. These results suggest that delta opioid receptor agonists exert their effects by opening K^+ channels.

To determine which K^+ channels are opened by opioid agonists, opioid-evoked K^+ currents have been studied under the voltage clamp condition. In the submucosal plexus with delta opioid receptors, delta opioids increase a K^+ current that has an inwardly rectifying voltage dependence [93,107]. In submocosal neurons, [Met^5]enkephalin increases the opening probability of background single K^+ channels of small (30–65 pS), intermediate (120–160 pS), and large (220–260 pS) conductance [108].

The spinal antinociceptive effect of DPDPE is blocked totally by a small conductance Ca^{2+}-activated K^+ channel blocker apamin [109]. The DPDPE-induced spinal antinociception is not inhibited by a ATP-sensitive K^+ channel blocker glyburide [109]. Like morphine, tetraethylammonium, 4-aminopyridine, and charybdotoxin are unable to block the effects of DPDPE [109]. These findings suggest that the modulation of apamin-sensitive K^+ channels appears to play a role in the DPDPE-induced antinociception in the spinal cord.

The ensuing studies of opioid actions on Ca^{2+} currents under the voltage clamp condition have proved that opioids exert many of their inhibitory effects by blocking voltage-dependent Ca^{2+} channels. Activation of delta opioid receptors has been found to reduce mostly N-type Ca^{2+} currents [110–117]. It should be noted that spinal antinociception induced by the delta opioid receptor agonist is potentiated by a selective N-type Ca^{2+} channel blocker omega-conotoxin GVIA, whereas the effect of mu opioid receptor agonist is not changed by this treatment [118]. Activation of delta opioid receptors sometimes reduces T-type Ca^{2+} currents [114,115]. The cloned delta opioid receptor expressed in GH3 cells has been shown to voltage-dependently couple through Gi protein to L-type Ca^{2+} channels [119]. In submucosal neurons, activation of the delta opioid receptors affects both inwardly rectifying K^+ channels and N-type Ca^{2+} channels in the same neuron [120]. Taken together, it is likely that activation of delta opioid receptors sometimes reduces N-type Ca^{2+} currents via the opening apamin-

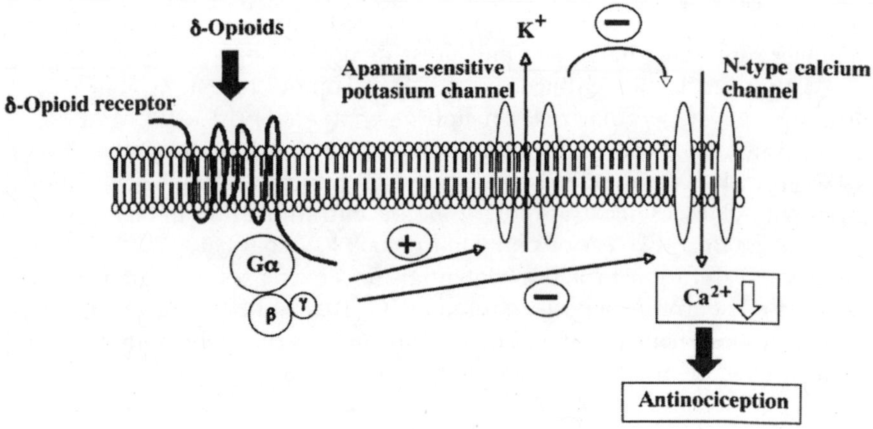

FIGURE 1 Role of ion channels in the delta opioid receptor agonist–induced spinal antinociception. Delta opioid receptor agonists acutely inhibit some neurons by increasing the conductance of an apamin-sensitive inwardly rectifying K^+ channel and decreasing an N-type Ca^{2+} channel-dependent inward current via activation of G_α and $G_{\beta\gamma}$. These changes, induced by activation of delta opioid receptor, lead to the delta opioid receptor agonist–induced spinal antinociception.

sensitive K^+ channel (small conductance Ca^{2+} activated K^+ channel), resulting in the expression of spinal antinociception (Fig. 1).

6 ROLE OF PROTEIN KINASE C IN SPINAL DELTA OPIOID RECEPTOR-MEDIATED ANTINOCICEPTION AND ITS TOLERANCE

It has been widely recognized that the activation of the mouse DOR-1 inhibits adenylyl cyclase, leading to the inhibition of cyclic AMP formation [121]. In addition to the inhibitory modulation through the delta opioid receptor function, delta opioid receptor agonists can also activate phospholipase C (PLC) via pertussis toxin (PTX)-sensitive G proteins. [122–125]. The activation of PLC produces at least two messengers—, inositol-1,4,5-triphosphate (IP_3), and diacylglycerol (DAG). The main effect of DAG is to activate protein kinase C (PKC); the effect of IP_3 is to release Ca^{2+} from intracellular stores. Indeed, delta opioid receptor agonists stimulate IP_3 formation, and so mobilize Ca^{2+} from intracellular stores, in NG108-15 cells [126]. In addition to stimulating IP_3 formation, delta opioid receptor agonists can activate PKC in *Xenopus* oocytes [127] and the mouse spinal cord [128]. It should be mentioned that ITH pretreatment with the PLC inhibitors neomycin and

U73122, which produce no changes in the basal tail flick latencies when they are injected alone, significantly attenuates the antinociception induced by ITH administration of [D-Ala^2]deltorphin II in mice [129]. The selective phosphatidylinositol-specific PLC inhibitor ET-18-OCH_3 inhibits the antinociception induced by ITH administration of [D-Ala^2]deltorphin II in a dose-dependent manner [129]. In mice undergoing treatment with LiCl, which impairs phosphatidylinositol synthesis, the antinociception induced by ITH administration of [D-Ala^2]deltorphin II can be significantly reduced. Coadministration of D-myo-IP_3 restores the [D-Ala^2]deltorphin II–induced antinociception in LiCl-pretreated mice [129]. It has been reported that both DPDPE and [D-Ala^2]deltorphin II increase the phosphoinositide metabolism in the mouse spinal cord [130]. These findings indicate a potential role for the PLC-IP_3 pathway in the expression of antinociceptive responses regulated by spinal delta opioid receptors in mice (Fig. 2).

In addition to stimulating IP_3 formation, delta opioid receptor agonists cause a concomitant activation of PKC through PLC [125]. PKC is a key regulatory enzyme that modulates both pre- and postsynaptic neuronal function, synthesis and release of neurotransmitters, and the regulation of receptors [131]. PKC has expanded into a family of closely related protein,

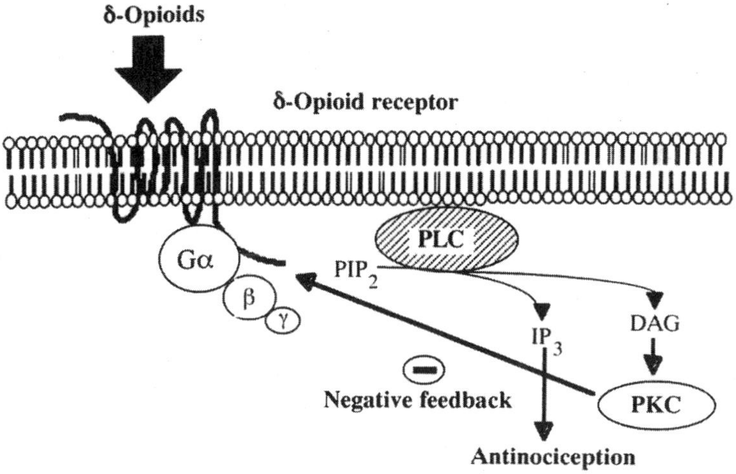

FIGURE 2 Schematic model of the intracellular signaling mechanisms including phosphatidylinositol-specific phospholipase C (PI-PLC) isoforms, inositol-1,4,5-triphosphate (IP_3) receptor, and protein kinase C (PKC) in the expression of delta opioid receptor agonist–induced spinal antinociception. In addition, PKC is considered to play a substantial role in an intracellular negative feedback action on the spinal delta opioid receptor–mediated antinociceptive pathway.

which can be subdivided and classified on the basis of certain structural and biochemical similarities [131]. Several PKC isoforms, especially conventional PKCs (cPKCs) including α, βI, βII, and γ that are Ca^{2+}-dependent and activated by both phosphatidylserine (PS) and DAG, have been identified in neurons of the spinal cord [132], and in each case immunocytochemistry has shown that they are concentrated in the superficial laminae of the dorsal horn [133,134]. It is of interest to note that ITH pretreatment with a potent and specific PKC inhibitor calphostin C, but not a highly selective PKA inhibitor KT5720, results in a dose-dependent enhancement of the [D-Ala^2]deltorphin II–induced antinociception [129]. Although a more extensive approach will be required, these findings suggest the possibility that the intracellular negative feedback action by PKC on the spinal delta opioid receptor–mediated antinociceptive pathway is necessary to maintain the basal level of pain sensitivity (Fig. 2).

A limiting factor in the clinical utilization of opioids for pain relief is that repeated administration leads to the development of tolerance to and physical dependence on opioids. At the cellular level, tolerance can be viewed as a form of persistent receptor desensitization associated with repeated drug administration. Phosphorylation of opioid receptors by protein kinases, especially PKC, is hypothesized to play a major role in this desensitization; hence, opioid receptors in tolerant and dependent states are thought to be highly phosphorylated [135–140].

Phorbol esters have a wide variety of pharmacological actions mediated through the activation of PKC [131,141–143]. Pei et al. [144] have reported that a phorbol ester stimulates phosphorylation of the delta opioid receptor in human embryonic kidney 293 cells, whereas a PKA activator, forskolin, has no such an effect. It is also proposed that PKC is involved in the functional uncoupling of the delta opioid receptor from G proteins in striatal membranes of young guinea pigs [145], *Xenopus* oocytes [127] and the membranes of the mouse spinal cord [128]. It is of interest to note that ITH pretreatment with a specific PKC activator PDBu at low doses, which injected alone does not affect the basal tail flick latency, produces a dose-dependent suppression of the antinociception induced by ITH administration of [D-Ala^2]deltorphin II [128]. The attenuation of ITH-administered [D-Ala^2]deltorphin II–induced antinociception by PDBu is reversed in a dose-dependent manner by ITH concomitant pretreatment with a specific PKC inhibitor, calphostin C [128]. In the binding experiment, incubation of the crude synaptic membrane of the spinal cord with PDBu causes a dose-dependent inhibition of the binding of [^{3}H][D-Ser^2,Leu^5]enkephalin-Thr^6 (DSLET), a delta opioid receptor ligand, [^{3}H]DSLET. Scatchard analysis of [^{3}H]DSLET binding reveals that PDBu at 10 mu M displays a 30% reduction in the number of [^{3}H]DSLET-binding sites with no significant change in affinity, compared with the nontreatment control [128]. These findings suggest the possibility that a loss of specific delta opioid receptor agonist–sensitive binding by the activation of PKC is involved in the PDBu-induced antinociceptive unresponsiveness to delta opioid receptor agonist in the mouse spinal cord.

An ITH injection of [D-Ala^2]deltorphin II produces an acute antinociceptive tolerance [128]. Concomitant pretreatment with lower doses of the PKC inhibitor calphostin C markedly prevents the development of acute tolerance to the ITH-administered [D-Ala^2]deltorphin II–induced antinociception. On the other hand, the PKA inhibitor KT5720, has no effect on the development of acute tolerance to [D-Ala^2]deltorphin II antinociception [128]. It is therefore likely that PKC, but not PKA, plays an important role in the process of homologous desensitization of the spinal delta opioid receptor–mediated antinociception.

7 TURNOVER OF DELTA OPIOID RECEPTOR TO PRODUCE SPINAL ANTINOCICEPTION

It has been proposed that the step that initiates receptor desensitization involves the phosphorylation of the receptor by protein kinases including PKC and G protein–coupled receptor kinases (GRKs), thereby promoting the recruitment of the cellular protein arrestin. Association of arrestin with the receptor enhances the uncoupled of the receptor from the respective G protein, thus blunting the signal transduction processes resulting in the receptor desensitization. The association of arrestin also appears to be critical in the agonist induced, clathrin-coated vesicle-mediated receptor internalization. The delta opioid receptor agonist induces rapid receptor desensitization [146,147]. The delta opioid receptors desensitize more rapidly than mu opioid receptors due to less efficient activation of arrestin [148]. A growing body of evidence suggests that regulatory events leading to desensitization, internalization, and recycling in functional state of delta opioid receptor involve phosphorylation by GRKs, internalization via clathrin-coated vesicles, and dephosphorylation by acid phosphatases [149]. Delta opioid receptor desensitization involves phosphorylation of both the carboxyl-terminal tail and the second intracellular loop that together leads to a more efficient activation of arrestin and thus faster desensitization. The delta opioid receptor internalization is intimately related to phosphorylation of Thr^{358} and Ser^{363} [147].

An ITH injection of [D-Ala^2]deltorphin II produces an acute tolerance to the antinociceptive effect of a subsequent ITH challenge of [D-Ala^2]deltorphin II. The acute tolerance is developed within 3 h after a single injection [150] (Fig. 3A). The suppression of [D-Ala^2]deltorphin II–induced antinociception lasts 3–9 h and completely subsides by 12 h [150]. It should be pointed

(A)

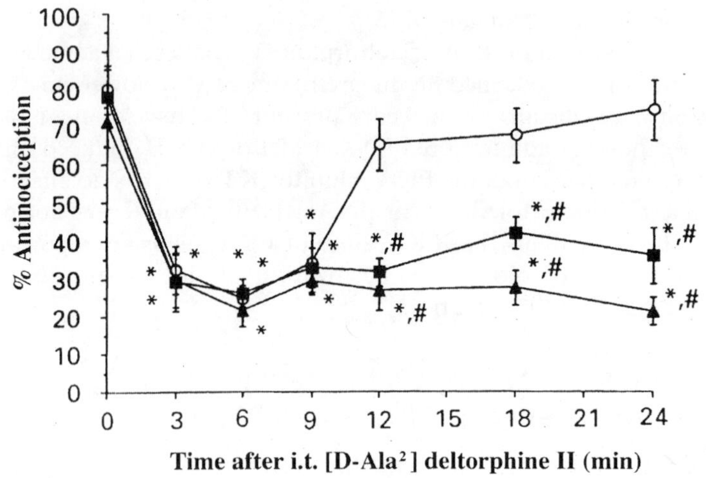
100
90
80
70
60
50
40
30
20
10
0
% Antinociception
0
3
6
9
12
15
18
21
24
Time after i.t. [D-Ala2] deltorphine II (min)
*,#

(B)

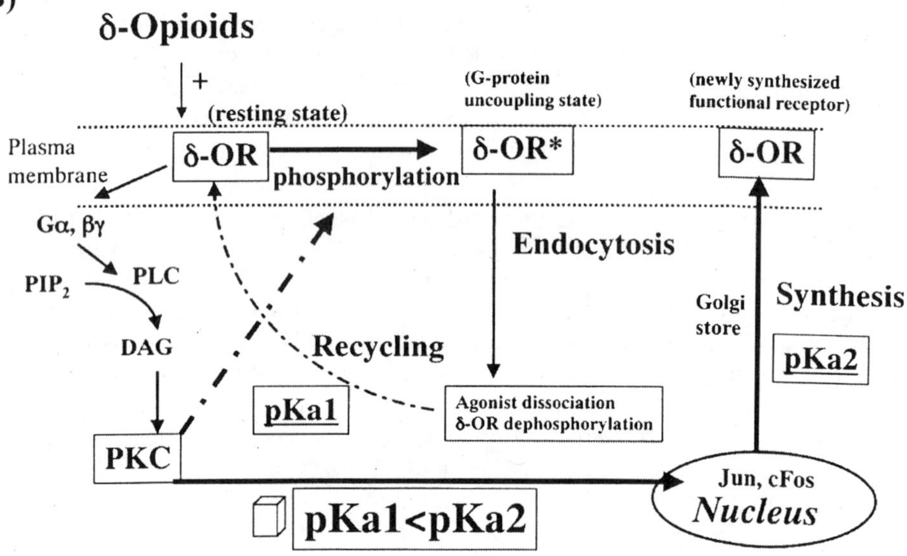
δ-Opioids
+
(resting state)
(G-protein uncoupling state)
(newly synthesized functional receptor)
Plasma membrane
δ-OR
phosphorylation
δ-OR*
δ-OR
Gα, βγ
PIP2
PLC
DAG
PKC
Endocytosis
Recycling
pKa1
Agonist dissociation
δ-OR dephosphorylation
Golgi store
Synthesis
pKa2
Jun, cFos
Nucleus
pKa1<pKa2

out that the inhibition of the biosynthesis of spinal delta opioid receptor protein by ITH pretreatment with antisense oligodeoxynucleotide against delta opioid receptor, but not mismatched oligodeoxynucleotide, prevents the recovery from acute tolerance to [D-Ala^2]deltorphin II-induced spinal antinociception in a dose-dependent manner [150]. However, ITH treatment with antisense oligodeoxynucleotide against delta opioid receptor cannot prevent the recovery from antinociceptive tolerance to either the mu opioid receptor agonist DAMGO or the kappa opioid receptor agonist U50,488H [150]. Under these conditions, ITH treatment with [D-Ala^2]deltorphin II significantly reduces the binding of [^{3}H]DSLET in membranes of the spinal cord at 3 h after treatment, but the binding returns to control levels by 24 h after treatment [150]. However, [^{3}H]DSLET binding in the spinal cord remained significantly reduces at 24 h if antisense oligodeoxynucleotide against delta opioid receptor is cotreated ITH with [D-Ala^2]deltorphin II [150]. These findings support the idea that a single stimulation of spinal delta opioid receptors induces a long-lasting desensitization of delta opioid receptor–mediated antinociception in the spinal cord, and this recovery from delta opioid receptor–mediated antinociceptive tolerance apparently depends on replenishment by newly synthesized delta opioid receptor protein rather than immediate reversal of delta opioid receptors in the spinal cord (Fig. 3B).

The inhibition of the biosynthesis of delta opioid receptor protein by ITH treatment with antisense oligodeoxynucleotide against delta opioid

---

FIGURE 3 Turnover of delta opioid receptor to produce spinal antinociception. (A) Time course of the change of ITH-administrated [D-Ala^2]deltorphin II–induced antinociception in mice pretreated ITH with [D-Ala^2]deltorphin II or a combination of [D-Ala^2]deltorphin II and antisense oligodeoxynucleotide to DOR-1 mRNA (DOR-1-AS). Groups of mice were pretreated ITH with [D-Ala^2]deltorphin II alone (○) or in combination with [D-Ala^2]deltorphin II and DOR-1-AS (low dose: ■ or high dose: ▲). Mice were then challenged ITH with [D-Ala^2]deltorphin II at different times after the first injection. The antinociception was measured 10 min after the [D-Ala^2]deltorphin II treatment at 0 h or 10 min after the second [D-Ala^2] deltorphin II injection at 3, 6, 9, 12, 15, 18, 21, or 24 h. A single injection was made at the time points. The point "0 h" indicates the value of a single injection of [D-Ala^2]deltorphin II or a combination of [D-Ala^2]deltorphin II and delta AS. $*P < .05$, vs. 0 h; $\#P < .05$, vs. [D-Ala^2]deltorphin II. (B) Schematic model of delta opioid receptor turnover to produce spinal antinociception. A single stimulation of spinal delta opioid receptors induces the desensitization of delta opioid receptor–mediated antinociception in the spinal cord, and this recovery from delta opioid receptor–mediated antinociceptive tolerance apparently depends on replenishment by newly synthesized delta opioid receptor protein (pKa2) rather than immediate reversal of delta opioid receptors (pKa1) in the spinal cord.

receptor causes a time-dependent attenuation of the delta opioid receptor mediated antinociception induced by ITH-challenged [D-Ala^2]deltorphin II [151]. The antinociception is not altered 1 day after the treatment with delta antisense oligodeoxynucleotide, suggesting that, initially, there are sufficient delta opioid receptors located in the cellular membrane for performing their pharmacological function [151]. However, the capacity of delta opioid receptors is progressively reduced after 2–3 days of ITH treatment with antisense oligodeoxynucleotide against delta opioid receptor, resulting in a significant reduction in the [D-Ala^2]deltorphin II–induced antinociception [151]. Interestingly, the attenuation of [D-Ala^2]deltorphin II–induced antinociception after ITH treatment with antisense oligodeoxynucleotide against delta opioid receptor is completely prevented by ITH treatment with NTB, which can selectively block spinal delta opioid receptors [151]. Furthermore, ITH cotreatment of antiserum to [Met^5]enkephalin, but not [Leu^5]enkephalin, β-endorphin, or dynorphin A [1–17], with ITH-treated delta antisense oligodeoxynucleotide blocks the attenuation of [D-Ala^2]deltorphin II–induced antinociception by ITH treated with antisense oligodeoxynucleotide against delta opioid receptor [151]. These findings indicate that the blockade of delta opioid receptors attenuates the turnover of spinal delta opioid receptors. This contention is further supported by the fact that ICV pretreatment with β-endorphin, which can subsequently release Met-enkephalin from the spinal cord [152], enhances the attenuation of the ITH-administered [D-Ala^2]deltorphin II–induced antinociception in mice copretreated ITH with antisense oligodeoxynucleotide against delta opioid receptor [152]. In contrast, thiorphan and bestatin, which inhibit the degradation of endogenously released [Met^5]enkephalin, are markedly enhanced by ITH treatment with bestatin or thiorphan. These findings give support to the hypothesis that the turnover of delta opioid receptors is selectively regulated by spontaneously released [Met^5]enkephalin.

Ultrastructural immunolabeling shows prominent vesicular localization of delta opioid receptor in the superficial layers of the rat spinal cord [23]. The intense vesicular localization of delta opioid receptor and the diversity and paucity of DOR-LI along the axonal plasmalemma in the delta opioid receptor provide further evidence for the slow or little resensitization of the delta opioid receptors stimulated by agonists and continuous supply or transport of the newly synthesized delta opioid receptor protein from the soma to the axon terminal in the spinal cord. This contention can be strongly supported by recent studies, indicating that targeting of the delta opioid receptor to lysosomes causes proteolytic downregulation through G protein–coupled receptor-associated sorting proteins, whereas recycling of the mu opioid receptor to the plasma membrane after endocytosis promotes rapid resensitization of signal transduction [153].

Conclusively, spinal delta opioid receptors undergo a conformational change by thc stimulation of spontaneously released [Met^5]enkephalin and are then internalized. It is likely that the internalization of spinal delta opioid receptors becomes an irreversible process. To maintain a normal physiological function, delta opioid receptors in the plasma membrane are constantly replenished by newly synthesized delta opioid receptor protein (Fig. 3B).

8 SUPRASPINAL ANTINOCICEPTION MEDIATED BY DELTA OPIOID RECEPTOR AGONIST AND INTERACTION BETWEEN MU AND DELTA OPIOID RECEPTORS IN THE BRAIN

The ability of delta opioid receptors in the spinal cord to mediate antinociception is well established but the ability of delta opioid receptors to mediate antinociception at the supraspinal level continues to be a subject of controversy. A preponderance of evidence seems to support supraspinal antinociception mediated by delta opioid receptors [27,42], however, the dispute still continues. Pharmacological evidence in favor of delta opioid–mediated antinociception in the brain can be summarized as: 1) the ability of highly selective delta opioid receptor agonists [35,50] to produce antinociceptive responses to chemical and thermal nociceptive stimuli in mice when give ICV; 2) the ability of selective delta opioid receptor antagonists to block these agonist effects; 3) the insensitivity of delta opioid receptor agonist-induced suparaspinal antinociception to mu opioid receptor antagonists; and 4) a lack of cross-tolerance between selective delta and mu opioid receptor agonists to produced supraspinal antinociception. On the contrary, a virtually absolute requirement from mu , but not delta, opioid receptor expression to allow expression of delta opioid receptor agonist–induced supraspinal antinociception has been proposed [59,73,154,155].

Mice lacking DOR-1 are markedly less sensitive to spinal antinociception induced by both DPDPE and [D-Ala^2]deltorphin II given ITH [59], indicating a major contribution of DOR-1 to spinal antinociception. However, both DPDPE and [D-Ala^2]deltorphin II retain their antinociceptive activity, with no reduction in potency, following ICV administration in DOR-1 knockout mice [59]. The retained antinociception induced by those agonists is only partially antagonized by NTI. Under these conditions, the nonpeptide delta opioid receptor agonist BW373U69 exhibits enhanced activity in DOR-1 mutant mice [59]. Such findings are unexpected in light of extensive pharmacological literature supporting antinociception mediated by the delta opioid receptor in the brain. In contrast, supraspinal antinociception induced by ICV administration of delta opioid receptor agonists is dramatically reduced in MOR-1 gene knockout mice in a gene dose-dependent fashion [154]. These findings suggest that the mu opioid receptor is necessary for delta opioid receptor agonist–induced supraspinal antinociception.

Although each type of opioid receptor can transduce its effects independently, evidence has been accumulating for the existence of cellular and/or molecular interactions between them. Thus, the cross-talk between mu and delta opioid receptors has been proposed on the basis of pharmacological studies demonstrating both competitive and noncompetitive changes in the binding of delta-selective ones [155,156]. Conversely, administration of delta opioid receptor antagonists, or of antisense oligodeoxynucleotide directed against DOR-1, is shown to reduce the development of tolerance to the antinociceptive effects of morphine [157–159]. Accordingly, DOR-1 knockout mice maintain mu opioid receptor–mediated antinociception but show a decrease in the development of tolerance to morphine [59]. In contrast, prolonged stimulation of neurons with morphine markedly increases recruitment of intracellular delta opioid receptors to the cell surface without any changes in protein levels of delta opioid receptors [160]. Recent evidence using transfected cell systems demonstrates direct molecular interactions between different members of the opioid receptor family, with reports if heterodimerization of the delta opioid receptor with either the mu [161,162] or kappa-opioid receptor [163]. These findings indicate that heterodimerization of delta and mu or kappa opioid receptors could account for the cross-modulation among opioid receptors. Furthermore, delta opioid receptors are considered to form heteromeric complexes with β_2-adrenergic receptor [164] and V2 vasopressin receptors [165]. It is therefore likely that the delta opioid receptor directly interacts with other receptors including mu opioid receptor involved in the modulation of supraspinal antinociception induced by delta opioid receptor agonists in the brain.

REFERENCES

1. Evans CJ, Keith DE, Morrison H, Magendzo K, Edwards RH. Science 1992; 258:1952–1955.
2. Kieffer BL, Befort K, Gaveriaux-Ruff C, Hirth CG. Proc Natl Acad Sci USA 1992; 89:12048–12052.
3. Chen Y, Mestek A, Liu J, Hurley JA, Yu L. Mol Pharmacol 1993; 44:8–12.
4. Chen Y, Mestek A, Liu J, Yu L. Biochem J 1993; 295:625–628.
5. Li S, Zhu J, Chen C, Chen Y-W, Deriel JK, Ashby B, Liu-Chen L-Y. Biochem J 1993; 295:629–633.
6. Meng F, Xie G-X, Thompson RC, Mansour A, Goldstein A, Watson SJ, Akil H. Proc Natl Acad Sci USA 1993; 90:9954–9958.

7. Nishi M, Takeshima H, Fukuda K, Kato S, Mori K. FEBS Lett 1993; 330:77–80.
8. Yasuda K, Raynor K, Kong H, Breder CD, Takeda J, Reisine T, Bell GI. Proc Natl Acad Sci USA 1993; 90:6736–6740.
9. Fukuda K, Kato S, Mori K, Nishi M, Takeshima H. FEBS Lett 1993; 327:311–314.
10. Wang J-B, Imai Y, Eppler CM, Gregor P, Spivak CE, Uhl GR. Proc Natl Acad Sci USA 1993; 90:10230–10234.
11. Wang J-B, Johnson PS, Persico AM, Hawkins AL, Griffin CA, Uhl GR. FEBS Lett 1994; 338:217–222.
12. Zastawny RL, George SR, Nguyen T, Cheng R, Tsatsos J, Briones-Urbina R, O'Dowd BF. J Neurochem 1994; 62:2099–2105.
13. Lewis RV, Stern AS, Kimura S, Rossier J, Stein S, Udenfriend S. Science 1980; 208:1459–1461.
14. Mansour A, Khachaturian H, Lewis ME, Akil H, Watson SJ. Trends Neurosci 1988; 11:308–314.
15. Nakanishi S, Inoue A, Kita T, Nakamura M, Chang AC, Cohen SN, Numa S. Nature 1979; 278:423–427.
16. Hughes J, Smith TW, Kosterlitz HW, Fothergill LA, Morgan BA, Morris HR. Nature 1975; 258:577–579.
17. Lord JA, Waterfield AA, Hughes J, Kosterlitz HW. Nature 1977; 267:495–499.
18. Bunzow JR, Saez C, Mortrud M, Bouvier C, Williams JT, Low M, Grandy DK. FEBS Lett 1994; 347:284–288.
19. Chen Y, Fan Y, Liu J, Mestek A, Tian M, Kozak CA, Yu L. FEBS Lett 1994; 347:279–283.
20. Mollereau C, Parmentier M, Mailleux P, Butour J-L, Moisand C, Chalon P, Caput D, Vassart G, Meunier J-C. FEBS Lett 1994; 341:33–38.
21. Monsour A, Khachaturian H, Lewis ME, Akil H, Watson SJ. J Neurosci 1987; 7:2445–2464.
22. Monsour A, Fox CA, Burke S, Meng F, Thompson RC, Akil H, Watson SJ. J Comp Neurol 1994; 350:412–438.
23. Cheng PY, Svingos AL, Wang H, Clarke CL, Jenab S, Beczkowska IW, Inturris CE, Pickel VM. J Neurosci 1995; 15:5976–5988.
24. Ribeiro-da-Silva A, Pioro EP, Cuello AC. J Neurosci 1991; 11:1068–1080.
25. Narita M, Funada M, Suzuki T. Pharmacol Ther 2001; 89:1–15.
26. Hutcheson DM, Matthes HW, Valjent E, Sanchez-Blazquez P, Rodriguez-Diaz M, Garzon J. Eur J Neurosci 2001; 13:153–161.
27. Heyman JS, Vaught JL, Raffa RB, Porreca F. Trends Pharmacol Sci 1988; 9:134–138.
28. Chang K-J, Rigdon GC, Howard JL, McNutt RW. J Pharmacol Exp Ther 1993; 267:852–857.
29. Heyman JS, Mulvaney SA, Mosberg HI, Porreca F. Brain Res 1987; 420:100–108.
30. Porreca F, Heyman JS, Mosberg HI, Omnaas JR, Vaught JL. J Pharmacol Exp Ther 1987; 241, 393–400.

31. Shaw JS, Miller L, Turnbull MJ, Gormley JJ, Morley JS. Life Sci 1982; 31: 1259–1262.
32. Mosberg HI, Hurst R, Hruby VJ, Gee K, Yamamura HI, Galligan JJ, Burks TF. Proc Natl Acad Sci USA 1983; 80:5871–5874.
33. James IF, Goldstein A. Mol Pharmacol 1984; 25:343–348.
34. Goldstein A, Naidu A. Mol Pharmacol 1989; 36:265–272.
35. Jiang Q, Takemori AE, Sultana M, Portoghese PS, Bowen WD, Mosberg HI, Porreca F. J Pharmacol Exp Ther 1991; 257:1069–1075.
36. Erspamer V, Melchiorri P, Falconieri-Erspamer G, Negri L, Corsi R, Severini C, Barra D, Simmaco M, Kreil G. Proc Natl Acad Sci USA 1989; 86:5188–5192.
37. Butelman ER, Negus SS, Gatch MB, Chang K-J, Woods JH. Eur J Pharmacol 1995; 277:285–287.
38. Dykstra LA, Schoenbaum GM, Yarbrough J, McNutt R, Chang K-J. J Pharmacol Exp Ther 1993; 267:875–882.
39. Negus SS, Butelman ER, Chang K-J, de Costa BR, Winger G, Woods JH. J Pharmacol Exp Ther 1994; 270:1025–1034.
40. Calderon SN, Rothman RB, Porreca F, Flippen-Anderson JL, McNutt RW, Xu H, Smith LE, Bilskey EJ, Davis P, Rice KC. J Med Chem 1994; 37:2125–2128.
41. Bilskey EJ, Calderon SN, Wang T, Bernstein RN, Davis P, Hruby J, McNutt RW, Rothman RB, Rice K, Porreca F. J Pharmacol Exp Ther 1995; 273:359–366.
42. Knapp RJ, Giovanna S, De Leon IA, Lee KB, Edsall SA, Waite S, Malatynska E, Varga E, Calderon SN, Rice KC, Rothman RB, Porreca F, Roeske WR, Yamamura HI, J Pharmacol Exp Ther 277:1284–1291.
43. Tseng LF, Narita M, Mizoguchi H, Kawai K, Mizusuna A, Kamei J, Suzuki T, Nagase H. J Pharmacol Exp Ther 1997; 280:600–605.
44. Cotton R, Giles MG, Miller L, Shaw JS, Timms D. Eur J Pharmacol 1984; 97: 331–332.
45. Portoghese PS, Sultana M, Takemori AE. Eur J Pharmacol 1988; 146:185–186.
46. Portoghese PS, Nagase H, MaloneyHuss KE, Lin CE, Takemori AE. J Med Chem 1991; 34:1715–1720.
47. Portoghese PS, Sultana M, Nagase H, Takemori AE. Eur J Pharmacol 1992; 218:195–196.
48. Schiller PW, Nguyen TM-D, Weltrowska G, Wilkes BC, Marsden BJ, Lemieux C, Chung NN. Proc Natl Acad Sci USA 1992a; 89:11871–11875.
49. Schiller PW, Weltrowska G, Nguyen TM-D, Wilkes BC, Chung NN, Lemieux C. J Med Chem 1992b; 35:3956–3961.
50. Mattia A, Vanderah T, Mosberg HI, Porreca F. J Pharmacol Exp Ther 1991; 258:583–587.
51. Mattia A, Farmer SC, Takemori AE, Sultana M, Portoghese PS, Mosberg HI, Bowen WD, Porreca F. J Pharmacol Exp Ther 1992; 260:518–525.
52. Stewart PE, Hammond DL. J Pharmacol Exp Ther 1993; 266:820–828.

53. Suzuki T, Tsuji M, Mori T, Misawa M, Endoh T, Nagase H. Life Sci 1995; 57: 155–168.
54. Tseng LF, Tsai JHH, Collins KA, Portoghese PS. Eur J Pharmacol 1995; 277:251–256.
55. Narita M, Tseng LF. Jpn J Pharmacol 1998; 76:233–253.
56. Hurley RW, Grabow TS, Tallarida RJ, Hammond DL. J Pharmacol Exp Ther 1999; 289:993–999.
57. Olianas MC, Onali P. J Pharmacol Exp Ther 1995; 275:1560–1567.
58. Hammond DL, Stewart PE, Littell L. J Pharmacol Exp Ther 1995; 274:1317–1324.
59. Zhu Y, King MA, Schuller AGP, Nitsche JF, Reidl M, Elde RP, Unterwald E, Pasternak GW, Pintar JE. Neuron 1999; 24:243–252.
60. Rossi GC, Su W, Leventhal L, Su H, Pasternak GW. Brain Res 1997; 753:176–179.
61. Porreca F, Mosberg HI, Hurst R, Hruby VJ, Burks TF. J Pharmacol Exp Ther 1984; 230:341–348.
62. Malmberg AB, Yaksh TL. J Pharmacol Exp Ther 1992; 263:264–275.
63. Dickenson AH, Sullivan AF, Knox R, Zajac JM, Roques BP. Brain Res 1987; 413:36–44.
64. Hope PJ, Fleetwood-Walker SM, Mitchell R. Br J Pharmacol 1990; 101:477–483.
65. Kalso EA, Sullivan AF, McQuay HJ, Dickenson AH. Eur J Pharmacol 1992; 216:97–101.
66. Duggan AW, Fleetwood-Walker SM. Opioids and sensory processing in the central nervous system. In: Handbook of Experimental Pharmacology, Vol. 104. Berlin: Springer-Verlag, 1993:731–771.
67. Go VLW, Yaksh TL. J Physiol 1987; 391:141–167.
68. Pohl M, Lombard MC, Bourgoin S, Carayon A, Benoliel JJ, Mauborgne A, Besson JM, Hamon M, Cesselin F. Neuropeptides 1989; 14:151–159.
69. Collin E, Mauborgne A, Bourgoin S, Chantrel D, Hamon M, Cesselin F. Neuroscience 1991; 44:725–731.
70. Ueda M, Sugimoto K, Oyama T, Kuraishi Y, Satoh M. Neuropharmacology 1995; 34:303–308.
71. Tseng LF, Collins KA, Kampine JP. Eur J Pharmacol 1994; 258:R1.
72. Narita M, Tseng LF. Eur J Pharmacol 1995; 284:185–189.
73. Matthes HWD, Maldonado R, Simonin F, Valverde O, Slowe S, Kitchen I, Befort K, Dierich A, Meur M, Dolle P, Tzavara E, Hanoune J, Roques BP, Kieffer BL. Nature 1996; 383:819–823.
74. Sora I, Takahashi N, Funada M, Ujike H, Revay RS, Donovan DM, Miner LL, Uhl GR. Proc Natl Acad Sci USA 1997; 94:1544–1549.
75. Tian MT, Broxmeyer HE, Fan Y, Lai ZN, Zhang SW, Aronica S, Cooper S, Bigsby RM, Steinmetz R, Engle SJ, Mestek A, Pollock JD, Lehman MN, Jansen HT, Ying M, Stambrook PJ, Tischfield JA, Yu L. J Exp Med 1997; 185:1517–1522.
76. Roy S, Barke RA, Loh HH. Brain Res Mol Brain Res 1998; 61:190–194.

77. Schuller AGP, King M, Zhang J, Bolan E, Pan Y-X, Morgan DJ, Czick ME, Chang A, Unterwald E, Pasternak GW, Pintar JE. Nat Neurosci 1999; 2:151–156.
78. Simonin F, Valverde O, Smadja C, Slowe S, Kitchen I, Dierich A, Le Meur M, Roques BP, Maldonado R, Kieffer BL. EMBO J 1998; 17:886–897.
79. Konig M, Zimmer AM, Steiner H, Holmes PV, Crawley JN, Brownstein MJ, Zimmer A. Nature 1996; 383:535–538.
80. Rubinstein M, Mogli JS, Japon M, Chan EC, Allen RG, Low MJ. Proc Natl Acad Sci USA 1996; 93:3995–4000.
81. Law PY. In: The Pharmacology of Opioid Peptides. Chur: Harwood Academic Publishers, 1995:109–130.
82. Narita M, Mizoguchi H, Tseng LF. Eur J Pharmacol 1996; 310:R1–R3.
83. Traynor JR, Nahorski SR. Mol Pharmacol 1995; 47:848–854.
84. Sim LJ, Selly DE, Childers SR. Proc Natl Acad Sci USA 1995; 92:7242–7246.
85. Sim LJ, Selly DE, Xiao R, Childers SR. Eur J Pharmacol 1996; 307:97–105.
86. Selly DE, Sim LJ, Xiao R, Liu Q, Childers SR. Mol Pharmacol 1997; 51:87–96.
87. Shimohira I, Tokuyama S, Himeno A, Niwa M, Ueda H. Neurosci Lett 1997; 237:113–116.
88. Narita M, Mizoguchi H, Narita M, Sora I, Uhl GR, Tseng LF. Br J Pharmacol 1999; 126:451–456.
89. Standifer KM, Pasternak GW. Cell Signal 1997; 9:237–248.
90. Standifer KM, Rossi GC, Pasternak GW. Mol Pharmacol 1996; 50:293–298.
91. Williams JT, Eagan TM, North RA. Nature 1982; 299:74–77.
92. Werz MA, MacDonald RL. J Pharmacol Exp Ther 1985; 234:49–56.
93. North RA, Williams JT, Surprenant A, Christie MJ. Proc Natl Acad Sci USA 1987; 84:5487–5491.
94. Triggle DJ. Neurotransmissions 1990; 6:1–5.
95. North RA. In: Handbook of Experimental Pharmacology. Amsterdam: Springer, 1992.
96. Aronsen JK. Biochem Pharmacol 1992; 43:11–14.
97. North RA, Katayama Y, Williams JT. Brain Res 1979; 165:66–67.
98. Pepper CM, Henderson G. Science 1980; 209:394–396.
99. Yoshimura M, North RA. Nature 1983; 305:529–530.
100. Loose MD, Kelly MJ. Brain Res 1990; 513:15–23.
101. Pan ZZ, Williams JT, Osborne PB. J Physiol 1990; 427:519–532.
102. Wuarin JP, Dudek FE. Neuroscience 1990; 36:291–298.
103. North RA, Williams JT. Br J Pharmacol 1983; 80:225–228.
104. Werz MA, MacDonald RL. Neurosci Lett 1983; 42:173–178.
105. Fan SF, Shen KF, Crain SM. Brain Res 1991; 558:166–170.
106. Fan SF, Shen KF, Crain SM. Brain Res 1993; 605:214–220.
107. Tatsumi H, Costa M, Schimerlik M, North RA. J Neurosci 1990; 10:1675–1682.
108. Shen KZ, North RA, Surprenant AI. J Physiol 1992; 445:581–599.
109. Welch SP, Dunlow LD. J Pharmacol Exp Ther 1993; 267:390–399.
110. MacDonald RL, Werz MA. J Physiol 1986; 377:237–249.

111. Tsunoo A, Yoshii M, Narahashi T. Proc Natl Acad Sci USA 1986; 83:9832–9836.
112. Gross RA, MacDonald RL. Proc Natl Acad Sci USA 1987; 84:5469–5473.
113. Surprenant A, shen KZ, North RA. J Physiol 1990; 431:585–608.
114. Eckert R, Trautwein W. Neurosci Lett 1991; 119:123–126.
115. Schroeder JE, Fischbach PS, Zheng D, McCleskey EW. Neuron 1991; 6:13–20.
116. Seward E, Hammond C, Henderson G. Proc R Soc Lond Ser B: Biol Sci 1991; 244:129–135.
117. Shen KZ, Surprenant A. Pflugers Archiv Eur J Physiol 1991; 418:614–616.
118. Lia EN, Prado WA. Acta Physiol Pharmacol Ther Latinoam 1999; 49:195–203.
119. Piros ET, Prather PL, Law PY, Evans CJ, Hales TG. Mol Pharmacol 1996; 50:947–956.
120. Huang LYM. Cellular Mechanisms of Excitatory and Inhibitory Actions of Opioids. Chur: Harwood Academic Publishers, 1995:131–149.
121. Nestler J. J Neurosci 1992; 12:2439–2450.
122. Miyamae T, Fukushima N, Misu Y, Ueda H. Eur Biochem Soc Lett 1993; 333:311–314.
123. Mutrhy KS, Makhlouf GM. Mol Pharmacol 1996; 50:870–877.
124. Sanchez-Blazquez P, Garzon J. J Pharmacol Exp Ther 1998; 285:820–827.
125. Tsu RC, Chan JCS, Wong TH. J Neurochem 1995; 64:2700–2707.
126. Smart D, Lambert DG. J Neurochem 1996; 66:1462–1467.
127. Ueda H, Miyamae T, Hayashi C, Watanabe S, Fukushima N, Sasaki Y, Iwamura T, Misu Y. J Neurosci 1995; 15:7485–7499.
128. Narita M, Mizoguchi H, Kampine JP, Tseng LF. Br J Pharmacol 1996; 118:1829–1835.
129. Narita M, Ohsawa M, Mizoguchi H, Aoki T, Suzuki T, Tseng LF. Neuroscience 2000; 99:327–331.
130. Sanchez-Blazquez P, Rodriguez-Diaz M, Frejo MT, Garzon J. Eur J Neurosci 1999; 11:2059–2064.
131. Nishizuka Y. FASEB J 1995; 9:484–496.
132. Way KJ, Chou E, King GL. Trends Pharmacol Sci 2000; 21:181–187.
133. Malmberg AB, Chen C, Tonegawa S, Basbaum AI. Science 1997; 278:279–283.
134. Martin WJ, Liu H, Wang H, Malmberg AB, Basbaum AI. Neuroscience 1999; 88:1267–1274.
135. Narita M, Feng YZ, Makimura M, Hoskins B, Ho IK. Eur J Pharmacol 1994; 271:543–545.
136. Narita M, Feng YZ, Makimura M, Hoskins B, Ho IK. Eur J Pharmacol 1994; 271:547–550.
137. Narita M, Makimura M, Feng YZ, Hoskins B, Ho IK. Brain Res 1994; 650:175–179.
138. Mayer DJ, Mao J, Price DD. Pain 1995; 61:365–374.
139. Mao J, Price DD, Phillips LL, Lu J, Mayer DJ. Brain Res 1995; 677:257–267.
140. Narita M, Narita M, Mizoguchi H, Tseng LF. Eur J Pharmacol 1995; 280:R1–R3.

141. Shearman MS, Sekiguchi K, Nishizuka Y. Pharmacol Rev 1989; 41:211–237.
142. Yang J, Tsien RW. Neuron 1993; 10:127–136.
143. Sasa M, Dimitrijevic W, Ryves J, Parker PJ, Rvans FJ. Mol Pharmacol 1995; 48:259–267.
144. Pei G, Kieffer BL, Lefkowitz RJ, Freedman NJ. Mol Pharmacol 1995; 48:173–177.
145. Fukushima N, Ueda H, Hayashi C, Katayama T, Miyamae T, Misu Y. Neurosci Lett 1994; 176:55–58.
146. Law PY, Kouhen OM, Solberg J, Wang W, Erickson LJ, Loh HH. J Biol Chem 2000; 275:32057–32065.
147. Kouhen OM, Wang G, Solberg J, Erickson LJ, Law PY, Loh HH. J Biol Chem 2000; 275:36659–36664.
148. Lowe JD, Celver JP, Gurevich VV, Chavkin C. J Biol Chem 2002; 277:15729–15735.
149. Hasubi A, Allouche S, Sichel F, Stanasila L, Massotte D, Landemore G, Polastron J, Jauzac P. J Pharmacol Exp Ther 2000; 293, 237–247.
150. Narita M, Mizoguchi H, Kampine JP, Tseng LF. Br J Pharmacol 1997; 120:587–592.
151. Narita M, Mizoguchi H, Nagase H, Tseng LF. Psychopharmacology 1997; 133:347–350.
152. Tseng LF, Narita M, Kampine JP. Eur J Pharmacol 1995; 287:169–172.
153. Whistler JL, Enquist J, Marley A, Fong J, Gladher F, Tsuruda P, Murray SR, von Zastrow M. Science 2002; 297:615–620.
154. Sora I, Funada M, Uhl GR. Eur J Pharmacol 1997; 324:R1–R2.
155. Hosohata Y, Vanderah TW, Burkey TH, Ossipov MH, Kovelowski GL, Sora I, Uhl GR, Zhang X, Rice KC, Roeske WR, Hruby VJ, Yamamura HI, Lai J, Porreca F. Eur J Pharmacol 2000; 388:241–248.
156. Gouardères C, Jhamandas K, Cridland R, Cros J, Quirion R, Zajac JM. Neuroscience 1993; 54:799–807.
157. Miyamoto Y, Portoghese PS, Takemori AE. J Pharmacol Exp Ther 1993; 264:1141–1145.
158. Bilsky EJ, Bernstein RN, Hruby VJ, Rothman RB, Lai J, Porreca F. J Pharmacol Exp Ther 1996; 277:491–501.
159. Kest B, Lee CE, McLemore GL, Inturrisi CE. Brain Res Bull 1996; 39:185–188.
160. Cahill CM, Morinville A, Lee M-C, Vincent J-P, Collier B, Beaudet A. J Neurosci 2001; 21:7598–7607.
161. George SR, Fan T, Xie Z, Tse R, Tam V, Varghese G, O'Dowd BF. J Biol Chem 2000; 275:26128–26135.
162. Gomes I, Jordan BA, Gupta A, Trapaidze N, Nagy V, Devi LA. J Neurosci 2000; 20:RC110(1-5).
163. Jordan BA, Devi LA. Nature 1999; 399:697–700.
164. Jordan BA, Trapaidze N, Gomes I, Nivarthi R, Devi LA. Proc Natl Acad Sci USA 2001; 98:343–348.
165. Klein U, Muller C, Chu P, Birnbaumer M, von Zastrow M. J Biol Chem 2001; 276:17442–17447.

20

Antidepressant-like Effects of Delta Opioid Receptor Agonists

Emily M. Jutkiewicz and James H. Woods
University of Michigan, Ann Arbor, Michigan, U.S.A.

1 POTENTIAL ANTIDEPRESSANT-LIKE EFFECTS OF DELTA OPIOID RECEPTOR AGONISTS

Endogenous opioids and exogenously administered delta opioid receptor agonists have been demonstrated to have antidepressant-like effects in animal models used to evaluate novel antidepressant compounds. This chapter reviews some of the previous research investigating the role of the opioid system, and more specifically the delta opioid receptor system, in clinical depression and in animal models used to study human depression and antidepressant treatments. In addition, this chapter discusses the interpretation of animal models of depression and their uses in the evaluation of new potential therapeutics.

2 THE RELATION BETWEEN DEPRESSION AND ANIMAL MODELS

Depression is an illness that alters mood, emotions, and physical conditions. Individuals suffering from depression can experience changes in physical characteristics such as changes in body weight and sleeping patterns, as well as persistent headaches, digestive disorders, and chronic pain. However, depression is best characterized by and diagnosed by the presence of some (but not all) symptoms of emotional distress, such as persistent sad mood, feelings of hopelessness, feelings of helplessness and worthlessness, decreased energy and fatigue, and thoughts of death and suicide. In the past few decades, many useful therapeutics have been developed for the treatment of this debilitating disease. Despite these advances, a significant portion of depressed patients are insensitive or unresponsive to these treatments. Therefore, it is important to develop new drugs and/or novel therapies to treat depression.

In addition to modifying the structures of successful antidepressants, it would be useful to investigate novel therapeutic approaches using nonhuman models. However, how does one study depression and actions of antidepressants in laboratory animals? Depression is based on human emotions and subjective reports of mood states. Can this be modeled in animals?

The American psychologist Martin Seligman [1] developed a theory for and a model of depression. This theory described depression as a disease that develops when an individual believes that he/she cannot control the outcome of life events. Seligman applied this theory to animal models and demonstrated that dogs, when exposed to stress, expressed "humanlike" depressive symptoms. This model was termed "learned helplessness" and was extensively studied in terms of psychological and physiological parameters that occur before, during, and after the induction of learned helplessness. Porsolt et al. [2] further developed this model to study the effects of antidepressants in order to develop a method for screening new therapeutic compounds.

The forced swim test [2,3], a model of learned helplessness or of behavioral despair, is conducted by forcing a rat or mouse to swim in an inescapable container filled with water. When placed in a swim tank, animals display escape-directed behaviors such as swimming, climbing, and diving. After an initial period of activity, animals become immobile or simply float in the water, using small movements to keep the head or nose above the water surface. The immobility (failure in the persistence of escape-directed behaviors) is frequently interpreted as "despair-like" behavior in which the subject has failed at escape attempts and adopts "hopeless" or "depressed" postures.

This interpretation of immobility in the forced swim test has been challenged numerous times since its creation. It was proposed that these behaviors are not emotional responses to stressful situations, but a learned response that occurs during extended or between successive swim exposures. The learned response might be considered an adaptive strategy used to conserve energy required for swimming and survival [for review see 4]. Based on these interpretations, it was suggested that antidepressants may decrease immobility by disrupting learning processes occurring between swim sessions. This issue will be addressed later in the chapter.

When considering animal models of depression and antidepressant tests, it is important to consider the validity of the model being used. There are three types of criteria used to evaluate animal models [5,6]. The first is face validity, referring to the degree of symptomology between the human clinical condition and the model. The second is predictive validity, referring to the concept that the model identifies compounds that are successful antidepressants in depressed patients. The third is construct validity, referring to the theoretical rationale of the model being either psychological or biological in nature. There are a number of models of depression that have been analyzed in terms of their validity [for reviews see 5–7]. According to Willner [5], the main model discussed in this chapter, the forced swim test, demonstrated good predictive validity for identifying compounds that are successful antidepressants in human depressed patients (as well as some face validity). In fact, there was a significant correlation between clinical potency and potency of antidepressants in the forced swim test in rats [5].

Although methods and drugs to treat depression have improved dramatically over the past 30 years, there is still no single cure or resolution for this disease. Animal models used to study human mental disorders should be designed and assessed according to certain criteria [8]. These criteria state that the model should resemble the human condition in terms of its etiology, biochemistry, symptomology, and treatment. However, this may be an unfair and unattainable model. Unless animal behaviors are massively anthropomorphized (which they commonly are), it would be impossible to equate human emotional conditions with those of animals, which may or may not exist. In addition, it is unreal to think that the depression condition is completely understood. Most likely, there is no "magic bullet" to treat depression and probably no single cause of this disease. Depression develops from a complex interaction of genetics, environment, and neurobiology that is just beginning to be comprehended. Taking this information into account, it may be impossible to model human depression in animals. Therefore, the best preclinical model used to study depression is one that identifies many known antidepressants of different classes and can be used easily to evaluate new drugs and to identify new potential targets.

In the quest for new therapeutics to treat depression, most compounds in development have similar mechanisms of action: increasing catecholamine or indolamine neurotransmission by blocking reuptake. These drugs block transporters that remove neurotransmitter from the synapse and transport it into the presynaptic cell from which it was released. These classes of drugs have improved the lives of many depressed people; however, many patients still suffer from this debilitating illness. A fair number of depressed patients are unresponsive to current drug treatments. In addition, depressed patients also have many compliance issues with the treatment regimens, and some patients find the side effects unbearable. Therefore, there is a strong necessity to find new drugs and to investigate new targets in order to treat patients as well as to learn more about the disease.

3 OPIOID SYSTEM AND DEPRESSION: NEW THERAPEUTIC TARGET

Most antidepressant therapies concentrate on the enhancement of aminergic neurotransmission. There are a number of other targets that may play a role in this disease, such as GABA, stress hormones, opioids, and others. Changes in opioid signaling have been associated with the antidepressant-like effects of electroconvulsive therapy (ECT); however, opioids are rarely discussed as a potential drug therapy. This chapter discusses the involvement of endogenous opioids in depression and opioid systems as potential opioid drug targets.

Opioids are known to alter mood states. For example, opiates such as morphine produce euphoria and pain relief. Prolonged use of and withdrawal from opiates produce depressive-like symptoms as well. Based on the mood-altering effects of opiates, the role of endogenous opiates in psychiatric symptoms of various diseases has been studied. In addition, endogenous opioids are believed to play a role in neuronal circuitry responsible for reward and pleasure. Therefore, it is thought that perhaps the anhedonia observed in depressed patients is due to dysregulation of endogenous opioids in neuronal reward circuitry.

A number of clinical studies have reported contradictory data about levels of endogenous β-endorphin levels in depressed patients. There was no difference in β-endorphin levels between depressed patients and normal controls [9]; however, β-endorphin was found to be elevated in patients with major depression [10], during the manic phase of manic-depressive patients [11], and in patients with primary affective disorders [12]. Another study demonstrated that β-endorphin did not correlate with the Hamilton Rating Scale for Depression, but high β-endorphin levels in depressed patients did correlate with severe anxiety, phobias, and obsessive-compulsive behaviors in these patients [13]. A more recent double-blind crossover randomized study demonstrated that depressed patients had lower β-endorphin levels than healthy controls, and patients with endogenous depression had even lower β-endorphin levels than patients with nonendogenous depression [14]. Interestingly, treatment with fluvoxamine improved depression scores on the Hamilton Rating Scale for Depression and increased β-endorphin levels. In addition, the tricyclic antidepressant doxepin improved pain and depression scores in patients diagnosed with a chronic pain condition and clinical depression in a randomized double-blinded study [15]. In this study, doxepin also increased nonspecific enkephalin-like activity, but not β-endorphin plasma levels.

In addition to known antidepressants increasing endogenous opioids, opioid ligands have also been administered to depressed patients to determine if opioid compounds have clinical efficacy to treat depression. The opioid ligand cyclazocine improved symptoms in severely depressed, chronically ill mental patients in an open clinical trial and in clinical trials with patients unresponsive to the tricyclic antidepressant imipramine [16]. Intravenous β-endorphin infusions improved mood in depressed patients in open case studies [17] and in depressed patients in a double-blind placebo-controlled study [18,19]. However, one study found a trend to improve depression scores in patients after acute and chronic β-endorphin infusions, but it was not significant [20].

ECT is one of the most successful and useful antidepressant therapies norepinephrine, available today, and alterations in endogenous opioid systems have been proposed to induce its therapeutic effects. ECT has been shown to change many systems, including serotonin, norepinephrine, and dopamine, although none of these changes can exclusively explain the therapeutic effects [for reviews see 21,22]. More recently, it was proposed that the antidepressant mechanism of ECT is determined by anticonvulsant effects [for review see 23]. The magnitude of change in seizure threshold is associated with therapeutic outcome. Changes in GABA-ergic functioning and endogenous opioids were proposed to play a role in the changing of seizure threshold and, therefore, to play a role in the antidepressant activity of ECT. ECT elevated plasma immunoreactive β-endorphin in depressed patients after single and repeated treatments [24,25]. Similarly, in preclinical studies, electroconvulsive shock (ECS) induced increases in β-endorphin as well as proenkephalin-derived peptides and increases in delta opioid receptors as measured by [^{3}H]DADLE binding and other opioid receptors [for review see 26]. Naloxone and ICI-174,864 antagonized the anticonvulsant effects of ECS that are thought to be the mechanism of therapeutic effects. Based on these findings, the opioid system may be a useful target for the development of new therapeutics.

4 OPIOIDS IN ANIMAL MODELS OF ANTIDEPRESSANT ACTIVITY

Opioid peptides have also been tested in animal models of depression and of antidepressant activity. Enkephalins and endorphins decreased immobility in the forced swim test and in the learned helplessness paradigm, demonstrating similar effects to known antidepressants [27,28]. It has also been demonstrated that preventing the breakdown of endogenous opioid peptides with enkephalinase inhibitors produces antidepressant-like effect in animal models. RB38A, a mixed inhibitor of enkephalinase, and RB38B, a selective inhibitor of endopeptidase EC 3.4.24.11, produced an antidepressant-like effect in the learned helplessness paradigm, and these effects were blocked by the opioid antagonist naloxone, indicating an opioid-mediated effect [28,29]. Similarly, the enkephalinase inhibitor BL-2401 elicited a naloxone-reversible antidepressant-like effect in the forced swim test in mice, indicating an opioid-mediated effect [30]. Although these studies demonstrated that the behavioral effects of the enkephalinase inhibitors were opioid-mediated, they did not speak to the different opioid receptor types. Baamonde et al. [31] and Tejedor-Real et al. [32] later demonstrated that the selective delta opioid antagonist naltrindole inhibited the antidepressant-like effects of RB101, a mixed enkephalinase inhibitor, in the learned helplessness paradigm in rats, demonstrating the antidepressant-like profile of RB101 is delta opioid receptor mediated. To further support this finding, they tested the selective delta opioid peptide agonist BUBU (Tyr-D.Ser-(*O-tert*-butyl)-Gly-Phe-Leu-Thr(O-Tet-butyl-OH) and found similar behavioral effects to the enkephalinase inhibitor [32]. These previous findings demonstrated that opioid receptor and, specifically, delta opioid receptor stimulation produces antidepressant-like behaviors in multiple animal models; however, the question of the role of endogenous opioid tone in depression and depressive-like symptoms remained. This point was addressed recently in a study that characterized the behavioral changes in delta opioid receptor–deficient mice (Oprd1-deficient) [33]. The delta opioid knockout mice demonstrated behaviors consistent with anxiogenic- and depressivelike responses, suggesting that delta opioid receptors and endogenous tone of the delta opioid system may play a role in mood states and depression.

More recent studies have investigated the effects of nonpeptidic delta opioid agonists in animal models of antidepressant activity. These nonpeptidic compounds are quite selective for the delta opioid receptor and are centrally active following peripheral administration. The nonpeptidic selective delta opioid agonists SNC80 and (+)BW373U86 decreased immobility in the forced swim test, indicating an antidepressant-like effect with these compounds [34–36]. The decrease in immobility was antagonized by the delta opioid antagonist naltrindole, demonstrating that these effects were mediated through the delta opioid receptor. Interestingly, the delta opioid agonists produce antidepressant-like profiles of action after a single injection, unlike the typical antidepressant compounds that usually require multiple administrations to produce an effect in many animal models of antidepressant activity. These data may imply that the antidepressant-like activity of the delta opioid agonists has a faster onset of action than usual antidepressant compounds. Animal models typically attempt to recreate conditions under which humans receive antidepressants and observe behavioral changes; therefore, antidepressants are frequently administered multiple times before testing. However, subchronic and chronic dosing regimens in animal models are not equivalent to human administration schedules, so this requirement may or may not be important to the observed behavioral changes. Either way, delta opioid agonists have immediate effects in the forced swim test and, therefore, may suggest that the drug has a faster onset of action than typical antidepressant drugs in this assay.

5 MECHANISM OF ACTION: STIMULANT ACTIVITY

The forced swim test and other assays that are used to identify antidepressant compounds detect compounds with different types of drug action—i.e., serotonin reuptake blockers, norepinephrine reuptake blockers, and atypical antidepressants. A major concern with these types of assays is the identification of false positive compounds. Traditionally, test compounds are evaluated in the forced swim test and in locomotor activity assays to test for stimulant activity. Stimulant compounds are considered false positives in the forced swim test since swimming is considered by some a form of locomotion. However, there are some compounds that increase dopaminergic signaling, such as nomifensine and bupropion, that had antidepressant-like effects in clinical and preclinical tests and demonstrated stimulant activity in some studies [37,38]. Interestingly, stimulant drugs are normally considered false positives in the forced swim test because they are not prescribed for depressed patients; however, no controlled studies have been conducted to test this assumption.

Like bupropion, the nonpeptidic delta opioid agonists SNC80 and BW373U86 increase locomotor activity [34,39]. Delta opioid agonist–induced locomotor stimulation was prevented by the delta opioid antagonist naltrindole [35,39] as well as by D1 and D2/D3 dopamine antagonists [39]. These results demonstrated that delta agonists display stimulant qualities most likely through an augmentation of the effects produced by the release of dopamine. Other studies, however, showed that SNC80 and BW373U86 failed to increased dopamine levels in the nucleus accumbens and actually decreased extracellular dopamine in the caudate putamen of the rat [40]. Clearly, there are some discrepancies of the circuitry involved on delta opioid signaling systems; however, the stimulant properties of nonpeptidic delta opioid agonists may confound the increased activity (i.e., decreased immobility) observed in the forced swim test, which is indicative of antidepressant-like activity. Broom et al. [35] attempted to address this issue by evaluating the antidepressant-like effects of nonpeptidic delta opioid agonists after the locomotor stimulation dissipated. This experiment demonstrated that 3h after an injection of SNC80, locomotor-stimulating effects were no longer observed; however, the antidepressant-like effect persisted. These data demonstrated that the antidepressant-like effects were not necessarily dependent on stimulant activity, but may coexist and have different durations of action.

The locomotor-stimulating properties of delta opioid agonists may be separated from the antidepressant-like effects by other measures as well. For example, it was demonstrated that tolerance to the behavioral effects of nonpeptidic delta agonists developed rapidly following a single injection of agonist in mice [41] and in rats [35]. Also, a profound tolerance to the locomotor-stimulating properties of the delta agonist SNC80 were observed after a single injection in rats [42]. Unlike convulsive and antidepressant-like effects, tolerance to antidepressant-like effects of (+)BW373U86 [35] and SNC80 [42] did not develop following a single administration. These data further support the concept that locomotor-stimulating and antidepressant-like properties may coexist but that stimulant activity is not required for antidepressant-like effects.

6 MECHANISM OF ACTION: CONVULSIVE ACTIVITY

In addition to locomotor activity, delta opioid agonists produce convulsions in mice [41,43], rats [34,35], and monkeys [44–46]. In the past, chemical-induced convulsions induced by camphor or pentylenetetrazol (Metrazol) were used as treatments for depression; today, however, ECT is the only convulsant therapy used because the treatment-induced effects are less unpleasant than those produced by chemical convulsants [47]. ECS was demonstrated to have antidepressant-like effects in the forced swim test in rats [3], and ECT is a very effective treatment for depression in humans. Based on these observations, it was proposed that delta opioid agonists produce antidepressant-like effects tough a convulsive- or electroconvulsive shock (ECS)-like mechanism of action [41].

Broom et al. [35] tested the hypothesis that convulsant activity was required for the antidepressant-like effects of the delta opioid agonist (+)BW373U86. In this study, midazolam, a short-acting benzodiazepine, was administered prior to the delta opioid agonist. Midazolam eliminated the observed convulsions, but did not alter the antidepressant-like effects of (+)BW373U86. These data demonstrated that convulsions were not required for the antidepressant-like effects of delta opioid agonists, suggesting that delta agonists do not produce these effects through an ECS-like mechanism. To further support this hypothesis, these studies demonstrated that tolerance to the convulsive effects develop after a single injection, but the antidepressant-like effects remain, similar to that observed with locomotor activity.

Similarly, convulsant effects can exist without the antidepressant-like effects. When the delta opioid agonist SNC80 was administered by rapid intravenous infusion (over 20 sec), some rats convulsed with a dose of 1.0 mg/kg SNC80; however, this dose did not decrease immobility in the forced swim test (Fig. 1). These results further demonstrate that the behavioral effects may coexist but convulsions alone do not produce antidepressant-like effects. Also, it was demonstrated that slow intravenous infusions of the delta opioid agonist SNC80 decreased and practically eliminated convulsant activity without altering the antidepressant-like effects of SNC80 [48] (Fig. 2). Again, these data suggest that it is possible to observe antidepressant-like activity without the presence of convulsions. Although multiple behavioral effects coexist, antidepressant-like effects appear to be a unique property of delta opioid receptor activation that is independent of convulsive and stimulantlike properties of these compounds.

Although these experiments analyzed the hypothesis that convulsant activity is required for the antidepressant-like effects of (+)BW373U86, it did not thoroughly address the question of whether seizure activity is required for the antidepressant-like effects of delta opioid agonists. The benzodiazepine midazolam was demonstrated to inhibit seizure activity [49,50]; however, seizure activity was not directly measured in the study by Broom et al. [35]. Interestingly, limited work has been done with the seizure activity of delta opioid agonists. One study demonstrated that the delta opioid agonist

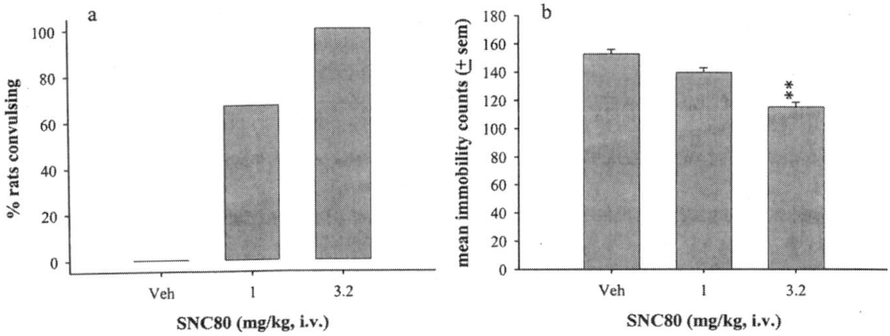

FIGURE 1 The convulsive (a) and antidepressant-like (b) effects of SNC80 administered by intravenous injection infused in 20 sec. During and immediately after the infusion, rats were observed for 20 min to watch for convulsive activity (a). Thirty minutes after the SNC80 infusion, rats were evaluated in the forced swim test (b). Counts of immobility were recorded as previously described [62]. $**P < .01$ as compared to vehicle by Dunnett's post hoc test.

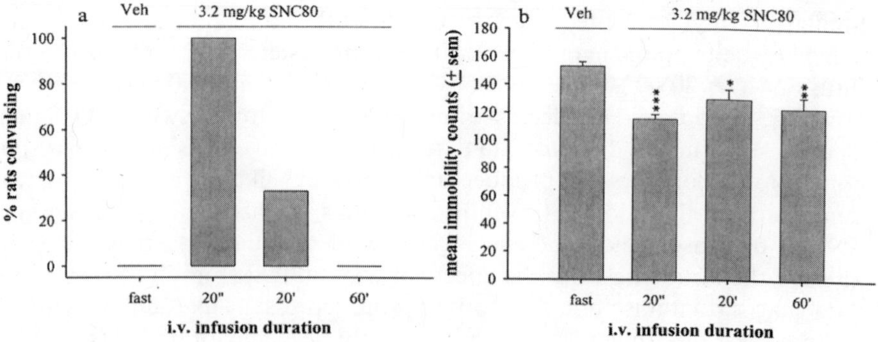

FIGURE 2 The convulsive (a) and antidepressant-like (b) effects of intravenous SNC80 administered with different speeds of infusion. SNC80 (3.2 mg/kg) was administered by a fast 20-sec infusion (20″), a 20-min infusion (20′), or a 60-min infusion (60′). SNC80 or vehicle 3.2 mg/kg was administered by different speeds of infusion; however, vehicle data from the fast (20″) infusion only is shown here. Immobility scores from rats receiving vehicle by different infusion speeds were not statistically different (data not shown). There were no statistical differences between immobility scores measured from rats that received 3.2 mg/kg SNC80 by different infusion speeds. $*P < .05$, $**P < .01$, $***P < .001$ as compared to vehicle by Dunnett's post hoc test.

BW373U86 increased hippocampal type 2 theta power in rat EEG measurements; however, this study did not test doses of BW373U86 that would produce convulsions in rats [51]. Therefore, it is difficult to determine the role of seizures in the antidepressant-like effects of delta opioid agonists. This concept must be studied in the future to determine the role of seizure activity or electroencephalographic activity in the antidepressant-like effects of delta opioid agonists.

Considering the fact that tolerance develops rapidly to the convulsant effects of delta opioid agonists but not to the antidepressant-like effects, it is interesting to consider the changes in ECT-induced seizures over time. During ECT, seizure activity usually terminates within a minute of initiation. With repeated ECT, the seizure threshold increases, and, as mentioned earlier, this is thought to be associated with or indicative of clinical efficacy. Endogenous opioids are thought to play a role in the increased seizure threshold, indicating that they may be natural anticonvulsant substances. Maintenance ECT is used to preserve the seizure threshold at an elevated level. This hypothesis of the mechanism of action of ECT may lend some insight into the antidepressant-like effects of delta opioid agonists. Delta opioid receptor activation alone and/or delta-induced convulsions may elevate endogenous opioid levels in a similar manner in which ECT elevates endogenous opioids, thus producing an antidepressant-like effect.

7 MECHANISM OF ACTION: IMPAIRMENTS IN LEARNING

Another concern with validity of the forced swim test is that learning and memory may play a role in the development of an immobile posture. For example, traditionally, the forced swim test in rats comprises a 2-day swim test. On the first day, rats are exposed to a 15-min habituation swim in which all rats are drug naïve. On day 2, rats are returned to the swim tank for a 5-min swim session and behavior is recorded during this second exposure. In the original interpretations of the forced swim test, the first swim exposure induced the despair-like behaviors, such that the despair or floating (immobility) was more readily observed on day 2. However, it was also suggested that rats became immobile faster on day 2 because they learned this adaptive response. Therefore, it was suggested that test drugs administered between day 1 and day 2 swim may decrease immobility on day 2 by interrupting learning and memory of the immobile behavior. For example, anisomycin, an antibiotic known to disrupt learning and memory and with no known antidepressant action, decreased immobility in the forced swim test [52]. However, much is known about the effects of antidepressants on learning and memory from a preclinical and clinical point of view. Some studies have demonstrated that tricyclic antidepressants, such as amitriptyline and imipramine, impaired memory and enhanced scopolamine-induced memory deficits [53]. On the other hand, fluoxetine [54] and mirtazepine [55] demonstrated memory-improving qualities and prevented scopolamine-induced memory deficits. Although these known antidepressant-like effects had different effects on learning and memory, these compounds all decreased immobility in the forced swim test, indicating an antidepressant-like effect. These results suggest that antidepressant-like effects in the forced swim test may not be dependent on the disrupting effects on learning and memory.

Another approach to consider is the removal of the habituation swim from the forced swim test, thus eliminating the possibility of learning between 2 days of swim. Some forced swim test experiments require only 1 day of swimming [34,35]. In this paradigm, rats were exposed to a 1-day 15-min forced swim test that was videotaped and later scored. On the day of swimming, rats that received vehicle injections demonstrated high levels of immobility, whereas the known antidepressant desipramine decreased immobility during the 1-day swim. These data demonstrate that the interpretation of "inducing despair" with repeated swim exposures may not be appropriate or accurate. Clearly, some antidepressants have effects in the forced swim test independent of "despair induction" or learning adaptive responses with 2 days of swim.

Since drugs that impair learning and memory may decrease immobility in the forced swim test, it is worthwhile to consider the ability of delta opioid agonists to alter these processes. A number of studies demonstrated that activation of delta opioid receptors impaired learning and memory in a number of behavioral assays. For example, [leu]enkephalin and [D-Pen^2,D-Pen^5]enkephalin (DPDPE) impaired acquisition of avoidance responding in mice [56], and the nonpeptidic delta opioid agonist BW373U86 increased errors in a repeated acquisition paradigm in squirrel monkeys [46] but had no significant effect on retention of a response chain 24 sec after acquisition [57].

Recently, the effects of the delta opioid agonist SNC80 in the forced swim test were compared with the effects of this delta opioid agonist on behaviors measured in the repeated-acquisition procedure [58]. These experiments showed that SNC80 increased errors in acquisition components, demonstrating an impairment in learning or acquiring a response pattern. The delta opioid agonist SNC80 also decreased immobility in the forced swim test, indicating an antidepressant-like effect. Doses that produced antidepressant-like effects were six to ten times lower than those required to produce deficits in learning acquisition. Although delta opioid agonists induce memory and learning impairments, this action alone cannot explain the antidepressant-like effects observed in the forced swim test.

8 SUMMARY

Experiments and data presented in this chapter demonstrate that delta opioid agonists have antidepressant-like effects in animal models used to measure antidepressant activity. The antidepressant-like effects can be separated from other behavioral effects produced by these compounds, such as locomotor stimulation, convulsions, and learning impairments. This separation lends validity to this potential target for depression by eliminating effects or sources that may produce false positives. These compounds should be tested in other models of antidepressant activity to confirm these findings in the forced swim test.

9 DRUG DEVELOPMENT CONSIDERATIONS: STRUCTURE-ACTIVITY RELATIONSHIP

To further develop delta opioid receptor agonists as potential therapeutics, more compounds should be developed and evaluated in the forced swim test and other models of antidepressant activity. This may allow for the identification of delta opioid agonists with even more selective behavioral effects. Indeed, there are a number of delta opioid agonists with different basic chemical structures that may prove useful.

The structural class of piperazinyl benzamides contains a number of SNC80-like delta opioid receptor ligands. These compounds have been

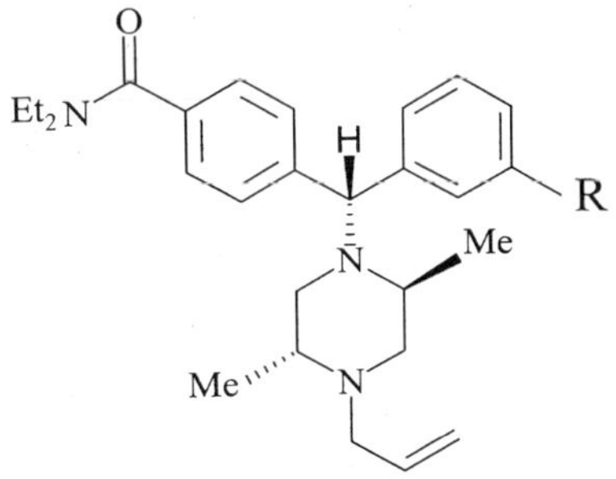

ligand	R group
SNC80	OMe
SNC86	OH
SNC162	H

FIGURE 3 Structures of SNC-related compounds.

studied in terms of radioligand binding affinities [e.g., 59]. The behavioral effects of three SNC-like compounds, SNC80, (+)BW373U86 (SNC86), and SNC162, were studied in the forced swim test and for their convulsive and locomotor-stimulating effects in rats (Fig. 3). In addition, these compounds were evaluated by intravenous and subcutaneous routes of administration.

By intravenous route of administration, SNC80 and (+)BW373U86 (SNC86) appeared very similar. Both compounds produced convulsions and

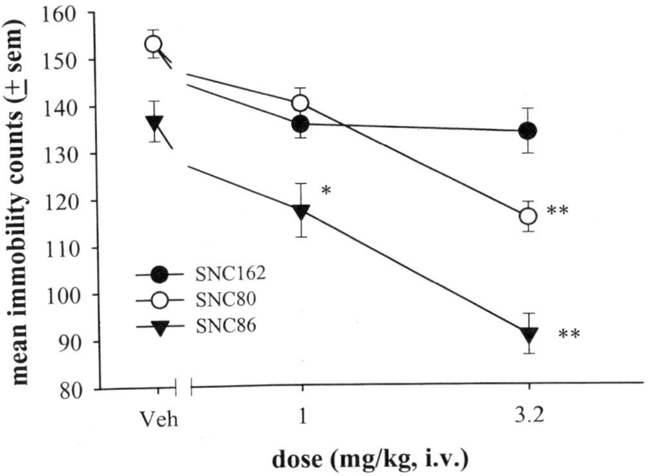

FIGURE 4 Antidepressant-like effects of SNC80, SNC86 [(+)BW373U86], and SNC162 (IV) measured in the forced swim test in rats. During and immediately after the drug unfusion, rats were observed for 20 min for convulsions. Thirty minutes after the drug infusion, rats were evaluated in the forced swim test. Counts of immobility were recorded as previously described [62]. $*P < .05$, $**P < .01$ as compared to vehicle by Dunnett's post hoc test.

demonstrate antidepressant-like effects in rats. (+)BW373U86 was slightly more potent for producing convulsions and antidepressant-like effects than SNC80. SNC162 produced convulsions in a manner similar to SNC80; however, it did not produce delta opioid receptor–mediated antidepressant-like effects when administered by the intravenous route (Fig. 4). These results demonstrated that SNC162 administered IV was three- to 10-fold less potent than SNC80 or (+)BW373U86 and further supported previous data demonstrating that convulsions alone were not sufficient to produce antidepressant-like effects. When these compounds are administered subcutaneously, SNC162 is at least 30-fold less potent than SNC80 or (+)BW373U86 (data not shown). This difference in potency is the largest difference reported. Stevenson et al. [60] reported an approximate sixfold difference between SNC80 and SNC162 in delta opioid discrimination in rats. In monkey studies, Brandt et al. [61] reported a threefold difference between SNC162 and SNC80 or (+)BW373U86 in monkeys trained to discriminate SNC80.

Interestingly, SNC162 did not produce long-lasting increases in locomotor activity (Fig. 5). (+)BW373U86 and SNC80 elevated locomotor activity, and the activity counts returned to baseline values approximately

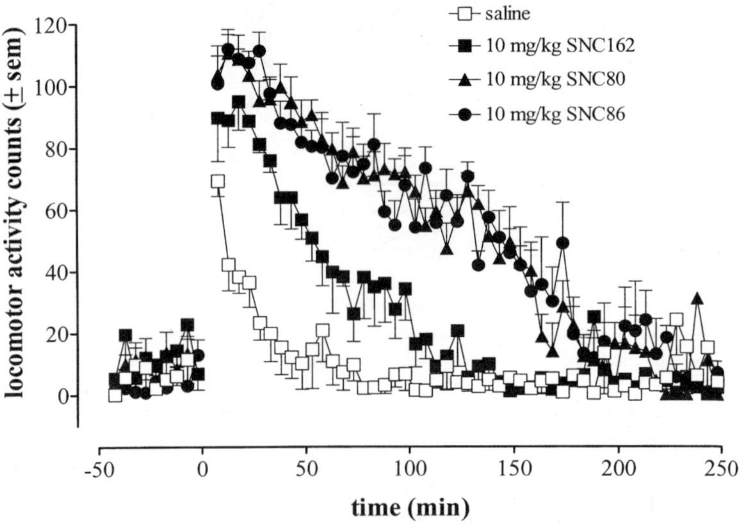

FIGURE 5 Rats were implanted with transmitters (Mini Mitter Co., Inc., Bend, OR, U.S.A.) in the peritoneal cavity that recorded changes in locomotor activity within the rats' home cages. Prior to injection, baseline activity counts were recorded (-50 to 0 min). Locomotor activity was recorded following a single injection (time 0) of vehicle, SNC80, SNC86 [(+)BW373U86], or SNC162.

3 hours after drug administration. However, SNC162-induced locomotor stimulation returned to baseline values after 1. These results suggested that SNC162 had a shorter duration of action, perhaps due to faster distribution, metabolism, and excretion. However, there is little information on the metabolism of SNC80-like compounds.

In conclusion, these data may provide information about the possibility of developing classes of delta opioid agonists that demonstrate potential therapeutic effects of delta opioid agonists, but do not demonstrate convulsions and locomotor-stimulating effects.

ACKNOWLEDGMENT

Research supported by USPHS grants DA00254, GM07767, and DA07267.

REFERENCES

1. Seligman MEP. Helplessness: On Depression, Development and Death. San Francisco: W.H. Freeman, 1975.
2. Porsolt RD, Le Pichon M, Jalfre M. Nature 1977; 266:730–732.
3. Porsolt RD, Anton G, Blavet N, Jalfre M. Eur J Pharmacol 1978; 47:379–391.
4. West AP. Prog Neuro-Psychopharmacol Biol Psychiatry 1990; 14:863–877.
5. Willner P. Psychopharmacology 1984; 83:1–16.
6. Willner P. Pharm Ther 1990; 45:425–455.
7. Willner P. Trends Pharmacol Sci 1991; 12:131–136.
8. McKinney WT Jr, Bunney WE. Arch Gen Psychiatry 1969; 21:240–248.
9. Naber D, Pickar D, Post RM, Van Kammen DP, Waters RN, Ballenger JC, Goodwin FK, Bunney WE. Am J Psychiatry 1981; 138:1457–1462.
10. Goodwin GM, Austin MP, Curran SM, Ross M, Murray C, Prentice N, Ebmeier KP, Bennie J, Carroll S, Dick H, Fink G. J Affect Disord 1993; 29:281–289.
11. Lindström LH, Widerlöv E, Gunne LM, Wahlström A, Terenius L. Acta Psychiatr Scand 1978; 57:153–164.
12. Genazzani AR, Petraglia F, Facchinetti F, Monittola C, Scarone S, Brambilla F. Neuropsychobiology 1984; 12:78–85.
13. Darko DF, Risch SC, Gillin JC, Golshan S. Am J Psychiatry 1992; 149:1162–1167.
14. Djurović D, Milić-Aškrabić J, Majkić-Singh N. Farmaco 1999; 54:130–133.
15. Hameroff SR, Cork RC, Scherer K, Crago R, Neuman C, Womble JR, Davis TP. J Clin Psychiatry 1982; 43:22–27.
16. Fink M, Simeon J, Itil TM, Freedman AM. Clin Pharmacol Ther 1969; 11(1): 41–48.
17. Kline NS, Li CH, Lehmann HE, Lajtha A, Laski E, Cooper T. Arch Gen Psychiatry 1977; 34:1111–1113.

18. Gerner RH, Catlin DH, Gorelick DA, Hui KK, Li CH. Arch Gen Psychiatry 1980; 37:642–647.
19. Gorelick DA, Catlin DH, Gerner RH. Mod Probl Pharmacopsychiatry 1981; 17:236–245.
20. Catlin DH, Gorelick DA, Gerner RH, Hui KK, Li CH. Adv Biochem Psychopharmacol 1980; 22:465–472.
21. Newman ME, Gur E, Shapira B, Lerer B. J ECT 1998; 14(3):153–171.
22. Mann JJ. J ECT 1998; 14(3):172–180.
23. Sackeim HA. J ECT 1999; 15(1):5–26.
24. Emrich HM, Hollt V, Kissling M, Fischler M, Laspe H, Heinemann H, Zerssen DV, Herz A. Pharmacopsychiatry 1979; 12:269–276.
25. Inturrisi CE, Alexopoulos G, Lipman R, Foley K, Roosier J. Ann NY Acad Sci 1982; 398:413–423.
26. Tortella FC, Long JB, Hong J-S, Holaday JW. Convuls Ther 1989; 5(3):261–273.
27. Kastin AJ, Scollan EL, Eensing RH, Schally AV, Coy DH. Pharmacol Biochem Behav 1978; 9:515–519.
28. Tejedor-Real P, Micó JA, Maldonado R, Roques BP, Gibert-Rahola J. Pharmacol Biochem Behav 1995; 52(1):145–152.
29. Tejedor-Real P, Micó JA, Maldonado R, Roques BP, Gibert-Rahola J. Biol Psychiatry 1993; 34:100–107.
30. Kita A, Imano K, Seto Y, Yakuo I, Deguchi T, Nakamura H. Jpn J Pharmacol 1997; 75:337–346.
31. Baamonde A, Daugé V, Ruiz-Gayo M, Fulga IG, Turcaud S, Fournié-Zaluski M-C, Roques BP. Eur J Pharmacol 1992; 216:157–166.
32. Tejedor-Real P, Micó JA, Smadja C, Maldonado R, Roques BP, Gibert-Rahola J. Eur J Pharmacol 1998; 354:1–7.
33. Filliol D, Ghozland S, Chluba J, Martin M, Matthes HWD, Simonin F, Befort K, Gaveriaux-Ruff C, Dietrich A, LeMeur M, Valverde O, Maldonado R, Kieffer BL. Nature 2000; 25:195–200.
34. Broom DC, Jutkiewicz EM, Folk JE, Traynor JR, Rice KC, Woods JH. Neuropsychopharmacology 2002; 26(6):744–755.
35. Broom DC, Jutkiewicz EM, Folk JE, Traynor JR, Rice KC, Woods JH. Psychopharmacology 2002; 164(1):42–48.
36. Broom DC, Jutkiewicz EM, Rice KC, Traynor JR, Woods JH. Jpn J Pharmacol 2002; 90(1):1–6.
37. Standford JA, Currier TD, Gerhardt GA. Pharmacol Biochem Behav 2002; 71:333–340.
38. Cooper BR, Hester TJ, Maxwell RA. J Pharmacol Exp Ther 1980; 215(1):127–134.
39. Spina L, Longoni R, Mulas A, Chang K-J, Di Chiara G. Behav Pharmacol 1998; 9:1–8.
40. Longoni R, Cadoni C, Mulas A, Di Chiara G, Spina L. Behav Pharmacol 1998; 9:9–14.
41. Comer SD, Hoenicke EM, Sable AI, McNutt RW, Chang K-J, De Costa BR, Mosberg HI, Woods JH. J Pharmacol Exp Ther 1993; 267(2):888–895.

42. Jutkiewicz EM, Rice KC, Traynor JR, Woods JH. FASEB, New Orleans, LA, 2002.
43. Hong EJ, Rice KC, Calderon S, Woods JH, Traynor JR. Analgesia 1998; 3:269–276.
44. Dykstra LA, Schoenbaum GM, Yarbroug J, McNutt R, Chang K-J. J Pharmacol Exp Ther 1993; 267:875–882.
45. Negus SS, Butelman ER, Chang K-J, De Costa B, Winger G, Woods JH. J Pharmacol Exp Ther 1994; 270:1025–1034.
46. Pakarinen ED, Woods JH, Moerschbaecher JM. J Pharmacol Exp Ther 1995; 272:552–559.
47. Fink M. Electroshock: Restoring the Mind. New York: Oxford University Press, 1999:87–91.
48. Jutkiewicz EM, Eller E, Rice KC, Woods JH, Behavioral effects in rats of a delta opioid agonist (SNC80) depend upon intravenous infusion rate. XIVth World Congress of Pharmacology. San Francisco, CA, USA, 2002.
49. Mandema JW, Tukker E, Danhof M. Br J Pharmacol 1991; 102:663–668.
50. Kubová H, Mocková M, Mares P. Physiol Res 1999; 48:491–500.
51. Marrosu F, Cozzolino A, Puligheddu M, Giagheddu M, Di Chiara G. Brain Res 1997; 776:24–29.
52. De Pablo JM, Parra A, Segovia S, Guillamón A. Physiol Behav 1989; 46:229–237.
53. Kumar S, Kulkarni KS. Indian J Exp Biol 1996; 34:431–435.
54. Nowakowska E, Chodera A, Kus K. Pol J Pharmacol 1996; 48:255–260.
55. Nowakowska E, Chodera A, Kus K. Pol J Pharmacol 1999; 51:463–469.
56. Schulteis G, Martinez JL Jr. Behav Neurosci 1988; 102(5):678–686.
57. Pakarinen ED, Faust WB, Moerschbaecher JM. Prog Neuro-Psychopharmacol Biol Psychiatry 1996; 20:883–898.
58. Jutkiewicz EM, Rice KC, Woods JH, Winsauer PJ. The delta-opioid receptor agonist SNC80 produces antidepressant-like effects in rats independent of its ability to disrupt learning. In press.
59. Knapp RJ, Santoro G, De Leon I, Lee K, Edsall S, Waite S, Malatynska E, Varga E, Calderon SN, Rice KC, Rothman RB, Porreca F, Roeske WR, Yamamura HI. J Pharmacol Exp Ther 1996; 277:1284–1291.
60. Stevenson GW, Canadas F, Gomez-Serrano M, Ullrich T, Zhang X, Rice KC, Riley AL. Pharmacol Biochem Behav 2002; 71:291–300.
61. Brandt MR, Negus SS, Mello NK, Furness S, Zhang X, Rice KC. J Pharmacol Exp Ther 1999; 290:1157–1164.
62. Rénéric JP, Lucki I. Psychopharmacology 1998; 136:190–197.

21

Mu-Delta Interactions In Vitro and In Vivo

Richard B. Rothman and Heng Xu
National Institute on Drug Abuse, National Institutes of Health, Baltimore, Maryland, U.S.A.

1 INTRODUCTION

Prior to the development of molecular biological methods, the delineation of receptor subtypes was made on the basis of various functional assays and ligand-binding studies, typically utilizing the discriminative power of selective ligands. With the advent of cloning technologies, molecular methods often led to the discovery of additional previously unrecognized receptor subtypes. This is particularly true for the serotonin, muscarinic and glutamate receptors. In the case of the opioid receptors, molecular cloning identified the three main types of opioid receptors previously delineated by pharmacological methods, mu, delta and kappa receptors. Interestingly, despite considerable effort, molecular researchers failed to clone receptors corresponding to various subtypes of opioid receptors identified by functional and ligand-binding studies. These include $delta_1$ and $delta_2$ receptors, mu_1 and mu_2 receptors, and the mu-delta opioid receptor complex. Thus, in some sense, molecular studies have failed to explain the rich complexity of opioid pharmacology. Possible explanations for these discrepant results are that as yet undiscovered genes code for these subtypes. An analogous situation occurred with histamine receptors [1]. The H1 and H2 receptors were cloned in 1991. However, the H3 receptor was not cloned until 1999 because its homology to known G protein receptors, including histamine receptors, was only 31%. Alternatively, some other mechanism such as splice variants, posttranslational modifications, or regulatory event might create new functional receptors from one (or more) of the basic three opioid receptor types.

The main purpose of this chapter is to review the evidence for delta-mu interactions in the context of recent research advances that reveal molecular mechanisms by which combinations of receptors might create "new" receptors with different pharmacological profiles.

2 "EARLY" IN VIVO STUDIES

The initial observations of delta-mu interactions were made by Vaught and Takemori [2], who showed that subeffective doses of [leu^5]enkephalin potentiated morphine antinociception. Lee et al. [3] confirmed these observations and also reported that subeffective doses of [met^5]enkephalin antagonized morphine antinociception. As reviewed in detail elsewhere [4], a number of studies conducted primarily by the group of Frank Porreca considerably extended these original observations. Some of these findings are summarized in Table 1. A key finding was that the delta antagonist, ICI174,864 [5] could block both the ability of subantinociceptive doses of [D-Pen^2,D-Pen^5]enkephalin (DPDPE*) to enhance morphine antinociception and the ability of subantinociceptive doses of [D-Ala^2,Met^5]enkephalinamide to decrease morphine antinociception [6]. Subsequent studies showed that certain opioid antagonists were able to block the modulatory effect of delta agonists on mu-mediated antinociception while other antagonists were not. For example, Porreca et al. [7] showed that pretreatment with the irreversible delta antagonist 5′-NTII blocked the positive and negative modulatory effects of delta agonists on mu-mediated antinociception. In contrast, pretreatment with the irreversible delta antagonist, DALCE, did not block the positive and negative modulatory effects of delta agonists on mu-mediated antinociception. Like 5′-NTII, the irreversible mu antagonist, beta-FNA, blocked the modulatory effects of delta agonists [6].

At the time these studies were conducted, two types of delta receptors were postulated: a delta receptor associated with a mu receptor, that together comprised the mu-delta opioid receptor complex, and a delta receptor

---

* Abbreviations: [D-Pen^2,D-Pen^5]enkephalin (DPDPE), [D-Pen^2,L-Pen^5]enkephalin (DPLPE), MeTyr-D-Ala-Gly-N(Et)-CH(CH_2-Ph)CH_2-N(CH_3)$_2$ (LY164929), N,N-diallyl-Tyr-Aib-Aib-Phe-Leu-OH (ICI174,864), [D-Ala^2,Leu^5,Thr^6]enkephalin (DTLET), [D-Ala^2,Leu^5,Ser^6]enkephalin (DSLET), [D-Ala^2,Met^5]enkephalin amide (DAMA), [D-ala^2-MePhe^4,gly-ol^5]enkephalin (DAMGO), N-phenyl-N-[1-(2-(4-isothiocyanato)phenethyl)-4-piperidinyl]propanamide-HCl (FIT).

TABLE 1 Summary of Modulatory Effects of Delta Agonists on Mu-Mediated Antinociception

Agent/activity	δ_{cx} (delta$_2$)	δ_{ncx} (delta$_1$)
DPDPE	Agonist	Agonist
	Positive modulation	Antinociception
[Leu^5]enkephalin	Agonist	Agonist
	Positive modulation	Antinociception
[Met^5]enkephalin	Inverse agonist	Agonist
	Negative modulation	Antinociception
DAMA	Inverse agonist	Agonist
	Negative modulation	Antinociception
ICI174,864	Antagonist	Antagonist
DALCE		Antagonist
Beta-FNA	Antagonist	
5′-NTII	Antagonist	

Source: Ref. 4.

not associated with the mu-delta opioid receptor complex. The delta site associated with the mu-delta opioid receptor complex was termed the delta$_{cx}$ site, and the delta site not associated with the mu-delta opioid receptor complex was termed the delta$_{ncx}$. In this formulation, compounds such as DPDPE produce antinociception via activation of delta$_{ncx}$ sites and positively modulate mu-mediated antinociception via activation of the delta$_{cx}$ site. Compounds like DAMA produce antinociception via activation of delta$_{ncx}$ sites and negatively modulate mu-mediated antinociception via inverse activation of the delta$_{cx}$ site. Antagonists such as beta-FNA and 5′-NTII, block the modulating effects of delta agonists via interactions with the delta$_{cx}$ site. As reviewed in detail elsewhere [8], several lines of pharmacological evidence led to the identification of delta$_1$ and delta$_2$ receptors. In light of data establishing DALCE and 5′-NTII as delta$_1$ and delta$_2$ antagonists, as well as other data, a consensus opinion developed that the delta$_{cx}$ and delta$_{ncx}$ sites are synonymous with the delta$_2$ and delta$_1$ receptors, respectively [8,9].

Importantly, another group of investigators, using different in vivo pharmacological models, independently concluded that a mu-delta opioid receptor complex provided the simplest explanation of their data [10–12]. Across three separate models of opioid receptor pharmacology, endotoxic shock, inhibition of fluorothyl-induced seizures, and inhibition of striatal cAMP levels, the Holaday group observed that pretreating rats with beta-FNA blocked the ability of the delta antagonist ICI174,864 to attenuate delta agonist–mediated responses. Working independently, Schoffelmeer and associates also presented evidence they interpreted as evidence for a mu-delta opioid receptor complex [13].

The hypothesis of a mu-delta opioid receptor complex is based on the in vivo work of the Porreca and Holaday groups and derives from two different types of observations: delta agonist–mu agonist interactions, and mu antagonist–delta antagonist interactions, respectively. A common theme to both sets of data is the ability of beta-FNA to block delta agonist modulation in the antinociception models and delta antagonist effects in the seizure, endotoxic shock, and cAMP models. As will be described below, beta-FNA also had definable effects on the mu-delta opioid receptor complex as studied with in vitro ligand binding methods.

3 EARLY LIGAND BINDING STUDIES

Initial observations that morphine was a noncompetitive inhibitor of [^{3}H][leu^5]enkephalin binding led us to suggest the concept of a mu-delta opioid receptor complex [14]. Since [^{3}H][leu^5]enkephalin could conceivably be labeling both mu and delta receptors, a key point in this and other studies was to distinguish between a two-site binding model and a noncompetitive (allosteric) model. The two-site model postulates that [^{3}H][leu^5]enkephalin labels mu and delta sites and that morphine appears to be a noncompetitive inhibitor of [^{3}H][leu^5]enkephalin binding because of its selectivity for the mu-binding site. The allosteric model, on the other hand, postulates that [^{3}H][leu^5]enkephalin labels a single binding site (the delta$_{cx}$ site) and that morphine decreases the B_{max} of the delta$_{cx}$ site via binding to an associated mu-binding site (the mu$_{cx}$ site). One approach to distinguishing these models is to deplete the membranes of delta receptors using an irreversible inhibitor. To do this, we used the irreversible delta receptor inhibitor (+)-trans-superfit [15–17].

Using this approach, Rothman and associates [17] characterized the interaction of a wide range of mu and delta compounds with delta$_{cx}$ sites present in membranes depleted of delta receptors with (+)-trans-superfit. In the data to be reviewed, the (+)-TSF-delta$_{cx}$ site refers to membranes pretreated with 1 mM (+)-trans-superfit and assayed with [^{3}H][D-Ala^2,D-Leu^5] enkephalin. The lysed-P2-delta$_{cx}$ site refers to lysed P2 membranes assayed with [^{3}H][D-Ala^2,D-Leu^5]enkephalin and 100 nM DPDPE to block delta receptors. This is a typical mu receptor assay. The lysed-P2-delta$_{ncx}$ site refers to lysed P2 membranes assayed with [^{3}H][D-Ala^2,D-Leu^5]enkephalin and 100 nM LY164929 to block mu receptors. This is a typical delta receptor assay. The (+)-TSF-mu site refers to membranes pretreated with 1 mM (+)-trans-superfit and assayed with [^{3}H]DAMGO. All assays took place in 10 mM TRIS-HCl, pH 7.4, 100 mM NaCl, 3 mM $MnCl_2$, 2 mM GTP, and a protease inhibitor cocktail.

As reported in Table 2, the observed ligand selectivity pattern for the lysed-P2-delta$_{ncx}$ (delta) and lysed-P2-delta$_{cx}$ (mu) sites were as expected. Compounds such as DPDPE were potent at the lysed-P2-delta$_{ncx}$ site and

TABLE 2 Interaction of Various Agents with Opioid Receptor Subtypes

Drug	(+)-TSF-δ_{cx} binding site[a]		Lysed-P2-δ_{cx} (μ) binding site[c]		Lysed-P2-δ_{ncx} (δ) Binding site[b]		(+)-TSF-μ (μ) Binding site[d]	
	IC_{50}	N	IC_{50}	N	IC_{50}	N	IC_{50}	N
DPDPE	49.6 ± 8.0	0.44	1523 ± 59	0.93	5.45 ± 0.91	0.93	1271 ± 24	0.91
Morphine	72.7 ± 13.1	0.49	204 ± 11	0.95	2010 ± 140	0.97	14.5 ± 0.4	0.97
LY164929	277 ± 45	0.24	4.70 ± 0.29	0.84	21829± 209	0.79	7.07 ± 0.26	1.00
Naloxone	40.2 ± 5.3	0.60	30.5 ± 2.1	0.78	110 ± 11	0.98	33.8 ± 1.8	0.85
DAMGO	97.9 ± 12.3	0.37	3.91 ± 0.15	0.92	2696 ± 169	0.94	5.12 ± 0.27	1.03
ICI174864	3002 ± 348	0.44	19582 ± 2252	1.19	297 ± 22	0.96	Not done	
DAMA	7.57 ± 0.58	0.91	4.00 ± 0.45	0.89	6.96 ± 0.70	0.96	1.32 ± .05	0.88
DADL	8.11 ± 0.40	1.06	14.5 ± 0.8	0.93	11.4 ± 0.7	0.98	Not done	
DTLET	1.97 ± 0.11	0.81	14.1 ± 0.87	1.05	1.38 ± 0.06	1.00	14.4 ± 0.7	0.97
DSLET	5.46 ± 0.37	0.83	24.0 ± 1.6	0.99	3.19 ± 0.22	1.10	23.6 ± 0.8	0.99
DPLPE	Not done	Not done	452 ± 20	0.99	6.16 ± 0.94	0.83	580 ± 11	0.91

Inhibition curves were generated by displacing [^{3}H][D-Ala^2,D-Leu^5]enkephalin (2 nM) or [^{3}H]DAMGO (2 nM) by eight concentrations of test drug. The data of three independent experiments were pooled (24 data points), and the IC_{50} (± SD) and slope factors (± SD) determined using MLAB. The SD values of the slope factors were <10% of the value. All assays took place in 10 mM TRIS-HCl, pH 7.4, 100 mM NaCl, 3 mM $MnCl_2$, 2 μM GTP, and a protease inhibitor cocktail.

[a] Membranes pretreated with 1 μM (+)-trans-superfit and assayed with [^{3}H][D-Ala^2,D-Leu^5]enkephalin.

[b] Lysed P2 membranes assayed with [^{3}H][D-Ala^2,D-Leu^5]enkephalin and 100 nM LY164929 (MeTyr-D-Ala-Gly-N(Et)-CH(CH_2-Ph)CH_2-N(CH_3)$_2$) to block mu receptors.

[c] Lysed P2 membranes assayed with [^{3}H][D-Ala^2,D-Leu^5]enkephalin and 100 nM DPDPE to block δ receptors.

[d] Membranes pretreated with 1 μM (+)-trans-superfit and assayed with [^{3}H]DAMGO.

Source: Ref. 17.

much weaker at the lysed-P2-delta$_{cx}$ site. Conversely, compounds such as DAMGO and LY164929 were considerably more potent at the lysed-P2-delta$_{cx}$ (mu) site than the lysed-P2-delta$_{ncx}$ (delta) site. Using membranes depleted of delta receptors, a mu-selective radioligand would be expected to generate a mu-like ligand selectivity pattern. Compounds such as DPDPE and DPLPE were weak, and compounds like morphine and DAMGO were potent at the (+)-TSF-mu site.

The results obtained for the (+)-TSF-delta$_{cx}$ site could be described as unexpected. The experiments targeting the (+)-TSF-delta$_{cx}$ site and the (+)-TSF-mu site were conducted under identical conditions and with identical membranes and differed only in the choice of radioligand. Nevertheless, DPDPE was remarkably more potent at the (+)-TSF-delta$_{cx}$ site (50 nM) than at the (+)-TSF-mu site (1271 nM). Conversely, mu-selective compounds such as DAMGO and LY164929 were less potent at the (+)-TSF-delta$_{cx}$ site than at the (+)-TSF-mu site. Subsequent experiments determined the effect of fixed concentrations of inhibitors on the K_d and B_{max} of [^{3}H][D-Ala^2,D-Leu^5]enkephalin binding to the (+)-TSF-delta$_{cx}$ site. As reported in Tables 3

TABLE 3 Effect of Various Agents on the B_{max} and K_d of the (+)-TSF-δ_{cx} Binding Site

Test drug	Concentration (nM)	B_{max} (fmol/mg protein) ±SD	Apparent K_d (nM)	Apparent K_i (nM)	Mechanism
Control	0	320 ± 16	1.66 ± 0.11		
DPDPE	300	212 ± 25	5.34*	135	Competitive
Morphine	500	118 ± 66*	3.81*	386	Competitive Noncompetitive
LY164929	500	178 ± 15*	2.69*	806	Competitive Noncompetitive
Naloxone	50	235 ± 22*	2.13		Noncompetitive
DAMGO	500	168 ± 16*	1.81		Noncompetitive
ICI174864	5000	402 ± 35	8.27*	1256	Competitive
DAMA	6.0	325 ± 52	10.88*	1.08	Competitive
DTLET	1.5	320 ± 27	4.19*	0.98	Competitive
DSLET	2.5	392 ± 19	3.94*	1.82	Competitive
DPLPE	20	312 ± 27	4.90*	10.0	Competitive

[^{3}H][D-Ala^2,D-Leu^5]enkephalin binding surfaces were generated in the absence (control) and presence of the indicated concentrations of test drugs using membranes pretreatred with (+)-trans-superfit. The data of two experiments were combined (36 data points) and fit to a one-site binding model for the best-fit estimates of the B_{max} and K_d, which are reported above. The K_i values were calculated using the equation $K_i = [\text{test drug}]/(K_d'/K_d - 1)$, where the K_d' and the K_d are the apparent K_d values in the presence and absence of test drug, respectively. $^*P < .01$ when compared to control (F-test). The SD of the K_d values were <10% of the K_d values.
Source: Ref. 17.

and 4, delta agents such as DPDPE, DAMA, DTLET, and DPLPE (among others), were competitive inhibitors at the (+)-TSF-delta$_{cx}$ site. In contrast, mu agents such as morphine, LY164929, DAMGO, naloxone, and oxymorphone (among others) were noncompetitive inhibitors at the (+)-TSF-delta$_{cx}$ site.

These findings serve to reinforce the point that [^{3}H][D-Ala^2,D-Leu^5]enkephalin appears to label a site traditionally called "mu", but at which mu ligands are noncompetitive inhibitors. Moreover, under the conditions we used the (+)-TSF-delta$_{cx}$ binding site has high to moderate affinity for delta ligands that exert modulatory effects on mu-mediated antinociception. These findings are consistent with the hypothesis of a mu-delta opioid receptor complex.

Studies with the irreversible mu antagonist beta-FNA provided a linkage between the in vitro ligand binding evidence for a mu-delta opioid receptor complex and the in vivo evidence described above [18]. As noted above, in vivo administration of beta-FNA blocks delta agonist modulation in the antinociception models and delta antagonist effects in the seizure, endotoxic shock, and striatal cAMP models. As reported by Rothman et al. [18], either pretreating membranes with beta-FNA or ICV administration of

TABLE 4 Effect of Additional Agents on the B_{max} and K_d of the (+)-TSF-δ_{cx} Binding Site

Test drug	Concentration (nM)	B_{max} (fmol/mg protein) ±SD	Apparent K_d (nM)	K_i (nm)	Mechanism
Control	0	237 ± 8	2.00 ± 0.10		
Oxymorphone	100	101 ± 6*	1.64 ± 0.09*		Noncompetitive
Levorphanol	30	108 ± 8*	3.30 ± 0.17		Noncompetitive
Dihydromorphine	200	94 ± 7*	1.46 ± 12*		Noncompetitive
(-)-Methadone	500	103 ± 7*	2.21 ± 0.14		Noncompetitive
(-)-U50488	10000	143 ± 12*	3.81 ± 0.27*	11049	Noncompetitive Competitive
[D-Ala^2] dynorphin-A$_{(1-13)}$ amide	5	144 ± 12*	3.22 ± 0.24		Noncompetitive
[MET^5]enkephalin	5	213 ± 13	4.77 ± 2.7*	3.61	Competitive
DAMA	5	198 ± 14	7.08 ± 0.33*	1.96	Competitive

[^{3}H][D-Ala^2,D-Leu^5]enkephalin binding surfaces were generated in the absence (control) and presence of the indicated concentrations of test drugs using membranes pretreated with (+)-trans-superfit. The data of two experiments were combined (36 data points) and fit to a one-site binding model for the best-fit estimates of the B_{max} and K_d, which are reported above. The K_i values were calculated using the equation K_i = [test drug]/($K_d'/K_d - 1$), where the K_d' and the K_d are the apparent K_d values in the presence and absence of test drug, respectively. *$P < .01$ when compared to control (F-test).

Source: Ref. 17.

beta-FNA selectively affected the $delta_{cx}$ binding site and increased the IC_{50} value twofold for naloxone binding to the $delta_{cx}$ site. Importantly, naloxone was a noncompetitive inhibitor at the $delta_{cx}$-binding site, in the absence and presence of beta-FNA pretreatment. Thus, the $delta_{cx}$ site is sensitive to the same agent that alters the function of the mu-delta opioid receptor complex, as defined on the basis of in vivo experiments.

4 RECENT PHARMACOLOGICAL STUDIES OF MU-DELTA INTERACTIONS

The studies reviewed above led investigators to test the hypothesis of mu-delta interactions in other experimental paradigms, including simpler cell-based systems. Perhaps the most striking observation to arise from these studies is that where mu-delta interactions are looked for, they are observed. For example, Schmidt et al. [19] reported that intra-accumbens injection of either a mu agonist alone or a delta agonist alone did not produce antinociception. However, coactivation of mu and delta receptors in rat nucleus accumbens was required to produce antinociception. Using undifferentiated SH-SY5Y cells that express both mu and delta receptors, Chen et al. [20] demonstrated, using isobolographic analysis of dose-response curve data, a synergistic interaction between mu and delta receptors. Activation of both mu and delta receptors synergistically increased the metabolic activity of cells, as indicated by the real-time extracellular acidification rate. Consistent with these findings, Martin and Prather [21] coexpressed mu and delta receptors in GH3 cells and reported evidence for mu-delta interactions at the level of both ligand binding and functional effects.

The work of Devi and associates [22] provided a potential molecular mechanism for mu-delta interactions and a physical basis to explain biochemical observations of a mu-delta opioid receptor complex [23,24]. Their work and the efforts of others [25] have clearly demonstrated that mu and delta receptors can form heterodimers, leading to synergistic interactions and the generation of novel "signaling units." This work, in other words, provides a molecular mechanism to explain how it is possible for three basic types of opioid receptors to form a greater number of pharmacologically defined opioid receptor subtypes.

5 SUMMARY

In vitro and in vivo studies conducted in the 1980s and 1990s clearly demonstrated synergistic interactions in vivo between mu and delta receptors that were most simply explained by a mu-delta opioid receptor complex. The failure to clone opioid receptors corresponding to the postulated mu-delta opioid receptor complex, as well as other opioid receptor subtypes defined with pharmacological methods, left this particular area of inquiry without molecular mechanisms to explain the findings. Receptor dimerization is a mechanism that can explain these aspects of opioid pharmacology. Future studies of this molecular mechanism may be able to provide a deeper understanding of how three opioid receptors generate the mu-delta opioid receptor complex and the in vivo mu-delta interactions observed in earlier studies.

REFERENCES

1. Hough LB. Genomics meets histamine receptors: new subtypes, new receptors. Mol Pharmacol 2001; 59(3):415–419.
2. Vaught JL, Takemori AE. A further characterization of the differential effects of leucine enkephalin, methionine enkephalin and their analogs on morphine-induced analgesia. J Pharmacol Exp Ther 1979; 211:280–283.
3. Lee NM, Leybn L, Chang J-K, Loh HH. Opiate and peptide interaction: effect of enkephalins on morphine analgesia. Eur J Pharmacol 1980; 68:181–185.
4. Rothman RB, Holaday JW, Porreca F. Allosteric coupling among opioid receptors: evidence for an opioid receptor complex. In: Herz A, ed. Handbook of Experimental Pharmacology. "Opioids." Vol. 104. Berlin: Springer-Verlag, 1992:217–237.
5. Cotton R, Giles MG, Miller L, Shaw JS, Tims D. ICI174,864: a highly selective anatgonist for the opioid delta-receptor. Eur J Pharmacol 1984; 97:331–332.
6. Heyman JS, Jiang Q, Rothman RB, Mosberg HI, Porreca F. Modulation of mu-mediated antinociception by delta agonists: characterization with antagonists. Eur J Pharmacol 1989; 169:43–52.
7. Porreca F, Takemori AE, Sultana M, Portoghese PS, Bowen WD, Mosberg HI. Modulation of mu-mediated antinociception in the mouse involves opioid delta-2 receptors. J Pharmacol Exp Ther 1992; 263:147–152.
8. Zaki PA, Bilsky EJ, Vanderah TW, Lai J, Evans CJ, Porreca F. Opioid receptor types and subtypes: the delta receptor as a model. Annu Rev Pharmacol Toxicol 1996; 36:379–401.
9. Xu H, Partilla JS, de Costa BR, Rice KC, Rothman RB. Differential binding of opioid peptides and other drugs to two subtypes of opioid $delta_{ncx}$ binding sites in mouse brain: further evidence for delta receptor heterogeneity. Peptides 1993; 14:893–907.
10. D'Amato R, Holaday JW. Multiple opiate receptors in endotoxic shock: evidence for delta involvement and mu-delta interactions in vivo. Proc Natl Acad Sci USA 1984; 81:2898–2901.
11. Tortella FC, Robles L, Holaday JW. The anticonvulsant effects of DADL are primarily mediated by activation of delta opioid receptors: interaction between delta and mu-receptor antagonists. Life Sci 1985; 37:497–503.
12. Holaday JW, Tortella FC, Maneckjee R, Long JB. In vivo interactions among opiate receptor agonists and antagonists. NIDA Res Monogr 1986; 71:173–188.
13. Schoffelmeer AN, Rice KC, Heijna MH, Hogenboom F, Mulder AH. Fentanyl isothiocyanate reveals the existence of physically associated mu- and delta-opioid

receptors mediating inhibition of adenylate cyclase in rat neostriatum. Eur J Pharmacol 1988; 149:179–182.

14. Rothman RB, Westfall TC. Morphine allosterically modulates the binding of 3H-leucine enkephalin to a particulate fraction of rat brain. Mol Pharmacol 1982; 21:538–547.
15. Kim C-H, Rothman RB, Jacobson AE, Mattson MV, Bykov V, Streaty RA, Klee WA, George C, Long JB, Rice KC. Probes for narcotic receptor mediated phenomena. 15. (3S,4S)-(+)-trans-3-methylfentanyl isothiocyanate, a potent site-directed acylating agent for the delta opioid receptors in vitro. J Med Chem 1989; 32:1392–1398.
16. Rothman RB, Mahboubi A, Bykov V, Kim C-H, Jacobson AE, Rice KC. Probing the opioid receptor complex with (+)-trans-SUPERFIT I Evidence that [D-Pen^2,D-Pen^5]enkephalin interacts with high affinity at the $*_{cx}$ binding site. Peptides 1991; 12:359–364.
17. Rothman RB, Mahboubi A, Bykov V, Kim C-H, de Costa BR, Jacobson AE, Rice KC. Probing the opioid receptor complex with (+)-trans-superfit. II Evidence that μ ligands are noncompetitive inhibitors of the δ_{cx} opioid peptide binding site. Peptides 1992; 13:1137–1143.
18. Rothman RB, Long JB, Bykov V, Jacobson AE, Rice KC, Holaday JW. Beta-FNA binds irreversibly to the opiate receptor complex: in vivo and in vitro evidence. J Pharmacol Exp Ther 1988; 247:405–416.
19. Schmidt BL, Tambeli CH, Levine JD, Gear RW. Mu/delta cooperativity and opposing kappa-opioid effects in nucleus accumbens-mediated antinociception in the rat. Eur J Neurosci 2002; 15(5):861–868.
20. Chen ZW, Yang K, Wang Y, Han JS. The metabolic evidence of synergistic interaction between DAMGO and DPDPE on undifferentiated SH-SY5Y cells. Neuroreport 2001; 12(4):845–849.
21. Martin NA, Prather PL. Interaction of co-expressed mu- and delta-opioid receptors in transfected rat pituitary GH(3) cells. Mol Pharmacol 2001; 59(4): 774–783.
22. Gomes I, Jordan BA, Gupta A, Trapaidze N, Nagy V, Devi LA. Heterodimerization of mu and delta opioid receptors: A role in opiate synergy. J Neurosci 2000; 20(22):RC110.
23. Schoffelmeer AN, Yao Y-H, Simon EJ. Cross-linking of human 125I-beta-endorphin to a mu-delta opioid receptor complex in rat striatum. Eur J Pharmacol 1989; 166:357–358.
24. Schoffelmeer AN, Yao YH, Gioannini TL, Hiller JM, Ofri D, Roques BP, Simon EJ. Cross-linking of human [^{125}I]beta-endorphin to opioid receptors in rat striatal membranes: biochemical evidence for the existence of a mu/delta opioid receptor complex. J Pharmacol Exp Ther 1990; 253:419–426.
25. Levac BA, O'Dowd BF, George SR. Oligomerization of opioid receptors: generation of novel signaling units. Curr Opin Pharmacol 2002; 2(1):76–81.

22

Delta Opioids and Immune Function

Richard J. Weber
University of Illinois College of Medicine at Peoria, Peoria, Illinois, U.S.A.

Ricardo Gomez-Flores
Universidad Autónoma de Nuevo León, San Nicolás de los Garza, NL, Mexico

1 INTRODUCTION

Immune functions are intricately regulated by complex physical interactions between cells of the immune system and release of autocrine or paracrine factors, i.e., interleukins or cytokines. An additional regulatory level involves components of the central nervous system (CNS), endocrine system (ES), and immune system (IS), which constitute a more extensive regulatory feedback loop subserving a myriad of physiological functions [1]. Such systems are known to share common receptors and effector molecules [2,3]. Receptor activation or alteration of the levels of endogenous or exogenous ligands can alter the immune function and possibly serve to maintain homeostasis in health and disease. Neurotransmitters, neuromodulators, neurohormones, and cytokines can affect the components of the immune system indirectly by stimulating the neuroendocrine axis or sympathetic nervous system or directly by interacting with receptors on leukocytes. Immunomodulation can therefore be achieved by directly influencing the immune system or indirectly modulating activity of the CNS or ES [4,5].

Since the first evidence of opioid receptors on cells of the immune system [6,7], other laboratories have provided pharmacological evidence of their presence on immunocytes such as granulocytes and monocytes [8]. Recent work has focused on characterization of these effects on immune function. The indirect effects opiate agonists elicit on leukocyte modulation through central pathways are known [4]. However, evidence for the direct modulation of opioids on the cells of the immune system has accumulated in recent years. Recent cloning of opioid receptors in immunocytes [9–13] and evidence for endogenous opioid peptide production in lymphocytes [14,15] and macrophages [16] has strengthened the theory of these proteins as autocrine/paracrine regulators of leukocyte activation, proliferation, and differentiation. Understanding the role these peptides play in immune function may provide clinical and therapeutic benefits to patients suffering immune abnormalities from immunodeficiencies including AIDS to autoimmune disorders. This paper will review the literature in respect to: 1) leukocyte opioid receptor protein structure, molecular genetics, binding data; 2) characterization and distribution of opioid receptors and the endogenous opioid peptides in cells of the immune system; 3) intracellular signaling mechanisms within immunocytes; and 4) direct effector functions of exogenous and endogenous opioids on cells of the immune system.

2 STRUCTURE AND EXPRESSION OF OPIOID RECEPTOR PROTEINS AND GENES IN LEUKOCYTES

Opioid receptors are also present on the surface of immunocompetent cells, and exhibit different patterns of ligand selectivity, stereoselectivity, saturability, and nanomolar affinity for opioids [3,17]. Presence of opioid receptors on cells of the immune system has been indirectly inferred by the use of mu-, delta-, and kappa-class receptor selective antagonists and agonists. In addition, opioid binding sites on lymphocytes and macrophages have been identified by opioids labeled with radioisotopes or fluorochromes, or by identifying the genes responsible for opioid-receptor expression [18]. Some endogenously produced neuroimmunomodulators are the delta opioid peptides met-enkephalin (Met-Enk; Try-Gly-Gly-Phe-Met-OH) and leu-enkephalin (Leu-Enk; Try-Gly-Gly-Phe-Leu-OH) at a ratio of 4:1, respectively [19,20]. They are more widely distributed in the brain than β-endorphins, being present in areas such as the hypothalamic nuclei, limbic structures, caudate-putamen, the brainstem, several layers of the dorsal horn, peripheral nerves, and the adrenal medulla.

Exogenous and endogenous opioids bind to receptors located on cells of the brain and immune system. These receptors, termed mu, kappa, and delta, are similar to classical G protein coupled receptors possessing an extracellular amino terminus, seven transmembrane domains connected by three extracellular and intracellular loops, and an intracellular carboxy terminus. An orphan opioid like receptor has been discovered exhibiting structural homology with the above receptors, but failing to bind classical opioid receptor agonists. Characterization of these receptors on cells of the immune system have been investigated showing structural homology with those receptors on neuronal cells [21–23].

Binding studies have determined the opioid receptor subtypes bind a wide range of agonists and antagonists with differences in specificity, affinity, and selectivity. Neutralizing antibodies specific for a an N-terminal sequence of the human kappa receptor [anti-kR-(33-52)] blocks KOR-specific agonist U50,488H-mediated immunosuppression of 1) *Staphylococcus aureus* Cowen strain I–induced B- and T-lymphocyte proliferation; 2) PHA-induced T-lymphocyte proliferation; and 3) *S. aureus* Cowen strain I–induced IgG synthesis. This suggests that this antibody does not bind to the effector portion of the KOR but inhibits agonist binding, and that the antiserum interacts with a site on the KOR important for ligand binding without intracellular second-messenger system modulation.

Recently, Sedqi et al. [24] were able to clone a delta opioid receptor complementary DNA by expression of cDNA library from activated thymocytes in Cos 7 cells, whose amino acid sequence was similar to the neural counterpart. Interestingly, they also observed that transcripts for kappa and mu opioid receptors were not detected in thymocytes. Furthermore, Gaveriaux et al. [25] demonstrated transcripts for the delta opioid receptor in T-lymphocyte, B-lymphocyte, and monocyte cell lines, as well as in murine splenocytes. However, they observed that the kappa opioid receptor transcript was only found in B-cell lines. These studies may suggest a selective expression of the delta opioid receptor in specific cells and tissues of the immune system and suggests specialized functions in different anatomical regions.

Characterization of the genomic organization of opioid receptors in immunocytes has been pursued to elucidate the regulation, processing and neuronal homology of these proteins. Analysis of the mu, delta, and kappa amino acid sequences show a high degree of amino acid homology with each other and neuronal receptors. Examination of the partial sequences of the delta, mu, and kappa receptors from monkey lymphocytes revealed nearly complete identity with their respective brain receptors. However, genetic differences have been revealed where the mu and delta subtypes possess three exon coding regions while the kappa receptor contains an additional exon 5′ of the translational start site [26]. A cDNA clone isolated from a PHA (phytohemagglutinin)-activated human lymphocyte population which encodes the opioid "orphan" receptor has complete homology to the clone isolated from human brain, but divergence at the 5′ untranslated region indicates tissue specificity of gene expression. Upon lymphocyte activation with PHA, a 10-fold induction of the orphan message was observed [27].

Distribution of the opioid receptor-like 1 (ORL-1) transcript has been observed in normal circulating human T-, B-, and monocytic cell lines by RT-PCR and RNAse Protection assay. These transcripts lack a 15-nucleotide stretch between exons 1 and 2 comprising the first intracellular loop but otherwise show similar distribution to the brain transcripts of the cortical areas, striatum, thalamus, and hypothalamus [28]. Low levels of delta opioid receptor (DOR) are consistently detected by RT-PCR from murine splenocytes with preferential expression in the T-cell fraction. Culturing cells without stimulation increased DOR levels and Con A reduced this effect [29]. However, delta opioid receptor expression is induced by concanavalin A in $CD4^+$ cells [30]. Activation of the DOR modulates calcium mobilization, IL-2 production, chemotaxis, and proliferation of T lymphocytes [29]. RT-PCR was also utilized to obtain cDNA clones from human T lymphocytes, which are nearly identical to the delta and kappa opioid receptor cDNA isolated from human brain and placenta, respectively [27]. The kappa receptor is present on immature mouse $CD4^+/CD8^+$ thymocytes [31], $CD4^+$ cells of human and monkey lymphocytes [12], human peripheral B-lymphocytes [25], and human fetal microglia exhibiting inhibition of HIV-1 gene expression upon KOR ligand binding [32]. Delta receptors are expressed in human and mouse T and B lymphocytes, human monocytes, mouse splenocytes [25], and unactivated murine thymocytes [24]. Mu opioid receptor mRNA is constitutively expressed in human microglial cells [33]. A naloxone-resistant β-endorphin binding site is present on murine peritoneal macrophages [34]. A naloxone-insensitive β-endorphin binding site is present on human promonocytic cell lines and murine peritoneal macrophages [34].

3 INTRACELLULAR SIGNALING MECHANISMS UPON OPIOID RECEPTOR LIGAND BINDING

Met-enkephalinamide, an endogenous delta selective opiate, is shown to modulate TCR/CD3-induced fluctuations in free intracellular calcium levels in human T lymphocytes, reinforcing the hypothesis of a direct immunomodulation by endogenous opioids [35]. Deltorphin and [D-Ala2,D-Leu5]enkephalin (DADLE) induced increases in intracellular calcium $[Ca^{2+}]$ were reversed in delta opioid receptor transfected Jurkat T cells with natrindole or pertussis toxin pretreatment. Also DADLE reduced cAMP production by 70%, which was reversed by pertussis toxin. Thus, delta opioid receptors are positively coupled to calcium mobilization pathways and negatively to cAMP production pathways that modulate T-cell modulation [36]. The mu and delta opioid receptors are coupled to the Gi subunit which has been implicated in the mediation of chemotaxis of HEK293 cells through the release of free beta-gamma subunits regardless of receptor internalization [37]. The mu receptor has been shown to increase the cAMP response to isoproterenol induction of β_2-adrenergic receptors in human mononuclear leukocytes [38]. Also, human CEMx174 lymphocytes increase expression of mitogen-activated protein kinase cascade proteins, which are important intermediates in signal transduction pathways. These effects were reversed by naloxone, a mu antagonist [39]. Met-enkephalin induced superoxide release from human neutrophils was associated with a dose-dependent increase in diacylglycerol (DAG) concentration and protein kinase C translocation to the neutrophil membrane accompanied by increases in intracellular calcium concentrations. The DAG degradation product, Tyr-Gly-Gly (TGG), inhibits suppression of superoxide anion release. However, this effect was not associated with an alteration in DAG concentration, pointing to an alternate pathway [40]. Delta opioid ligation to its specific agonist, deltorphin, enhances IL-2 cytokine transcription in human Jurkat T cells by facilitating an increase in the AP-1 component of the NF-AT/AP-1 transcription factor. Despite an increase in intracellular calcium levels, the effects of deltorphin ligation to the DOR occurred via a mechanism independent of calcineurin [41].

Coupling of opioids with their receptors on leukocytes will activate membrane G proteins that transmit signals through two major pathways, the adenylate cyclase and the phospholipase C systems. The activation of these pathways induces protein phosphorylation enabling cells to respond rapidly to diverse signals in the extracellular environment. Protein phosphorylation is crucial for a number of cellular processes such as intermediate metabolism, cytoskeletal architecture, cell adhesion, gene expression, and cell cycle progression. Protein phosphorylation induces second messengers such as inositol phosphates, diacylglycerol, Ca^{2+}, arachidonic acid, and transcription of new proteins. On this respect, Sharp et al. [36] have shown that treatment of a transfected human T-cell line with DADLE ([D-Ala_2,D-Leu_5]enkephalin) decreased forskolin-mediated intracellular cAMP levels, but increased intracellular Ca^{2+} levels (i[Ca^{2+}]). In the macrophage, opioid-mediated changes in second messengers have been also implicated. Met-Enk was reported to activate Ca^{2+} influx and to increase cAMP and cGMP intracellular levels in rat peritoneal macrophages—effects (except cAMP levels) reversed by naloxone

[42]. Activation of the G proteins Gi2, Gi3, Go, and Gs have been associated with modulating intracellular signaling induced by delta opioid ligands [43]. Stimulation of Gi proteins may lead to Rac- and Cdc42-dependent regulation of c-Jun N-terminal kinases, which belong to the mitogen-activated protein kinases (MAPKs) [44].

4 CHARACTERIZATION OF ENDOGENOUS OPIOIDS IN THE IMMUNE SYSTEM

Originally, the endogenous opioids enkephalin, dynorphin, and endorphin were considered to be specific products of the nervous system, but cells of the immune system have recently been shown to express these peptides. These mediators stem from three different prohormones: proenkephalin, proopiomelanocortin (POMC), and prodynorphin, encoded by three separate genes. All opioids start with the same sequence, Tyr-Gly-Gly-Phe, followed by methionine or leucine. Transcription of the POMC gene has been demonstrated in stimulated macrophages and lymphocytes. Proenkephalin A mRNA has been expressed in thymocytes, bone marrow cells, and splenocytes. Localized inflammation of a rat's hindpaw elicits an accumulation of β-endorphin-releasing lymphocytes, suggesting the immune system is a key factor of inflammatory pain management [45]. A secreted peptidase from activated murine $CD4^+$ and $CD8^+$ cells metabolizes endorphin into biologically active and possibly immunoregulatory peptides [30]. Endorphin binds to Jurkat potassium channel proteins, displacing channel inhibitors charybdotoxin and tetraethylammonium, relieving the cell of calcium ion flux inhibition [46]. Endorphin levels are positively correlated to NK- and T-cell activity in mistletoe lectin treated breast carcinoma patients [47]. Increases in immunocyte-derived endorphin were seen in rats injected intracerebroventricularly with IL-1 or stress induced by foot intermittent foot shock. These increases led to decreased NK cell activity and lymphocyte proliferation, which were blocked by a corticosteroid releasing hormone antagonist, despite elevated corticosteroid levels [48]. Aging is associated with the increase in intracellular endorphin concentrations in resting and fresh cultured peripheral blood mononuclear cells (PBMCs), the intracellular decrease, and the increased release of the endogenous opioid from mitogen induced PBMCs from young and old subjects. These data suggest that changes in Ca^{2+} homeostasis modulation may be involved in these cells [49]. It is shown that synthesis of the endogenous opioid precursor proenkephalin A in human peripheral T cells and monocytes is induced through stimulation of CD2/CD28 receptors and LPS treatment, respectively. Lack of enkephalin production resulted in inhibition of monocytic IL-6 production acting via membrane opioid receptors [50]. Signal transduction pathways involving calcium seem to be mediated by met-enkephalin in human T-cells [35]. Endogenous enkephalins are expressed in fetal but not newborn murine thymocytes, suggesting their possible role in T-cell differentiation and maturation [51]. The endopeptidase carboxypeptidase M, with numerous immunological substrates including enkephalin, serves as a marker for macrophage maturation [52].

5 DIRECT EFFECTS OF ENDOGENOUS AND EXOGENOUS OPIOIDS ON IMMUNOCYTES

Mu, delta and kappa receptors are independent in terms of antinociception. Activation of analgesia by stimulating non-mu receptors suggests action of other opioid agonists, mainly of delta and kappa types. Particularly, the delta receptor may be relevant for opioid therapy; studies on animals have revealed that opioids selective for the delta receptor are as potent analgesics as morphine by acting at spinal and supraspinal sites [53,54]. In addition, delta opioids have been shown not to depress respiratory function. In this regard, the highly selective delta opioid receptor agonist (+)-4-[(alpha R)-alpha-((2S, 5R)-4-allyl-2, 5-dimethyl-1-piperazinyl)-3-methoxybenzyl]-N, N-diethylbenzamide (SNC 80) was observed to stimulate rather than depress respiratory rate in sheep. In this respect, Nowak et al. in 1998 [55] and Nelson et al. in 2000 [56] reported that the delta opioids SNC 80 and DPDPE, respectively, did not alter NK cell, lymphocyte, and macrophage functions following ICV administration. Therefore, delta opioids have a great clinical potential compared with morphine, because of their high potency as analgesics [57,58], reduced induction of respiratory distress [59], and lack of immune alterations [55,56] or induction of immunopotentiation [60]. Furthermore, House et al. in 1996 [61] showed that in vitro exposure to delta opioid receptor agonists resulted in significant immunostimulation.

Activation of opiate delta receptors has been shown to influence immune function. With regard to enkephalins, Met-Enk is more active than Leu-Enk in modulating immune responses. Met-Enk action is considered bimodal since administration of high doses suppresses, whereas at low doses it enhances immune function [62]. In addition, enkephalins have been shown to enhance survival time of tumor-bearing mice, and NK and T-cell activity. Therefore, by increasing activity of T and NK cells it would be possible to enhance host resistance to viral and tumor challenge. Enkephalins are known to affect a wide range of immune parameters both in rodents and humans. In lymphocytes, Met-Enk and delta-selective opioid analogues were reported to stimulate human peripheral lymphocytes [63], LPS-induced IL-6, and TNF-α production by macrophages [61,64,65], superoxide and/or hydrogen peroxide production by macrophages of different sources [66], IFN-γ production by human mononuclear cells [67], and suppress the production of T-lymphocyte chemotactic factor [68] and LPS-induced TNF-α production by macrophages [64]. Similarly, deltorphin was shown to inhibit LPS-mediated nitrite production by the cell line J774 [69]. Met-Enk has been also reported to increase cytolytic T lymphocytes (CTL) activity [70] and B-cell proliferation [71], and Met-Enk and Leu-Enk were observed to increase or decrease NK cell activity depending upon concentration of peptides [72], potentiate LPS-induced IL-1 production by macrophages [73], reduce macrophage [74] or granulocyte-macrophage [75] colony formation, and potentiate macrophage tumoricidal activity [76].

Similarly, exogenous delta opioids such as DPDPE (D-Pen_2,D-Pen_5-enkephalin) were shown to increase IL-2 and IL-4 production, NK cell activity, and LPS-induced IL-6 production by macrophages [61]. However, B-cell proliferation [71], IgM and IgG production [77], and LPS-mediated nitrite production by the cell line J774 [69] were reported to be suppressed by DPDPE. In contrast, SNC 80, a novel nonpeptidic delta opioid ligand, was found to potentiate in vitro and ex vivo TNF-α and nitric oxide production by peritoneal macrophages [60]. In addition, Sharp et al. [78] demonstrated that the delta opioids deltorphin and SNC-80 concentration dependently inhibited the production of p24 antigen, an index of HIV-1 expression, probably acting on delta opioid receptors on T-cells [79,80]. These in vitro findings supported the anti-HIV-1 property of deltorphin and SNC 80 ligands which may have therapeutic potential for treating patients with acquired immunodeficiency syndrome [81].

The direct effects opioid and opioidlike peptides exhibit on cells of the immune system is both varied and, in some instances, contradictory, depending on which receptor subtype is being studied. Mu and kappa receptors generally affect immunofunction in a suppressive manner, where delta receptors tend to express immunostimulation [82–85]. However, selected delta antagonists have shown to elicit suppressive effects on B-cell proliferation, NK cell activity, and T-helper cell cytokine production [86]. The alteration of leukocyte function via opioid receptors will be discussed highlighting specific cell subtype immunomodulation. Endorphin shows a inhibitory effect on splenocyte proliferation through central and peripheral autocrine/paracrine pathways [87].

5.1 Macrophages

Macrophages serve a number of immunological functions including cytokine production, phagocytosis of foreign organisms, and chemotaxis to sites of inflammation. Opioid receptors present on these cells modulate these activities in vivo. The subcutaneous injection of morphine, administered prior to

LPS stimulation, decreased the expression of TNF-α in murine macrophages. This effect was reversible by naloxone and dose dependent, indicating an opioid receptor stimulation event. Despite evidence leading to a direct effect, suppression of the sympathetic nervous system blocks morphine's inhibitory results [88]. *Staphylococcus aureus*, Cowen strain I (SAC) induced inhibition of IL-6 production in intact PBMCs [89]. The inhibitory affects of opioids also affect the production of nitric oxide (NO) by macrophages as well. Among the many biological activities of NO, limiting the proliferative response of lymphocytes is critical to immune function. Acute central administration of morphine or the mu-selective agonist DAMGO is followed by an increase in production of NO in macrophages leading to suppression of T-cell proliferation in vivo [90]. Conversely, the mu receptor agonist DAGO and morphine inhibited NO production in murine macrophages before LPS induction in vitro. However, morphine was unable to inhibit NO production after LPS administration, suggesting the production of NO synthase is affected, not the activity [69]. Morphine inhibits chemotaxis [33] and stimulates phagocytosis [91] of microglial cells of the brain, and these results were reversed by the mu-selective antagonist β-funaltrexamine.

The production of IFN-γ and uptake of thymidine were increased in human mononuclear cells by delta receptor selective agonist deltorphin and delta-selective antagonist naltrindole, suggesting nonclassical opioid receptor mediation [92]. Suppression of M-CSF-induced proliferation of murine BAC 1.2F5 macrophages shows partial delta receptor mediation (20% of binding sites were reversible by naloxone, and RT-PCR revealed identical homology to the brain receptor). The other 80% of binding sites are not reversible by naloxone, suggesting a possible novel receptor [74].

Opioid modulation of macrophage function is also seen inhibiting LPS-induced IL-1 and TNF-α production using the kappa agonist U50,488H in the murine P388D1 macrophage cell line. However, the above results were not shown in the RAW 264.7 murine macrophage cell line, indicating variations of opioid responsiveness among subpopulations of cells [93]. These results were completely reversed using the kappa antagonist norbinaltoriphimine (norBNI), indicating kappa receptor stimulation [94]. Dynorphin has been shown to upregulate the expression of HIV-1 in a chronically infected promonocytic cell line [95]. HIV-1-infected microglial cells downregulate viral expression in response to kappa receptor agonists U50,593 and U69,593. These antiviral effects were blocked by the kappa antagonist nor-BNI [32].

5.2 T-Lymphocytes

T-lymphocytes are the backbone of the immune system responsible for antigen presentation, cytotoxic effects, and cytokine release. Any attenuation of T-cell function can lead to widespread immunological suppression increasing a propensity for tumor metastases or infection, such as HIV-1 or tuberculosis. Early clinical observations of immune dysfunction in addicts was initially attributed to dirty needles and impure drug preparations until it was shown heroin addicts have decreased peripheral T-cell proliferation responsiveness to classic lymphocyte mitogens [96]. Evidence of opioid binding sites on human T-cells through naloxone binding strengthened the premise for direct immunomodulation via opioids [97]. The inclination of heroin addicts of contracting AIDS has advanced the premise of morphine's action as a cofactor in HIV-1 infection. Morphine's involvement in the mechanisms of HIV-1 infection has been shown to inhibit Sendai and Newcastle virus-induced IFN production by lymphocytes, inhibit HIV-1 protein-induced lymphocyte proliferative responses, and promote apoptosis of normal lymphocytes. Inhibition of interferon production was reversed by naloxone, implicating opioid receptor mediation [98]. Conversely, delta opioid receptor ligation on human T-cells increased IL-2 production [41,99], stimulated transformation [100], and enhanced rosette formation through upregulation of CD2 receptor [6,99]. It has been shown β-endorphin stimulates the uptake of polyamines (implicated in protein synthesis stimulation, RNA synthesis and translation) in human T-cells independent of naloxone blockade [101]. Acute exposure to morphine suppresses mouse cytotoxic T-cell activity, in a naltrexone [102] and β-funaltrexamine [103] reversing manner. In vivo treatment with morphine and methadone produces a down-regulation of opioid receptor density on B- and T-lymphocytes. This contradicts classic opioid receptors in the nervous system, which downregulate only after chronic morphine exposure [104]. Furthermore, activation of human T-cells with IL-2 or PHA induces a high-affinity, saturable, and displaceable mu binding site [105].

5.3 NK Cells

Natural killer cells along with the macrophages and neutrophils form the major part of the innate immunity. NK cell cells are able to attack and lyse targeted cells in a MHC class I–independent manner without prior exposure or priming. Cancer cells downregulating MHC class I serve as important targets in tumor suppression. Additionally, NK cells aid in inflammation through the secretion of gamma interferon in response to IL-12 and TNF-α. Both chronic and acute administration of morphine affects NK cytotoxicity or cytokine secretion, indirectly rather than through opioid receptors present on the cells.

5.4 B Cells

B-cell effects in immune function are necessary in targeting foreign pathogens and proteins for elimination by the rest of the immune system. The B-cells produce and secrete (as differentiated plasma cells) antibodies, which neutralize, opsonize, and activate complement. B-cells are produced and matured in the bone marrow, then circulate the blood and lymph systems ready to bind foreign antigens. The bound antigen is internalized by the IgM receptor, broken down into smaller peptides, and then attached to MHC class II molecules for presentation to T-helper cells within the lymph nodes. Ligation to the TCR bound to the same antigen determinant results in IL-4 secretion by the T-helper cell and activation of the B-cell into an antibody secreting plasma cell.

Met-enkephalin has been shown to downregulate immunoglobulin production of IgG2a, IgG3, and IgM and in a naloxone-sensitive mechanism. Its precursor protein, pro-enkephalin, induced no change in antibody production; however, enhanced IgG1 and IgG2a yield 6 h post-LPS/DxS activation. The necessity for increased concentrations of MENK at 6- and 24-h and PENK at 6 h postactivation suggests opioid receptor expression and/or affinity decreases as a function of activation time length [77]. Similarly, mu, delta, and kappa receptor-selective agonists are able to inhibit *Staphylococcus aureus* Cowen strain I (Sac)–induced IgG production by human B-cells, and these effects are reversed by the appropriate receptor antagonist [89]. Alteration of growth hormone receptor gene expression and human GH binding has been shown in human IM-9 cells upon morphine administration. Growth hormone receptor mRNA levels and GH protein binding increased at 3 and 12 h post drug addition, respectively. These effects are abolished by naloxone, implicating a classical opioid receptor mediation [106]. In addition to antibody and GH receptor expression, endogenous opioids are chemotactic for neutrophils, monocytes, and T-cells [107–109] and B-cells [110] via naloxone-sensitive opioid receptors.

The opioid system can clearly influence responses to nociceptive and antigenic stimuli at all levels and stages (from peripheral nociceptors to the cerebral cortex and from the precursors of immunocompetent cells to mature effector cells). In most experimental and clinical studies, opioid-mediated analgesia has been associated with immunosuppression. It is well recognized that both endogenous and exogenous opioids have the potential to alter the immune system. Cloning of opioid receptors of neural and immune origins has facilitated defining the molecular events involved in opioid-mediated immunomodulation. CNS mediated indirect effects of opioids suppress many parameters of immune function through actions at mu receptors in the periaqueductal gray matter of the mesencephalon [4,111–113]. These observations have shown that central opioid receptor activation involves both the hypothalamic pituitary adrenal (HPA) axis [114] and adrenergic pathways [115,116] in suppressing NK activity following acute morphine administration. Eisenstein et al. have also demonstrated that opioids directly affect cellular and humoral immune functions though classical opioid receptors [117–120]. It is now clear that endogenous opioid peptides and exogenous opioid alkaloids are also capable of modulating the immune function by directly acting on opioid receptors on the surface of leukocytes. The significance of opioid involvement with the immune system is evidenced by the presence of opioid receptors on inflammatory cells, which may result in modulation of cellular activity after activation [121]. Involvement of opiate/opioid signaling has been observed in pathologies caused by the human immunodeficiency virus, substance abuse, parasitism, and the diffuse inflammatory response associated with surgery [121]. Opioids act like cytokines, and both types of molecules share many properties including paracrine, autocrine, and endocrine sites of action, functional redundancy, pleiotropy, and effects that are both dose and time dependent [122].

Novel opioid compounds have been synthesized that have analgesic capacity, but lack immunosuppressive effects or even potentiates immune function [60,123–126]. In this respect, Nowak et al. [123] recently reported that the delta opioid agonist SNC 80 did not alter NK cell, lymphocyte, and macrophage functions following ICV administration [123]. Furthermore, IV administration of SNC 80 was associated with ex vivo immunopotentiation, following an activating challenge.

REFERENCES

1. Jankovic BD. Immunol Lett 1979; 1:145–146.
2. Blalock JE. Chem Immunol 1992; 52:1–24.
3. Weber RJ, Pert CB. In: Muller EE Genazzani AR, eds. Central and Peripheral Endorphins: Basic and Clinical Aspects. New York: Raven Press, 1984:35–42.
4. Weber RJ, Pert A. Science 1989; 245:188–190.
5. Weber RJ, Pert A. In: Smith B, Adelman G, eds. Encyclopedia of Neuroscience. Boston: Birkhauser, 1999.
6. Wybran J, Appelboom T, Famaey JP, Govaerts A. J Immunol 1979; 123:1068–1070.
7. McDonough RJ, Madden JJ, Falek A, Shafer DA, Pline M, Gordon D, Bokos P, Kuehnle JC, Mendelson J. J Immunol 1980; 125:2539–2543.
8. Falke NE, Fischer EG, Martin R. Cell Biol Int Rep 1985; 9:1041–1047.
9. Belkowski SM, Zhu J, Liu-Chen LY, Eisenstein TK, Adler MW, Rogers TJ. J Neuroimmunol 1995; 62:113–117.
10. Chuang LF, Chuang TK, Killam KFJ, Chuang AJ, Kung HF, Yu L, Chuang RY. Biochem Biophys Res Commun 1994; 202:1291–1299.
11. Chuang LF, Chuang TK, Killam KF Jr, Qiu Q, Wang XR, Lin JJ, Kung HF, Sheng W, Chao C, Yu L. Biochem Biophys Res Commun 1995; 209:1003–1010.

12. Chuang TK, Killam KF Jr, Chuang LF, Kung HF, Sheng WS, Chao CC, Yu L, Chuang RY. Biochem Biophys Res Commun 1995; 216:922–930.
13. Sedqi M, Roy S, Ramakrishnan S, Elde R, Loh HH. Biochem Biophys Res Commun 1995; 209:563–574.
14. Blalock JE, Smith EM. Proc Natl Acad Sci USA 1980; 77:5972–5974.
15. Harbour-McMenamin D, Smith EM, Blalock JE. Infect Immun 1985; 48:813–817.
16. Lolait SJ, Lim AT, Toh BH, Funder JW. J Clin Invest 1984; 73:277–280.
17. Sibinga NE, Goldstein A. Annu Rev Immunol 1988; 6:219–249.
18. Gomez-Flores R, Weber RJ. In: Plotnikoff, Faith, Murgo, Good, eds. Cytokines: Stress and Immunity. Boca Raton: CRC Press 1999; 281–314.
19. Cox BM. Life Sci 1982; 31:1645–1658.
20. Jankovic BD. Ann NY Acad Sci 1994; 741:1–38.
21. Evans CJ, Keith DEJ, Morrison H, Magendzo K, Edwards RH. Science 1992; 258:1952–1955.
22. Kieffer BL, Befort K, Gaveriaux-Ruff C, Hirth CG. Proc Nat Acad Sci USA 1992; 89:12048–12052.
23. Wang JB, Johnson PS, Wu JM, Wang WF, Uhl GR. J Biol Chem 1994; 269: 25966–25969.
24. Sedqi M, Roy S, Ramakrishnan S, Loh HH. J Neuroimmunol 1996; 65:167–170.
25. Gaveriaux C, Peluso J, Simonin F, Laforet J, Kieffer B. FEBS Lett 1995; 369: 272–276.
26. Carr DJJ, Rogers TJ, Weber RJ. Proc Soc Exp Biol Med 1996; 213:248–257.
27. Wick MJ, Minnerath SR, Roy S, Ramakrishnan S, Loh HH. Brain Res Mol Brain Res 1995; 32:342–347.
28. Peluso J, Laforge KS, Matthes HW, Kreek MJ, Kieffer BL, Gaveriaux-Ruff C. J Neuroimmunol 1998; 81:184–192.
29. Sharp BM, Shahabi N, McKean D, Li MD, McAllen K. J Neuroimmunol 1997; 78:198–202.
30. Miller B. J Immunol 1996; 157:5324–5328.
31. Bidlack JM, Lawrence DM, Ignatowski TA. Adv Exp Med Biol 1996; 402:13–22.
32. Chao CC, Gekker G, Hu S, Sheng WS, Shark KB, Bu DF, Archer S, Bidlack JM, Peterson PK. Proc Natl Acad Sci USA 1996; 93:8051–8056.
33. Chao CC, Hu S, Shark KB, Sheng WS, Gekker G, Peterson PK. J Pharmacol Exp Ther 1997; 281:998–1004.
34. Woods JA, Shahabi NA, Sharp BM. Life Sci 1997; 60:573–586.
35. Sorensen AN, Claesson MH. Life Sci 1998; 62:1251–1259.
36. Sharp BM, Shahabi NA, Heagy W, McAllen K, Bell M, Huntoon C, McKean DJ. Proc Natl Acad Sci USA 1996; 93:8294–8299.
37. Neptune ER, Bourne HR. Proc Nat Acad Sci USA 1997; 94:14489–14494.
38. Pende A, Ioverno A, Musso NR, Vergassola C, Lotti G. Biomed Pharmacother 1995; 49:33–37.
39. Chuang LF, Killam KFJ, Chuang RY. J Biol Chem 1997; 272:26815–26817.
40. Haberstock H, Marotti T, Banfic H. Neuropeptides 1996; 30:193–201.

41. Hedin KE, Bell MP, Kalli KR, Huntoon CJ, Sharp BM, McKean DJ. J Immunol 1997; 159:5431–5440.
42. Foris G, Medgyesi GA, Hauck M. Mol Cell Biochem 1986; 69:127–137.
43. McKenzie FR, Milligan G. Biochem J 1990; 267:391–398.
44. Kam AY, Chan AS, Wong YH. J Neurochem 2003; 84:503–513.
45. Cabot PJ, Carter L, Gaiddon C, Zhang Q, Schafer M, Loeffler JP, Stein C. J Clin Invest 1997; 100:142–148.
46. Millar DB, Mazorow DL, Hough C. J Neuroimmunol 1997; 78:8–18.
47. Heiny BM, Albrecht V, Beuth J. Anticancer Res 1998; 18:583–586.
48. Panerai AE, Sacerdote P, Bianchi M, Manfredi B. Int J Clin Pharmacol Res 1997; 17:115–116.
49. Manfredi B, Clementi E, Sacerdote P, Bassetti M, Panerai AE. Peptides 1995; 16:699–706.
50. Kamphuis S, Eriksson F, Kavelaars A, Zijlstra J, van de Pol M, Kuis W, Heijnen CJ. J Neuroimmunol 1998; 84:53–60.
51. Linner KM, Quist HE, Sharp BM. Endocrinology 1996; 137:857–863.
52. Krause SW, Rehli M, Andreesen R. Immunol Rev 1998; 161:119–127.
53. Sullivan AF, Dickenson AH, Roques BP. Br J Pharmacol 1989; 98:1039–1049.
54. Jiang Q, Takemori AE, Sultana M, Portoghese PS, Bowen WD, Mosberg HI, Porreca F. J Pharmacol Exp Ther 1991; 257:1069–1075.
55. Nowak DM, Ansell I, Hjelle JT, Ross JA, Miller-Hjelle MA, Dobbie JW, Dombrink-Kurtzman MA. Adv Perit Dial 1998; 14:158–163.
56. Nelson CJ, Schneider GM, Lysle DT. Brain Behav Immun 2000; 14:170–184.
57. Calderon SN, Rothman RB, Porreca F, Flippen-Anderson JL, McNutt RW, Xu H, Smith LE, Bilsky EJ, Davis P, Rice KC. J Med Chem 1994; 37:2125–2128.
58. Bilsky EJ, Calderon SN, Wang T, Bernstein RN, Davis P, Hruby VJ, McNutt RW, Rothman RB, Rice KC, Porreca F. J Pharmacol Exp Ther 1995; 273: 359–366.
59. Negus SS, Gatch MB, Mello NK, Zhang X, Rice K. J Pharmacol Exp Ther 1998; 286:362–375.
60. R Gomez-Flores, KRN Vietti, WJ Dunn, R Tamez-Guerra, RJ Weber (2003). Selective lymphocyte activation and inhibition of in vitro tumor cell growth by novel morphinans. (Unpublished.)
61. House RV, Thomas PT, Bhargava HN. Peptides 1996; 17:75–81.
62. Jankovic BD, Maric D. Ann NY Acad Sci 1987; 496:115–125.
63. Hucklebridge FH, Hudspith BN, Lydyard PM, Brostoff J. Immunopharmacol 1990; 19:87–91.
64. Bian TH, Wang XF, Li XY. Zhongguo Yao Li Xue Bao 1995; 16:449–451.
65. Zhong F, Li XY, Yang SL, Stefano GB, Fimiani C, Bilfinger TV. Int J Cardiol 1998; 64:S53–S59.
66. Radulovic J, Dimitrijevic M, Laban O, Stanojevic S, Vasiljevic T, Kovacevic-Jovanovic V, Markovic BM. Peptides 1995; 16:1209–1213.
67. Brown SL, Van Epps DE. Cell Immunol 1986; 103:19–26.
68. Brown SL, Van Epps DE. J Immunol 1985; 134:3384–3390.

69. Iuvone T, Capasso A, D'Acquisto F, Carnuccio R. Biochem Biophys Res Commun 1995; 212:975–980.
70. Carr DJ, Klimpel GR. J Neuroimmunol 1986; 12:75–87.
71. Gang L, Fraker PJ. Acta Pharmacol Sin 1989; 10:216–221.
72. Oleson DR, Johnson DR. Brain Behav Immun 1988; 2:171–186.
73. Si-Xun Y, Xiao-Yu L. Acta Pharmacol Sin 1989; 10:266.
74. Roy S, Ramakrishnan S, Loh HH, Lee NM. Eur J Pharmacol 1991; 195:359–363.
75. Krizanac-Bengez L, Boranic M, Testa NG, Marotti T. Biomed Pharmacother 1992; 46:367–373.
76. Cameron DJ. Jpn J Exp Med 1987; 57:31–39.
77. Das KP, Hong JS, Sanders VM. J Neuroimmunol 1997; 73:37–46.
78. Sharp BM, Gekker G, Li MD, Chao CC, Peterson PK. Biochem Pharmacol 1998; 56:289–292.
79. Sharp BM, McAllen K, Gekker G, Shahabi NA, Peterson PK. J Immunol 2001; 167:1097–1102.
80. Sharp BM, Li MD, Matta SG, McAllen K, Shahabi NA. Ann NY Acad Sci 2000; 917:764–770.
81. Sharp BM, Roy S, Bidlack JM. J Neuroimmunol 1998; 83:45–56.
82. Bhargava HN, House RV, Thorat SN, Thomas PT. Brain Res 1995; 690:121–126.
83. Caroleo MC, Arbitrio M, Melchiorri D, Nistico G. Neuroimmunomodulation 1994; 1:141–147.
84. Mazumder S, Nath I, Dhar MM. Immunol Lett 1993; 35:33–38.
85. Rogers TJ, Taub DD, Eisenstein TK, Geller EB, Adler MW. NIDA Res Monogr 1991; 105:82–88.
86. House RV, Thomas PT, Kozak JT, Bhargava HN. Neurosci Lett 1995; 198: 119–122.
87. Panerai AE, Manfredi B, Granucci F, Sacerdote P. J Neuroimmunol 1995; 58: 71–76.
88. Bencsics A, Elenkov IJ, Vizi ES. J Neuroimmunol 1997; 73:1–6.
89. Morgan EL. J Neuroimmunol 1996; 65:21–30.
90. Schneider GM, Lysle DT. Adv Exp Med Biol 1996; 402:81–88.
91. Peterson PK, Gekker G, Hu S, Sheng WS, Molitor TW, Chao CC. Adv Neuroimmunol 1995; 5:299–309.
92. Noviello L, Papadia S, Capobianchi MR, Negri L. Eur J Pharmacol 1997; 332:R1–R2.
93. Belkowski SM, Alicea C, Eisenstein TK, Adler MW, Rogers TJ. J Pharmacol Exp Ther 1995; 273:1491–1496.
94. Alicea C, Belkowski S, Eisenstein TK, Adler MW, Rogers TJ. J Neuroimmunol 1996; 64:83–90.
95. Chao CC, Gekker G, Hu S, Sheng WS, Portoghese PS, Peterson PK. Biochem Pharmacol 1995; 50:715–722.
96. Brown SM, Stimmel B, Taub RN, Kochwa S, Rosenfield RE. Arch Intern Med 1974; 134:1001–1006.

97. Madden JJ, Donahoe RM, Zwemer-Collins J, Shafer DA, Falek A. Biochem Pharmacol 1987; 36:4103–4109.
98. Nair MP, Schwartz SA, Polasani R, Hou J, Sweet A, Chadha KC. Clin Diagn Lab Immunol 1997; 4:127–132.
99. Bajpai K, Singh VK, Dhawan VC, Haq W, Mathur KB, Agarwal SS. Immunopharmacology 1997; 35:213–220.
100. Bajpai K, Singh VK, Agarwal SS, Dhawan VC, Naqvi T, Haq W, Mathur KB. Int J Immunopharmacol 1995; 17:207–212.
101. Ientile R, Ginoprelli T, Cannavo G, Romeo S, Macaione S. Life Sci 1997; 60: 1545–1551.
102. Carpenter GW, Breeden L, Carr DJJ. Int J Immunopharmacol 1995; 17:1001–1006.
103. Carpenter GW, Carr DJ. Immunopharmacol 1995; 29:129–140.
104. Patrini G, Massi P, Ricevuti G, Mazzone A, Fossati G, Mazzucchelli I, Gori E, Parolaro D. J Pharmacol Exp Ther 1996; 279:172–176.
105. Madden JJ, Ketelsen D, Whaley WL, Donahoe RM, Oleson D. Adv Exp Med Biol 1995; 373:37–40.
106. Henrohn D, Le Greves P, Nyberg F. Mol Cell Endocrinol 1997; 135:147–152.
107. Epps Van DE, Saland L. J Immunol 1984; 132:3046–3053.
108. Ruff MR, Wahl SM, Mergenhagen S, Pert CB. Neuropeptides 1985; 5:363–366.
109. Foris G, Medgyesi GA, Nagy JT, Varga Z. Ann NY Acad Sci 1987; 496:151–157.
110. Manfreda SE, Dunzendorfer S, Schratzberger P, Buratti T, Reinisch N, Kahler CM, List WF, Wiedermann CJ. Anesth Analg 1998; 86:670–672.
111. Gomez-Flores R, Suo JL, Weber RJ. Brain Behav Immun 1999; 13:212–224.
112. Gomez-Flores R, Weber RJ. Immunopharmacology 2000; 48:145–156.
113. Liang-Suo J, Gomez-Flores R, Weber RJ. Life Sci 2002; 71:2595–2602.
114. Freier DO, Fuchs BA. J Pharmacol Exp Ther 1994; 270:1127–1133.
115. Carr DJ, Mayo S, Gebhardt BM, Porter J. J Neuroimmunol 1994; 53:53–63.
116. Carr DJ, Carpenter GW, Garza HH Jr, Baker ML, Gebhardt BM. Adv Exp Med Biol 1995; 373:131–139.
117. Eisenstein TK, Meissler JJ Jr, Rogers TJ, Geller EB, Adler MW. J Pharmacol Exp Ther 1995; 275:1484–1489.
118. Rojavin M, Szabo I, Bussiere JL, Rogers TJ, Adler MW, Eisenstein TK. Life Sci 1993; 53:997–1006.
119. Taub DD, Eisenstein TK, Geller EB, Adler MW, Rogers TJ. Proc Natl Acad Sci USA 1991; 88:360–364.
120. Rahim RT, Meissler JJ Jr, Cowan A, Rogers TJ, Geller EB, Gaughan J, Adler MW, Eisenstein TK. Int Immunopharmacol 2001; 1:2001–2009.
121. Stefano GB, Scharrer B, Smith EM, Hughes TKJ, Magazine HI, Bilfinger TV, Hartman AR, Fricchione GL, Liu Y, Makman MH. Crit Rev Immunol 1996; 16:109–144.

122. Bidlack JM. Clin Diagn Lab Immunol 2000; 7:719–723.
123. Nowak JE, Gomez-Flores R, Calderon SN, Rice KC, Weber RJ. J Pharmacol Exp Ther 1998; 286:931–937.
124. Riley ME, Ananthan S, Weber RJ. Adv Exp Med Biol 1998; 437:183–187.
125. ME Hicks, S Ananthan, RJ Weber, unknown 2001.
126. Hicks ME, Gomez-Flores R, Wang C, Mosberg H, Weber RJ. Life Sci 2001; 68:2685–2694.

23

Delta Opioids and Substance Abuse

S. Stevens Negus
McLean Hospital, Harvard Medical School, Belmont, Massachusetts, U.S.A.

1 INTRODUCTION

Alcohol and drug abuse are pervasive public health problems that affect millions of people at an estimated annual cost to society of nearly $250 billion in the United States alone [1,2]. The potential relationship between delta opioid systems and substance abuse is of practical interest for at least two reasons. First, as reviewed in this volume, novel and selective delta opioid ligands are being developed and evaluated for a wide range of clinical applications, and the abuse liability of these compounds may influence the conditions under which they can be made available for clinical use. In one historically famous example, heroin was originally developed and marketed at the end of the 19th century as an alternative to morphine for the treatment of cough. Despite its effectiveness as an antitussant, heroin's abuse potential quickly became apparent, and because of its widespread abuse, clinical use of heroin was ultimately prohibited [3]. Preclinical procedures are now available to assess the abuse liability of novel drugs before they are approved for clinical use, and data from these procedures are routinely used by the U.S. Food and Drug Administration to determine the degree of control that will be exercised over the distribution of new drugs.

TABLE 1 Drugs Commonly Used to Examine the Relationship Between Delta Opioid Systems and Substance Abuse[a]

	Agonists	Antagonists
Peptides	Methinonine enkephalin	ICI174864
	Leucine enkephalin	DALCE (∂1)
	DSLET	
	DPDPE (∂1)	
	Deltorphin II (∂2)	
Nonpeptides	SNC80	Naltrindole
	SNC162	7-Benzylidenenaltrexone (BNTX; ∂1)
	TAN-67	Naltriben (∂2)
		Naltrindole isothiocyanate (NTII; ∂2)

[a] In some cases, common abbreviations are also provided. Some compounds have been associated with delta-1 and delta-2 opioid receptor subtypes, and notations are provided to identify drugs used to define these subtypes.

A second reason for interest in the relationship between delta opioid systems and substance abuse is the possibility that delta opioids may modulate the effects of alcohol or other abused drugs and may be useful in the treatment of some forms of alcohol or drug abuse. In contrast to the great strides made during the last 50 years in the pharmacological treatment of some other mental health disorders, relatively few medications are available for the treatment of substance abuse, and many of these medications have limited efficacy. The development of novel drug abuse treatment medications is an active area of investigation, and delta opioids are one class of drugs under consideration for this purpose. This chapter will address these issues in two sections. The first section reviews data that speak to the abuse liability of delta opioids, and in particular, of delta opioid agonists. The second section reviews data pertaining to the potential utility of delta opioids, and in particular delta opioid antagonists, for the treatment of alcohol and drug abuse. The relationship of delta opioid systems and substance abuse has also been reviewed previously [4–8], and Table 1 shows drugs commonly used to conduct these studies.

2 ABUSE LIABILITY OF DELTA OPIOID AGONISTS

Substance abuse is defined by maladaptive patterns of drug-taking behavior that produce symptoms to indicate that the individual continues use of the substance despite significant substance-related problems [9]. The goal of preclinical abuse liability evaluations is to predict the likelihood that some drug will maintain these maladaptive patterns of drug-taking behavior in a significant number of people. Several experimental procedures have been developed to examine the abuse-related effects of drugs, and the effects of delta agonists in these procedures will be described.

2.1 Reinforcing Effects of Delta Opioid Agonists in Drug Self-Administration Studies

Drug-taking behavior in humans can be considered as an example of operant behavior, in which the delivery of a stimulus (in this case, the drug) is contingent on the performance of certain behaviors (obtaining, preparing, and administering the drug). By definition, drugs that increase the probability of the behaviors that lead to their delivery are considered to act as reinforcers of behavior and to produce reinforcing effects. These reinforcing effects can be most directly evaluated using drug self-administration procedures, in which drug delivery is made contingent on behavior under simplified conditions that are typically arranged to maximize the likelihood of identifying a drug's reinforcing effects [10]. For example, subjects may be placed alone in an operant conditioning chamber and allowed to emit a simple and quantifiable response such as a lever press to receive intravenous drug injections. Drugs that maintain higher rates of responding than their vehicles are considered to be reinforcing. There is a high, though not perfect, correlation between drugs that produce reinforcing effects in self-administration assays and drugs that are abused by humans, indicating that these procedures serve as a good predictor of a drug's abuse potential [10–12]. In view of the functional similarity between drug self-administration and substance abuse and the good predictive validity of self-administration procedures for identifying drugs with high abuse liability, data from drug self-administration studies often serve as a cornerstone in the preclinical evaluation of a drug's abuse liability.

The endogenous opioid peptides methionine and leucine enkephalin were used in the seminal studies that first proposed the existence of delta opioid receptors [13,14], and these were the first delta agonists examined in drug self-administration studies [15,16]. Both compounds were self-administered into the cerebral ventricles by rats [15]. A later study went on to show that drug-naïve rats could be trained to self-administer methionine enkephalin directly into the nucleus accumbens [16]. Figure 1 (left panel) shows that the methionine enkephalin self-administration dose-effect curve had an inverted-U shape, which is typical of many self-administered drugs [17]. Doses of 350 and 500 pmol methionine enkephalin maintained the highest rates of self-administration, and these rates were five to six times higher than rates maintained by vehicle. The neuroanatomical specificity of this effect was suggested by the finding that reliable self-administration was observed only in those rats with verified cannula placements in the nucleus accumbens; rats

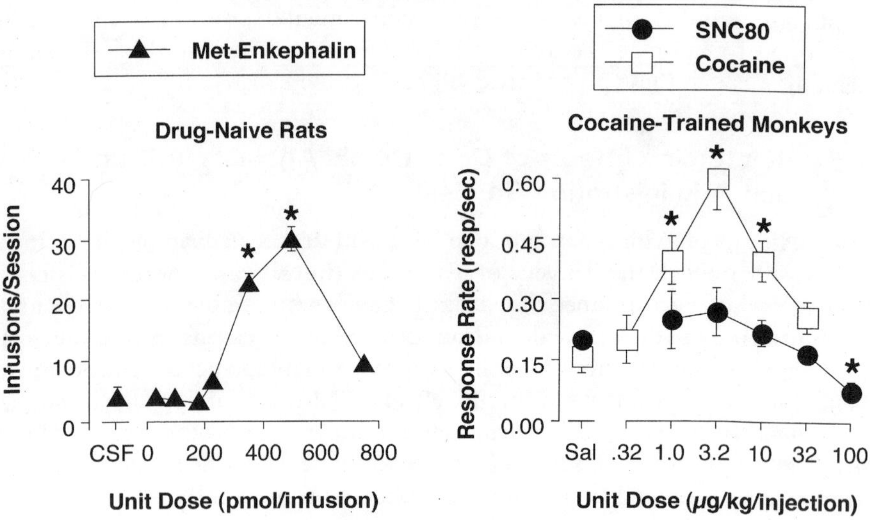

FIGURE 1 Reinforcing effects of delta opioid agonists in drug self-administration studies. (*Left*) The reinforcing effect of the prototype peptidic delta agonist methionine enkephalin (met-enkephalin) administered intracerebrally into the nucleus accumbens in rats. Drug-naïve rats reliably acquired self-administration of centrally administered met-enkephalin, and the highest rates of self-administration were maintained by 350 and 500 pmol/inf met-enkephalin. (From Ref. 16.) (*Right*) Responding maintained by intravenous delivery of cocaine and the non-peptidic delta agonist SNC80 in rhesus monkeys initially trained to self-administer cocaine. Cocaine maintained self-administration at rates greater than those maintained by saline in these animals, but SNC80 did not. During availability of the highest dose of SNC80 (100 μg/kg/inj), rates of responding were lower than those maintained by saline. (From Ref. 29.) CSF, artificial cerebrospinal fluid; Sal, saline.

with cannula placements outside the nucleus accumbens did not self-administer any dose of methionine enkephalin. Other data provided additional evidence to suggest that methionine enkephalin served as an effective reinforcer. First, when optimal doses of methionine enkephalin were available, rats responded at much higher rates on the active lever than on a concurrently available inactive lever. Second, when the active and inactive levers were switched, response rates rapidly increased on the new active lever and decreased on the new inactive lever. Finally, when the number of lever presses required to produce an infusion was increased from one press (fixed ratio 1 [FR 1] schedule of reinforcement) to five presses (FR 5), rats increased the number of lever presses accordingly to sustain similar levels of methionine enkephalin intake.

Taken together with the results of the earlier study [15], these findings by Goeders et al. [16] provided compelling evidence to indicate that centrally administered methionine enkephalin served as an effective reinforcer in rats. However, the role of delta receptors in mediating these effects was not clear, because methionine and leucine enkephalin have low selectivity for delta as opposed to mu receptors [18]. As a result, it is possible that the reinforcing effects of the enkephalins demonstrated in these studies were mediated by mu opioid receptors. This concern was at least partially addressed by a later study, which found that the much more selective delta opioid peptide DPDPE was self-administered directly into the ventral tegmental area [19]. However, even DPDPE may produce mu receptor–mediated effects under some conditions [20,21], and as a result, additional studies would be required to confirm that the reinforcing effects of DPDPE were mediated by delta receptors.

The studies described above with peptidic delta agonists exemplify a general trend in behavioral science to study the behavioral effects of peptidic compounds following central administration. This trend reflects, at least in part, the assumption that peptidic compounds do not distribute as well as some nonpeptides across the blood-brain barrier, and hence will display either low potency or be inactive following peripheral administration. Moreover, the behavioral effects of peptidic delta agonists have been examined almost exclusively in rodents, perhaps because technical difficulties have discouraged widespread use of central routes of administration in other species, such as nonhuman primates. Such studies can provide important information about the neurobiology of delta agonist effects in rodents. Given the current technologies of drug delivery, though, it is unlikely that peptidic delta agonists would be abused by central routes of administration. Rather, because all abused drugs are currently self-administered systemically, it is especially relevant to examine the abuse-related effects of systemically administered delta opioids. In addition, there are species differences in the density, distribution and ligand-binding characteristics of opioid receptors [22,23] and in the abuse-related behavioral effects of some opioids [cf. 12,24]. Accordingly, studies in non-human primates provide a valuable complement to studies in rodents for the prediction of the abuse liability of novel compounds. The recent development of nonpeptidic delta agonists such as BW373U86 and SNC80 [25,26] has greatly facilitated evaluation of behavioral effects, including abuse-related effects, produced by systemically administered delta agonists in nonhuman primates as well as in rodents.

In stark contrast to results with centrally administered peptidic delta agonists in rats, intravenous administration of the nonpeptidic delta agonists BW373U86 and SNC80 did not maintain self-administration in drug-experienced rhesus monkeys previously trained to self-administer other drugs of abuse [27–29]. In a study that examined the reinforcing effects of SNC80, for example, monkeys were implanted with intravenous catheters and initially trained to self-administer cocaine under a FR 30 schedule during daily sessions lasting 100 min. Following training, different doses of cocaine (0.32–10 μg/kg/inj) and SNC80 (1.0–100 μg/kg/inj) were substituted for the maintenance dose of cocaine. Each test dose of cocaine and SNC80 was substituted for 1 day. Figure 1 (right panel) shows that cocaine dose-dependently maintained drug self-administration under these conditions, and 3.2 μg/kg/inj cocaine maintained peak rates of self-administration approximately four times higher than rates maintained by saline. In contrast, no dose of SNC80 maintained self-administration at rates higher than those maintained by saline, and the highest dose of SNC80 (100 μg/kg/inj) actually maintained lower rates than those maintained by saline. Furthermore, analysis of patterns of responding during each 100-min test session indicated that all doses of SNC80 produced saline-like patterns of extinction responding, characterized by high rates of responding during the first quartile of the test session and very low response rates during the remainder of the session. Taken together, these results indicated that SNC80 did not produce reinforcing effects in rhesus monkeys under conditions in which cocaine did function as a reinforcer. In agreement with these findings with SNC80, the structurally related but less selective nonpeptidic delta agonist BW373U86 also failed to maintain self-administration in rhesus monkeys initially trained to self-administer either cocaine or the mu opioid agonist alfentanil [27,28].

The determinants of the contrasting results described above are not well understood. One possibility is that peptidic delta agonists such as DPDPE have greater reinforcing effects than the nonpeptidic delta agonists SNC80 and BW373U86. Although these compounds are all classified as delta agonists, functional evidence suggests that they are not interchangeable. For example, the peptidic delta agonists DPDPE and deltorphin II increased extracellular dopamine levels in the nucleus accumbens of rats, a neurochemical effect associated with many abused drugs [30–33]. BW373U86 and SNC80, in contrast, did not increase dopamine levels in the nucleus accumbens [34]. Other differences between peptidic and nonpeptidic delta agonists have also been reviewed [6], and some of these other differences will be discussed below (see Sec. 2.3).

In addition to these differences in the types of delta agonists used, these studies also differed in numerous environmental and subject-related parameters, including 1) route of delta agonist administration (central vs. systemic), 2) schedule of reinforcement (FR $\leq$ 5 vs. FR $\geq$ 30), 3) use of acquisition of self-administration in drug-naïve subjects versus maintenance of self-administration in drug-experienced subjects as a means of evaluating reinforcement, and 4) species of subject (rat vs. rhesus monkey). A better understanding of the role of these variables will require further research. Overall, results from drug self-administration studies suggest that activation of delta opioid receptors in selected brain areas by peptidic agonists may be sufficient to produce reinforcing effects in rats, although additional studies would be required to confirm that these reinforcing effects were in fact mediated by delta opioid receptors. Systemically administered nonpeptidic delta agonists have not been shown to produce reinforcing effects in rhesus monkeys.

2.2 Other Effects of Delta Agonists Related to Reinforcement

Two other behavioral procedures have been used extensively to investigate delta agonist effects that may be related to reinforcement. One of these procedures, place conditioning, typically uses an experimental apparatus that consists of two or more compartments separated by removable barriers [35,36]. During the training phase, subjects are confined to one compartment after vehicle administration and are confined to a different compartment after administration of a dose of a test drug. During the testing phase, the barrier is removed, and the untreated subject is free to explore both compartments. The primary dependent variable is the time spent in each compartment during the test phase. Drugs are considered to produce a place preference if the subject spends more time in the drug-paired compartment than in the vehicle-paired compartment, and many abused drugs produce conditioned place preferences [35,36].

Centrally administered peptidic delta agonists have usually been found to produce dose-dependent place preferences in rats and mice [37–40]. Moreover, studies with receptor-selective antagonists suggest that these place conditioning effects are mediated by delta receptors. One study, for example, found that place preferences produced by the mu agonist DAMGO could be blocked by the mu-selective antagonist CTOP but not by the delta antagonist ICI174864, whereas DPDPE-induced place preferences could be blocked by ICI174864, but not by CTOP [38]. Other studies provided evidence to suggest that place conditioning effects could be mediated by different delta receptor subtypes in mice [39,40]. Specifically, the effects of DPDPE in mice were blocked by the putative delta-1 receptor antagonist BNTX but not by the putative delta-2 receptor antagonist naltriben. Conversely, the effects of another peptidic delta agonist, deltorphin II, were blocked by naltriben but not by BNTX. These results were interpreted to suggest that DPDPE and deltorphin II produced place conditioning effects by acting at delta-1 and delta-2 receptors, respectively [39,40]. In parallel with these findings with centrally administered peptidic delta agonists, systemic administration of the nonpeptidic delta agonists SNC80 and BW373U86 also produced place preferences in rats, and these effects could be blocked by the delta-selective antagonist naltrindole [34,41]. However, another nonpeptidic delta agonist, TAN-67, failed to produce place preferences in mice [42]. In addition, a recent study found that deltorphin II failed to produce place preferences in mu opioid receptor knockout mice, which suggested that mu receptors were necessary for deltorphin II–induced place preferences in these mice [43]. These latter results challenge the notion that delta receptor activation is sufficient to mediate delta agonist-induced place preferences.

Reinforcement-related effects of delta agonists have also been examined in procedures that assess drug effects on responding for electrical brain stimulation. In this procedure, electrodes are implanted into a target brain site, such as the ventral tegmental area or the medial forebrain bundle at the level of the lateral hypothalamus, and stimulation of this electrode is contingent on some simple behavior, such as pressing a response lever [44]. Responding is usually evaluated at multiple frequencies of electrical stimulation, and techniques have been developed to measure threshold frequencies necessary to maintain responding or to determine the complete function that relates response rate to stimulation frequency. Many drugs of abuse facilitate responding for electrical brain stimulation, as demonstrated by a decrease in the frequency threshold for responding or a leftward shift in the rate-frequency curves [44].

Administration of the peptidic delta agonists DPDPE or DSTBULET into the ventral tegmental area, ventral pallidum, centromedial caudate, or nucleus accumbens facilitated electrical brain stimulation in rats [45–49]. These results agree with the finding that centrally administered peptidic delta agonists maintained drug self-administration and produced place preferences in rats. However, the receptor mechanisms of delta agonist effects on electrical brain stimulation have not been systematically examined. In the only study to investigate the effects of a delta-selective antagonist, naltrindole dose-dependently blocked DPDPE-induced facilitation of electrical brain stimulation, but naltrindole also blocked the effects of the mu-selective antagonist DAMGO [49]. Further research will be required to confirm that the effects of peptidic delta agonists on responding for electrical brain stimulation are actually mediated by delta opioid receptors. Moreover, the effects of systemically administered nonpeptidic delta agonists on electrical brain stimulation have not been reported.

2.3 Discriminative Stimulus Effects of Delta Agonists

In addition to producing reinforcing effects, drugs may also produce discriminative stimulus effects [50,51]. In the terms of operant conditioning, a discriminative stimulus signals the presence or absence of a behavior-reinforcer contingency. For example, subjects may be trained to press one lever for food reinforcement following injection of some dose of a training drug, and to press a different lever for food reinforcement following injection of vehicle. The role of a drug's discriminative stimulus effects in generating drug-taking behavior is not clear. Whereas nearly all drugs that produce reinforcing effects also have abuse liability, many drugs that produce discriminative stimulus effects are not abused. Thus, the demonstration that laboratory animals or humans can be trained to discriminate a drug does not provide a useful prediction of the drug's abuse liability. However, a drug's discriminative stimulus effects do display a high degree of pharmacological selectivity, and this characteristic can be exploited for the purpose of abuse liability testing. Specifically, test drugs that mimic the discriminative stimulus effects of a training drug often share some pharmacological mechanism of action with the training drug and produce many other effects in common with the training drug. Thus, if a test drug shares discriminative stimulus effects with a known drug of abuse, then the test drug may also have abuse liability.

Pigeons, rats, and rhesus monkeys have all been trained to discriminate delta agonists from vehicle in drug discrimination procedures [52–56]. In pigeons, both centrally administered DPDPE and systemically administered BW373U86 served as the training drug, whereas SNC80 served as the training drug in both rats and rhesus monkeys. In all cases, the discriminative stimulus effects of the delta agonist training drug were antagonized by the delta-selective antagonist naltrindole. These results demonstrate that delta agonists produce robust, delta receptor–mediated discriminative stimulus effects. However, studies conducted to date suggest that the discriminative effects of peptidic and nonpeptidic delta agonists may differ. For example, in pigeons trained to discriminate centrally administered DPDPE from vehicle, systemic administration of the nonpeptidic delta agonist BW373U86 produced only partial substitution [53]. Similarly, in pigeons trained to discriminate systemically administered BW373U86, centrally administered BW373U86 substituted completely for the training stimulus, but centrally administered DPDPE and DSLET did not [52]. It has been suggested that these results may reflect differences in either the selectivity of the delta agonists employed or the neuroanatomical distribution of peptidic and nonpeptidic delta agonists delivered by different routes of administration [52,53].

No known drugs of abuse have been found to substitute consistently for any delta agonist training drug. Although a relatively limited range of drugs has been tested, mu opioid agonists such as DAMGO, morphine, and fentanyl, the stimulant cocaine, and the phencyclidine-like dissociative anesthetic ketamine usually failed to substitute for delta agonist training drugs in most or all subjects [52–56]. For example, Figure 2 (left panel) shows that cocaine failed to substitute for SNC80 in rhesus monkeys trained to discriminate SNC80 from saline [55]. Similarly, morphine and cocaine failed to

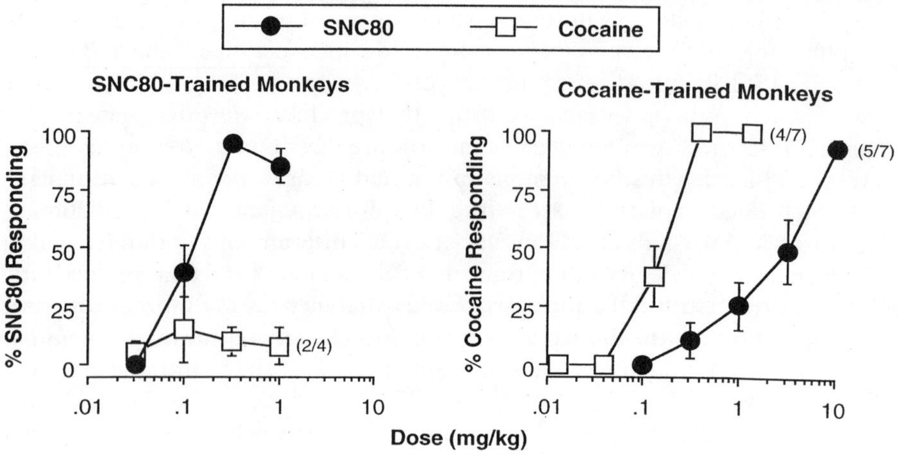

FIGURE 2 Comparison of the discriminative stimulus effects of SNC80 and cocaine in rhesus monkeys. (*Left*) The effects of intramuscular SNC80 and cocaine in rhesus monkeys trained to discriminate SNC80 from saline. SNC80 produced a dose-dependent increase in responding on the SNC80-appropriate response key, whereas cocaine did not. (From Ref. 55.) (*Right*) The effects of intramuscular SNC80 and cocaine in rhesus monkeys trained to discriminate cocaine from saline. Both SNC80 and cocaine produced dose-dependent increases in responding on the cocaine-appropriate response key. However, much higher doses of SNC80 were required to produce cocaine-appropriate responding than SNC80-appropriate responding in these monkeys. (From Ref. 29.) Figures in parentheses show proportion of monkeys responding at the highest doses of each drug. Response rates were eliminated in the remaining monkeys.

substitute for DPDPE in pigeons trained to discriminate centrally administered DPDPE from saline [53]. The discriminative stimulus effects of delta agonists have also been examined in subjects trained to discriminate known drugs of abuse. With some exceptions described in detail below, selective delta agonists usually did not substitute for mu agonist training drugs [27,57–59] or for cocaine as a training drug [27,60–62] in rats or nonhuman primates. In general, these results support the conclusion that the discriminative stimulus effects of delta agonists differ from the effects of several known drugs of abuse. One implication of these findings is that delta agonists would not be expected to display mu agonist-like, cocaine-like, or ketamine-like abuse liability.

There are, however, two notable exceptions to these general findings. First, BW373U86 has been found to share discriminative stimulus effects with mu agonists in pigeons. For example, several mu agonists produced high levels of substitution in pigeons trained to discriminate BW373U86 from vehicle [52,54], and similarly, BW373U86 substituted for the mu agonist fentanyl in four of five pigeons trained in a three-way discrimination among fentanyl, the kappa agonist bremazocine, and saline [63]. Antagonism studies with naltrindole and naloxone suggested that these overlapping effects of BW373U86 and mu agonists were mediated by delta and mu receptors, respectively [54,63]. Thus, activation of either delta or mu receptors may produce similar discriminative stimulus effects in pigeons. The second exception is that both peptidic and nonpeptidic delta agonists produced high levels of substitution for a cocaine training stimulus in at least some studies. Centrally administered DPLPE and deltorphin II (but not DPDPE) substituted completely for cocaine in rats trained to discriminate cocaine from saline [64,65]. Similarly, systemically administered TAN-67 produced partial (56–70%) substitution for both cocaine and methamphetamine in rats [65,66], and Figure 2 (right panel) shows that systemically administered SNC80 produced complete substitution in five of seven rhesus monkeys trained to discriminate cocaine from saline [29]. Antagonism studies demonstrated that the cocainelike effects of both DPLPE in rats and SNC80 in monkeys could be blocked by naltrindole, which further suggests that these effects were mediated by delta opioid receptors [29]. Moreover, although SNC80 did not substitute for cocaine in squirrel monkeys, it did produce a dose-dependent and naltrindole-reversible enhancement of the discriminative stimulus effects of cocaine [62]. These results suggest that there may be some overlap between the discriminatory stimulus effects of stimulants such as cocaine and methamphetamine and some delta agonists. However, as noted above and shown in Figure 2 (left panel), cocaine did not substitute for either DPDPE or SNC80 in subjects trained to discriminate these delta agonists [53,55,62], so any overlap between cocaine- and delta agonist-induced discriminative stimulus effects is incomplete. Moreover, the relevance of this overlap for the prediction of the reinforcing effects and abuse liability of delta agonists is not clear. Although SNC80 substituted for the discriminative stimulus effects of cocaine in five of seven monkeys, it did not maintain self-administration in rhesus monkeys trained to self-administer cocaine [29].

2.4 Tolerance and Physical Dependence

Tolerance is defined as a decrease in the effect of a drug following preexposure to that drug. Physical dependence is defined by the emergence of abstinence signs following withdrawal from some regimen of drug treatment. Tolerance and physical dependence are neither necessary nor sufficient for substance abuse [9], and some abused drugs (e.g., cocaine) produce relatively subtle signs of either tolerance or physical dependence. Nonetheless, tolerance and physical dependence are included as possible criteria for the clinical diagnosis of substance abuse [9], and the ability of drug exposure to produce either tolerance or physical dependence may influence a drug's abuse liability. Specifically, tolerance to a drug's effects may lead to increased drug consumption, because the user may become less sensitive to drug effects that normally limit consumption. For example, heroin addicts may be able to tolerate very high doses of heroin that would kill a drug-naïve user. It has also been argued that physical dependence may contribute to drug abuse, because a physically dependent user may consume drug not only in response to the drug's positive reinforcing effects, but also to avoid presumably aversive signs of withdrawal.

Tolerance develops rapidly following acute or chronic treatment with peptidic or nonpeptidic delta agonists [67–71]. For example, Figure 3 shows dose-effect curves for SNC80 administered alone or 24 h after pretreatment with SNC80 (1.0–10 mg/kg) in an assay of schedule-controlled responding for food reinforcement in rhesus monkeys [71]. Under control conditions, SNC80

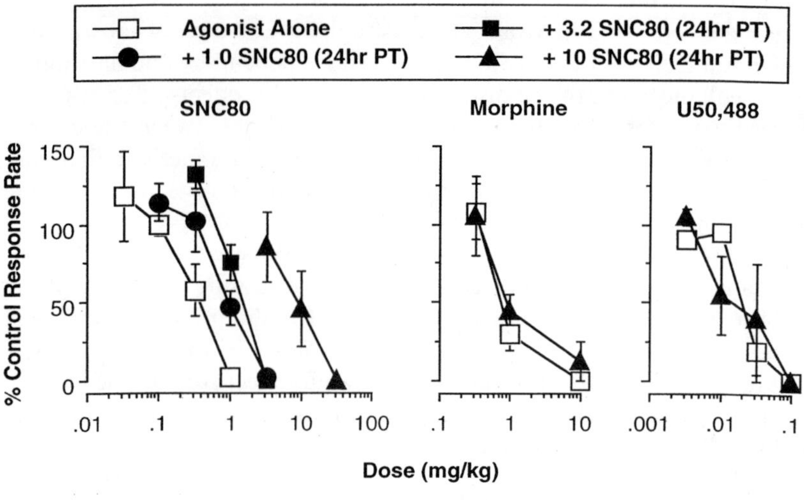

FIGURE 3 Acute tolerance to the rate-decreasing effects of SNC80 in rhesus monkeys. (*Left*) The effects of intramuscular SNC80 administered either alone or 24 h after pretreatment with SNC80 (1.0–10.0 mg/kg) in an assay of schedule-controlled responding for food reinforcement in rhesus monkeys. (*Center and right*) The effects of the mu agonist morphine and the kappa agonist U50,488 administered alone or after SNC80 pretreatment. SNC80 dose-dependently produced acute tolerance to its own rate-decreasing effects, as indicated by rightward shifts in the SNC80 dose-effect curve. These effects were pharmacologically selective, in that the effects of morphine and U50,488 were not altered by SNC80 pretreatment. (From Ref. 71.)

produced a dose-dependent decrease in response rates in this procedure, and 24-h pretreatment with SNC80 produced acute tolerance to this effect, as evidenced by dose-dependent rightward shifts in the SNC80 dose-effect curve. Figure 3 (center and right panels) shows that SNC80 pretreatment did not alter the effects of the mu agonist morphine or the kappa agonist U50,488, indicating that tolerance was pharmacologically selective for delta opioid agonists in this study. Finally, chronic treatment with SNC80 for several weeks produced a more profound tolerance than acute treatment [71]. Other studies have demonstrated tolerance to the antinociceptive effects [67,68] and endocrine effects (increased corticosterone levels) [70] of DPDPE in mice and to the convulsant effects of BW373U86 in mice [69]. Although tolerance has been demonstrated to all delta agonist effects studied to date, it should be recognized that studies with other drug classes have shown that tolerance can develop to different degrees and at different rates to different drug effects. For example, tolerance developed more readily to the antinociceptive effects of mu opioid agonists than to their respiratory depressant, constipating, and pupillary-constricting effects [72–74]. Further research will be required to provide a more comprehensive understanding of tolerance to delta agonist effects and to assess the implications of tolerance for the abuse liability of delta agonists.

In contrast to the profound tolerance that has been found to develop to delta agonist effects, there is little evidence to suggest that delta agonist treatment produces physical dependence [43,67,70,71,75,76]. Withdrawal from central administration of the peptidic delta agonists DPDPE or deltorphin II produced fewer and less severe signs of abstinence in rodents than withdrawal from mu agonists [43,67,70,75]. Moreover, any physical dependence that developed to these peptidic delta agonists may have been mediated by their actions at mu receptors. For example, withdrawal from chronic treatment with deltorphin II produced mild but measurable abstinence signs in wild-type mice but not in mu receptor knockout mice, suggesting that mu receptors were necessary for even mild physical dependence to develop to deltorphin II [43]. Withdrawal from chronic treatment with BW373U86 in rats or SNC80 in monkeys produced no evidence of abstinence signs [71,76].

2.5 Summary

No delta selective opioids are currently available for human use, and as a result, the degree to which delta agonists might have abuse liability is unknown. Data from preclinical studies provide a mixed picture. The strongest evidence suggestive of abuse liability comes from studies showing that centrally administered peptidic delta agonists were self-administered, produced conditioned place preferences, facilitated responding for electrical brain stimulation, partially substituted for the discriminative stimulus effects of cocaine, and produced tolerance and mild physical dependence in rats and mice. However, it remains possible that mu opioid receptors mediated at least some of these effects. Systemic administration of nonpeptidic delta agonists produced place preferences in rats, partial substitution for the discriminatory stimulus effects of mu agonists in pigeons and cocaine in monkeys, and tolerance in monkeys. However, nonpeptidic delta agonists did not maintain self-administration in monkeys or produce evidence of physical dependence in rats or monkeys. Overall, these data are consistent with the conclusion that central delta opioidergic systems may be capable of mediating at least some abuse-related effects in rodents. However, systemic administration of currently available nonpeptidic delta agonists does not produce effects suggestive of high abuse liability in nonhuman primates. Additional studies are required to asses the abuse-related effects of delta opioid agonists under a wider array of conditions.

3 UTILITY OF DELTA OPIOID ANTAGONISTS FOR THE TREATMENT OF SUBSTANCE ABUSE

On the basis of clinical experience with the treatment of opioid abuse, two general categories of substance abuse treatment medications have been proposed [17,77]. "Agonist" medications produce some effects in common with the abused drug and may produce tolerance to and prevent withdrawal from the abused drug. Methadone is the prototype agonist medication for the treatment of opioid abuse [78], and other examples of agonist medications include formulations of nicotine for the treatment of tobacco dependence and clinical trials of amphetamine for the treatment of psychostimulant abuse [79–81]. "Antagonist" medications, in contrast, pharmacologically block the effects of the abused drug. Naltrexone is the prototype antagonist medication for the treatment of opioid dependence [82], and mecamylamine and dopamine antagonists have been evaluated as candidate antagonist medications for the treatment of tobacco and psychostimulant abuse, respectively [79,80,83,84]. In general, agonist medications have proven to be more effective than antagonist medications, in large part because they maintain better compliance.

The investigation of delta opioids as candidate treatments for substance abuse is predicated on the hypothesis that opioid systems in general, and delta opioid systems in particular, may mediate some abuse-related effects of known drugs of abuse. Within this conceptual framework, delta agonists might be expected to reproduce delta receptor–mediated effects of abused drugs and serve as candidate agonist medications. Conversely, delta antagonists might attenuate delta receptor–mediated effects of abused drugs and serve as candidate antagonist medications. To date, the vast majority of studies to address these possibilities have examined the effects of delta antagonists on the abuse-related effects of abused drugs, and this section will review those studies.

3.1 Opioid Abuse

Prevailing evidence indicates that the reinforcing and other abuse-related effects of abused opioids are mediated primarily by mu opioid receptors [4,8]. For example, the reinforcing effects of heroin and other opioids were antagonized by the mu-selective opioid antagonists quadazocine, β-funaltrexamine, and clocinnamox [85–89]. In comparison to the effects of mu antagonists, delta antagonists are often relatively ineffective in altering the abuse-related effects of abused opioids. For example, delta-selective doses of naltrindole did not alter heroin or morphine self-administration in rats [87,90] or the discriminative stimulus effects of fentanyl in pigeons [63] or morphine in rats [57]. Similarly, the peptidic delta antagonist ICI174864 had no effect on a morphine-induced place preference in rats [37], and a range of delta antagonists did not alter morphine-induced place preferences in CXBK mice [91]. On the other hand, naltrindole, the putative delta-1 receptor antagonist BNTX and the putative delta-2 antagonist naltriben attenuated morphine place preferences in ddY mice, and the delta agonist TAN-67 enhanced morphine place preferences [42,91]. Also, very high doses of naltrindole (≥10 mg/kg) attenuated some abuse-related effects of opioids [49,63,87], suggesting a possible role for delta receptors. However, naltrindole is only marginally selective for delta over mu receptors [92,93], and these high naltrindole doses may have blocked mu as well as delta receptors.

More compelling evidence to suggest a role for delta receptors in opioid reinforcement comes from a recent report that the irreversible delta antagonist NTII, which is reputed to be selective for delta-2 receptors, antagonized the reinforcing effects of heroin in rats [94]. In this study, complete heroin self-administration dose-effect curves were determined daily. NTII (10–40 nmol ICV) produced dose-dependent and long-lasting rightward and downward shifts in the heroin self-administration dose-effect curve, and up to 17 days was required for heroin self-administration to recover to baseline levels. Importantly, the selectivity of these NTII doses for delta-2 receptors was supported by additional studies, which showed that NTII also antagonized the antinociceptive effects of the putative delta-2 agonist deltorphin, but not those of heroin, the mu agonist DAMGO or the putative delta-1 receptor agonist DPDPE. Moreover, these doses of NTII had minimal effects on cocaine self-administration, suggesting that NTII-induced decreases in heroin self-administration did not result from nonselective decreases in all operant behavior (e.g., due to sedation). Overall, these results were interpreted to suggest that blockade of delta-2 receptors decreased the reinforcing effects of heroin, and that delta-2 antagonists may be useful in the treatment of opioid abuse [94].

Another intriguing theme of research has examined the role of delta opioid receptors in the development and expression of physical dependence on mu opioid agonists. Regarding the expression of physical dependence, delta antagonists such as naltrindole usually display low potency in precipitating various signs of withdrawal in subjects chronically treated with morphine [95–97]. For example, the high doses of naltrindole that precipitated signs of withdrawal in morphine-treated monkeys [95,96] were similar to doses that produced mu antagonist effects in untreated monkeys and higher than doses that produced selective delta antagonist effects [27,29]. These findings suggest that any naltrindole-precipitated withdrawal signs likely reflected mu antagonist effects of high naltrindole doses, and further suggest that delta receptors play little role in mediating signs of mu agonist withdrawal. In agreement with this conclusion, the nonpeptidic delta agonists BW373U86 and SNC80 did not reverse somatic signs of withdrawal in morphine-dependent rats and monkeys [76,98]. However, naltrindole was more potent in precipitating a withdrawallike place aversion than in producing somatic withdrawal signs in morphine-treated rats [99]. These latter results were interpreted to suggest that delta receptors may mediate some effects of morphine withdrawal that contribute to place conditioning in rats.

In contrast to these findings that suggest a limited role for delta receptors in the *expression* of physical dependence on morphine, numerous studies over the past decade have implicated delta receptors in the *development* of physical dependence [76,100–105]. In these experiments, the delta opioid is not administered at the conclusion of morphine treatment to precipitate withdrawal, but rather is administered together with morphine. The severity of physical dependence is then measured by quantifying withdrawal signs precipitated by a nonselective antagonist such as naloxone, and a reduction in naloxone-precipitated withdrawal signs is interpreted as a decrease in the severity of physical depencence. One study, for example, examined naloxone-precipitated withdrawal in mice treated with morphine administered either alone or together with a relatively low dose of naltrindole (10 pmol, ICV) [100]. Withdrawal-induced jumping was greater in mice treated only with morphine than in mice treated with both morphine and naltrindole. Importantly, the dose of naltrindole used in this study did not block the acute antinociceptive effects of the mu-selective agonist DAMGO, suggesting that naltrindole's attenuation of withdrawal signs in morphine-treated mice could not be attributed to any mu antagonist effects of naltrindole [100]. Subsequent studies by this group and others suggested that blockade of delta-2 receptors was more effective than blockade of delta-1 receptors in preventing morphine-induced physical dependence [106–109]. Some of signs of naloxone-precipitated withdrawal may be more sensitive than others to modulation by delta opioid antagonists. For example, withdrawal-associated salivation was usually blocked by delta antagonists [76, 101,104], whereas withdrawal-associated ptosis was not [76,102,104]. Moreover, one study found that withdrawal-associated salivation in morphine-dependent rats was the only one of 13 withdrawal signs reduced by naltrindole in rats, whereas some other signs (wet dog shakes, forelimb tremor, and teeth chattering) were reduced in a naltrindole-reversible manner by the delta agonist BW373U86, and other signs were not affected by delta agonists or antagonists [76].

Delta receptors may also modulate the development of tolerance to some effects of morphine. Both naltrindole [100,104] and an antisense probe for delta receptors [103] attenuated the development of tolerance to the antinociceptive effects of morphine in rodents, and delta receptor knockout mice also failed to show tolerance to the antinociceptive effects of morphine [110]. However, naltrindole did not affect tolerance to morphine-induced respiratory depression in rats [104], and another study found that naltrindole did not affect antinociceptive tolerance to morphine in rats, although the peptidic delta antagonist TIPPψ was effective [101]. Overall, these findings suggest that delta receptors may play a prominent but complex role in the development of morphine physical dependence and tolerance. Additional research will be required to extend these provocative results and examine such issues as the ability of delta antagonists to modulate physical dependence and tolerance that develop to mu agonists other than morphine in species other than rodents. Also, the implications of these findings for the abuse liability of morphinelike opioids remains to be determined.

One final set of data has addressed the possible connection between delta receptors and opioid abuse in humans. In a population of 103 German heroin abusers, a correlation was identified between a polymorphism in the delta receptor gene and the likelihood of heroin abuse [111]. However, two subsequent studies in larger populations failed to replicate this finding [112,113].

3.2 Alcohol Abuse

The potential utility of delta antagonists for the treatment of alcohol abuse has excited considerable interest over the last decade for at least two reasons. First, numerous studies suggest that the abuse-related effects of alcohol are mediated, at least in part, by endogenous opioid systems [114]. Second, the relatively nonselective opioid antagonist naltrexone has acknowledged clin-

TABLE 2 Summary of Delta Opioid Antagonist Effects on the Reinforcing Effects of Ethanol

Delta antagonist (dose/route)	Species[a]	Procedure (reinforcers[b]	H_2O Dep[c]	Effect on EtOH SA[d]	Selectivity[e]	Ref.
Studies showing a selective delta antagonist–induced decrease in ethanol self-administration						
ICI174864 (0.5–3 mg/kg IP)	HAD rats	2-Bottle choice 10% EtOH/water	Yes	Decrease	>3-fold	118
Naltrindole (10 mg/kg SC)	C57 mice	2-Bottle choice 12% EtOH/water	No	Decrease	Not reported	121
ICI174864 (3–8 mg/kg SC)	P rats	2-Bottle choice 10% EtOH/water	No	Decrease	>2.7-fold	119
Naltrindole (5–20 mg/kg IP)	P rats	2-Bottle choice 10% EtOH/water	No	Decrease	>4-fold	119
Naltriben (0.375–6 mg/kg IP)	P rats	2-Bottle choice 10% EtOH/water	No	Decrease	>1-fold	122
Naltriben (0.375–6 mg/kg IP)	P rats	2-Bottle choice 10% EtOH + Sacc/Sacc	No	Decrease	>1-fold	122
Naltriben (0.375–6 mg/kg IP)	P rats	2-Bottle choice 10% EtOH + Quin/Quin	No	Decrease	>1-fold	122
ICI174864 (0.1 mg/kg IP)	SD rats	2-Bottle choice 5% EtOH/water	Yes	Decrease	>1-fold	120
Naltrindole (0.1–3 mg/kg IP)	SD rats	2-bottle choice 5% EtOH/water	Yes	Decrease	>3-fold	120
Naltriben (0.6–4 mg/kg SC)	P rats	2-Operandum operant 10% EtOH/Sacc	No	Decrease	4.4-fold	123
Naltrindole (3–30 μg ICV)	W rats	2-Operandum operant 10% EtOH/water	No	Decrease	>1-fold	124
Naltrindole (0–500 ng IC)	W rats	2-Operandum operant 10% EtOH/water	No	Decrease[f]	Not reported	124

Naltrindole (3–30 μg ICV)	AA rats	2-operandum operant 10% EtOH/water	No	Decrease	>3-fold	124
Studies showing no effect or a nonselective delta antagonist–induced decrease in ethanol self-administration						
ICI174864 (1.5–3 μg ICV)	AA rats	2-Bottle choice 10% EtOH/water	No	No effect	—	126
Naltrindole (15 mg/kg IP)	P rats	2-Bottle choice 10% EtOH + Sacc/Sacc	No	Decrease	No selectivity	119
Naltrindole (1–5 mg/kg IP)	AA rats	2-Bottle choice 10% EtOH/water	No	No effect	—	127
Naltrindole (1 mg/kg)	AA rats	2-Bottle choice 10% EtOH/water	No	No effect	—	127
Naltrindole (5–15 mg/kg SC)	W rats	2-Bottle choice 6% EtOH/water	No	No effect	—	128
Naltrindole (1–3.2 mg/kg IM)	Rh monk	2-Operandum operant 1 or 2% EtOH/water	No	No effect	—	125
Naltrindole (0.3–30 mg/kg)	C57 mice	1-Operandum operant 12% EtOH or water	No	Decrease	No selectivity	129

[a] Abbreviations: AA-Alko, Alcohol (alcohol-preferring rats); HAD, high alcohol drinking; P, alcohol-preferring rats; SD, Sprague-Dawley; W, Wistar.

[b] Two types of procedures were used. In 2-bottle choice procedures, two bottles containing different solutions were concurrently available. In operant procedures, one or two operanda were available, and subjects could respond on these opperanda to obtain the reinforcer solutions. Abbreviations: Quin, quinine solution; Sacc, Saccharin solution.

[c] Indicates whether or not subjects were water deprived prior to self-administration tests.

[d] Effect on ethanol self-administration.

[e] Selectivity calculated as (lowest dose to decrease self-administration of alternative *or* highest dose tested ÷ lowest dose to significantly decrease EtOH self-administration).

[f] In this study, naltrindole decreased ethanol self-administration following intracranial administration of naltrindole into the basolateral amygdala and nucleus accumbens but not into the ventral tegmental area.

ical utility in the treatment of alcohol abuse [115–117]. These findings have stimulated research to investigate the potential role of delta receptors, as well as mu and kappa receptors, in mediating the clinically useful effects of naltrexone.

Table 2 summarizes preclinical studies that have investigated the effects of delta antagonists on the reinforcing effects of alcohol (ethanol). The results of approximately half of these studies have been interpreted to suggest that delta antagonists may reduce the reinforcing effects of alcohol and may be useful in the treatment of alcohol abuse. In the seminal study to implicate delta receptors in the reinforcing effects of alcohol, water-deprived rats bred to consume large quantities of alcohol (high alcohol drinking, or HAD rats) were given concurrent access to one bottle containing 10% ethanol and another bottle containing water during daily 30-min sessions [118]. Under baseline conditions, rats consumed slightly more water than ethanol. Systemic administration of the peptidic delta antagonist ICI174864 (0.5–3.0 mg/kg, IP) dose-dependently decreased ethanol consumption without altering water consumption. A higher dose of ICI174864 (i.e., 10 mg/kg) was not tested owing to concerns about toxicity. In this study, then, ICI174864 decreased consumption of 10% ethanol at doses approximately three- to 10-fold lower than those expected to produce toxicity [118]. Other studies also found that systemically administered ICI174864 [119,120], systemically administered naltrindole and naltriben [119–123], and centrally administered naltrindole [124] selectively decreased either ethanol drinking or operant responding for ethanol (see Table 2).

Although these findings are provocative, their implications both for the role of delta receptors in ethanol reinforcement and for the clinical promise of delta antagonists as alcohol abuse treatment medications should be regarded with caution for several reasons. First, the pharmacokinetics and pharmacodynamics of the delta antagonists employed must be considered in interpreting the results. ICI174864 is a peptidic delta antagonist that does not distribute well across the blood-brain barrier. The pharmacology of this compound after systemic administration has not been thoroughly examined, and it is not clear that decreases in alcohol consumption produced by systemic administration of this compound result from a selective antagonism of central delta opioid receptors. Pharmacokinetic issues are usually less problematic in studies with nonpeptidic delta antagonists such as naltrindole. Nonetheless, the limited pharmacological selectivity of these compounds for delta receptors must still be taken into account, and the possibility that high doses of these antagonists may produce nondelta effects (e.g., mu antagonist effects) must be considered. In rhesus monkeys, for example, systemically administered naltrindole produces delta-selective antagonist effects at doses up to 1.0–3.2 mg/kg, and these naltrindole doses failed to alter ethanol self-administration in rhesus monkeys given concurrent access to 10% ethanol and water in an operant procedure [125]. Higher naltrindole doses may have decreased ethanol self-administration, but higher doses were not tested because of their low selectivity. These results were interpreted to suggest that delta receptors did not mediate the reinforcing effects of ethanol in monkeys [125].

Second, the behavioral selectivity of delta antagonists for reducing consumption of ethanol in comparison to consumption of other reinforcers may depend on such procedural factors as the nature of the alternative reinforcer. For example, one revealing study examined the effects of naltrindole on consumption of 10% ethanol under two conditions in alcohol-preferring (P) rats [119]. Under one set of conditions, rats had limited daily access to a 10% solution of ethanol in water from one bottle and 24 h/d access to water from a second bottle. Naltrindole (5–20 mg/kg) produced a dose-dependent and selective decrease in ethanol consumption without altering total daily consumption of water. Under the second set of conditions, though, rats had concurrent limited daily access to a solution of saccharin + 10% ethanol and to a saccharin solution without ethanol. Only water was available for the remainder of each day. Under these conditions, naltrindole produced a nonselective decrease in consumption of both saccharin + ethanol and saccharin alone. These results were interpreted to suggest that delta opioid receptors may mediate the reinforcing effects of both ethanol and sweet solutions, but while this may be the case, medications that decrease the reinforcing effects of a wide range of reinforcers may have limited clinical utility. As noted in Table 2, many other studies also found that delta antagonists either did not alter ethanol self-administration or nonselectively decreased ethanol self-administration [126–129]. Moreover, alcohol self-administration was actually increased in delta opioid receptor knockout mice [130]. The inconsistent effects of delta antagonists on alcohol-maintained behavior suggest that delta antagonists may be effective under only a limited range of conditions, and additional research will be necessary to identify those conditions under which delta antagonists are most likely to be effective.

Third, even when delta antagonists do produce a behaviorally selective decrease in ethanol consumption, this selective decrease usually occurs over a relatively narrow dose range in comparison to the effects of naltrexone or naloxone. For example, both naloxone and the putative delta-2 receptor antagonist naltriben produced a selective decrease in responding for 10% ethanol in P rats choosing between ethanol and a saccharin solution in a two-choice operant task [123]. However, naloxone decreased ethanol-maintained responding at a very low dose of 0.00325 mg/kg, and even doses as high as 0.75 mg/kg failed to alter saccharin-maintained responding. Thus, naloxone selectively decreased ethanol-maintained responding across a >200-fold range of doses. By comparison, the lowest dose of naltriben to decrease ethanol-maintained responding was 0.9 mg/kg, and a dose of 4.0 mg/kg significantly decreased saccharin-maintained responding. Thus, naltriben was selective only across a 4.4-fold range of doses. These findings suggest that even if existing delta antagonists do selectively decrease alcohol consumption under some conditions, they may have a narrower margin of selectivity and safety than naltrexone or naloxone.

A final concern regarding the literature that describes delta antagonist effects on alcohol consumption is that no study has examined delta antagonist effects on complete alcohol self-administration dose-effect curves, and few studies have examined the effects of chronic delta receptor antagonism. It is well established that candidate substance abuse treatment medications can have different effects depending on the unit dose or concentration of the self-administered drug [17]. Moreover, most substance abuse treatment medications (including naltrexone for alcohol abuse [115–117]) are administered chronically for at least several days and sometimes up to several years, and the effects of medications can change over time. As a result, we have argued that preclinical assessment of medications should ideally examine the effects of chronic medication treatment on complete drug self-administration dose-effect curves [17]. Such data are not yet available to describe opioid antagonist effects on ethanol self-administration.

In addition to the results summarized in Table 2, delta antagonists have also been reported to alter some other abuse-related effects of alcohol. At a neurochemical level, naltrindole blunted the ethanol-induced increase in extracellular dopamine levels in the nucleus accumbens of Sprague-Dawley rats [32]. Naltrindole also blocked this effect of the delta agonist deltorphin II but not of cocaine, suggesting some selectivity in naltrindole's effects [32]. At a behavioral level, naltrindole dose-dependently blocked both ethanol-induced place preferences in Sprague-Dawley rats exposed to stressors [131,132] and reinstatement of extinguished ethanol self-administration in Wistar rats [133]. Finally, naltrindole enhanced conditioned taste aversion to low ethanol concentrations in P rats, and it was suggested that this effect may contribute to delta antagonist–induced decreases in ethanol drinking [134]. However, neither naltrindole nor naltriben altered the discriminative stimulus effects of ethanol in rats or mice [129,135,136].

3.3 Cocaine and Other Stimulants

The effects of delta antagonists on abuse-related effects of cocaine and other stimulants have been examined with inconsistent results. In assays of drug self-administration, delta antagonists usually decreased cocaine self-administration in at least some subjects under some conditions [28,90,94,137]. For example, in rhesus monkeys responding for cocaine and food pellets under a second-order schedule of reinforcement, chronic treatment with naltrindole (0.1–3.2 mg/kg/d) decreased cocaine self-administration without altering food-maintained responding in three of four monkeys tested [28]. However, these effects appeared to be influenced by several independent variables. First, one monkey was completely insensitive to the effects of naltrindole, suggesting that there may be individual differences in the sensitivity of cocaine reinforcement to delta receptor antagonism. One study in rats also reported dramatic individual differences in naltrindole effects [90], and another study in rats found virtually no effect of naltrindole on cocaine self-administration [137]. Second, responding maintained by low unit doses of cocaine in monkeys was more sensitive than responding maintained by higher cocaine doses to suppression by naltrindole. The putative delta-2 antagonist NTII also decreased low-dose but not high-dose cocaine self-administration in rats [137]. Third, intermediate doses of naltrindole often had greater effects than lower or higher naltrindole doses on cocaine self-administration in monkeys, so the effects of naltrindole on cocaine self-administration were not monotonically related to naltrindole dose. This agrees with the finding that intermediate doses of naltrindole were also most effective in modifying sensitization to the locomotor effects of cocaine in rats [138]. Finally, the effects of naltrindole on cocaine self-administration in monkeys were not replicable upon repeated determinations, suggesting that any effects of naltrindole on cocaine self-administration diminished over time. Taken together, these results indicate that naltrindole has unreliable effects on the reinforcing effects of cocaine in drug self-administration procedures.

Results with delta antagonists have also been inconsistent in studies of other abuse-related effects of cocaine. Naltrindole attenuated cocaine-induced place preferences [139,140] and cocaine-induced facilitation of electrical brain stimulation [141]. Cocaine-induced place preferences were also blocked by the putative delta-2 antagonist naltriben, but not by the delta-1 antagonist BNTX, and these results were interpreted to suggest that delta-2 receptors may be especially important in modulating these abuse-related effects of cocaine [140]. However, other studies found that naltrindole did not block cocaine-induced place preferences [137], and naltrindole also failed to alter the discriminative stimulus effects of cocaine in rats or monkeys [28,61,62]. Also, the importance of delta-2 receptors in mediating abuse-related effects of cocaine is challenged by the finding that the putative delta-2 antagonist NTII produced only modest effects on cocaine self-administration in rats [94]. Finally, naltrindole did not alter the ability of cocaine to increase extracellular dopamine levels in the nucleus accumbens in rats [32].

Few studies have examined the effects of delta opioids on abuse-related effects of other psychostimulants. Naltrindole and naltriben, but not BNTX, attenuated both methamphetamine-induced place preferences and the discriminative stimulus effects of low methamphetamine doses in rats; however, naltrindole did not alter the discriminative stimulus effects of a higher dose of methamphetamine [140,142]. A single dose of 3 mg/kg naltrindole was also reported to block facilitation of electrical brain stimulation by a single dose of methylenedioxymethamphetamine (MDMA) [143]. Finally, a recent study found that both morphine and TAN-67 prevented mecamylamine-induced place aversions in rats chronically-treated with nicotine. These results were interpreted to suggest that both mu and delta agonists may attenuate some aversive effects associated with nicotine withdrawal [144].

3.4 Summary

Research on the utility of delta opioids for the treatment of substance abuse is predicated on the hypothesis that delta opioid systems may contribute to the effects of some abused drugs, especially opioids, alcohol, and stimulants such as cocaine. Most of this research has focused on the ability of delta opioid antagonists to block the effects of abused drugs. In general, delta antagonists have produced modest and inconsistent changes in the reinforcing effects and other abuse-related effects of opioids, alcohol, and stimulants. The most promising effects have been produced by the putative delta-2 antagonists NTII and naltriben, and further research with these compounds in particular may be warranted. It will also be useful to identify conditions under which delta antagonists are most likely to attenuate the effects of abused drugs. Any enthusiasm for the use of delta antagonists in the treatment of drug dependence should be tempered by two considerations. First, even when delta antagonists have been found to attenuate effects of abused drugs, it is not clear that delta antagonists are superior to the nonselective and clinically available antagonists naltrexone and naloxone. Second, antagonist medications have been less successful than agonist medications for the treatment of substance abuse. In view of this latter point, it may be of interest in the future to evaluate the potential utility of delta agonists as candidate drug abuse treatment medications.

ACKNOWLEDGMENTS

The author would like to thank Nancy K. Mello, Ph.D., and Glenn W. Stevenson, Ph.D., for their comments on an earlier version of this review. Supported by RO1 DA-11460 from the National Institute on Drug Abuse.

REFERENCES

1. National Institute on Drug Abuse. NIH Publication 98-4327, 1998.
2. Substance Abuse and Mental Health Services Administration. Office of Applied Studies, NHSDA Series H-13, DHHS Publication No. (SMA) 01-3549, Rockville, MD, 2001.
3. Courtwright DT. In: Musto DF, ed. One Hundred Years of Heroin. Westport, CT: Auburn House, 2002:3–19.
4. Negus SS, Dykstra LA. In: Watson RW, ed. Biochemistry and Physiology of Substance Abuse, Vol 1. Boca Raton: CRC Press, 1989:211–242.
5. Self DW, Stein L. Pharmacol Toxicol 1992; 70:87–94.
6. Negus SS, Picker MJ. CNS Drug Rev 1996; 2:52–74.
7. Herz A. Can J Physiol Pharmacol 1998; 76:252–258.
8. Narita M, Funada M, Suzuki T. Pharmacol Ther 2001; 89:1–15.
9. American Psychiatric Association. Diagnostic and Statistical Manual of Mental Disorders (4th ed, Text Revision). Washington, DC: American Psychiatric Association, 2000.
10. Brady JV, Griffiths RR, Hienz RD, Ator NA, Lukas SE, Lamb RJ. In: Bozarth MA, ed. Methods of Assessing the Reinforcing Effects of Abused Drugs. New York: Springer-Verlag, 1987:48–86.
11. Woods JH. Prog Neuro-Psychopharmacol Biol Psychiatry 1983; 7:577–584.
12. Collins RJ, Weeks JR, Cooper MM, Good PI, Russell RR. Psychopharmacology 1984; 82:6–13.
13. Hughes J, Smith TW, Kosterlitz HW, Fothergill LA, Morgan BA, Morris HR. Nature 1975; 258:577–579.
14. Lord JA, Waterfield AA, Hughes J, Kosterlitz HW. Nature 1977; 26:495–499.
15. Belluzzi JD, Stein L. Nature 1977; 266:556–558.
16. Goeders NE, Lane JD, Smith JE. Pharmacol Biochem Behav 1984; 20:451–455.
17. Mello NK, Negus SS. Neuropsychopharmacology 1996; 14:375–424.
18. Wood PL, Charleson SE, Lane D, Hudgin RL. Neuropharmacology 1981; 20:1215–1220.
19. Devine DP, Wise RA. J Neurosci 1994; 14:1978–1984.
20. Fraser GL, Pradhan AA, Clarke PBS, Wahlestedt C. J Pharmacol Exp Ther 2000; 295:1135–1141.
21. Hosohata Y, Vanderah TW, Burkey TH, Ossipov MH, Kovelowski CJ, Sora I, Uhl GR, Zhang X, Rice K, Roeske WR, Hruby VJ, Yamamura HI, Lai J, Porreca F. Eur J Pharmacol 2000; 388:241–248.
22. Mansour A, Khachaturian H, Lewis ME, Akil H, Watson SJ. Trends Neurosci 1988; 11:308–314.
23. Yeardon M, Kitchen I. J Pharm Pharmacol 1988; 40:736–739.
24. Woods JH, Young AM, Herling S. Fed Proc 1982; 41:221–227.
25. Chang K-J, Rigdon GC, Howard JL, McNutt RW. J Pharmacol Exp Ther 1993; 267:852–857.
26. Calderon SN, Rothman RB, Porreca F, Flippen-Anderson JL, McNutt RW, Xu H, Smith LE, Bilsky EJ, Davis P, Rice KC. J Med Chem 1994; 37:2125–2128.

27. Negus SS, Butelman ER, Chang KJ, DeCosta BR, Winger G, Wood JH. J Pharmacol Exp Ther 1994; 270:1025–1034.
28. Negus SS, Mello NK, Portoghese PS, Lukas SE, Mendelson JH. J Pharmacol Exp Ther 1995; 273:1245–1256.
29. Negus SS, Gatch MB, Mello NK, Zhang X, Rice K. J Pharmacol Exp Ther 1998; 286:362–375.
30. Spanagel R, Herz A, Shippenberg TS. J Neurochem 1990; 55:1734–1740.
31. Longoni R, Spina L, Mulas A, Carboni E, Garau L, Melchiorri, Di Chiara G. J Neurosci 1991; 11:1565–1576.
32. Acquas E, Meloni M, DiChiara G. Eur J Pharmacol 1993; 230:239–241.
33. Devine DP, Leone P, Pocock D, Wise RA. J Pharmacol Exp Ther 1993; 266: 1236–1246.
34. Longoni R, Cadoni C, Mulas A, Di Chiara G, Spina L. Behav Pharmacol 1998; 9:9–14.
35. Mucha RF, van der Kooy D, O'Shaughnessy M, Bucenieks P. Brain Res 1982; 243:91–105.
36. Van der Kooy D. In: Bozarth MA, ed. Methods of Assessing the Reinforcing Properties of Abused Drugs. New York: Springer-Verlag, 1987:229–240.
37. Shippenberg TS, Bals-Kubic R, Herz A. Brain Res 1987; 436:234–239.
38. Bals-Kubic R, Shippenberg TS, Herz A. Eur J Pharmacol 1990; 175:63–69.
39. Suzuki T, Tsuji M, Mori T, Misawa M, Nagase H. Psychopharmacology 1996; 124:211–218.
40. Suzuki T, Tsuji M, Mori T, Ikeda H, Misawa M, Nagase H. Brain Res 1997; 744:327–334.
41. Morales L, Perez-Garcia C, Alguacil LF. Br J Pharmacol 2001; 133:172–178.
42. Suzuki T, Tsuji M, Mori T, Misawa M, Endoh T, Nagase H. J Pharmacol Exp Ther 1996; 279:177–185.
43. Hutcheson DM, Matthes HW, Valjent E, Sanchez-Blazquez P, Rodriguez-Diaz M, Garzon J, Kieffer BL, Maldonado R. Eur J Neurosci 2001; 13:153–161.
44. Wise RA. Annu Rev Neurosci 1996; 19:319–340.
45. Jenck F, Gratton A, Wise RA. Brain Res 1987; 423:34–38.
46. Heidbreder CA, Gewiss M, Lallemand S, Roques BP, De Witte P. Neuropharmacology 1992; 31:293–298.
47. Johnson PI, Stellar JR. Neuropharmacology 1994; 33:1171–1182.
48. Johnson PI, Goodman JB, Condon R, Stellar JR. Psychopharmacology 1995; 120:195–202.
49. Duvauchelle CL, Fleming SM, Kornetsky C. Eur J Pharmacol 1996; 316:137–143.
50. Holtzman SG. Drug Alcohol Depend 1985; 14:263–282.
51. Overton DA. In: Bozarth MA, ed. Methods of assessing the reinforcing properties of abused drug. New York: Springer-Verlag, 1987:291–340.
52. Comer SD, McNutt RW, Chang KJ, DeCosta BR, Mosberg HI, Woods JH. J Pharmacol Exp Ther 1993; 267:866–874.
53. Jewett DC, Mosberg HI, Woods JH. Psychopharmacology 1996; 127:225–230.

54. Picker MJ, Cook CD. Behav Pharmacol 1998; 9:319–328.
55. Brandt MR, Negus SS, Mello NK, Furness MS, Zhang X, Rice KC. J Pharmacol Exp Ther 1999; 290:1157–1164.
56. Stevenson GW, Canadas F, Gomez-Serrano M, Ullrih T, Zhang X, Rice K, Riley AL. Pharmacol Biochem Behav 2002; 71:283–292.
57. Stevenson GW, Canadas F, Zhang X, Rice KC, Riley AL. Pharmacol Biochem Behav 2000; 66:851–856.
58. Easterling KW, Holtzman SG. Brain Res Bull 2001; 56:545–551.
59. Platt DM, Rowlett JK, Spealman RD. J Pharmacol Exp Ther 2001; 29:760–767.
60. Spealman RD, Bergman J. Behav Pharmacol 1994; 5:21–31.
61. Broadbent J, Gaspard TM, Dworkin SI. Pharmacol Biochem Behav 1995; 51: 379–385.
62. Rowlett JK, Spealman RD. Psychopharmacology 1998; 140:217–224.
63. Negus SS, Morgan D, Cook CD, Picker MJ. Psychopharmacology 1996; 126: 199–205.
64. Ukai M, Mori E, Kameyama T. Eur J Pharmacol 1993; 231:143–144.
65. Suzuki T, Mori T, Tsuji M, Maeda J, Kishimoto Y, Misawa M, Nagase H. Eur J Pharmacol 1997; 324:21–29.
66. Suzuki T, Kishimoto Y, Ozaki S, Narita M. Eur J Pain 2001; 5(suppl A):63–65.
67. Kovacs GL, Nyolczas N, Krivan M, Gulya K. Eur J Pharmacol 1988; 150: 347–353.
68. Suh HH, Tseng LF. Life Sci 1990; 46:759–765.
69. Comer SD, Hoenicke EM, Sable AI, McNutt RW, Chang KJ, DeCosta BR, Mosberg HI, Woods JH. J Pharmacol Exp Ther 1993; 267:888–895.
70. Gonzalvez ML, Vargas ML, Milanes MV. Gen Pharmacol 1994; 25:719–723.
71. Brandt MR, Furness MS, Rice KC, Fischer BD, Negus SS. J Pharmacol Exp Ther 2001; 299:629–637.
72. Ling GS, Paul D, Simantov R, Pasternak GW. Life Sci 1989; 45:1627–1636.
73. Paronis CA, Woods JH. J Pharmacol Exp Ther 1997; 282:355–362.
74. Brandt MR, France CP. J Pharmacol Exp Ther 2000; 294:166–178.
75. Cowan A, Zhu XZ, Mosberg HI, Omnaas JR, Porreca F. J Pharmacol Exp Ther 1988; 246:950–955.
76. Lee PH, McNutt RW, Chang KJ. J Pharmacol Exp Ther 1993; 267:883–887.
77. Rothman RB, Glowa JR. Mol Neurobiol 1995; 10:1–19.
78. Lowinson JH, Payte JT, Salsitz E, Joseph H, Marion IJ, Dole VP. In: Lowinson JH, Ruiz P, Millman RB, Langrod JG, eds. Substance Abuse: A Comprehensive Textbook. Baltimore: Williams and Wilkins, 1997:405–415.
79. Hurt RD. Nicotine Tob Res 1999; 1(suppl 2):S175–S179.
80. Haustein KO. Int J Clin Pharmacol Ther 2000; 38:273–290.
81. Grabowski J, Rhoades H, Schmitz J, Stotts A, Daruzska LA, Creson D, Moeller FG. J Clin Psychopharmacol 2001; 21:522–526.
82. Greenstein RA, Fudala PJ, O'Brien CP. In: Lowinson JH, Ruiz P, Millman RB, Langrod JG, eds. Substance Abuse: A Comprehensive Textbook. Baltimore: Williams and Wilkins, 1997:415–425.

83. Grabowski J, Rhoades H, Silverman P, Schmitz J, Stotts A, Creson D, Bailey R. J Clin Psychopharmacol 2000; 20:305–310.
84. Soyka M, DeVry J. Eur Neuropsychopharmacol 2000; 10:325–332.
85. Bertalmio AJ, Woods JA. J Pharmacol Exp Ther 1989; 251:455–460.
86. Winger G, Skjoldager P, Woods JH. J Pharmacol Exp Ther 1992; 261:311–317.
87. Negus SS, Henriksen SJ, Mattox A, Pasternak GW, Portoghese PS, Takemori AE, Weinger MG, Koob GF. J Pharmacol Exp Ther 1993, 1245–1252.
88. Martin TJ, Dworkin SI, Smith JE. J Pharmacol Exp Ther 1995; 272:1135–1140.
89. Zernig G, Lewis JW, Woods JH. Psychopharmacology 1997; 129:233–242.
90. Reid LD, Glick SD, Menkens KA, French ED, Bilsky EJ, Porreca F. Neuroreport 1995; 6:1409–1412.
91. Suzuki T, Yoshiike M, Mizoguchi H, Kamei J, Misawa M, Nagase H. Jpn J Pharmacol 1994; 66:131–137.
92. Portoghese PS, Sultana M, Takemori AE. Eur J Pharmacol 1988; 146:185–186.
93. Emmerson PJ, Liu M-R, Woods JH, Medzihradsky F. J Pharmacol Exp Ther 1994; 271:1630–1637.
94. Martin TJ, Kim SA, Cannon DG, Sizemore GM, Bian D, Porreca F, Smith JE. J Pharmacol Exp Ther 2000; 294:975–982.
95. Aceto MD, Bowman ER, Harris LS, May EL. NIDA Res Monogr 1990; 95: 578–631.
96. France CP, de Costa BR, Jacobson AE, Rice KC, Woods JH. J Pharmacol Exp Ther 1990; 252:600–604.
97. Madonado R, Negus SS, Koob GF. Neuropharmacology 1992; 31:1231–1241.
98. Jacobson AE. NIDA Res Monogr 1997; 174:323–337.
99. Funada M, Schutz CG, Shippenberg TS. Eur J Pharmacol 1996; 300:17–24.
100. Abdelhamid EE, Sultana M, Portoghese PS, Takemori AE. J Pharmacol Exp Ther 1991; 258:299–303.
101. Fundytus ME, Schiller PW, Shapiro M, Weltrowska G, Coderre TJ. Eur J Pharmacol 1995; 286:105–108.
102. Suzuki T, Tsuji M, Mori T, Misawa M, Nagase H. Life Sci 1995; 57:PL247–PL252.
103. Kest B, Lee CE, McLenmore GL, Inturrisi CE. Brain Res Bull 1996; 39:185–189.
104. Hepburn MJ, Little PJ, Gingras J, Kuhn CM. J Pharmacol Exp Ther 1997; 281:1350–1356.
105. Suzuki T, Ikeda H, Tsuji M, Misawa M, Narita M, Nagase H. Life Sci, 1997.
106. Miyamoto Y, Portoghese PS, Takemori AE. J Pharmacol Exp Ther 1993; 264:1141–1145.
107. Miyamoto Y, Portoghese PS, Takemori AE. J Pharmacol Exp Ther 1993; 265:1325–1327.
108. Miyamoto Y, Bowen WD, Portoghese PS, Takemori AE. J Pharmacol Exp Ther 1994; 270:37–39.

109. Suzuki T, Tsuji M, Mori T, Ikeda H, Misawa M, Nagase H. Pharmacol Biochem Behav 1997; 57:293–299.
110. Zhu Y, King MA, Schuller AGP, Nitsche JF, Reidl M, Elde RP, Unterwald E, Pasternak GW, Pintar JE. Neuron 1999; 24:243–252.
111. Mayer P, Rochlitz H, Rauch E, Rommelspacher H, Hasse HE, Schmidt S, Hollt V. Neuroreport 1997; 8:2547–2550.
112. Franke P, Nothen MM, Wang T, Neidt H, Knapp M, Lichtermann D, Weiffenbach O, Mayer P, Hollt V, Propping P, Maier W. Am J Med Genet 1999; 88:462–464.
113. Xu K, Liu XH, Nagarajan S, Gu XY, Goldman D. Am J Med Genet 2002; 110:45–50.
114. Herz A. Psychopharmacology 1997; 129:99–111.
115. O'Malley SS, Jaffe AJ, Chang G, Scottenfeld RS, Meyer RE, Rounsaville B. Arch Gen Psychiatry 1992; 49:881–887.
116. Volpicelli JR, Alterman AI, Hayashida M, O'Brien CP. Arch Gen Psychiatry 1992; 49:876–880.
117. O'Brien CP, Volpicelli LA, Volpicelli JR. Alcohol 1996; 13:35–39.
118. Froehlich JC, Zweifel M, Harts J, Lumeng L, Li TK. Psychopharmacology 1991; 103:467–472.
119. Krishnan-Sarin S, Jing SL, Kurtz DL, Zwiefel M, Portoghese PS, Li TK, Froehlich JC. Psychopharmacology 1995; 120:177–185.
120. Franck J, Lindholm S, Raaschou P. Alcohol Clin Exp Res 1998; 22:1185–1189.
121. Le AD, Poulos CX, Quan B, Chow S. Brain Res 1993; 630:330–332.
122. Krishnan-Sarin S, Portoghese PS, Li TK, Froehlich JC. Pharmacol Biochem Behav 1995; 52:153–159.
123. June HL, McCane SR, Zink RW, Portoghese PS, Li TK, Froehlich JC. Psychopharmacology 1999; 147:81–89.
124. Hyytia P, Kiianmaa K. Alcohol Clin Exp Res 2001; 25:25–33.
125. Williams KL, Woods JH. Alcohol Clin Exp Res 1998; 22:1634–1639.
126. Hyytia P. Pharmacol Biochem Behav 1993; 45:697–701.
127. Honkanen A, Vilamo L, Wegelius K, Sarviharju M, Hyytia P, Korpi ER. Eur J Pharmacol 1996; 304:7–13.
128. Stromberg MF, Casale M, Volpicelli L, Volpicelli JR, O'Brien CP. Alcohol 1998; 15:281–289.
129. Middaugh LD, Kelley BM, Groseclose CH, Cuison ERJ. Pharmacol Biochem Behav 2000; 65:145–154.
130. Roberts AJ, Gold LH, Polis I, McDonald JS, Filliol D, Kieffer BL, Koob GF. Alcohol Clin Exp Res 2001; 25:1249–1256.
131. Matsuzawa S, Suzuki T, Misawa M, Nagase H. Brain Res 1998; 803:169–177.
132. Matsuzawa S, Suzuki T, Misawa M, Nagase H. Eur J Pharmacol 1999; 368:9–16.
133. Ciccocioppo R, Martin-Fardon R, Weiss F. Neuropsychopharmacology 2002; 27:391–399.
134. Froehlich JC, Badia-Elder NE, Zink RW, McCullough DE, Portoghese PS. J Pharmacol Exp Ther 1998; 287:284–292.

135. Spanagel R. Pharmacol Biochem Behav 1996; 54:645–664.
136. Mhatre MC, Carl K, Garrett KM, Holloway FA. Pharmacol Biochem Behav 2000; 66:701–706.
137. De Vries TJ, Babovic-Vuksanovic D, Elmer G, Shippenberg TS. Psychopharmacology 1995; 120:442–448.
138. Heidbreder C, Goldberg SR, Shippenberg TS. Eur J Pharmacol 1993; 243: 123–127.
139. Menkens K, Bilsky EJ, Wild KD, Portoghese PS, Reid LD, Porreca F. Eur J Pharmacol 1992; 19:345–346.
140. Suzuki T, Mori T, Tsuji M, Misawa M, Nagase H. Life Sci 1994; 55:PL339–PL344.
141. Reid LD, Hubbell CL, Glaccum MB, Bilsky EJ, Portoghese PS, Porreca F. Life Sci 1993; 52:PL67–PL71.
142. Suzuki T, Mori T, Tsuji M, Misawa M, Nagase H. Eur J Pharmacol 1997; 331: 1–8.
143. Reid LD, Hubbell CL, Tsai J, Fishkin MD, Amedola CA. Pharmacol Biochem Behav 1996; 53:477–480.
144. Ise Y, Narita M, Nagase H, Suzuki T. Psychopharmacology 2000; 151:49–54.

24

Delta Opioid Receptors in the Gastrointestinal Tract

DeWayne Townsend IV and David R. Brown
University of Minnesota, St. Paul, Minnesota, U.S.A.

1 INTRODUCTION

Opium has been employed historically for the alleviation of dysentery and diarrheal disorders [1]. Although the principal opiate alkaloids morphine and codeine were chemically isolated in the 19th century and employed as antidiarrheal medications, the worldwide use of antidiarrheal preparations of opium and opium tinctures, such as paregoric, has continued into the 21st century. With the development of many synthetic and semisynthetic opioid drugs in the twentieth century, newer antidiarrheal opioids have been identified and introduced into modern medical practice. These include the peripherally acting opioid agonists diphenoxylate and loperamide, which were first shown in the 1970s to be clinically effective in limiting chronic diarrhea [2,3]. Inhibitors of neutral endopeptidase (EC 3.4.24.11), such as the thiorphan derivative acetorphan, have been found to possess antidiarrheal efficacy after oral administration that may be comparable with that of loperamide in decreasing acute watery diarrheas in children and adults [4,5]. By inhibiting this enzyme, acetorphan diminishes the degradation of one family of endogenous opioid peptides, the enkephalins, in the intestinal wall. Opioid antidiarrheal drugs are also used in the treatment of diarrhea-predominant inflammatory bowel disease [6] and to increase intestinal transit time in the short bowel syndrome [7].

In addition to their antidiarrheal activity, opioids can produce significant constipating effects. This is of particular concern in the management of cancer pain, for which long-term opioid administration is a cornerstone of therapy [8,9]. It remains a problem in postoperative patients as well, because prolonged ileus significantly extends recovery times [10]. Recently, opioid antagonists that are either poorly absorbed orally or excluded from the central nervous system (CNS) have been shown to selectively inhibit the constipating actions while sparing the analgesic actions of opioids mediated at CNS sites [11,12].

The antidiarrheal and constipating actions of opioids have been attributed to their ability to inhibit intestinal motility and active transepithelial ion transport. Their actions, as discussed below, are mediated by local opioid receptors in the enteric nervous system (ENS) as well as through opioid receptors in spinal and supraspinal CNS sites. Myenteric neurons of the guinea pig ileum express mu opioid receptors at relatively high densities, and the longitudinal muscle–myenteric plexus preparation (LMMP) preparation from this species has become a widely used, classical bioassay to assess the actions of mu opioid receptor ligands. Indeed, it was instrumental in the discovery of endogenous opioid peptides and opioid receptor heterogeneity [13]. In addition, the guinea pig intestine has been the major template on which the chemical and functional neuroanatomy of the intestinal tract has been based over the past two decades [14]. Morphine, codeine, and other opioid agonists with a preference for mu-type opioid receptors are clearly potent and effective antidiarrheal drugs in modern medical practice. It is therefore not surprising that of the three opioid receptor types, emphasis in gastrointestinal pharmacology has been placed on mu opioid receptors. However, the guinea pig ileum is somewhat unique in its response to morphine [15].

Recent investigations have revealed that the delta opioid receptor also is involved in the modulation of gastrointestinal functions, particularly in mammals other than the guinea pig. In this chapter, we will discuss the presence and roles of this opioid receptor type in mediating the actions of opioid drugs and their endogenous counterparts in the digestive tract.

2 LOCALIZATION OF DELTA OPIOID RECEPTORS IN THE DIGESTIVE TRACT

2.1 Localization of Receptor mRNA Expression

In several studies designed to examine the pattern of delta opioid receptor expression in the gut, messenger RNA transcripts encoding the delta opioid receptor have been isolated from intestinal segments and subjected to amplification and analysis through the reverse transcriptase-polymerase chain reaction. *Delta* opioid receptor transcripts have been amplified from the wall of the rat intestine [16] and in smooth muscle–myenteric plexus and mucosa–submucosal plexus preparations from the porcine ileum [17]. In developmental studies with mice, mRNA transcripts for kappa and mu opioid receptors are expressed in the intestinal epithelium (kappa) and enteric neurons (kappa, mu) in early to midgestation embryonic stages, in advance of delta opioid receptor expression [18]. These results mirror an earlier study in the developing guinea pig small intestine which demonstrated that stereospecific binding sites for the nonselective opioid receptor ligand [^{3}H]diprenorphine were detectable at 25 days of gestation, at a time when enteric neurons could first be observed [19].

In rats, the chronic, intracerebroventricular (ICV) administration of an antisense oligodeoxynucleotide targeting the second exon of rat delta opioid receptor mRNA decreased the inhibitory action of SNC80 on colonic propulsion; injection of an oligonucleotide with a mismatched sequence did not alter the activity of this delta opioid agonist [20].

Intestinal inflammation may increase the expression of enteric opioid receptors (see below), and the potency of delta opioid agonists such as DPDPE ([D-Pen^{2,5}]enkephalin) to decrease intestinal transit and mucosal permeability is enhanced in the inflamed mouse gut. The intraperitoneal administration of antisense oligodeoxynucleotides to "knock down" enteric delta opioid receptor expression is associated with decreases in the inhibitory effects of DPDPE on transit and permeability, particularly in the inflamed intestine [21].

2.2 Localization of Delta Opioid Receptor Immunoreactivity

Subsequent to the molecular cloning and cDNA sequence analysis of the delta opioid receptor, antireceptor antibodies were raised against synthetic peptides based on the deduced amino acid sequence of this receptor. To date, there have been a limited number of studies using antibodies directed toward the amino terminus of the cloned murine receptor to assess the distribution of delta opioid receptorlike immunoreactivity in the gastrointestinal tract, and these have been confined so far to the porcine small intestine. They have shown the presence of delta opioid receptor–immunoreactive neurons and fibers in both the myenteric and submucosal plexuses and as well as in myenteric neurons maintained in primary culture [22, 23]. Receptor-like immunoreactivity in neuronal cell bodies appears to be localized in the cytoplasm and is likely to be trafficked to nerve terminals. These neurons are coimmunoreactive for the acetylcholine-synthesizing enzyme choline acetyltransferase, suggesting that they are cholinergic neurons. This is perhaps not surprising as cholinergic neurons, identified by choline acetyltransferase immunoreactivity, are abundant in the ENS of many species examined [24–27]. Moreover, delta opioid receptor–immunoreactive neurons and fibers frequently coexpress immunoreactivities for other substances commonly associated with primary afferent neurons, such as the vanilloid VR1 receptor [28]. A small subset of delta opioid receptor–positive neurons in the porcine ileal myenteric plexus coexpress immunoreactivity for the kappa opioid receptor [22,28,29]. Immunoreactivity to the amino terminus of the mu opioid receptor could not be detected in porcine enteric neurons, although it was present in neurons from the porcine hypothalamus and myenteric plexus of the guinea pig ileum [22,23]. However, mu and kappa opioid receptor immunoreactivities were localized in neurons within the rat intestine [30]. These disparate findings may imply that species differences exist in the expression of opioid receptors on enteric neurons.

The presence of preprodynorphin- and preproenkephalin-derived peptides in enteric neurons has been detected by immunohistochemical methods as well. The distribution of the enkephalins is of particular interest as these peptides are thought to be a class of endogenous ligands for delta opioid receptors [31]. Indeed, they bind preferentially to the cloned delta opioid receptor relative to the other recombinant opioid receptor types [32]. Enkephalinlike immunoreactivity is present in neurons both intrinsic and extrinsic to the intestinal wall and is localized in nerve fibers along the entire length of the digestive tract, from esophagus to colon, in a wide variety of species including rats, guinea pigs, hamsters, cats, pigs, and humans [33–39]. In general, enkephalin-immunoreactive neurons are present in myenteric ganglia, but are rare or absent in submucosal ganglia. Unfortunately, there have been no reports of the anatomical interrelationships between enkephalinergic nerves and postsynaptic target cells expressing immunoreactivity for delta opioid receptors.

Immunoreactivity towards the frog dermal peptide [D-Ala^2]deltorphin I, a selective ligand for delta opioid receptors, is reportedly present in high abundance in goblet cells of the rat jejunal mucosa, from which it may be secreted into the intestinal lumen [40]. Peptide-positive epithelial cells can be detected in the late fetal period [41]. Deltorphinlike immunoreactivity is not detectable in the enteric nervous system, and its precise chemical identity and functional significance remain to be determined.

2.3 Distribution and Characteristics of Specific Delta Opioid–Binding Sites

Quantitative receptor autoradiography is one technique that has been used in assessing the distribution and pharmacological characteristics of delta opioid receptor receptors in subregions of the stomach and intestine. Although specific binding sites can be visualized in intestinal subregions, this technique offers relatively low anatomical resolution of binding sites. Through this approach, specific binding sites for the mixed mu and delta opioid receptor ligand [^{3}H]DADLE ([D-Ala^2, D-Leu^5]enkephalin) have been localized in guinea pigs and rats to the muscularis mucosae, submucosal plexus, and circular muscle of the gastric fundus; the mucosa and submucosal regions of the gastric antrum and corpus; the duodenal mucosa; and to the rat, but not guinea pig, ileal mucosa [42]. In the human sigmoid colon, specific [^{3}H]DADLE binding sites are distributed in neurons and fibers within the myenteric plexus [43]. Specific binding sites for [^{3}H]DPDPE have been visualized over the villous and crypt epithelium of the rat small intestine [44].

Radioligand binding to membrane homogenates isolated from intestinal tissues has also been employed to assess the presence of delta opioid binding sites. Some degree of tissue localization can be obtained by isolating membranes from dissected layers of the intestine and using differential centrifugation to isolate specific cellular membranes. This technique was first used to study specific [^{3}H]naloxone binding sites present in membrane homogenates from a LMMP preparation of the guinea pig ileum [45]. Submucosal neural fractions from guinea pig small intestine express high-affinity binding sites for [^{3}H]DADLE [46]. Displacement of the nonselective opioid receptor ligand [^{3}H]diprenorphine by ligands selective for the each opioid receptor type has been examined in neuronally enriched membranes isolated from the canine ileum. *Delta* and *mu* opioid binding sites, as measured by displacement of [^{3}H]diprenorphine binding by DPDPE and [*N*-Me-Phe^3, D-Pro^4]morphiceptin (PL017) respectively, each represent ~40% of the opioid receptor population in the myenteric synaptosomes, with the remaining 20% of specific [^{3}H]diprenorphine binding representing kappa opioid–binding sites [47]. In canine submucosal synaptosomes, these sites are predominately mulike; i.e., 64% of total, high-affinity [^{3}H] diprenorphine binding was displaced by PL017, and 21% was displaced by DPDPE [48].

Saturation binding analyses with [^{3}H]diprenorphine and the selective delta, mu, and kappa opioid receptor ligands [^{3}H]naltrindole, [^{3}H]DAMGO ([D-Ala^2,Me-Phe^4,Gly^5]enkephalin-ol), and [^{3}H]U-69,593 have been conducted under Na^+-free conditions in membrane homogenates from both the submucosal and myenteric plexuses from five segments of the porcine intestinal tract. [^{3}H]Naltrindole bound to specific sites in these neuronal homogenates with K_Ds ranging from 0.02 to 0.1 nM and B_{max} values ranging from 8 to 60 fmol/mg protein. These [^{3}H]naltrindole binding sites predominated in neuronal homogenates from both the small intestine and colon relative to the other opioid receptor–selective radioligands employed [49]. These results suggest that, at least in the porcine intestinal tract, delta opioid receptors are the predominant opioid receptor type present in the ENS.

In addition to these studies, which focused on opioid binding sites associated with enteric neurons, there are reports suggesting that delta opioid receptors may exist on intestinal smooth muscle cells [50,51] and epithelial cells [52,53]. In other studies, however, specific opioid binding sites were observed neither in membrane fractions containing 5′-nucleotidase activity, which is specific to smooth muscle cells [47] nor in mucosal epithelial cells [49,54,55]. Functional opioid receptors are clearly expressed by enteric neurons, but it is premature to rule out the existence of extraneuronal opioid receptors in the intestine. Opioid binding sites, particularly on the many types of nonneuronal cells in the intestinal mucosa, may be difficult to detect owing to species or segmental differences in receptor expression as well as several methodological variables.

2.4 Electrophysiological Actions of Delta Opioid Agonists on Enteric Neurons

Mu-and kappa-type opioid receptors are functionally expressed in myenteric neurons of the guinea pig ileum, and agonists of these receptors decrease myenteric neurotransmission albeit through different cellular mechanisms [56,57]. Methionine and leucine enkephalins inhibit spontaneous firing of myenteric neurons from the guinea pig intestine, but it is likely that their effects are mediated by mu opioid receptors [58]. Although there is some functional evidence that delta opioid receptors may reside on myenteric neurons [59,60], this issue has not been thoroughly resolved by electrophysiological analyses to date.

Through interactions with delta opioid receptors, opioid agonists such as DADLE, [D-Ser^2,Leu^5]enkephalin-Thr (DSLET) and DPDPE mediate an increase in membrane K^+ conductance and inhibit inward Ca^{2+} currents in submucosal neurons from the guinea pig cecum [61–63]. Their effects are inhibited by selective delta opioid antagonists, such as ICI174, 864 [61]. Because DAMGO and normorphine produce neither of these effects, it appears that functional mu opioid receptors are not expressed in these cells.

3 FUNCTIONAL ROLES FOR DELTA OPIOID RECEPTORS IN THE GASTROINTESTINAL TRACT

The body of literature concerning the functional roles of delta opioid receptors in the digestive tract is somewhat complicated by the results of studies that employed opioid agonists or antagonists with relatively poor selectivity for opioid receptor types (e.g., capable of interacting with both delta and mu opioid receptors within a narrow concentration range). Such ligands include methionine or leucine enkephalin and peptidase-resistant enkephalin derivatives with such as [D-Ala^2, Met^5]enkephalinamide, which act as mixed mu and delta opioid agonists. To more clearly delineate those actions of opioids that are mediated by delta opioid receptors, we will mainly discuss those studies meeting one or more of the following criteria: 1) the use of agonists such as DPDPE, deltorphin II, and SNC80 with high delta opioid receptor selectivity; 2) the use of a delta opioid receptor–selective antagonist, such ICI-174, 864 and naltrindole; or 3) the use of poorly selective opioid agonists, such as the enkephalin pentapeptides, in biological preparations devoid of mu opioid receptors, such as those pretreated with mu opioid antagonists or that are receptor gene knockout models and which therefore do not respond to morphine or other, more selective mu opioid agonists. The reader is referred to some earlier review articles for additional discussions of the actions of opiate alkaloids and endogenous opioid peptides in the digestive tract [64–69].

3.1 Esophagus

Esophageal motility is decreased by opioid agonists in both the proximal striated and distal smooth muscle portions. This effect appears to be mediated by mu opioid receptors in rats [70] or kappa opioid receptors in guinea pigs [71].

In the dog, electrically evoked contractions of the lower esophageal sphincter are inhibited by DPDPE, a response interpreted to indicate the involvement of delta opioid receptors [72]. On the other hand, DADLE increases lower esophageal sphincter pressure in the opossum. As the mu opioid agonists meperidine and buprenorphine both reduce sphincter pressure, the opposite effect of DADLE is probably mediated by delta opioid receptors [73].

3.2 Stomach

3.2.1 Cytoprotection

In the rat stomach, opioid agonists have been shown to prevent the formation of gastric ulcers in response to acidified ethanol or other irritants. This action is mediated through delta opioid receptors present in both the ENS and CNS [74,75] and is not due to a reduction in gastric acid output [76]. Agonist stimulation of delta opioid receptors in the brainstem appears to increase vagal outflow to the stomach, resulting in the release of cytoprotective nitric oxide and prostaglandins from vagal efferent neurons [77,78]. Interestingly, both naloxone and naltrindole also prevent the gastroprotective action of clonidine, an α_2-adrenergic agonist [79]. Although the nature of the antagonistic action was not defined, this result suggests that there may be an interaction between supraspinal α_2-adrenergic and delta opioid receptors in neural pathways mediating the protective effect of this drug on the gastric mucosa.

3.2.2 Acid Secretion

In rats, delta opioid agonists such as DPDPE and [D-Ala^2]deltorphin II appear to have little or no effect on gastric acid secretion [76,80], although one research group reported that acid output was inhibited by a methione-enkephalin derivative putatively selective for delta opioid receptors [81]. In parietal cells isolated from the guinea pig stomach, the mixed mu and delta opioid agonist DADLE increased histamine-stimulated acid secretion. This suggests that opioid receptors, possibly of the delta type, may be expressed by these cells [82]. However, it is not known whether this effect can be antagonized by naloxone or a selective delta opioid antagonist. In the canine stomach, both morphine and DADLE inhibit acid secretion stimulated by 2-deoxy-D-glucose in a gastric fistula model. The action of DADLE is inhibited by the delta opioid antagonist naltrindole, indicating the involvement of delta opioid receptors in this effect. Neither morphine nor DADLE alters basal acid production in a vagotomized Heidenhain pouch preparation in the dog, but they can increase acid output stimulated by pentagastrin in this preparation [83,84]. The delta opioid receptor may also mediate the inhibitory effects of enkephalins on postprandial gastrin secretion in dogs [85]. These findings imply that both mu and delta opioid receptors may mediate the effects of opioids on gastric acid secretion in this species.

3.2.3 Motor Function

In the rat, the δ-opioid agonist [D-Ala^2]deltorphin II does not alter gastric emptying after its peripheral or central administration [80]. In cats however, DPDPE reduces gastric contractions induced by vagal stimulation, and its effect is prevented by the δ-opioid antagonist ICI-174, 864 [86]. In dogs, a role for δ-opioid receptors in the regulation of smooth muscle contractility in the gastric fundus and antrum is not well defined. On the one hand, DPDPE has no effect on gastric smooth muscle relaxation in response to feeding after its peripheral (IV) or central (ICV) administration, whereas morphine and DAMGO enhance gastric relaxation in this paradigm [87]. On the other hand, contractions of the gastric antrum are increased by the addition of the delta opioid receptor–selective antagonist ICI174,864, suggesting that endogenous opioids are released in this tissue and stimulate δ-opioid receptors to depress antral contractions [88]. In addition, delta opioid agonists inhibit contractions of the canine pylorus both in intact animals [89] and isolated muscle strips [90]. Based on these few studies, it appears that delta opioid receptors may exert an inhibitory influence on gastric motility and emptying, depending on the nature of the stimulus initiating gastric motor alterations and the animal species examined.

3.3 Small and Large Intestine

3.3.1 Motor Function

Although the mu opioid receptor appears to be the most significant receptor type mediating the inhibitory actions of opioids on the electrically evoked contractions of the LMMP preparation from the guinea pig ileum, the delta opioid receptor also plays a significant role in the local modulation of smooth muscle contractility in the small intestines of many mammalian species [68]. In the canine ileal circular smooth muscle, for example, it has been shown that both delta and mu opioid receptors mediate opioid agonist–induced increases in contraction [91]. This increased motor activity occurs secondary to an attenuation of inhibitory junctional potentials measured within smooth muscle cells [91,92]. Similar electrophysiological actions of delta opioid agonists have been observed in the primate jejunum [93]. Electrically stimulated contractions of smooth preparations from the rat, mouse, and rabbit small intestine are inhibited by delta opioid receptor agonists. Both kappa and delta opioid receptors appear to mediate the inhibitory actions of opioids on field-stimulated contractions of a circular muscle–myenteric plexus preparation from porcine ileum; these studies suggest the possibility of an interaction between these two opioid receptors [22]. In the isolated rabbit jejunum, a short (5 min) exposure to nanomolar concentrations of deltorphin I is associated with acute dependence as manifested by the appearance of smooth muscle contractions during naloxone-precipitated withdrawal. The dependence phenomenon was likely mediated by delta opioid receptors because it was selectively prevented by naltrindole [94].

Many intestinal functions are mediated through enteric neural reflexes, and delta opioid receptors appear to be present in these reflex pathways. In the rat ileum, the delta opioid receptor agonist DSLET decreases the rate of peristaltic contractions [95]. This effect may be secondary to an increased latency in the contractile response as described for DPDPE in this tissue preparation [96]. DPDPE inhibits reflex contractile responses of the guinea pig intestine to volume distention by raising the stimulus threshold required to evoke contraction [97]. However, this effect is probably not mediated by conventional delta opioid receptors, as SNC80 is considerably weaker than DPDPE in inhibiting peristaltic contractions in this preparation and its effects are not attenuated by naltrindole [98].

Despite evidence for a role of delta opioid receptors in modulating mechanical activity in isolated intestinal preparations, there have been few studies to confirm their role in vivo. These receptors appear to play a minor role in modulating basal motility in the rat and dog small intestine [99,100]. However, it is possible that the function of delta opioid receptors in intestinal motor function may be unmasked in states of inflammation, volume distention, or increased enteric neural activity. Indeed, the antimotility potency of delta opioid agonists is increased in the inflamed bowel [101].

As in the small intestine, delta opioid receptors may mediate a portion of the antimotility effects of opioids on the colon. For example, delta opioid agonists such as DADLE, DSLET, and DPDPE delay colonic transit in cats and rabbits [102,103]. In the human colon, the enkephalins and their derivatives DADLE and DPDPE decrease the amplitude of inhibitory junction potentials in circular smooth muscle [104]. This finding may indicate that these agonists act on presynaptic delta opioid receptors to reduce inhibitory neurotransmission to colonic myocytes. A similar effect of DPDPE on inhibitory junction potentials was observed in the guinea pig colon which was attributed to a decrease in N-type calcium conductance in inhibitory nerve terminals [105]. Smooth muscle contractions induced by electrical field stimulation are inhibited by delta opioid agonists in the large intestines of humans [106,107], cats [108], guinea pigs [107], rats [109,110], and mice [110]. Neurogenic contractions of the feline colonic longitudinal muscle, evoked by pelvic nerve stimulation, are inhibited by either leucine or methionine enkephalin as well as DPDPE, and these agonists were considerably more potent than mu or kappa opioid agonists. Moreover, their actions were antagonized by ICI174,864. These studies suggest that inhibitory delta opioid receptors mediate a reduction in sacral parasympathetic outflow to the distal colon in this species [108]. In sum, delta opioid receptors may mediate a reduction in the release of both inhibitory and excitatory neurotransmitters as they appear to do in the small intestine, depending on the species and colonic segment examined, with the net effect of decreasing colonic propulsion.

Opioid agonists also affect intestinal motility after their administration into the spinal cord and brain [111]. ICV injection of selective delta opioid agonists including [D-Ala^2]deltorphin II and DPDPE slows colonic transit and inhibits castor oil–induced diarrhea in rodents [112,113], but does not affect small intestinal transit [113–115]. In further support of a mechanism involving delta opioid receptors, the inhibitory effect of SNC80 on colonic motility is prevented in rats pretreated ICV with an antisense oligonucleotide targeted to the rat delta opioid receptor [20].

After intrathecal injection, delta opioid agonists slow transit in the small intestine [116,117] and inhibit diarrhea [114,118]. Based on these studies, it appears that delta opioid receptors in the spinal cord, unlike those in the brain, are associated with opioid-induced decreases in small intestinal motility. Although delta opioid receptors in the CNS may mediate opioid antidiarrheal activity through decreases in small and large intestinal propulsion [114,119,120], peripheral delta opioid receptors clearly contribute to this drug effect [113,120]. Lengthened intestinal transit time may increase contact time between luminal fluid and the colonic mucosa, and facilitate absorption of osmolytes and water.

3.3.2 Active Ion Transport

The columnar epithelial cells of the intestinal mucosa actively absorb and secrete extracellular ions, nutrients, and water. The active secretion of ions by these cells with an accompanying fluid flux acts to dilute and purge microorganisms or toxins in the bowel; promotes the transfer of secretory immunoglobulin A, antimicrobial defensin peptides, and mucin into intestinal mucus and the gut lumen; and, by affecting intraluminal pH, may alter the growth characteristics of enteric microflora [121]. Mucosal secretion is modulated by several enteric neurotransmitters, as well as inflammatory mediators released by mucosal mast cells that may affect transport indirectly through their ability to stimulate enteric neurons [122].

The intestinal antisecretory effects of opioids are now generally assumed to be an important component in the antidiarrheal action of opioids [123]. In experimental animals, opioids, including morphine and enkephalin analogues, decrease intestinal secretion induced by prostaglandin E, cholera toxin, and other secretagogues [69,124]. Codeine and loperamide, both considered to be mu opioid agonists, are very effective antidiarrheal drugs and markedly slow intestinal transit, yet they do not enhance water and electrolyte absorption when given to human subjects [125,126]. Perhaps this is not unexpected, as the antisecretory, rather than pro-absorptive, action of these drugs would be more closely associated with antidiarrheal activity, but this issue this has not been assessed thoroughly in human subjects.

Most studies of the local actions of opioids on the intestinal mucosa have utilized muscle-stripped sheets of ileal mucosa with attached submucosa mounted in Ussing flux chambers. Peptidase-resistant enkephalin derivatives such as DPDPE decrease short-circuit current, an electrical measure of active transepithelial ion transport, across isolated mucosal sheets from the guinea pig ileum [46,127,128], rabbit ileum [129–131], mouse jejunum [132], and pig distal jejunum/ileum [133]. This effect, which occurs after the application of opioid agonists to the serosal aspect of epithelial sheets, is due to an increase in electroneutral salt absorption and a decrease in electrogenic chloride secretion [46,132,133]. In contrast to enkephalin derivatives, opiate alkaloids have limited effects on active transepithelial transport of ions [69]. Pretreatment of tissues with axonal conduction blockers, such as tetrodotoxin or saxitoxin abolishes opioid effects on ion transport, suggesting that they are mediated by opioid receptors situated on submucosal neurons present in mucosal preparations. Indeed, short-circuit current elevations evoked by transmural electrical stimulation of submucosal neurons are attenuated by the serosal administration of selective delta opioid agonists, and agonist action is prevented by selective delta opioid antagonists [23,133]. In isolated sheets of the porcine ileal mucosa-submucosa, a tissue preparation that does not appear to contain functional mu or kappa opioid receptors, the delta opioid receptor mediating the antisecretory effect of opioid agonists has been extensively characterized in mucosal sheets subjected to transmural electrical stimulation [23,134]. The pharmacological characteristics of this receptor appear to differ from those of the cloned delta opioid receptor in the following respects: 1) peptidic delta and mu opioid agonists, including DPDPE, deltorphin II, and DAMGO, exhibited similar antisecretory potencies; 2) these peptidic opioid agonists were considerably more potent than selective kappa opioid agonists or nonpeptidic mu or delta opioid agonists such as morphine, loperamide, and SNC80; and 3) the mixed-acting mu/delta opioid antagonist 7-benzylidenenaltrexone (BNTX) appeared to be more potent than the more selective delta opioid antagonists naltriben or TIPP (Tyr-1,2,3,4-tetrahydroisoquinoline-Phe-Phe-OH) or the selective mu opioid antagonist CTOP (D-Phe-Cys-Tyr-D-Trp-Orn-Thr-Pen-Thr-NH_2) in reducing DPDPE potency [23].

Opioid agonists, administered into the cerebral ventricles of anesthetized rodents and conscious dogs, increase basal fluid absorption or decrease fluid accumulation evoked by cholera toxin in small intestinal loops [135–138]. Experiments with drugs selective for each opioid receptor type revealed that these effects are mediated predominantly by central mu opioid receptors with little involvement of delta opioid receptors [137–139].

3.3.3 Mucosal Defense and Repair Processes

Intestinal inflammation may influence opioid pharmacodynamics, and conversely, opioids acting through neural pathways responsive to inflammatory mediators appear to have antiinflammatory activity. Inflammation in peripheral locations increases the analgesic potency of delta opioid agonists after their peripheral or central administration, a phenomenon attributed to an inflammation-induced increase in opioid receptor expression at local sites of inflammation and in the CNS [140]. In acute or chronic inflammation of the murine intestine produced by chemical irritants such as croton oil, delta opioid agonists exhibit increased potency in slowing gut transit and decreasing mucosal permeability to the nonpermeant marker ^{51}Cr-EDTA [101,141].

In mucosal sheets from porcine ileum, the delta opioid agonist DPDPE inhibits saxitoxin-sensitive elevations in neurogenic ion transport evoked by histamine [142], tryptase-like enzymes [143], serotonin [144], kallidin [145], and type I hypersensitivity [142]. These effects of DPDPE are inhibited by naltrindole. In contrast, elevations in neurogenic ion transport occurring secondary to an immediate hypersensivity reaction in the guinea pig ileal mucosa are augmented by DPDPE, indicating that the neuromodulatory actions of opioids on active mucosal transport evoked by inflammation or anaphylaxis may depend on the species examined [146].

Opioids have been shown to modulate both innate and acquired immunity through direct and indirect actions on leukocyte function [147]. The gastrointestinal tract is the largest immunologic organ in the body, but studies of opioid action on intestinal immune function are rare. Morphine has been shown to inhibit mucosal immune responses to the potent mucosal adjuvant cholera toxin [148,149]. Although delta opioid receptors have a role in immune function, their role in mucosal immunity is largely unknown. Naltrindole has been reported to suppress immunoglobulin production in Peyer's patch lymphocytes, but the mechanisms underlying this effect await further investigation [150].

In the rabbit, delta opioid receptors may have a role in protecting the enterohepatic system from ischemic injury [151]. This effect may be related to the presynaptic inhibition of mesenteric arteriolar constriction and consequent preservation of blood flow by delta opioid agonists [152]. However, vasoactive effects of opioids mediated by delta opioid receptors are not observed in the rat [153,154]. *Delta*-opioid receptors have a similar protective action in the myocardium, which is described in detail elsewhere in this volume.

3.4 Gallbladder

The inhibitory action of opioids on gallbladder emptying has been known for many years [155]. This effect appears to be mediated by all three opioid receptor types in gallbladder myocytes in the dog [156] and in excitatory neurons innervating gallbladder smooth muscle in the guinea pig [157]. In the latter species, opioid agonists selective for each opioid receptor type are extremely potent in inhibiting excitatory neurotransmission in the gallbladder [157]. In guinea pigs, ligation of the bile duct is associated with a dramatic decrease in the potency of morphine to inhibit electrically induced twitches of LMMP strips, a tissue preparation containing mu and kappa opioid receptors. Likewise, in cholestatic mice, the potency of SNC80 in inhibiting electrically evoked contractions of the isolated vas deferens, a tissue containing delta opioid receptors, is substantially reduced [158]. Restricted bile flow is associated with increased circulating levels of preproenkephalin-derived peptides, and it is been hypothesized that the excess endogenous opioids downregulate their cognate receptors. A logical extension of this idea would be that cholestasis might be associated with decreased delta opioid receptor expression in the gastrointestinal tract.

4 PERSPECTIVES ON THE ROLE OF DELTA OPIOID RECEPTORS IN THE GASTROINTESTINAL TRACT

Opium alkaloids and endogenous opioid peptides have profound effects on the digestive tract, and delta opioid receptors play a significant role in mediating many opioid actions in this system. The ENS, a particularly important site of opioid action, represents an excellent biological model for the study of opioid receptors because it represents a complex neuronal network with high plasticity and relatively well defined and quantifiable outputs. However, our understanding of the neuronal organization of the mammalian intestine is at present narrowly confined to the guinea pig, but is rapidly developing for a few other species, such as the mouse, rat, and pig. Electrophysiological studies of opioid action have also been limited to cavine enteric neurons, and studies of delta opioid receptor distribution in the ENS have generally been limited to the canine and porcine gut. Thus, the cellular sites and mechanisms underlying the pronounced species differences reported for some of the local actions of opioids in the gastrointestinal tract have not been completely identified. *Delta* opioid receptors present in the CNS, as well as the ENS, mediate opioid effects on some aspects of digestive function, but the sites of opioid action in the spinal cord and brain remain to be precisely localized. Finally, there is a paucity of investigations designed to rigorously characterize the pharmacological properties and functional roles of delta opioid receptors in the alimentary canal, despite their potentially important role in mucosal protection and repair processes. We believe that as the knowledge base in this area of scientific endeavor expands in the twenty-first century, increasing attention will be focused on the delta opioid receptor as a novel drug target in the digestive tract and other peripheral organs.

REFERENCES

1. Scarborough J. In: Porter R Teich M, eds. Drugs and Narcotics in History. Cambridge: Cambridge University, Press 1995:4–23.
2. Bitar J, Najjar SS, Asfour RY. Arch Dis Child 1970; 45:190–192.
3. Demeulenaere L, Verbeke S, Muls M, Reyntjens A. Curr Ther Res Clin Exp 1974; 16:32–39.

4. Matheson AJ, Noble S. Drugs 2000; 59:829–835.
5. Prado D. Scand J Gastroenterol 2002; 37:656–661.
6. Camilleri M. Gastroenterology 2001; 120:652–668.
7. Platell CF, Coster J, McCauley RD, Hall JC. World J Gastroenterol 2002; 8:13–20.
8. Meuser T, Pietruck C, Radbruch L, Stute P, Lehmann KA, Grond S. Pain 2001; 93:247–257.
9. Mancini I, Bruera E. Support Care Cancer 1998; 6:356–364.
10. Taguchi A, Sharma N, Saleem RM, Sessler DI, Carpenter RL, Seyedsadr M, Kurz A. N Engl J Med 2001; 345:935–940.
11. Schmidt WK. Am J Surg 2001; 182:27S–38S.
12. Foss JF. Am J Surg 2001; 182:19S–26S.
13. Hughes J, Smith TW, Kosterlitz HW, Fothergill LA, Morgan BA, Morris HR. Nature 1975; 258:577–580.
14. Furness JB. J Auton Nerv Syst 2000; 81:87–96.
15. Daniel E. In: Bertaccini G, ed. Gastrointestinal Motility II: Endogenous and Exogenous Agents. Heidelberg: Springer Verlag 1982:295–322.
16. Wittert G, Hope P, Pyle D. Biochem Biophys Res Commun 1996; 218:877–881.
17. Brown DR, Poonyachoti S, Osinski MA, Kowalski TR, Pampusch MS, Elde RP, Murtaugh MP. Dig Dis Sci 1998; 43:1402–1410.
18. Zhu Y, Hsu MS, Pintar JE. J Neurosci 1998; 18:2538–2549.
19. Gintzler AR, Rothman TP, Gershon MD. Brain Res 1980; 189:31–48.
20. Negri L, Broccardo M, Lattanzi R, Melchiorri P. Br J Pharmacol 1999; 128: 1554–1560.
21. Pol O, Valle L, Puig MM. Eur J Pharmacol 2001; 428:127–136.
22. Poonyachoti S, Portoghese PS, Brown DR. J Pharmacol Exp Ther 2001; 297: 69–77.
23. Poonyachoti S, Portoghese PS, Brown DR. J Pharmacol Exp Ther 2001; 297: 672–679.
24. Furness JB, Costa M, Eckenstein F. Neurosci Lett 1983; 40:105–109.
25. Porter AJ, Wattchow DA, Brookes SJ, Schemann M, Costa M. Gastroenterology 1996; 111:401–408.
26. Sang Q, Young HM. Anat Rec 1998; 251:185–199.
27. Nakajima K, Tooyama I, Yasuhara O, Aimi Y, Kimura H. J Chem Neuroanat 2000; 18:31–40.
28. Poonyachoti S, Kulkarni-Narla A, Brown DR. Cell Tissue Res 2002; 307:23–33.
29. Kulkarni-Narla A, Brown DR. Neurosci Lett 2001; 308:153–156.
30. Bagnol D, Mansour A, Akil H, Watson SJ. Neuroscience 1997; 81:579–591.
31. Kosterlitz HW. Proc R Soc Lond B Biol Sci 1985; 225:27–40.
32. Mansour A, Hoversten MT, Taylor LP, Watson SJ, Akil H. Brain Res 1995; 700:89–98.
33. Furness JB, Costa M, Miller RJ. Neuroscience 1983; 8:653–664.
34. Elde R, Hokfelt T, Johansson O, Terenius L. Neuroscience 1976; 1:349–351.

35. Linnoila RI, DiAugustine RP, Miller RJ, Chang KJ, Cuatrecasas P. Neuroscience 1978; 3:1187–1196.
36. Schultzberg M, Hokfelt T, Nilsson G, Terenius L, Rehfeld JF, Brown M, Elde R, Goldstein M, Said S. Neuroscience 1980; 5:689–744.
37. Porcher C, Jule Y, Henry M. J Histochem Cytochem 2000; 48:333–344.
38. Polak JM, Bloom SR, Sullivan SN, Facer P, Pearse AG. Lancet 1977; 1:972–974.
39. Bagnol D, Henry M, Cupo A, Jule Y. J Auton Nerv Syst 1997; 64:1–11.
40. Fujimiya M, Okumiya K, Renda T, Kimura H, Maeda T. Peptides 1994; 15: 1095–1100.
41. Matsui J, Fujimiya M, Matsui S, Amakata Y, Renda T, Kimura H, Maeda T. J Histochem Cytochem 1994; 42:1377–1381.
42. Nishimura E, Buchan AM, McIntosh CH. Gastroenterology 1986; 91:1084–1094.
43. James S, Hoyle CH, Burnstock G, Jass JR, Jeffrey IJ, Lennard-Jones JE. Eur J Pharmacol 1987; 142:185–186.
44. Dashwood MR, Sykes RM, Thomson CS. Prog Clin Biol Res 1990; 328:165–169.
45. Creese I, Snyder SH. J Pharmacol Exp Ther 1975; 194:205–219.
46. Schulzke JD, Fromm M, Riecken EO, Reutter W. Eur J Clin Invest 1990; 20: 182–191.
47. Allescher HD, Ahmad S, Kostka P, Kwan CY, Daniel EE. Am J Physiol 1989; 256:G966–G974.
48. Ahmad S, Allescher HD, Manaka H, Manaka Y, Daniel EE. Am J Physiol 1989; 256:G957–G965.
49. Townsend D, Brown DR. J Pharmacol Exp Ther 2002; 300:900–909.
50. Zhang L, Gu ZF, Pradhan T, Jensen RT, Maton PN. Am J Physiol 1992; 262:G461–G469.
51. Kuemmerle JF, Makhlouf GM. Am J Physiol 1992; 263:G269–G276.
52. Lang ME, Davison JS, Bates SL, Meddings JB. J Physiol (Lond) 1996; 497: 161–174.
53. Nano JL, Fournel S, Rampal P. Pflugers Arch 2000; 439:547–554.
54. Binder HJ, Laurenson JP, Dobbins JW. Am J Physiol 1984; 247:G432–G436.
55. Gaginella TS, Rimele TJ, Wietecha M. J Physiol (Lond) 1983; 335:101–111.
56. Cherubini E, Morita K, North RA. Br J Pharmacol 1985; 85:805–817.
57. Cherubini E, North RA. Proc Natl Acad Sci USA 1985; 82:1860–1863.
58. North RA, Williams JT. Nature 1976; 264:460–461.
59. Egan TM, North RA. Science 1981; 214:923–924.
60. Gintzler AR, Hyde D. Proc Natl Acad Sci USA 1984; 81:2252–2254.
61. Mihara S, North RA. Br J Pharmacol 1986; 88:315–322.
62. Tatsumi H, Costa M, Schimerlik M, North RA. J Neurosci 1990; 10:1675–1682.
63. Surprenant A, Shen KZ, North RA, Tatsumi H. J Physiol (Lond) 1990; 431: 585–608.
64. Kromer W. Pharmacol Rev 1988; 40:121–162.
65. Bueno L, Fioramonti J. Baillieres Clin Gastroenterol 1988; 2:123–139.

66. Kromer W. Dig Dis 1990; 8:361–373.
67. Konturek SJ. Am J Gastroenterol 1980; 74:285–291.
68. De Luca A, Coupar IM. Pharmacol Ther 1996; 69:103–115.
69. Brown DR, Miller RJ. In: Handbook of Physiology. Neurohormonal Control of Fluid and Electrolyte Transport in Intestinal Mucosa. Bethesda, MD: American Physiological Society, 1991:527–589.
70. Storr M, Geisler F, Neuhuber WL, Schusdziarra V, Allescher HD. Neurogastroenterol Motil 2000; 12:441–448.
71. Kamikawa Y, Shimo Y. Br J Pharmacol 1983; 78:693–699.
72. Barnette MS, Grous M, Manning CD, Callahan JF, Barone FC. Eur J Pharmacol 1990; 182:363–368.
73. Rattan S, Goyal RK. J Pharmacol Exp Ther 1983; 224:391–397.
74. Gyires K, Ronai AZ, Toth G, Darula Z, Furst S. Life Sci 1997; 60:1337–1347.
75. Gyires K, Ronai AZ. J Pharmacol Exp Ther 2001; 297:1010–1015.
76. Fox DA, Burks TF. J Pharmacol Exp Ther 1988; 244:456–462.
77. Gyires K, Mullner K, Ronai AZ. J Physiol (Paris) 2001; 95:189–196.
78. Ronai AZ, Gyires K, Barna I, Mullner K, Reichart A, Palkovits M. Brain Res 2002; 947:90–99.
79. Gyires K, Mullner K, Furst S, Ronai AZ. J Physiol (Paris) 2000; 94:117–121.
80. Improta G, Broccardo M. Neuropharmacology 1994; 33:977–981.
81. Glavin GB, Pinsky C, Hall AM. Life Sci 1990; 46:1075–1079.
82. Kromer W, Skowronek B, Stark H, Netz S. Pharmacology 1983; 27:298–304.
83. Intorre L, Mengozzi G, Vanni E, Grassi F, Soldani G. Eur J Pharmacol 1993; 243:265–272.
84. Soldani G, Del Tacca M, Mengozzi G, Bernardini C, Bartolini D. Eur J Pharmacol 1985; 117:295–301.
85. Kostritsky-Pereira A, Woussen-Colle MC, De Graef J. Arch Int Physiol Biochim 1984; 92:19–26.
86. Okamoto T, Kurahashi K, Fujiwara M. Br J Pharmacol 1988; 95:329–334.
87. Gue M, Junien JL, Bueno L. J Pharmacol Exp Ther 1989; 250:1006–1010.
88. Holle GE, Steinbach E. Dig Dis Sci 2002; 47:1027–1033.
89. Allescher HD, Ahmad S, Daniel EE, Dent J, Kostolanska F, Fox JE. Am J Physiol 1988; 255:G352–G360.
90. Bayguinov O, Sanders KM. Br J Pharmacol 1993; 108:1024–1030.
91. Fox-Threlkeld JE, Daniel EE, Christinck F, Hruby VJ, Cipris S, Woskowska Z. J Pharmacol Exp Ther 1994; 268:689–700.
92. Bauer AJ, Szurszewski JH. J Physiol (Lond) 1991; 434:409–422.
93. Bauer AJ, Sarr MG, Szurszewski JH. Gastroenterology 1991; 101:970–976.
94. Valeri P, Morrone LA, Romanelli L. Br J Pharmacol 1992; 106:39–44.
95. Coupar IM. J Pharm Pharmacol 1995; 47:643–646.
96. Allescher HD, Storr M, Piller C, Brantl V, Schusdziarra V. Neuropeptides 2000; 34:181–186.
97. Waterman SA, Costa M, Tonini M. Br J Pharmacol 1992; 106:1004–1010.
98. Shahbazian A, Heinemann A, Schmidhammer H, Beubler E, Holzer-Petsche U, Holzer P. Br J Pharmacol 2002; 135:741–750.

99. Pinnington J, Wingate DL. Life Sci 1982; 31:2217–2219.
100. Tavani A, Petrillo P, La Regina A, Sbacchi M. J Pharmacol Exp Ther 1990; 254:91–97.
101. Puig MM, Pol O. J Pharmacol Exp Ther 1998; 287:1068–1075.
102. Krevsky B, Cowan A, Maurer AH, Butt W, Fisher RS. Life Sci 1991; 48:1597–1602.
103. Pairet M, Ruckebusch Y. Life Sci 1984; 35:1653–1658.
104. Hoyle CH, Kamm MA, Burnstock G, Lennard-Jones JE. J Physiol (Lond) 1990; 431:468–478.
105. Zagorodnyuk V, Maggi CA. Br J Pharmacol 1994; 112:1077–1082.
106. Chamouard P, Rohr S, Meyer C, Baumann R, Angel F. Eur J Pharmacol 1994; 262:33–39.
107. Mako E, Ronai AZ, Adam G, Juhasz G, Ritter L, Lestar B, Crunelli V. J Physiol (Paris) 2000; 94:135–138.
108. Kennedy C, Krier J. Br J Pharmacol 1987; 92:291–298.
109. Scheurer U, Wenger F, Caliezi A, Drack E, Varga L, Halter F. J Pharmacol Exp Ther 1990; 252:1324–1330.
110. Menzies JR, Glen T, Davies MR, Paterson SJ, Corbett AD. Eur J Pharmacol 1999; 385:217–223.
111. Burks TF. Gastroenterology 1978; 74:322–324.
112. Broccardo M, Improta G. Eur J Pharmacol 1992; 218:69–73.
113. Shook JE, Lemcke PK, Gehrig CA, Hruby VJ, Burks TF. J Pharmacol Exp Ther 1989; 249:83–90.
114. Burks TF, Fox DA, Hirning LD, Shook JE, Porreca F. Life Sci 1988; 43: 2177–2181.
115. Galligan JJ, Mosberg HI, Hurst R, Hruby VJ, Burks TF. J Pharmacol Exp Ther 1984; 229:641–648.
116. Porreca F, Burks TF. J Pharmacol Exp Ther 1983; 227:22–27.
117. Heyman JS, Williams CL, Burks TF, Mosberg HI, Porreca F. J Pharmacol Exp Ther 1988; 245:238–243.
118. Lemcke PK, Shook JE, Burks TF. Eur J Pharmacol 1991; 193:109–115.
119. Broccardo M, Improta G, Tabacco A. Eur J Pharmacol 1998; 342:247–251.
120. Shook JE, Pelton JT, Hruby VJ, Burks TF. J Pharmacol Exp Ther 1987; 243: 492–500.
121. Hecht G. Am J Physiol 1999; 277:C351–C358.
122. Perdue MH, McKay DM. Am J Physiol 1994; 267:G151–G165.
123. Powell DW. Gastroenterology 1981; 80:406–408.
124. Coupar IM. Life Sci 1987; 41:917–925.
125. Schiller LR, Davis GR, Santa Ana CA, Morawski SG, Fordtran JS. J Clin Invest 1982; 70:999–1008.
126. Schiller LR, Santa Ana CA, Morawski SG, Fordtran JS. Gastroenterology 1984; 86:1475–1480.
127. Kachur JF, Miller RJ, Field M. Proc Natl Acad Sci USA 1980; 77:2753–2756.
128. Vinayek R, Brown DR, Miller RJ. Eur J Pharmacol 1983; 94:159–161.

129. Dobbins J, Racusen L, Binder HJ. J Clin Invest 1980; 66:19–28.
130. Berschneider HM, Martens H, Powell DW. Gastroenterology 1988; 94:127–136.
131. Meyer G, Botta G, Tavani A, Algeri S, Cremaschi D. J Pharm Pharmacol 1989; 41:494–495.
132. Sheldon RJ, Riviere PJ, Malarchik ME, Moseberg HI, Burks TF, Porreca F. J Pharmacol Exp Ther 1990; 253:144–151.
133. Quito FL, Brown DR. J Pharmacol Exp Ther 1991; 256:833–840.
134. Townsend D, Portoghese PS, Brown DR. FASEB J 2002; 16:A297.
135. Brown DR, Miller RJ. Eur J Pharmacol 1983; 90:441–444.
136. Brown DR, Miller RJ. J Pharmacol Exp Ther 1984; 231:114–119.
137. Primi MP, Fargeas MJ, Bueno L. Regul Pept 1988; 21:107–115.
138. Jiang Q, Sheldon RJ, Porreca F. J Pharmacol Exp Ther 1990; 253:784–790.
139. Quito FL, Brown DR. Neuropeptides 1989; 14:39–44.
140. Stein C, Machelska H, Schafer M. Z Rheumatol 2001; 60:416–424.
141. Pol O, Ferrer I, Puig MM. J Pharmacol Exp Ther 1994; 270:386–391.
142. Poonyachoti S, Brown DR. J Neuroimmunol 2001; 112:89–96.
143. Green BT, Bunnett NW, Kulkarni-Narla A, Steinhoff M, Brown DR. J Pharmacol Exp Ther 2000; 295:410–416.
144. Green BT, Brown DR. Eur J Pharmacol 2002; 451:185–190.
145. Green BT, Calvin A, O'Grady SM, Brown DR. J Pharmacol Exp Ther 2003; 305:733–739.
146. Poonyachoti S, Brown DR. Eur J Pharmacol 1999; 379:81–85.
147. McCarthy L, Wetzel M, Sliker JK, Eisenstein TK, Rogers TJ. Drug Alcohol Depend 2001; 62:111–123.
148. Dinari G, Ashkenazi S, Marcus H, Rosenbach Y, Zahavi I. Digestion 1989; 44:14–19.
149. Peng X, Cebra JJ, Adler MW, Meissler JJ Jr, Cowan A, Feng P, Eisenstein TK. J Immunol 2001; 167:3677–3681.
150. Carr DJ, Radulescu RT, deCosta BR, Rice KC, Blalock JE. Life Sci 1990; 47:1059–1069.
151. Tubbs RJ, Porcaro WA, Lee WJ, Blehar DJ, Carraway RE, Przyklenk K, Dickson EW. Acad Emerg Med 2002; 9:555–560.
152. Illes P, Ramme D, Busse R. Naunyn Schmiedebergs Arch Pharmacol 1987; 335:701–704.
153. Ralevic V, Rubino A, Burnstock G. J Pharmacol Exp Ther 1994; 268:772–778.
154. Li YJ, Duckles SP. Eur J Pharmacol 1991; 204:323–328.
155. Sacchetti G, Roncoroni L, Mandelli V, Rocca F, Magni E. Eur J Clin Pharmacol 1976; 10:127–131.
156. Severi C, Grider JR, Makhlouf GM. Life Sci 1988; 42:2373–2380.
157. Guarraci FA, Pozo MJ, Palomares SM, Firth TA, Mawe GM. Gastroenterology 2002; 122:340–351.
158. Dehpour AR, Rastegar H, Jorjani M, Roushanzamir F, Joharchi K, Ahmadiani A. J Pharmacol Exp Ther 2000; 293:946–951.

25

Cardioprotection and Delta Opioid Receptors

Garrett J. Gross and Hemal H. Patel
Medical College of Wisconsin, Milwaukee, Wisconsin, U.S.A.

Ryan M. Fryer
Abbott Laboratories, Abbott Park, Illinois, U.S.A.

Jo El J. Schultz
University of Cincinnati, Cincinnati, Ohio, U.S.A.

1 INTRODUCTION

Ischemic preconditioning (IPC) is a cardioprotective phenomenon in which a single or multiple brief period of coronary artery occlusion results in an increased tolerance to a subsequent more prolonged period of ischemia [1]. This phenomenon occurs in two phases temporally—an acute or early phase in which the cardioprotective effect lasts for 1–3 h, and a delayed phase or second window of protection which reappears 24–48 h after the acute phase and may last for 72 h [2].There has been considerable interest in the triggers and/or mediators of this phenomenon since identification of a mechanism responsible might lead to a powerful new therapeutic approach to treating patients with ischemic heart disease at risk for a myocardial infarction. Adenosine, bradykinin, and opioid receptors have all been identified as potential targets for drug development since occupation and activation of all three of these G protein–coupled receptors have been shown to trigger IPC in a number of animal models and species [2]. The purpose of this chapter then will be to focus on the evidence which suggests that delta opioid receptor activation is an important component of both early and delayed IPC and potential mechanisms by which delta opioid receptor agonists produce a direct cardioprotective effect on the cardiac myocyte. We will also discuss possible clinical implications of this approach to cardioprotection.

2 OPIOID PEPTIDES AND RECEPTORS IN THE HEART

Opioid receptors involved in regulating cardiovascular function have been localized to various regions of the central nervous system and peripherally to cardiac myoctyes and autonomic nerve endings [3,4]. Myocardial binding studies have shown the presence of delta and kappa opioid receptors on rat ventricular myocytes [5,6]. Ventura et al. [7] demonstrated that kappa and delta receptors are present on the sarcolemmal membranes of rat hearts. These receptors appeared to be involved in the regulation of contractility in rat myocytes; however, functional mu receptors were notably absent. In a similar vein, Krumins et al. [8] found kappa and delta opioid receptors in adult rat hearts but no mu receptors in either atrial or ventricular tissue. Finally, Wittert et al. [9] did not detect any mu opioid receptor gene expression in rat hearts; however, the delta opioid receptor transcript appeared to be the predominant subtype. Based on these findings, the effects of IPC and opioid agonists to elicit cardioprotection are most likely the result of delta or kappa opioid receptor activation and the studies performed thus far indicate that this is indeed the case, with the predominance of evidence supporting a major role for the delta receptor.

There is evidence to support the existence of an endogenous opioid receptor system in the heart that may contribute to functional changes in the normal and diseased myocardium. It has been demonstrated that delta opioid receptor activation attenuates adrenergic responses [10], inhibits vagal-induced bradycardia [11], suppresses baroreceptor function [12], and results in a decrease in cardiac performance, perhaps via its antiadrenergic effects [13]. Kappa receptor stimulation has also been shown to be involved in arrhythmogenesis [14] and changes in cardiac function [15], and to inhibit norepinephrine release by an action on noradrenergic nerves [16]. The lack of mu receptor activity in the myocardium agrees with the evidence that the cardiac myocyte lacks these receptors [8,9].

Similar to opioid receptors, endogenous stores of opioid peptides have been found in the heart [17,18]. The heart has the ability to synthesize all three major types of opioid peptides including enkephalins, endorphins, and dynorphin [19,20]. Howells et al. [21] showed that preproenkephalin mRNA was the highest in rat ventricular tissue as compared to other organs, including the brain. Low et al. [22] demonstrated that proenkephalin, the precursor to the enkephalins, was associated with polyribosomes in the myocardium, which suggests that this tissue has translational capability.

A number of investigators have demonstrated that opioid peptides are released during stress into the peripheral circulation [23,24]. Myocardial ischemia has been demonstrated to result in the synthesis and release of opioid peptides such as met and leu-enkephalin [25,26]. Paradis et al. [27] suggested that increasing concentrations of enkephalins in rat ventricular tissue during ischemia may be a mechanism by which the heart protects itself from increased amounts of catecholamines, which are also released during ischemia. Thus, the release and/or synthesis of endogenous opioids might be a compensatory mechanism to minimize infarct size development. These observations and those obtained from our studies implicating endogenous opioids in triggering and mediating the protective effects of IPC [28] all suggest that endogenous opioids may serve an autocrine function possibly through release from myocytes during ischemia and an interaction with myocardial opioid receptors to limit cardiac cell injury, perhaps by reducing Ca^{2+} overload in the stressed myocyte.

3 DELTA OPIOIDS AND CYTOPROTECTION IN MULTIPLE ORGANS

Several early papers suggested that endogenous opioids protected several organs including the heart from hypoxic or ischemic injury [29,30]. Most notably, Mayfield and D'Alecy [30,31] found that several brief intermittent periods of hypoxia produced a phenotype in which mice exposed to a subsequent more prolonged period of hypoxia were resistant to its deleterious effects and survived longer than a control group not subjected to a previous hypoxic preconditioning protocol. Additionally, these investigators found that this adaptive response was mediated via the delta opioid receptor. Further support for a cardioprotective role of opioids in the heart was provided by Maslov et al. [32], who showed that delta opioid receptor stimulation was antiarrhythmic in the rat heart. Finally, there is intriguing evidence that hibernating animals possess a hibernation-inducing trigger (HIT) that appears to share many characteristics with delta opioid receptor activation [33] in protecting organs from stressful insults such as hypoxia. Both this HIT factor and a delta opioid agonist [D-Ala^{2}-D-Leu^{5}]enkephalin (DADLE) have been shown to enhance organ survival time and produce tissue preservation prior to organ transplantation [34]. Taken together, this evidence supports an important role for the delta opioid system to serve an endogenous protective mechanism that is activated during stress-induced situations such as ischemia or hypoxia.

4 DELTA OPIOIDS AND EARLY PRECONDITIONING

The first evidence to demonstrate that opioids were involved in early IPC was obtained by Schultz et al. [28] in intact rat hearts. These investigators showed that the nonselective opioid antagonist, naloxone, completely blocked the infarct size–reducing effect of IPC either when administered prior to IPC or after IPC but prior to the index ischemic period (Fig. 1). These results suggested that opioids served as both a trigger and mediator of IPC in rat hearts. Similarly, Chien and Van Winkle [35] found that the active enantiomer of naloxone [(−)naloxone] blocked the effect of IPC in rabbit hearts whereas

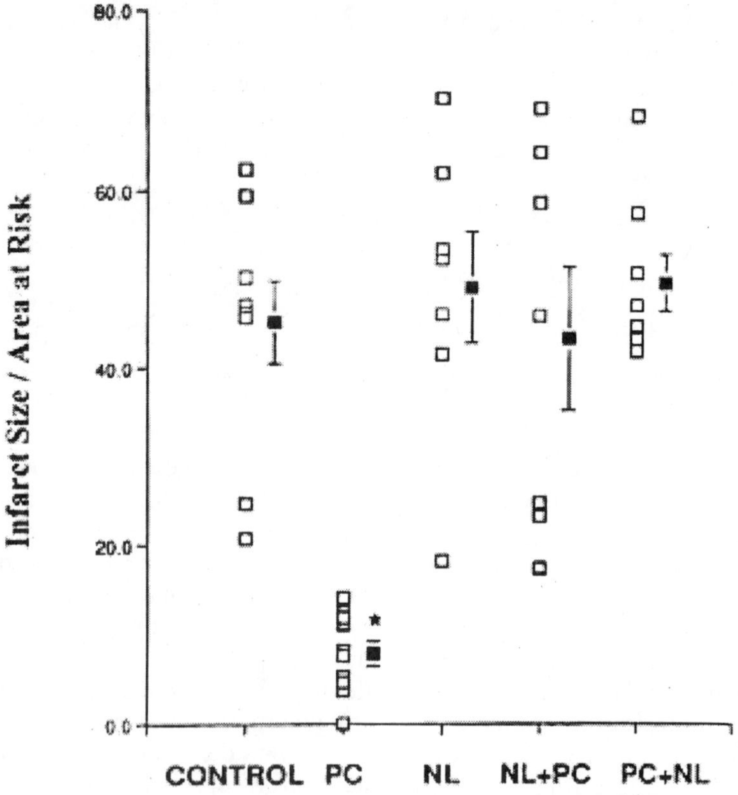

FIGURE 1 Infarct size (IS) expressed as a percent of the area at risk (AAR) in intact rat hearts subjected to vehicle (control), ischemic preconditioning (PC), or naloxone (NL) in the absence of PC, NL treatment prior to PC (NL + PC), and NL treatment after PC (PC + NL) but before the index ischemic period. The filled squares are the mean ± SE of each group. *$P < .05$ vs. the control group. (From Ref. 28.)

the inactive enantiomer, (+)-naloxone, did not block IPC. These data suggested that an opioid receptor was mediating the effect of endogenous opioids in IPC.

More recent data [36] also demonstrated that thiorphan, an enkephalinase inhibitor, was able to produce cardioprotection in a nonpreconditioned heart and that this effect was blocked by naloxone. These data suggested that by preventing the rapid breakdown of endogenous enkephalins released during ischemia it was possible to produce cardioprotection. Furthermore, Takashi et al. [37] undertook a study in isolated rabbit cardiomyocytes to determine which opioid peptides were primarily responsible for the cardioprotection observed following IPC. These investigators used an isolated myocyte model exposed to simulated ischemia and/or hypoxia in the presence or absence of IPC or administration of endogenous opioid peptides. IPC was produced by 15 min of simulated ischemia and 15 min of reoxygenation prior to 180 min of simulated ischemia. The results indicated that IPC limited cell death, an effect blocked by naloxone. Similarly, Met^5-enkephalin (ME), Leu^5-enkephalin, and ME-Arg-Phe produced cardioprotection whereas β-endorphan was not effective. These results suggested that the enkephalins are the most likely triggers of IPC in the rabbit heart.

The observation that the enkephalins are the endogenous opioids responsible for the cardioprotective effect of IPC is not surprising since it has been previously demonstrated by several investigators that large amounts of endogenous opioids are released during ischemia [38,39]. In fact, Weil et al. [40] suggested that the left ventricle of the rat heart behaved like an endocrine organ that supplied the body with endogenously released enkephalins. In agreement with this hypothesis, recent evidence obtained from our laboratory [41] and that of Dickson et al. [42] suggests that endogenous opioids released from the intestine or heart following mesenteric or coronary artery occlusion and reperfusion can also produce a cardioprotective effect at a distance in the heart and jejunum, a phenomenon known as remote preconditioning . In both studies, the protective effect in the heart and jejunum was blocked by the opioid receptor antagonist naloxone. Thus, this system seems to be uniquely poised to serve an endogenous protective role throughout the body, with the heart serving as a major supplier of opioid peptides during a variety of stressful situations.

5 CARDIOPROTECTIVE EFFECT OF EXOGENOUS OPIOIDS

The initial study to suggest that nonpeptide opioid agonists also mimic IPC was reported by Schultz et al. [43] in the rat heart. These investigators found that exposure of the heart to three brief (5 min) infusions of morphine (100 μg/

kg each) interspersed with three 5-min periods of no drug resulted in a reduction in infarct size nearly similar to that observed with IPC. The effect of morphine and IPC was blocked by naloxone and glibenclamide, a nonselective K_{ATP} channel antagonist. These results were the first to suggest that both IPC and morphine were acting via an opioid receptor-K_{ATP} channel–linked mechanism to produce cardioprotection in the intact rat heart.

Interestingly, morphine is primarily considered to have selective effects on the mu opioid receptor for its analgesic effects; however, there is also evidence that it possesses effects on delta or kappa opioid receptors and that crosstalk can occur between mu and delta opioid receptors [44]. To test the hypothesis that the cardioprotective effects of IPC and morphine were acting via a delta opioid receptor, Schultz et al. [45] administered the selective delta receptor antagonist naltrindole to rats prior to IPC or morphine infusion. In both instances, the cardioprotective effects of morphine and IPC were completely abolished at a dose of naltrindole that had no effect by itself on infarct size in nonpreconditioned rat hearts. These data clearly suggest that both IPC and morphine are exerting their cardioprotective effects via the delta opioid receptor in the intact rat heart.

Although the results of these studies suggested that morphine and other endogenous opioids are producing their effects in the cardiomyocyte, it is possible that these compounds could be producing some of their cardioprotective effects on other cells, bloodborne mediators or nerves innervating the intact heart. Therefore, Liang and Gross [46] addressed this question by performing studies in an isolated embryonic chick myocyte model in culture. These investigators found that a 5-min exposure of the myocytes to morphine (1 μM) produced a cardioprotective effect similar to that observed following a hypoxic preconditioning stimulus. The effect of morphine was concentration dependent and was blocked by naloxone (0.1–10 μM) and BNTX (7-benzylidenenaltrexone), a $delta_1$ receptor–selective antagonist. These results are in agreement with those of Schultz et al. [43,45] and provide direct evidence that morphine is exerting its infarct size reducing effect on the cardiomyocte via a $delta_1$ opioid receptor. Furthermore, these cardioprotective effects of morphine were blocked by glibenclamide, a nonselective K_{ATP} channel antagonist, and 5-hydroxydecanoic acid (5HD), a putative selective mitochondrial K_{ATP} channel antagonist [46].

6 EVIDENCE FOR A DELTA OPIOID RECEPTOR SUBTYPE BEING RESPONSIBLE FOR OPIOID-INDUCED CARDIOPROTECTION

Although our initial studies with IPC and morphine suggest that the beneficial effect of these interventions was occurring via a delta opioid receptor, both delta and kappa receptors are present on cardiomyocytes, and Xia et al. [47] and Wu et al. [48] demonstrated that kappa receptors mediate cardioprotection in rat myocytes. With this in mind, Schultz et al. [49] undertook a more detailed study to determine the opioid receptor subtype responsible for IPC and morphine-induced cardioprotection in the intact rat heart. Two doses of BNTX, a $delta_1$-selective antagonist or naltriben, a selective $delta_2$ receptor antagonist were administered prior to IPC. BNTX produced a dose-related reduction in the protective effect of IPC, whereas naltriben had no effect. Furthermore, the protective effect of IPC was not blocked by the selective mu or kappa receptor antagonists β-funaltrexamine or nor-binaltorphimine (nor-BNI), respectively. These results suggested that the $delta_1$ opioid receptor plays a major role in IPC in the rat heart. Similar results have been obtained by Tsuchida et al. [50] and Aitchison et al. [51] in rat and human myocardium, respectively. In contrast, Wang et al. [52] studied the role of kappa and delta opioid receptors in mediating IPC in a perfused rat heart preparation and found that both kappa and delta receptors mediated IPC-induced infarct size reduction, whereas only activation of kappa receptors produced a reduction in the incidence of arrhythmias. Thus, these interesting findings and those previously mentioned [47,48] suggest that more work needs to be done to determine the importance of the kappa receptor in opioid-induced cardioprotection.

7 SIGNALING PATHWAYS MEDIATING OPIOID-INDUCED CARDIOPROTECTION

Initial studies by Schultz et al. [43,45,49] and Liang and Gross [46] indicated that IPC and morphine-induced preconditioning were working via the $delta_1$ opioid receptor and the mitochondrial K_{ATP} (mitoK_{ATP}) channel to produce cardioprotection. However the signaling pathway between the $delta_1$ receptor and the mitoK_{ATP} channel remained unknown. An initial event in the signaling pathway of IPC and pharmacological preconditioning produced by several agonists of G protein–coupled receptors appears to be the generation of reactive oxygen species (ROS). As far as opioids are concerned, a recent study by McPherson and Yao [53] demonstrated that morphine produced a cardioprotective effect in chick myocytes that could be blocked by pretreatment with BNTX, a selective $delta_1$ opioid antagonist, 2-mercaptopropionyl glycine (2-MPG), a thiol reductant and ROS scavenger, and by 5HD, the selective mitoK_{ATP} channel antagonist. Similar results have been obtained with morphine, acetylcholine, and bradykinin in isolated rabbit hearts by Pain et al. [54]. Taken together, these results strongly support the idea that ROS are an integral part of the trigger phase of acute IPC and pharmacological preconditioning.

To further address the signaling pathways involved in opioid-induced cardioprotection, Schultz et al. [55] determined the involvement of a Gi/o protein in mediating $delta_1$-induced cardioprotection produced by the selective nonpeptide $delta_1$ opioid agonist, TAN-67. Pretreatment with pertussis toxin for 48 h prior to TAN-67 administration completely blocked its cardioprotective effect as well as that to IPC, suggesting that a Gi/o protein is intimately involved in the cardioprotection produced by these two interventions. Subsequently, Miki et al. [56] found that morphine produced a cardioprotective effect in isolated rabbit hearts which was blocked by pretreatment with chelerythrine, a protein kinase C (PKC) inhibitor at a concentration that had no effect on infarct size in the absence of morphine. More recently, Fryer et al. [57] extended these findings to the intact rat heart and showed that the protective effect of TAN-67 to reduce infarct size was blocked by chelerythrine and GF 109203X, two selective PKC inhibitors with different binding sites, and that TAN-67 produced a selective translocation of the PKC-delta isoform to the mitochondria.

Furthermore, Fryer and colleagues [57] showed that the translocation of PKC-delta and the cardioprotective effect of TAN-67 were all blocked by

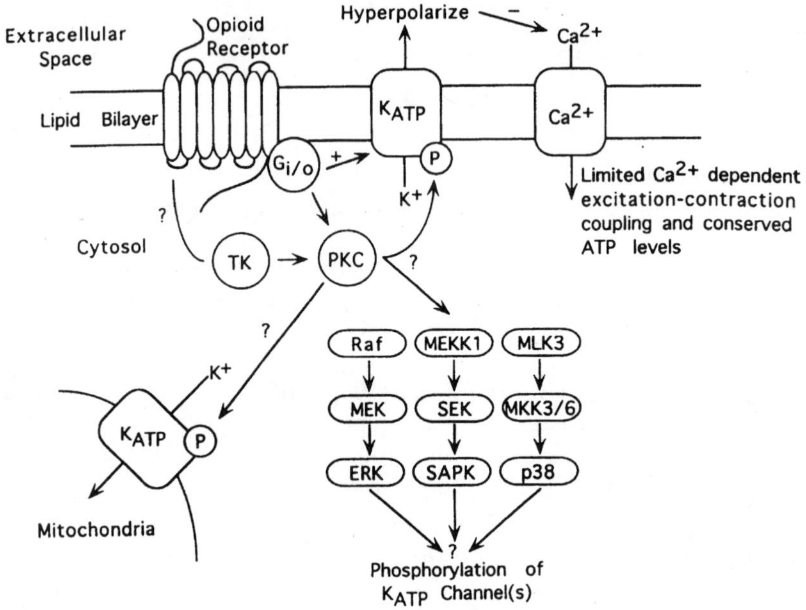

FIGURE 2 A schematic diagram of some of the major pathways thought to be involved in acute opioid-induced cardioprotection.

pretreatment with rottlerin, a selective blocker of PKC-delta. Fryer et al. [58,59] further showed the importance of tyrosine kinase (TK) and the mitogen-activated protein kinase (MAPK) cascade in opioid-induced cardioprotection. These authors found that genestein and lavendustin A, two TK inhibitors [58], and the MEK-1 inhibitor PD 098059 [59], which inhibits extracellular regulated kinase (ERK 1/2) phosphorylation and activation blocked opioid-induced cardioprotection produced by TAN-67. Taken together, these data suggest that TK, ERK 1/2, and PKC-delta are integral components of acute opioid-induced cardioprotection.

A final series of experiments were performed in our laboratory to determine the role of the sarcolemmal K_{ATP} (sarcK_{ATP}) channel and the mitoK_{ATP} channel in TAN-67-induced cardioprotection in the intact rat heart [60]. Administration of the selective sarcK_{ATP} inhibitor HMR 1098 prior to TAN-67 did not significantly block the cardioprotection produced by TAN-67. However, pretreatment with 5HD, the selective mitoK_{ATP} channel blocker, completely abolished TAN-67-induced cardioprotection. These data clearly suggest that delta$_1$ opioid receptor–induced infarct size reduction is mediated by the mito K_{ATP} channel in rats. A summary of the major signaling components involved in acute opioid-induced preconditioning is schematically depicted in Figure 2.

8 OPIOID-INDUCED DELAYED CARDIOPROTECTION

Baxter et al. [61] were the first to demonstrate the phenomenon of delayed preconditioning in rabbit hearts 24–48 h following IPC or the administration of the adenosine A_1 receptor agonist CCPA, and this group and others [62,63] subsequently showed that this delayed reduction in infarct size was associated with a translocation of PKC to the nucleus and the transcription of cardioprotective molecules such as heat shock proteins and inducible nitric oxide synthase (iNOS). Although opioids had not been studied in the context of delayed IPC, Ventura et al. [64] demonstrated that kappa opioid receptor stimulation resulted in a translocation of PKC to the nucleus and an increased synthesis of opioid peptides. Gustein et al. [65] also demonstrated that opioids induced an activation of ERK and p38 MAPK, similar to our findings, which also might result in increased gene transcription and translation. Based on these intriguing findings, Fryer et al. [66] found that adminstration of TAN-67, the nonpeptide delta$_1$ agonist, produced a significant reduction of infarct size 24–48 h following its administration to conscious rats (Fig. 3). These effects of TAN-67 were blocked by pretreatment with BNTX, the selective delta$_1$ antagonist, and by glibenclamide and 5HD administered 24–48 h after TAN-67. These results suggested that like adenosine, opioids are also capable of producing a delayed cardioprotective effect mediated by the mitoK_{ATP} channel.

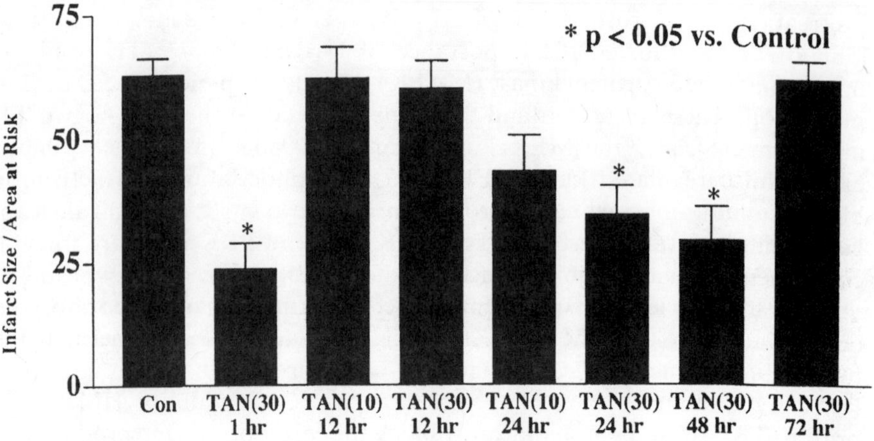

FIGURE 3 Infarct size (IS) expressed as a percent of the area at risk (AAR) in rats treated with 10 or 30 mg/kg of TAN-67, either 1, 12, 24, 48, or 72 h before subjecting the hearts to 30 min of ischemia and 2 h of reperfusion. A 1-h pretreatment with TAN-67 produced a significant reduction in IS/AAR. Pretreatment with both doses of TAN-67 12 h prior to ischemia/reperfusion or low-dose TAN-67 24 h prior to ischemia had no significant effect on IS/AAR. However, pretreatment with the large dose of TAN-67 24–48 h prior to ischemia/reperfusion significantly reduced IS/AAR. This cardioprotective effect was lost following 72 h of pretreatment. All values are the mean ± SE. $*P < .05$ vs. control. (From Ref. 66.)

Interestingly, Wu et al. [48] observed that kappa opioid receptor stimulation induced an early and late window of cardioprotection in isolated rat myocytes exposed to simulated metabolic-induced ischemia and reoxygenation, an effect blocked by a PKC antagonist. Thus, depending on the model and conditions, it appears that both delta and kappa opioid receptors are capable of eliciting acute and delayed cardioprotection. More recently, Patel et al. [67] expanded on the original observations of Fryer et al. [66] and studied the mechanisms responsible for delayed cardioprotection following the administration of several nonpeptide selective delta receptor agonists in our intact rat model of infarction. Interestingly, Patel et al. [67] discovered that significantly lower doses of two nonpeptide delta agonists, BW373U86 and TAN-67, produced a delayed cardioprotective effect at ~0.1 mg/kg given IV to conscious rats. Surprisingly, the salubrious effects at this low dose of agonist were not completely antagonized by the delta opioid selective antagonists, naltrindole, or BNTX. In addition, naloxone, which blocks all three major opioid receptors, did not completely block their cardioprotective effects. However, pretreatment with the ROS scavenger 2-MPG on the first day of drug administration completely abolished the cardioprotective effects of all three delta agonists 24 h later, which suggests that ROS production is an essential triggering factor similar to that previously reported for acute IPC and pharmacological preconditioning [53,54].

These results also suggest that nonpeptide opioid agonists such as TAN-67 or BW373U86 may have cellular actions independent from their known effects on classical opioid receptors, a possibility also suggested by the work of Zhu et al. [68] in delta receptor knockout mice. These investigators showed that BW373U86 actually produced an enhanced analgesic effect in delta receptor knockout mice, which suggested the possibility of a second deltalike opioid system in the central nervous system that is different from the classical

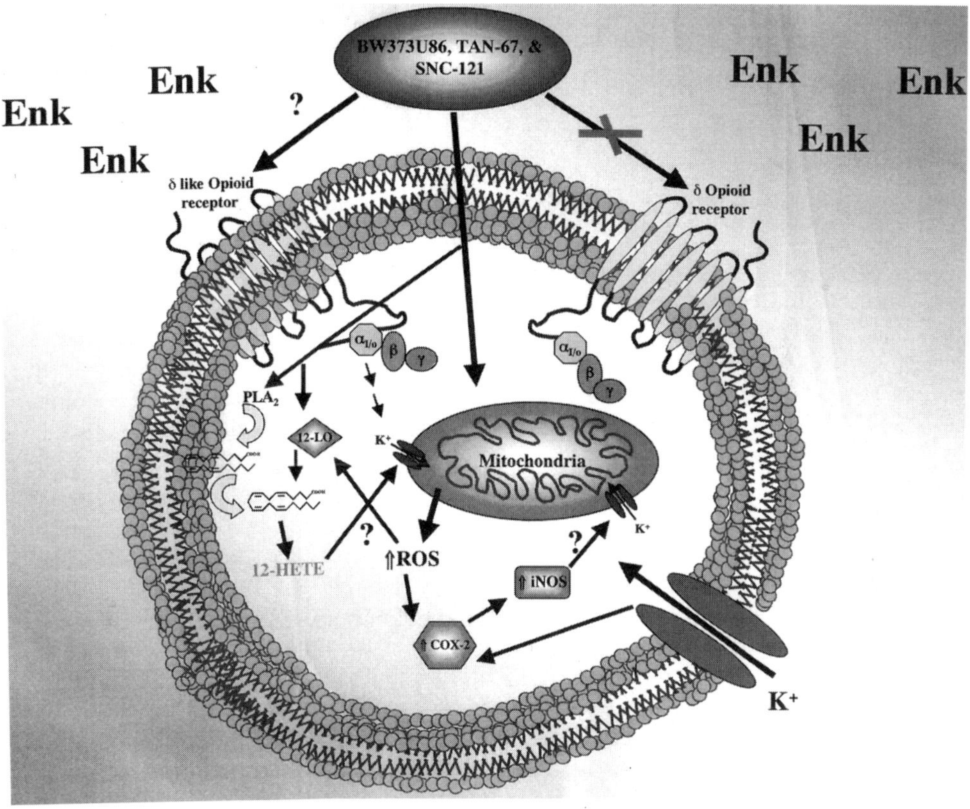

FIGURE 4 A schematic diagram of some of the major pathways thought to be involved in delayed opioid-induced cardioprotection.

system. Whether this also exists in the heart requires further study in the genetic knockout model.

Patel et al. [69] further characterized the signaling pathway involved in the triggering and mediator phase of delayed opioid-induced cardioprotection by studying the role of the sarcK_{ATP} and mitoK_{ATP} channel. Interestingly, these investigators found that the sarcK_{ATP} channel triggered the protective signal and that the mitoK_{ATP} channel mediated the beneficial effect observed 24 h later. These data are the first to suggest that the sarcK_{ATP} channel plays an important role in triggering both IPC and opioid-induced delayed preconditioning. There is also accumulating evidence obtained from our laboratory and that of Bolli and colleagues (unpublished data) which suggest that iNOS, cyclo-oxygenase 2 (COX-2), and 12-lipoxygenase (12-LO) are three additional downstream mediators in opioid-induced delayed cardioprotection. A schematic depicting the signaling events thought to be involved in opioid-induced delayed preconditioning is illustrated in Fig. 4.

9 DELTA OPIOID RECEPTORS AND CARDIOPROTECTION IN MAN

Although these studies performed in animal models of ischemia/reperfusion injury are intriguing, it is necessary to determine if the opioid receptor plays an important cardioprotective role in human myocardium. In this regard, Tomai et al. [70] showed that naloxone blocked the adaptation to ischemia in humans undergoing balloon angioplasty (PTCA). Similarly, Xenopoulos et al. [71] showed that intracoronary morphine mimics IPC, as assessed by changes in the ST segment of the electrocardiogram in patients undergoing PTCA. Preliminary results of Bell et al. [72] have demonstrated that opioids appear to mimic IPC in human right atrial trabeculae via a delta opioid receptor mechanism and via mitoK_{ATP} channels. In this same study, they demonstrated by RT-PCR that delta receptors are present in human atrial and ventricular muscle. These results suggest a possibility that delta opioid agonists may have clinical potential for the therapy of acute or chronic myocardial ischemia in man.

Another area of cardiovascular medicine where opioids may find a use is in the area of organ transplantation. Bolling et al. [73,74] have shown that the delta opioid agonist DADLE protects hearts that have been subjected to 18 h of cold storage at 4°C or 2 h of global ischemia in the presence of a standard cardioplegic solution. Subsequently, Kevelaitus et al. [75] demonstrated that activation of delta opioid receptors resulted in an improved recovery of function in cold-stored rat hearts similar to that of IPC. This group also showed that this cardioprotective effect was mediated downstream by the K_{ATP} channel.

10 CONCLUSIONS

In summary, the results from a number of animal and human studies indicate that stimulation of delta opioid receptors produces a consistent acute as well as a delayed cardioprotective effect and that this effect is mediated via signaling pathways similar to those responsible for acute and delayed IPC. Based on these fascinating findings, future animal and clinical studies are warranted with more selective and efficacious opioid agonists to determine the clinical potential of this approach in treating the ischemic myocardium. Since many opioids are already approved for the clinical treatment of pain, a long period of drug development may not be necessary before it is possible to bring this novel cardioprotective concept to the bedside.

ACKNOWLEDGMENTS

The authors wish to thank Anna Hsu and Jeannine Moore for their outstanding technical assistance in performing many of the experiments in the authors' laboratory. The authors also wish to acknowledge the financial support of NIH grant HL 08311 and the generous supply of drugs furnished by Dr. H. Nagase, Toray Industries, Inc., Kanagawa, Japan.

REFERENCES

1. Murry CE, Jennings RB, Reimer KA. Circulation 1986; 74:1124–1136.
2. Cohen MV, Baines CP, Downey JM. Annu Rev Physiol 2000; 62:79–109.
3. May CJ, Dashwood MR, Whitehead CJ, Mathias CJ. Br J Pharmacol 1989; 98:903–913.
4. Zunkin RS, Zunkin SR. Life Sci 1981; 29:2681–2690.
5. Zhang WM, Jin WQ, Wong TM. J Mol Cell Cardiol 1996; 28:1547–1554.
6. Zimlichman R, Gefel D, Eliahou H, Matas Z, Rosen B, Glass S, Ela C, Eilam Y, Vogel Z, Barg J. Circulation 1996; 93:1020–1025.
7. Ventura C, Bastagli L, Bernardi P, Campus CM, Capogrossi MC. Biochim Biophys Acta 1989; 987:69–74.
8. Krumins SA, Faden AI, Feuerstein G. Biochem Biophys Res Commun 1985; 127:120–128.
9. Wittert G, Hope P, Pyle D. Biochem Biophys Res Commun 1996; 218:877–881.
10. Xiao RP, Pepe S, Spurgeon HA, Capogrossi MC, Lakatta EG. Am J Physiol 1997; 272:H797–H805.
11. Caffrey JL, Mateo Z, Napier LD, Gaugl JF, Barron BA. Am J Physiol 1995; 268:H848–H855.
12. Giles TD, Sander GE, Rice JC, Quiroz AC. Peptides 1987; 8:609–612.
13. Wenzlaff H, Stein B, Teschemacher H. Naunyn Schmiederbergs Arch Pharmacol 1998; 358:360–366.

14. Sitsapesan R, Parratt JR. Br J Pharmacol 1989; 97:795–800.
15. Sheng JZ, Wong NS, Wang HX, Wong TM. Am J Physiol 1997; 272:C560–C564.
16. Gu H, Barron BA, Gaugl JF, Caffrey JL. Am J Physiol 1992; 263:H153–H161.
17. Barron BA, Gu H, Gaugl JF, Caffrey JL. J Mol Cell Cardiol 1992; 24:6–7.
18. Caffrey JL, Boluyt MO, Younes A, Barron BA, O'Neill L, Crow MT, Lakatta EG. J Mol Cell Cardiol 1994; 26:701–711.
19. McLaughlin PJ, Wu Y. Dev Dyn 1998; 211:153–163.
20. Millington WR, Rosenthal DW, Unal CB, Nyquist-Battie C. Cardiovasc Res 1999; 43:107–116.
21. Howells RD, Kilpatrick DL, Bailey LC, Noe M, Udenfriend S. Proc Natl Acad Sci USA 1986; 83:1960–1963.
22. Low KG, Allen RG, Melner MH. Mol Endocrinol 1990; 4:1408–1415.
23. Akil H, Watson SJ, Young E, Lewis ME, Kachaturian H, Walker JM. Annu Rev Neurosci 1984; 7:223–255.
24. Howlett T, Tomlin S, Ngahfoong L. BMJ 1984; 28:1950–1952.
25. Maslov LN, Lishmanov YB. Clin Exp Pharmacol Physiol 1995; 22:812–816.
26. Eliassion T, Mannheimer C, Waagstein F, Andersson B, Bergh CH, Augustinsson LE, Hedner T, Larson G. Cardiology 1998; 89:170–177.
27. Paridis P, Dumont M, Belichard P, Rouleau JL, Lemaire S, Brakier-Gingras L. Biochem Cell Biol 1992; 70:593–598.
28. Schultz JJ, Gross GJ. Pharmacol Ther 2001; 89:123–137.
29. Chien S, Oeltgen PR, Diana JN, Salley RK. J Thorac Cardiovasc Surg 1994; 107:964–967.
30. Mayfield KP, D'Alecy LG. Brain Res 1992; 582:226–231.
31. Mayfield KP, D'Alecy LG. J Pharmacol Exp Ther 1994; 268:74–77.
32. Maslov LN, Krylatov AV, Lismanov YB. Bull Exp Biol Med 1996; 121:20–21.
33. Oeltgen PR, Nilekani SP, Nichols PA, Spurrier WA, Su TP. Life Sci 1988; 43:1565–1574.
34. Oeltgen PR, Horton ND, Bolling SF, Su TP. Ann Thorac Surg 1996; 61:1488–1492.
35. Chien GL, Van Winkle DM. J Mol Cell Cardiol 1996; 28:1895–1900.
36. Hsu AK, Patel HH, Gross GJ. FASEB J 2002; 16:10.
37. Takashi Y, Wolff RA, Chien GL, Van Winkle DM. Am J Physiol 1999; 277, H2442–H2450.
38. Oldroyd KG, Harvey K, Gray CE, Beastall GH, Cobbe SM. Br Heart J 1992; 67:230–235.
39. Falcone C, Guasti L, Ochan M, Codega S, Tortorici M, Angoli L, Bergamaschi R, Montemartini C. J Am Coll Cardiol 1993; 22:1614–1620.
40. Weil J, Eschenhagen T, Fleige G, Mittman C, Orthey E, Scholz H. Am J Physiol 1998; 275:H378–H384.
41. Patel HH, Moore J, Hsu AK, Gross GJ. J Mol Cell Cardiol 2002; 34:1–7.
42. Dickson EW, Tubbs RJ, Porcaro WA, Lee WJ, Blehar DJ, Carraway RE, Darling CE, Przyklenk K. Am J Physiol 2002; 283:H22–H28.

43. Schultz JJ, Hsu AK, Gross GJ. Circ Res 1996; 78:1100–1104.
44. Traynor JR, Elliott J. Trends Pharmacol Sci 1993; 14:84–86.
45. Schultz JJ, Hsu AK, Gross GJ. J Mol Cell Cardiol 1997; 29:2187–2195.
46. Liang BT, Gross GJ. Circ Res 1999; 84:1396–1400.
47. Xia Q, Zhang WM, Shen YL, Wong TM. Life Sci 1996; 58:1307–1313.
48. Wu S, Li HY, Wong TM. Circ Res 1999; 84:1388–1395.
49. Schultz JJ, Hsu AK, Gross GJ. Circulation 1998; 97:1282–1289.
50. Tsuchida A, Miura T, Tanno M, Nozowa Y, Kita Y, Shimamoto K. Cardiovasc Drugs Ther 1998; 12:365–373.
51. Aitchison KA, Baxter GF, Awan MM, Smith RM, Yellon DM, Opie LH. Basic Res Cardiol 2000; 95:1–10.
52. Wang GY, Wu S, Pei JM, Yu XC, Wong TM. Am J Physiol 2001; 280:H384–H391.
53. McPherson BC, Yao Z. Circulation 2001; 103:290–295.
54. Pain T, Yang XM, Critz SD, Yankun Y, Nakano A, Liu GS, Heusch G, Cohen MV, Downey JM. Circ Res 2000; 87:460–466.
55. Schultz JJ, Hsu AK, Nagase H, Gross GJ. Am J Physiol 1998; 274:H909–H914.
56. Miki T, Cohen MV, Downey JM. Mol Cell Biochem 1998; 186:3–12.
57. Fryer RM, Schultz JJ, Hsu AK, Gross GJ. Am J Physiol 1999; 276:H1229–H1235.
58. Fryer RM, Schultz JJ, Hsu AK, Gross GJ. Am J Physiol 1998; 275:H2009–H2015.
59. Fryer RM, Pratt PF, Hsu AK, Gross GJ. J Pharmacol Exp Ther 2001; 296:647–654.
60. Fryer RM, Hsu AK, Nagase H, Gross GJ. J Pharmacol Exp Ther 2000; 294:451–457.
61. Baxter GF, Marber MS, Patel VC, Yellon DM. Circulation 1984; 90:2993–3000.
62. Baxter GF, Goma FJ, Yellon DM. Br J Pharmacol 1995; 115:222–224.
63. Bolli R, Bhatti ZA, Tang XL, Qui Y, Zhang Q, Guo Y, Jadoon AK. Circ Res 1997; 81:42–52.
64. Ventura C, Pintus G, Vaona I, Bennardini F, Pinna G, Tadolini B. J Biol Chem 1995; 270:30115–30120.
65. Gustein HB, Rubie EA, Mansour A, Akil H, Woodgett JR. Anesthesiology 1997; 87:1118–1126.
66. Fryer RM, Hsu AK, Eells JT, Nagase H, Gross GJ. Circ Res 1999; 84:846–851.
67. Patel HH, Hsu AK, Gross GJ. J Mol Cell Cardiol 2001; 33:1455–1465.
68. Zhu Y, King MA, Schullar AGP, Nitsche JF, Reidl M, Elde RP, Unterwald E, Pasternak GW, Pintar JE. Neuron 1999; 24:243–252.
69. Patel HH, Hsu AK, Peart JN, Gross GJ. Circ Res 2002; 91:186–188.
70. Tomai F, Crea F, Gaspardone A, Versaci F, Ghini AS, Ferri C, Desideri G, Chiariello L, Gioffre PA. J Am Coll Cardiol 33:1863–1869.
71. Xenopoulos NP, Lessar M, Bolli R. J Am Coll Cardiol 1998; 65:65A.

72. Bell SP, Sack MN, Patel A, Opie LH, Yellon DM. Circulation 1998; 98:1486.
73. Bolling SF, Su TP, Childs KF, Ning XH, Horton N, Kilgore K, Oeltgen PR. Transplantation 1997; 63:42–52.
74. Bolling SF, Tramontini NL, Kilgore KS, Su TP, Oeltgen PR, Harlow HH. Ann Thorac Surg 1997; 64:623–627.
75. Kevelaitus E, Peynet J, Mouas C, Launay JM, Menasche P. Circulation 1999; 99:3079–3085.

26

The Delta Opioid Receptor and Brain Pain-Modulating Circuits

Mary M. Heinricher
Oregon Health and Science University, Portland, Oregon, U.S.A.

Howard L. Fields
University of California at San Francisco, San Francisco, California, U.S.A.

1 INTRODUCTION

Although it has long been recognized that the prototypic opioid analgesic morphine acts in the central nervous system to produce analgesia, the underlying mechanisms have only recently been elucidated at the level of neural circuits. Mu opioid receptors have received the most attention. Multiple lines of evidence now demonstrate that the basis for the analgesic actions of drugs acting at the mu opioid receptor is recruitment of a brainstem pain modulatory network that includes the midbrain periaqueductal gray (PAG) and rostral ventromedial medulla (RVM) [1,2]. The role of the kappa receptor in supraspinal mechanisms of pain modulation is also reasonably well understood. However, contrary to general expectations regarding the effects of opioids, activation of kappa receptors in brainstem pain modulatory circuits often exerts a functional antagonism of supraspinal mu opioid analgesia [3]. By comparison, analysis of the role of supraspinal delta receptors in pain modulation is less developed. This review will consider the evidence relating to the function of the delta receptor in brainstem pain modulatory circuitry.

2 BRAINSTEM PAIN MODULATING CIRCUITRY: THE PAG-RVM SYSTEM

In brain, mu opioid mediated antinociception is mediated by activation of a brainstem pain-modulating system with links in PAG and RVM, which includes the nucleus raphe magnus and adjacent reticular formation at the level of the facial nucleus [1,2]. Indeed, progress in our understanding of that circuitry has been greatly facilitated by the availability of selective mu opioid receptor ligands. Direct focal application of morphine or mu-selective ligands such as DAMGO into the PAG or the RVM produces potent antinociception, an effect mimicked by electrical stimulation at the same sites. The antinociceptive effect of PAG or RVM mu opioid microinjection or electrical stimulation is due in large part to activation of a descending inhibitory influence that suppresses spinal nociceptive transmission. The PAG does not itself have a major direct projection to the dorsal horn, and its modulatory influence is relayed through the RVM. The RVM projects via the spinal dorsolateral funiculus to terminate densely and specifically in spinal cord dorsal horn laminae involved in nociceptive transmission. Although the details of how the RVM interfaces with dorsal horn nociceptive processing are not known, there is no question that this system can influence processing of the nociceptive signal in the dorsal horn, quite likely as early as the first central synapse. Sensory information can reach the PAG-RVM axis directly, via spinoreticular and spinomesencephalic projections, and indirectly, via higher structures such as the amygdala that have strong reciprocal connections with the PAG. There is also evidence that the PAG/RVM system modulates supraspinal processing through ascending projections. However, supraspinal mechanisms of nociception are in general much more difficult to study than spinal circuitry. As a consequence, ascending modulation has received relatively little attention.

The antinociceptive effects of mu opioid agonists within the RVM require activation of a class of neurons termed "off-cells" that exert a net inhibitory effect on nociception [4–6]. Mu opioid activation of off-cells is indirect, and triggered by disinhibition. In addition to the indirect activation of off-cells, mu opioids inhibit another class of neurons in RVM, termed "on-cells." This inhibition is direct, but suppression of on-cell discharge is not sufficient to produce behavioral analgesia. Opioid inhibition of on-cells may however assume an important role in inflammatory pain states, in which these neurons likely contribute to hyperalgesia [7].

Mu, delta, and kappa opioid receptors have been demonstrated anatomically in both PAG and RVM [8–11]. Mu receptors are apparently the most prominent, particularly in the PAG. Delta receptor binding in the PAG and RVM is detectable, although not prominent [12,13].

Immunohistochemical studies of delta receptor localization in this system are inconsistent. Cahill et al. [14] report moderate levels of delta labeling in the PAG, and low levels in the RVM. These authors find that labeling is primarily on cell bodies and dendrites in both regions. In contrast, Wessendorf and colleagues [15] find that the delta receptor is localized almost entirely to axon terminals. The basis for this discrepancy is not apparent, although Cahill and colleagues [14] suggest that the immunogenic properties of somatodendritic and axonal delta receptors may be different. There are clear functional implications for either scenario. If the receptor were restricted to terminals, one would predict that delta agonists would influence synaptic transmission within the PAG and RVM but have no postsynaptic effect. Conversely, somatodendritic localization would lead one to expect postsynaptic actions of these compounds. Unfortunately, electrophysiological studies to date have not provided a resolution of this question, with little support for either pre- or postsynaptic effects of delta agonism in PAG or RVM [16–19]. Mu receptor activation has both pre- and postsynaptic effects in both PAG and RVM [16–18,20,21].

The identification of endogenous opioid peptides within the PAG and RVM [22–24] is significant because of evidence that these peptides contribute to the modulatory action of this system. Indeed, exogenous opioid analgesics such as morphine likely mimic some aspects of the endogenous circuitry. Thus, electrical stimulation produced analgesia from the PAG or RVM can be attenuated by naloxone given systemically or at a site downstream from the stimulation (e.g., in the spinal cord or in the RVM when stimulation is delivered to the PAG). Although opioid mediation of the effects of PAG or RVM stimulation apparently depends on the exact stimulation site, stimulation parameters and behavioral test, these data demonstrate that the PAG-RVM system employs endogenous opioids (in addition to numerous other neurotransmitters and neuromodulators) to produce antinociception.

3 ANTINOCICEPTIVE EFFECTS OF EXOGENOUS DELTA RECEPTOR AGONISTS WITHIN THE PAG-RVM SYSTEM

When applied via the cerebral ventricles, $delta_1$ and $delta_2$ receptor agonists produce measurable behavioral analgesia in rat and mouse, although $delta_2$ effects appear to be more robust, especially in rat [25–29]. However, the site of action of drugs given intracerebroventricularly (ICV) is not known. Indeed, drugs given by this route are likely to gain access to multiple sites, which may interact in an unpredictable fashion. A microinjection approach is required to identify individual sites that have the circuitry needed to support any antinociceptive effects of opioid receptor activation. Although microinjection of mu opioid receptor agonists into the PAG or RVM produces potent antinociception, only the RVM supports strong delta-mediated antinociception.

A number of groups have found that microinjection of $delta_1$ or $delta_2$ agonists into the PAG has no effect on nociceptive responses [25,30,31], although there is one report that microinjection of the $delta_2$ agonist deltorphin into the PAG produces a modest increase in latency on the tail flick response [32]. This failure of delta agonists to produce an antinociceptive effect when applied to the PAG is consistent with electrophysiological data. Working in the slice or with dissociated neurons, Christie and colleagues have shown that delta receptor agonists, unlike mu agonists, do not hyperpolarize PAG neurons or inhibit synaptic transmission in this region [16,17,19,21,33].

Thus, although delta binding has been reported in the PAG and some consequences of delta activation can be measured biochemically [34,35], effects on neurons have not been documented electrophysiologically, and there is no behavioral evidence that this receptor is linked to analgesia. The role or roles of delta receptors within the PAG therefore remain something of a puzzle. One interesting possibility is that the delta receptor is important in other aspects of defense that are known to be mediated by the PAG. Keay and colleagues [36] have focused on the integration of antinociception, cardiovascular control, and behavioral inhibition within the PAG as part of defense. They find that microinjection of the $delta_1$ agonist DPDPE within the ventrolateral PAG has a depressor effect, although, as already noted, DPDPE does not produce antinociception in this region. A significant role for the delta receptor in cardiovascular aspects of defense is further supported by the observation that microinjection of the delta antagonist naltrindole into the caudal midline medulla (at the level of the inferior olive) attenuates the hypotension and bradycardia produced by hemorrhage [37].

Most investigators now agree that focal application of delta agonists with the RVM, especially $delta_2$ agonists produces measurable hypoalgesia. Thus, microinjection of deltorphin in the RVM results in at least modest analgesia on measures of thermal nociception (tail flick, hotplate, paw withdrawal) and on the formalin test [25,38–42]. There is less consensus as to whether the $delta_1$ agonist DPDPE has an antinociceptive effect in the RVM. Early reports indicated that DPDPE was ineffective when applied within the RVM [25,32]. However, more recent studies demonstrate mild to substantial analgesic effects of DPDPE on both hotplate and tail flick tests [40–43].

Current evidence therefore demonstrates that activation of $delta_1$ and $delta_2$ receptors within the RVM produces antinociception. One aspect of this work that remains unclear is whether the behavioral test of nociception is a critical variable. Ossipov and colleagues [25] proposed that the effect of delta agonism in the RVM is primarily due to local or ascending modulation, since they found that deltorphin and DPDPE have greater efficacy on the supra-spinally mediated hotplate test than on the tail flick, which can be elicited in spinalized rodents. In contrast, Thorat and Hammond [40] reported in a similar study that the tail flick was more sensitive than the hotplate test to both DPDPE and deltorphin. They suggested that some factor other than the CNS level mediating the behavioral nocifensive response, likely stimulus intensity, might be the relevant variable. Consistent with this latter idea, the antinociceptive effect of RVM deltorphin microinjection on the formalin test was completely blocked by lesions of the dorsolateral funiculus, which would include the axons of descending projections from the RVM to the dorsal horn. RVM deltorphin microinjection also reduced formalin-evoked increases in Fos-like immunoreactivity in the dorsal horn, and this reduction, like antinociception, was attenuated by lesions of the dorsolateral funiculus [42]. These data indicate that activation of $delta_2$ receptors in the RVM, like activation of mu receptors, produces analgesia in large part by recruiting descending inhibition. It is thus probably not too surprising that $delta_2$ agonists administered simultaneously in the RVM and at the spinal cord interact in a synergistic fashion to produce antinociception, as mu agonists show a similar spinal synergy. It is not entirely clear whether $delta_1$ receptor agonists have a synergistic, or merely additive, effect when given simultaneously in the RVM and spinal cord [39,41,43].

As noted above, it is now well established that the antinociceptive effects of mu agonists within the RVM require activation of a class of neurons called off-cells. The concomitant suppression of on-cell firing may contribute, but it is not by itself sufficient to produce antinociception. In contrast to this understanding of mu-mediated antinociception, we have very little knowledge of the neural basis for the antinociceptive action of delta agonists within the RVM. Given that delta agonists produce behavioral analgesia, one might expect that these compounds would have the net effect of activating off-cells and/or inhibiting on-cells through either a pre- or postsynaptic mechanism. However, although several groups now report that microinjection of the $delta_1$ agonist DPDPE into the RVM produces significant analgesia in the awake behaving rat [41,43], this compound is reported to have no effect on RVM neurons recorded in a slice preparation obtained in rat [18]. (Interestingly, a subset of RVM neurons recorded in the guinea pig slice were hyperpolarized by DPDPE. Unfortunately, it is not known whether DPDPE produces analgesia when applied within the RVM in this species.)

Infusion of deltorphin into the RVM in lightly anesthetized rats has only been studied in a paradigm in which the infusion produced small (1–2 sec) changes in tail flick latency [44]. Under those conditions, identified on- and off-cells showed only modest changes in firing pattern, with a small increase in the latencies of neuronal responses and no change in ongoing firing. It was suggested that the delayed responses during noxious stimulation reflected a strictly presynaptic action of deltorphin on the inputs to both classes of RVM neurons, an explanation consistent with immunohistochemical studies of DOR-1 localization on terminals in the RVM [15]. However, DOR-1 is apparently localized to the somatodendritic regions when different antibodies are used [14]. Moreover, it must be emphasized that the minimal effects of deltorphin on RVM neuronal firing in this experiment are entirely comparable to those observed following infusion of the selective mu agonist DAMGO in animals in which the microinjection failed to produce complete inhibition of the tail flick reflex [45]. Thus, mechanistic conclusions drawn from this experiment with deltorphin should be viewed with some caution. In future studies, it will be critical to characterize the responses of RVM neurons to delta receptor agonists under conditions of full antinociceptive effectiveness, as well as to identify direct and indirect effects of delta agonists within the RVM, and to define mu and delta interactions on individual neurons.

4 INTERACTIONS OF MU AND DELTA RECEPTORS WITHIN THE PAG-RVM SYSTEM

The idea that interactions between mu and delta contribute to behavioral and physiological effects of opioid agonists has been considered for over a decade. Evidence for mu/delta cooperativity has been obtained using molecular, cellular, and behavioral approaches [46–52]. However, models of mu/delta interaction have not yet been considered in detail from a circuit perspective. Thus, given that both mu and delta agonists can produce antinociception following focal application within the RVM, the question arises whether mu and delta actions in this system are coordinated or functionally independent. This issue has been addressed from two perspectives.

The first approach considers whether the antinociceptive effects of exogenous mu agonists applied within the PAG or RVM are potentiated by exogenous delta agonists. Evidence to date indicates that mu and delta agonists likely interact in a supra-additive fashion within the PAG-RVM system as a whole; that is, cooperativity appears to be a distributed property of the circuit, and does not involve a co-localization of receptors on a single neuron [32]. This conclusion is based on the observation that coadministration of a mu and delta receptor agonist within the PAG or RVM does not result in a greater than additive effect, whereas administration of different agonists at the two sites simultaneously generates a greater than expected antinociception. Thus, the effect of PAG DAMGO is enhanced in the presence of RVM deltorphin, and that of RVM DAMGO is similarly potentiated in the presence of PAG deltorphin. In contrast, the interaction is at most effect-additive (i.e., not synergistic) when the different agonists were applied simultaneously at the same site. These authors therefore argued that the interaction of mu- and delta-mediated effects within the PAG-RVM system is a property of the circuit as a whole, and does not reflect an interaction within an individual region or on individual neurons.

Other investigators have asked the related, but distinct, question of whether endogenous opioids working through the delta receptor are engaged when the pain-modulating circuit is activated by mu agonists. At one level, the answer to this question seems to be that delta receptor activation is not required for the antinociceptive effects of mu agonists, since the antinociceptive effects of acutely administered morphine or DAMGO given systemically or ICV are not attenuated by delta receptor antagonists or knockdown [50–56]. This conclusion was confirmed with the knockout approach: animals deficient in preproenkephalin or DOR-1 exhibit a full analgesic response to morphine [57,58]. There is thus little support for the idea that endogenous delta transmission has a necessary role in mu-mediated antinociception when drugs are given systemically or ICV. The same conclusion is reached when the focus is restricted to the RVM. Administration of the delta antagonist naltrindole or the more selective $delta_2$ antagonist naltriben into the RVM does not shift the dose-response relation for the mu agonist DAMGO applied at the same site [38,59]. That is, mu agonists maintain full antinociceptive effectiveness within the RVM when delta receptors in the same region are blocked, consistent with the lack of a mu/delta synergy when agonists are coadministered within the RVM [32].

The above data indicate that focal application of an exogenous mu agonist within either the PAG or RVM does not induce local release of endogenous delta agonists. In contrast, application of mu agonists within the PAG recruits both mu- and delta-mediated processes within the RVM [60,61], and stimulation within the RVM similarly recruits delta systems at the level of the spinal cord [62]. In the same way, activation of mu receptors in the amygdala, a major forebrain input to the PAG, engages both mu and $delta_2$ receptor activation in the PAG to produce antinociception [63–65]. These findings indicate that endogenous ligands for the delta receptor are an important element in the communication between different levels of the PAG-

RVM modulatory system, but do not play a critical role in the response to mu receptor activation within the PAG or RVM itself, or when the mu agonist is given systemically or ICV so that multiple levels of the system are activated in concert.

A complementary question is whether the antinociceptive effects of delta agonists require an action at the mu receptor. Studies in mice lacking the mu receptor are not entirely consistent, but suggest that the supraspinal antinociceptive effects of DPDPE are slightly depressed or even blocked (although this apparently depends on the nociceptive assay), whereas deltorphin is no less effective in the long-term absence of the mu receptor [66–72]. The data obtained using this genetic approach are entirely consistent with pharmacological observations, in that reports of the effects of selective mu receptor antagonism on DPDPE antinociception are mixed, whereas deltorphin antinociception is unaffected [28,73]. Studies focused specifically on the RVM show no effect of the selective mu antagonism on the antinociceptive effect of deltorphin applied at the same site [38,59], which again underscores the importance of viewing delta/mu interactions within the context of an identified circuit.

5 PHYSIOLOGICAL RECRUITMENT OF ENDOGENOUS DELTA RECEPTOR LIGANDS

The physiological and behavioral effects seen following administration of exogenous ligands for the delta receptor point to potential roles for delta receptor activation, but such data do not permit the conclusion that the receptor actually plays that role under physiological or pathophysiological conditions. Exogenous agonists may produce actions that differ in a number of ways from those produced by endogenous ligands released by the natural adequate stimulus. Exogenous drug may influence cell populations that would not normally be exposed simultaneously to endogenous opioids. The temporal pattern of ligand availability may be very different. Finally, the lack of normal comodulators, positive and negative, may greatly influence the observed outcome. Thus, analysis of the physiological functions of the delta receptor requires tests of the effects of pharmacological antagonists.

Investigators taking this approach have shown that delta receptor antagonists applied within the PAG or RVM have no effect on baseline nociceptive responding [40,59,61,63,74]. The lack of effect of delta receptor blockade in the PAG-RVM system is consistent with work using systemic or ICV administration of delta receptor antagonists, and was subsequently confirmed in mice deficient in the delta receptor or preproenkephalin [57, 58]. It is thus clear that endogenous ligands for the delta receptor, like those for the mu receptor, do not maintain an ongoing modulation of nociceptive processing in normal animals.

The challenge then is to identify the mechanisms that stimulate endogenous opioid systems acting through the delta receptor. An obvious candidate is stress. The interaction of stress and endogenous opioid systems is well documented [75], and a possible role for delta activation in stress-induced analgesia has received significant attention. Mu, delta, and kappa opioid receptors have all been implicated in supraspinal mechanisms of stress-induced analgesia in various paradigms. A number of groups have demonstrated that administration of delta antagonists via the cerebral ventricles interferes with swim stress analgesia [53,76]. A role for delta activation in swim stress effects is maintained in mice lacking the mu receptor [77]. Mouse lines bred for low swim stress analgesia also show reduced antinociception in response to mu and delta receptor agonists, a behavioral change that is apparently not due to reduced mu or delta opioid binding [78]. The late analgesia produced by multiple tail shocks is also blocked by ICV administration of the delta agonist naltrindole, but not by the mu-selective antagonist CTOP. In contrast, kappa, but not delta, receptors are implicated in this effect at the spinal cord [79,80]. There are few studies focused specifically on the PAG or RVM, but it appears that RVM mu, but not delta, receptor activation is required for the analgesia associated with conditioned fear [74]. Delta and mu receptors in the RVM are required for the antinociceptive effect of transcutaneous electrical stimulation (TENS) in arthritic animals, with high-frequency TENS mediated by endogenous ligands acting at the delta receptor, and low-frequency TENS mediated by ligands acting at the mu receptor [38].

An intriguing role for delta-mediated effects in nociceptive modulation has been demonstrated in regulation of the response to a novel environment. Animals exposed to a novel environment demonstrate a suppression of nociceptive responding. Although this antinociception itself does not require an opioid synapse, habituation of the antinociception is retarded by the delta antagonist naltrindole given ICV [81]. Thus, the opioid action in this case is not to produce analgesia. Rather, it contributes to the regulation of the analgesic response.

6 PLASTICITY IN SUPRASPINAL DELTA OPIOID SYSTEMS INVOLVED IN NOCICEPTIVE MODULATION

It has long been recognized that the effectiveness of opioid analgesics is dynamically regulated, with increased potency in inflammatory pain states and decreased potency with chronic opioid use (the phenomenon of tolerance). Increased opioid potency in inflammatory pain states seems to be at least in part due to a time-dependent recruitment of supraspinal delta opioid systems. In an important series of studies, Hurley and Hammond [59,82] demonstrated an increased antinociceptive potency for the mu agonist DAMGO as well as for the $delta_2$ agonist deltorphin microinjected into the RVM that develops over a period of 2 weeks after inflammatory injury (complete Freund's adjuvant, CFA). Importantly, the leftward shift of the dose-response curve for both drugs can be seen when testing the uninjured paw contralateral to the injured paw, which suggests that the RVM itself has undergone a change in sensitivity. These authors also showed that the enhanced potency of RVM DAMGO is reversed by delta receptor blockade within the RVM. These data thus demonstrate a novel interaction of mu and delta receptor–mediated effects within the RVM itself that develops with inflammation. Hurley and Hammond [59] suggest that this may be due to formation of functional mu/delta receptor complexes following inflammation [51,52].

Of equal interest, delta receptor blockade in the RVM produced hyperalgesia in the uninflamed paw in these studies and evoked other behaviors indicating a general hyperalgesia [59]. This observation demonstrates an ongoing delta-mediated antinociceptive "tone" in the chronically inflamed animals. Again, this delta-mediated process is recruited over a period of several weeks, and is not seen in normal animals or at early time points after inflammation in the CFA [59] or arthritis models [38]. The ongoing delta modulation may reflect increased release of endogenous delta ligands in the brainstem with inflammation. Met-enkephalin release is increased in the PAG after inflammation [83], and met-enkephalin levels are increased in several brainstem areas, including the RVM [59]. An additional possibility, complementary to an increase in available ligand, is that there is an increase in the targeting of the delta receptor to the cell membrane, as has been demonstrated in the dorsal horn of the spinal cord following inflammation [84]. At the circuit level, inflammation may reduce the influence of antiopioid peptides, such as cholecystokinin, which under normal conditions could mask or inhibit delta signaling within the PAG-RVM system [85,86]. Whatever the mechanism, these findings suggest that the delta receptor may be a particularly useful target for developing drugs for treating inflammatory pain. However, it must be noted that similar increases in met-enkephalin release and delta receptor availability are seen with acute and chronic administration of morphine [83,87,88]. Given that delta receptor blockade or deficiency reduces the tolerance that would normally result from chronic administration of morphine or mu agonists [58,89–91], it may be that the net effect of a prolonged increase in delta transmission over time is to functionally antagonize morphine's antinociceptive effects.

Additional evidence for plasticity in delta-mediated responses was recently provided in work using mice deficient in the delta receptor [58]. Surprisingly, the supraspinal antinociceptive effects of DPDPE and deltorphin were maintained in these animals, although the same drugs given intrathecally showed almost complete loss of effectiveness. The supraspinal antinociceptive actions of the delta compounds in the DOR-1 knockout mice were sensitive to naltrexone and naltrindole given ICV, but not to β-FNA or the kappa antagonist norBNI. This indicated that the antinociception produced by supraspinal DPDPE and deltorphin in these animals was not mediated by a mu or kappa receptor, and suggested the presence of a novel "secondary delta system" that is either unmasked or upregulated in the absence of DOR-1.

7 SUMMARY AND CONCLUSIONS

Behavioral studies now document that delta receptor agonists, like mu opioid agonists, exert their supraspinal antinociceptive effects in large part by activating the PAG-RVM pain-modulating system. The overlap of mu and delta actions within the PAG-RVM system raises important questions about the mechanisms underlying their interactions within this circuitry, and these would be best addressed within a framework provided by known properties of the circuit. The neural basis for delta actions within the RVM is still poorly understood, and it will be important to determine the extent to which mu- and delta-mediated actions intersect at the level of individual neural elements and cell populations within this region. It is also critical to recognize that the actions of endogenous and exogenous opioid ligands vary with the brain region involved and the status of the circuit as a whole. Delta-mediated transmission within the PAG-RVM system likely contributes to at least some forms of environmental analgesia, including stress-induced analgesia. Finally, the evidence for plasticity in delta-mediated antinociceptive modulation, especially in inflammatory pain models, suggests that the delta receptor may be a useful therapeutic target in some specific pain states.

REFERENCES

1. Fields HL, Basbaum AI. In: Wall PD, Melzack R, eds. Central Nervous Mechanisms of Pain Modulation. Edinburgh: Churchill Livingston, 1999:309–329.
2. Heinricher MM, Morgan MM. In: Stein C, ed. Supraspinal Mechanisms of Opioid Analgesia. Cambridge: Cambridge University Press, 1999:46–69.
3. Pan ZZ. Trends Pharmacol Sci 1998; 19:94–98.
4. Heinricher MM, McGaraughty S, Farr DA. Pain 1999; 81:57–65.

5. Heinricher MM, McGaraughty S, Tortorici V. J Neurophysiol 2001; 85:280–286.
6. Heinricher MM, Schouten JC, Jobst EE. Pain 2001; 92:129–138.
7. Kincaid W, Xu M, Kim CJ, Neubert MJ, Roberts WJ, Heinricher MM. Soc Neurosci Abstr 2001.
8. Mansour A, Khachaturian H, Lewis ME, Akil H, Watson SJ. Trends Neurosci 1988; 11:308–314.
9. Pierce TL, Wessendorf MW. J Chem Neuroanat 2000; 18:181–207.
10. Mansour A, Fox CA, Burke S, Akil H, Watson SJ. J Chem Neuroanat 1995; 8: 283–305.
11. Gutstein HB, Mansour A, Watson SJ, Akil H, Fields HL. Neuroreport 1998; 9: 1777–1781.
12. Mansour A, Khachaturian H, Lewis ME, Akil H, Watson SJ. J Neurosci 1987; 7:2445–2464.
13. Gouarderes C, Tellez S, Tafani JA, Zajac JM. Synapse 1993; 13:231–240.
14. Cahill CM, McClellan KA, Morinville A, Hoffert C, Hubatsch D, O'Donnell D, Beaudet A. J Comp Neurol 2001; 440:65–84.
15. Kalyuzhny AE, Arvidsson U, Wu W, Wessendorf MW. J Neurosci 1996; 16: 6490–6503.
16. Chieng B, Christie MJ. Br J Pharmacol 1994; 113:121–128.
17. Chieng B, Christie MJ. Br J Pharmacol 1994; 113:303–309.
18. Pan ZZ, Williams JT, Osborne PB. J Physiol 1990; 427:519–532.
19. Connor M, Christie MJ. J Physiol 1998; 509:47–58.
20. Osborne PB, Vaughan CW, Wilson HI, Christie MJ. J Physiol 1996; 490:383–389.
21. Vaughan CW, Christie MJ. J Physiol 1997; 498:463–472.
22. Finley JC, Maderdrut JL, Petrusz P. J Comp Neurol 1981; 198:541–565.
23. Khachaturian H, Lewis ME, Watson SJ. J Comp Neurol 1983; 220:310–320.
24. Skinner K, Basbaum AI, Fields HL. Neuroreport 1997; 8:2995–2998.
25. Ossipov MH, Kovelowski CJ, Nichols ML, Hruby VJ, Porreca F. Pain 1995; 62:287–293.
26. Mattia A, Vanderah T, Mosberg HI, Porreca F. J Pharmacol Exp Ther 1991; 258:583–587.
27. Negri L, Improta G, Lattanzi R, Potenza RL, Luchetti F, Melchiorri P. Br J Pharmacol 1995; 116:2931–2938.
28. Fraser GL, Pradhan AA, Clarke PB, Wahlestedt C. J Pharmacol Exp Ther 2000; 295:1135–1141.
29. Roerig SC, Fujimoto JM. J Pharmacol Exp Ther 1989; 249:762–768.
30. Smith DJ, Perrotti JM, Crisp T, Cabral ME, Long JT, Scalzitti JM. Eur J Pharmacol 1988; 156:47–54.
31. Bodnar RJ, Williams CL, Lee SJ, Pasternak GW. Brain Res 1988; 447:25–34.
32. Rossi GC, Pasternak GW, Bodnar RJ. Brain Res 1994; 665:85–93.
33. Connor M, Schuller A, Pintar JE, Christie MJ. Br J Pharmacol 1999; 126: 1553–1558.
34. Garzon J, Martinez-Pena Y, Sanchez-Blazquez P. Eur J Neurosci 1997; 9: 1194–1200.

35. Rodriguez-Diaz M, Garzon J, Sanchez-Blazquez P. Life Sci 1998; 62:PL253–PL258.
36. Keay KA, Crowfoot LJ, Floyd NS, Henderson LA, Christie MJ, Bandler R. Brain Res 1997; 762:61–71.
37. Henderson LA, Keay KA, Bandler R. Neuroreport 2002; 13:729–733.
38. Kalra A, Urban MO, Sluka KA. J Pharmacol Exp Ther 2001; 298:257–263.
39. Grabow TS, Hurley RW, Banfor PN, Hammond DL. Pain 1999; 83:47–55.
40. Thorat SN, Hammond DL. J Pharmacol Exp Ther 1997; 283:1185–1192.
41. Kovelowski CJ, Bian D, Hruby VJ, Lai J, Ossipov MH, Porreca F. Brain Res 1999; 843:12–17.
42. Kovelowski CJ, Ossipov MH, Hruby VJ, Porreca F. Pain 1999; 83:115–122.
43. Hurley RW, Grabow TS, Tallarida RJ, Hammond DL. J Pharmacol Exp Ther 1999; 289:993–999.
44. Harasawa I, Fields HL, Meng ID. Pain 2000; 85:255–262.
45. Heinricher MM, Morgan MM, Tortorici V, Fields HL. Neuroscience 1994; 63: 279–288.
46. Traynor JR, Elliott J. Trends Pharmacol Sci 1993; 14:84–86.
47. Jiang Q, Mosberg HI, Porreca F. Eur J Pharmacol 1990; 186:137–141.
48. Heyman JS, Vaught JL, Mosberg HI, Haaseth RC, Porreca F. Eur J Pharmacol 1989; 165:1–10.
49. Adams JU, Tallarida RJ, Geller EB, Adler MW. J Pharmacol Exp Ther 1993; 266:1261–1267.
50. Porreca F, Takemori AE, Sultana M, Portoghese PS, Bowen WD, Mosberg HI. J Pharmacol Exp Ther 1992; 263:147–152.
51. Gomes I, Jordan BA, Gupta A, Trapaidze N, Nagy V, Devi LA. J Neurosci 2000; 20:RC110.
52. Jordan BA, Devi LA. Nature 1999; 399, 697–700.
53. Vanderah TW, Wild KD, Takemori AE, Sultana M, Portoghese PS, Bowen WD, Hruby VJ, Mosberg HI, Porreca F. J Pharmacol Exp Ther 1993; 267:449–455.
54. Bilsky EJ, Bernstein RN, Hruby VJ, Rothman RB, Lai J, Porreca F. J Pharmacol Exp Ther 1996; 277:491–501.
55. Calcagnetti DJ, Holtzman SG. Pharmacol Biochem Behav 1991; 38:185–190.
56. Sluka KA, Deacon M, Stibal A, Strissel S, Terpstra A. J Pharmacol Exp Ther 1999; 289:840–846.
57. Nitsche JF, Schuller AG, King MA, Zengh M, Pasternak GW, Pintar JE. J Neurosci 2002; 22:10906–10913.
58. Zhu Y, King MA, Schuller AG, Nitsche JF, Reidl M, Elde RP, Unterwald E, Pasternak GW, Pintar JE. Neuron 1999; 24:243–252.
59. Hurley RW, Hammond DL. J Neurosci 2001; 21:2536–2545.
60. Kiefel JM, Rossi GC, Bodnar RJ. Brain Res 1993; 624:151–161.
61. Hirakawa N, Tershner SA, Fields HL. Neuroreport 1999; 10:3125–3129.
62. Hammond DL, Donahue BB, Stewart PE. Brain Res 1997; 765:177–181.
63. Tershner SA, Helmstetter FJ. Brain Res 2000; 865:17–26.
64. Pavlovic ZW, Cooper ML, Bodnar RJ. Brain Res 1996; 741:13–26.

65. Helmstetter FJ, Bellgowan PS, Poore LH. J Pharmacol Exp Ther 1995; 275: 381–388.
66. Matthes HW, Smadja C, Valverde O, Vonesch JL, Foutz AS, Boudinot E, Denavit-Saubie M, Severini C, Negri L, Roques BP, Maldonado R, Kieffer BL. J Neurosci 1998; 18:7285–7295.
67. Hutcheson DM, Matthes HW, Valjent E, Sanchez-Blazquez P, Rodriguez-Diaz M, Garzon J, Kieffer BL, Maldonado R. Eur J Neurosci 2001; 13:153–161.
68. Fuchs PN, Roza C, Sora I, Uhl G, Raja SN. Brain Res 1999; 821:480–486.
69. Sora I, Funada M, Uhl GR. Eur J Pharmacol 1997; 324:R1–R2.
70. Hosohata Y, Vanderah TW, Burkey TH, Ossipov MH, Kovelowski CJ, Sora I, Uhl GR, Zhang X, Rice KC, Roeske WR, Hruby VJ, Yamamura HI, Lai J, Porreca F. Eur J Pharmacol 2000; 388:241–248.
71. Loh HH, Liu HC, Cavalli A, Yang W, Chen YF, Wei LN. Brain Res Mol Brain Res 1998; 54:321–326.
72. Qiu C, Sora I, Ren K, Uhl G, Dubner R. Eur J Pharmacol 2000; 387:163–169.
73. Heyman JS, Mulvaney SA, Mosberg HI, Porreca F. Brain Res 1987; 420:100–108.
74. Foo H, Helmstetter FJ. Pain 1999; 83:427–431.
75. Drolet G, Dumont EC, Gosselin I, Kinkead R, Laforest S, Trottier JF. Prog Neuropsychopharmacol Biol Psychiatry 2001; 25:729–741.
76. Killian P, Holmes BB, Takemori AE, Portoghese PS, Fujimoto JM. J Pharmacol Exp Ther 1995; 274:730–734.
77. LaBuda CJ, Sora I, Uhl GR, Fuchs PN. Brain Res 2000; 869:1–5.
78. Kest B, Jenab S, Brodsky M, Sadowski B, Belknap JK, Mogil JS, Inturrisi CE. Brain Res 1999; 816:381–389.
79. Watkins LR, Wiertelak EP, Maier SF. Brain Res 1992; 582:10–21.
80. Watkins LR, Wiertelak EP, Grisel JE, Silbert LH, Maier SF. Brain Res 1992; 594:99–108.
81. Spreekmeester E, Rochford J. Psychopharmacology (Berl) 2000; 148:99–105.
82. Hurley RW, Hammond DL. J Neurosci 2000; 20:1249–1259.
83. Williams FG, Mullet MA, Beitz AJ. Brain Res 1995; 690:207–216.
84. Cahill CM, Morinville A, Hoffert C, O'Donnell D, Beaudet A. Pain 2003; 101: 199–208.
85. Vanderah TW, Lai J, Yamamura HI, Porreca F. Neuroreport 1994; 5:2601–2605.
86. Stanfa L, Dickenson A, Xu XJ, Wiesenfeld-Hallin Z. Trends Pharmacol Sci 1994; 15:65–66.
87. Cahill CM, Morinville A, Lee MC, Vincent JP, Collier B, Beaudet A. J Neurosci 2001; 21:7598–7607.
88. Nieto MM, Wilson J, Cupo A, Roques BP, Noble F. J Neurosci 2002; 22:1034–1041.
89. Kest B, Lee CE, McLemore GL, Inturrisi CE. Brain Res Bull 1996; 39:185–188.
90. Hepburn MJ, Little PJ, Gingras J, Kuhn CM. J Pharmacol Exp Ther 1997; 281: 1350–1356.
91. Abdelhamid EE, Sultana M, Portoghese PS, Takemori AE. J Pharmacol Exp Ther 1991; 258:299–303.

Index